Flow and Transport Properties of Unconventional Reservoirs 2018

Flow and Transport Properties of Unconventional Reservoirs 2018

Special Issue Editors

Jianchao Cai
Zhien Zhang
Qinjun Kang
Harpreet Singh

MDPI • Basel • Beijing • Wuhan • Barcelona • Belgrade

Special Issue Editors
Jianchao Cai
China University of Geosciences
China

Zhien Zhang
The Ohio State University
USA

Qinjun Kang
Los Alamos National Laboratory
USA

Harpreet Singh
The University of Texas at Austin
USA

Editorial Office
MDPI
St. Alban-Anlage 66
4052 Basel, Switzerland

This is a reprint of articles from the Special Issue published online in the open access journal *Energies* (ISSN 1996-1073) from 2018 to 2019 (available at: https://www.mdpi.com/journal/energies/special_issues/unconventional_reservoirs2018)

For citation purposes, cite each article independently as indicated on the article page online and as indicated below:

LastName, A.A.; LastName, B.B.; LastName, C.C. Article Title. *Journal Name* **Year**, *Article Number*, Page Range.

ISBN 978-3-03921-116-6 (Pbk)
ISBN 978-3-03921-117-3 (PDF)

Contents

About the Special Issue Editors

Jianchao Cai received his B.Sc. in Physics from Henan Normal University in 2005, and earned an MSc and a Ph.D. in Condensed Matter Physics from Huazhong University of Science and Technology in 2007 and 2010, respectively. He joined the Institute of Geophysics and Geomatics at the China University of Geosciences (Wuhan) in 2010. From July 2013 to July 2014, he acted as a Visiting Scholar at the University of Tennessee-Knoxville, USA. He currently is a professor of Geological Resources and Geological Engineering (since 2015). Furthermore, he is the founder and Editor-in-Chief of *Advances in Geo-Energy Research* and serves as an Associate/Guest Editor for other international journals. Dr. Cai focuses on the petrophysical characterization and micro-transport phenomena in porous media, as well as fractal theory and its application. He has published more than 110 peer-refereed journal articles, three books, and numerous book chapters.

Zhien Zhang is currently a postdoctoral researcher in the William G. Lowrie Department of Chemical and Biomolecular Engineering at Ohio State University. His research interests include advanced processes and materials (i.e., membranes) for CO2 capture, CCUS processes, gas separation, and gas hydrates. Dr. Zhang has published 80+ journal papers and 10+ editorials in high-impact journals, including Renewable and Sustainable Energy Reviews. He has written three Hot Papers (top 0.1%) and 10 Highly Cited Papers (top 1%). He also serves as an Editor or Guest Editor for several international journals, including *Applied Energy, Fuel, and Journal of Natural Gas Science and Engineering.* He currently works as a Visiting Professor at the University of Cincinnati.

Qinjun Kang is a senior scientist at the Earth and Environmental Sciences (EES) Division of Los Alamos National Laboratory (LANL). His current research focuses on the modeling and simulation of transport and interfacial processes in porous media at the pore (nano/meso) scale, and on multiscale models bridging different scales. His work is applied to problems in a broad range of engineering and science disciplines, including the geologic storage of carbon dioxide and nuclear waste, conventional and unconventional hydrocarbon exploration and production, the fate and transport of underground contaminants, and the engineering of energy storage and conversion devices (e.g., fuel cells and batteries). Dr. Kang has coauthored more than 100 publications, which have been cited nearly 6000 times. He has been invited to give 20+ talks at various international and national conferences, universities, national laboratories, and industries, and has organized/chaired sessions at numerous conferences. He is currently a member of the Editorial Board or an Associate Editor for multiple journals, and has served as a reviewer for over 50 journals and numerous funding agencies, including DOE and NSF. Dr. Kang has been a top publisher in the EES Division at LANL since 2010.

Harpreet Singh is a postdoctoral research associate at the National Energy Technology Laboratory, Morgantown, West Virginia. He is a reservoir engineer by training and holds Master's and Ph.D. degrees in petroleum engineering from the University of Texas at Austin. He has published 23 manuscripts, 17 of them as the first author, and authored one chapter in a book published by Elsevier. He currently serves as an Associate Editor for *Advances in Geo-Energy Research*, and has been a Guest Editor for journals such as *Fuel* and *Energies*. He has served as a reviewer for over 20 journals.

Editorial

Recent Advances in Flow and Transport Properties of Unconventional Reservoirs

Jianchao Cai [1,*], Zhien Zhang [2], Qinjun Kang [3] and Harpreet Singh [4]

1 Institute of Geophysics and Geomatics, China University of Geosciences, Wuhan 430074, China
2 William G. Lowrie Department of Chemical and Biomolecular Engineering, The Ohio State University, Columbus, OH 43210, USA; zhienzhang@hotmail.com
3 Computational Earth Science Group, Earth and Environmental Sciences Division, Los Alamos National Laboratory, Los Alamos, NM 87545, USA; qkang@lanl.gov
4 National Energy Technology Laboratory, Morgantown, WV 26505, USA; harpreet.singh@netl.doe.gov
* Correspondence: caijc@cug.edu.cn

Received: 4 March 2019; Accepted: 13 May 2019; Published: 16 May 2019

Abstract: As a major supplement to conventional fossil fuels, unconventional oil and gas resources have received significant attention across the globe. However, significant challenges need to be overcome in order to economically develop these resources, and new technologies based on a fundamental understanding of flow and transport processes in unconventional reservoirs are the key. This special issue collects a series of recent studies focused on the application of novel technologies and theories in unconventional reservoirs, covering the fields of petrophysical characterization, hydraulic fracturing, fluid transport physics, enhanced oil recovery, and geothermal energy.

Keywords: unconventional reservoirs; petrophysical characterization; fluid transport physics

---

1. Introduction

Unconventional reservoirs, such as shale, coal, and tight sandstone reservoirs, are complex and highly heterogeneous, generally characterized by low porosity and ultralow permeability. Additionally, the strong physical and chemical interactions between fluids and pore surfaces further lead to the inapplicability of conventional approaches for characterizing fluid flow in these porous reservoir rocks [1]. Therefore, new theories, techniques, and geophysical and geochemical methods are urgently needed to characterize petrophysical properties, fluid transport, and their relationships at multiple scales for improving production efficiency from unconventional reservoirs.

Petrophysical characterization covers the study of the physical and chemical properties of rock and its interactions with fluids, which has many applications in different industries, especially in the oil and gas industries. The key parameters studied in petrophysics are lithology, porosity, water saturation, permeability, and density. Petrophysical characterization is the basis for understanding the special properties of unconventional reservoirs.

Fluid transport physics in micropore structures and macro-reservoirs covers a wide range of research studies including hydrocarbon extraction, geosciences, environmental issues, hydrology, and biology. Implementing reliable methods for the characterization of fluid transport at multiple scales is crucial in many fields, especially in unconventional reservoirs and rocks.

Hydraulic fracturing is currently considered as one of the most important stimulation methods in the oil and gas industry, which significantly improves the productivity of the wells and the overall recovery factor, especially for low-permeability reservoirs, such as shale-gas and tight-gas reservoirs. Problems that are associated with unconventional oil and gas production in hydraulic fracturing operations include aqueous phase trapping, diversion mechanisms of fracture networks, and fluid incompatibility with the formation.

This collection associated with the special issue in *Energies* emphasizes fundamental innovations and gathers 21 recent papers on novel applications of new techniques and theories in unconventional reservoirs.

2. Overview of Work Presented in This Special Issue

The papers published in this special issue present new advancements in the characterization of porous media and the modeling of multiphase flow in porous media. These studies are classified into five categories.

The first category focuses on petrophysical characterization. By means of a set of experiments including scanning electron microscopy, mercury intrusion capillary pressure, X-ray diffraction, and nuclear magnetic resonance measurements, Xu et al. [2] characterized the pore structure of a tight oil reservoir in Permain Lucaogou formation of Jimusaer Sag and further performed a consecutive prediction for its pore structures. The pore types of this formation were mainly divided into four categories, and the capillary pressure curve and the T2 distribution data were analyzed in depth.

A matrix–fracture interaction model was developed by Liu et al. [3] to investigate the transient response of coal deformation and permeability to the temporal and spatial variations of effective stresses under mechanically unconstrained conditions. The impacts of fracture properties, initial matrix permeability, injection processes, and confining pressure were separately evaluated through the developed model.

Base on a low-pressure nitrogen adsorption experiment and fractal theory, Li et al. [4] studied the characteristics of nanopore structure in shale, tight sandstone, and mudstone, with an emphasis on the relationships between pore structure parameters, mineral compositions, and fractal dimensions. The relationships among average pore diameter, Brunner–Emmet–Teller specific surface area, pore volume, porosity, and permeability were also discussed.

Ma et al. [5] introduced the local force to define the interactions between the matrix and the fracture and derived a set of partial differential equations to define the full coupling of rock deformation and gas flow both in the matrix and fracture systems. Permeability evolution profiles during unconventional gas extraction were obtained by solving the full set of cross-coupling formulations.

A comprehensive experiment, including petrophysical measurements (porosity and permeability), pore structure measurements (low-field nuclear magnetic resonance and carbon dioxide/nitrogen adsorption), geochemical measurements (vitrinite reflectance, pyrolysis, and residual analysis), and petrological analysis (X-ray diffraction, thin section, scanning electron microscopy, and isothermal adsorption measurement), was designed by Fan et al. [6] to explore the influential and controlling factors of the gas adsorption capacity.

By using the data from casting thin section and mercury intrusion capillary pressure experiments, Sha et al. [7] investigated the pore structure characterization, permeability estimation, and fractal characteristics of Carboniferous carbonate reservoirs.

The second category focuses on fluid transport at multiple scales. Based on Swartzendruber equation and conformable derivative approach, as well as the modified Hertzian contact theory and fractal geometry, Lei et al. [8] developed a novel nonlinear flow model for tight porous media, which manifests the most important fundamental controls on low-velocity nonlinear flow. According to this model, the average flow velocity in tight porous media is a function of microstructural parameters of the pore space, rock lithology, and differential order, as well as hydraulic gradients and threshold hydraulic gradients. Moreover, the relationships between average flow velocity and effective stress, the rougher pore surfaces, and rock elastic modulus were further discussed.

Chen et al. [9] proposed a novel model for characterizing boundary layer thickness and fluid flow at microscales, which has a wide range of applications proved mathematically. Based on this model, the effects of fluid–solid interaction on flow in microtubes and tight formation were analyzed in depth.

Two different productivity models, the steady-state productivity model of shale horizontal wells with volume fracturing and the transient productivity calculation model of fractured wells, were derived by Zeng et al. [10]. The former considered the multiscale flowing states, shale gas desorption, and diffusion, while the latter combined the material balance equation. Furthermore, the horizontal well productivity prediction and the analysis of influencing factors were carried out.

In order to describe the pressure-transient behaviors in shale gas reservoirs in a way that considers the stimulated reservoir volume region with anomalous diffusion and fractal features, an improved analytical model was established by Tao et al. [11] through introducing the time-fractional flux law. Base on this model, the influences of relevant parameters, such as fractal-anomalous diffusion, stress sensitivity, absorption, and Knudsen diffusion, on the pressure-transient response were further analyzed through sensitivity analysis.

By introducing an improved pseudopotential multirelaxation-time lattice Boltzmann method, Wang et al. [12] simulated the fluid flow in a microfracture. The effects of contact angles, driving pressure, and the liquid–gas density ratio on the slip length were discussed.

Based on the dual-media theory and discrete-fracture network models, Ren et al. [13] built a mathematical flow model of a stimulated reservoir volume fractured horizontal well with multiporosity and multipermeability media. The differences of flow regimes between triple-porosity, dual-permeability and triple-porosity, triple-permeability models were identified. Moreover, the productivity contribution degree of multimedium was analyzed.

Tang et al. [14] summarized the flow law in shale gas reservoirs and established a three-dimensional composite model, which uses dual media to describe matrix-natural microfractures and utilizes discrete media to describe artificial fractures. The production of multisection fractured horizontal wells in a rectangular shale gas reservoir was described, considering multiscale flow mechanisms in the matrix, such as gas desorption, the Klinkenberg effect, and gas diffusion.

The third category focuses on hydraulic fracturing. By means of the extended finite element method, Wang et al. [15] investigated the diversion mechanisms of a fracture network in tight formations with frictional natural fractures. The effects of some key factors, for example, the location of natural fracture, the intersection angle between natural fracture and hydro-fracture, the horizontal stress difference, and the fluid viscosity on the mechanical diversion behavior of the hydro-fracture, were analyzed in detail.

Kamal et al. [16] developed a new smart fracturing fluid system mainly consisting of a water-soluble polymer and chelating agent, which can be either used for proppant fracturing (high pH) or acid fracturing (low pH) operations in tight as well as conventional formations. The optimal conditions and concentration of this fracturing fluid system were determined by performing thermal stability, rheology, Fourier transform infrared spectroscopy, and core flooding experiments.

By measuring the solution viscosity, Tang et al. [17] investigated the effects of hydrophobic chain, spacer group, concentration, temperature, and addition of nano-MgO on the viscosity of sulfonate Gemini surfactant solution. Moreover, their micellar microstructures were observed by Cryo-SEM. Further, the thickening mechanism of sulfonate Gemini surfactant was investigated by correlating the relationship between solution viscosity and its microstructure.

The fourth category focuses on enhanced oil recovery. A novel depletion laboratory experimental platform and its evaluation method for a tight oil reservoir were developed by Chen et al. [18] to effectively measure the oil recovery and pressure propagation over pressure depletion. On this platform, under different temperatures, formation pressure coefficients, and oil property conditions, the recovery factor as well as the real-time monitoring of the pressure propagation in the process of reservoir depletion were measured to reveal the drive mechanism and recovery factor of tight oil reservoir depletion.

Lyu et al. [19] applied the nuclear magnetic resonance technique to explore the spontaneous imbibition mechanism and the oil displacement recovery by imbibition in tight sandstones under all face open boundary conditions. The distribution of remaining oil and the effect of microstructures on imbibition were analyzed.

Through three groups of core displacement experiments with cores containing different clay mineral compositions, Jiang et al. [20] studied the effect of different clay mineral compositions on low-salinity water flooding. Additionally, the properties of the effluent were determined in different flooding stages, and the mechanism of enhanced oil recovery effect of low-salinity water flooding was analyzed.

The fifth category focuses on geothermal energy. Based on hydrogeochemical and isotopic constraints, the deep circulation of the groundwater flow system was surveyed by Long et al. [21] to elucidate the origin of the geothermal fluids and the source of solutes and to discern the mixing and hydrogeochemical alteration. The conceptual models and mechanisms for the deep circulation of the groundwater flow system were further discussed.

Combining the fracture continuum method and genetic algorithm, a well-placement optimization framework was proposed by Zhang et al. [22] to address the optimization of the well-placement for an enhanced geothermal system. The optimization efficiency and effect of this framework were further analyzed.

3. Conclusions

Many researchers around the world from different areas, ranging from natural sciences to engineering fields, have been working on the characterization of petrophysical properties for unconventional reservoirs, fluid transport at multiscales, and technologies for the efficient development of unconventional resources. The aim of this special issue is to provide new technologies and theories of characterizing petrophysical properties, fluid transport, and their relationships at multiple scales in unconventional reservoirs. Clearly, the studies covered by this special issue will be helpful to the economic and effective development of unconventional oil and gas resources.

Author Contributions: The authors contributed equally to this work.

Acknowledgments: The guest editors would like to acknowledge MDPI for the invitation to act as the guest editors of this special issue in "*Energies*" with the kind cooperation and support of the editorial staff. The guest editors are also grateful to the authors for their inspiring contributions and the anonymous reviewers for their tremendous efforts. The first guest editor, J.C., would like to thank the National Natural Science Foundation of China for supporting his series of studies on flow and transport properties in porous media. H.S. acknowledges the support in part by an appointment to the National Energy Technology Laboratory Research Participation Program, sponsored by the U.S. DOE and administered by the Oak Ridge Institute for Science and Education.

Conflicts of Interest: The authors declare no conflicts of interest.

References

1. Cai, J.; Hu, X. *Petrophysical Characterization and Fluids Transport in Unconventional Reservoirs*; Elsevier: Amsterdam, The Netherlands, 2019; p. 352.
2. Xu, Z.; Zhao, P.; Wang, Z.; Ostadhassan, M.; Pan, Z. Characterization and consecutive prediction of pore structures in tight oil reservoirs. *Energies* **2018**, *11*, 2705. [CrossRef]
3. Liu, X.; Sheng, J.; Liu, J.; Hu, Y. Evolution of coal permeability during gas injection—From initial to ultimate equilibrium. *Energies* **2018**, *11*, 2800. [CrossRef]
4. Li, X.; Gao, Z.; Fang, S.; Ren, C.; Yang, K.; Wang, F. Fractal characterization of nanopore structure in shale, tight sandstone and mudstone from the Ordos basin of china using nitrogen adsorption. *Energies* **2019**, *12*, 583. [CrossRef]
5. Ma, X.; Li, X.; Zhang, S.; Zhang, Y.; Hao, X.; Liu, J. Impact of local effects on the evolution of unconventional rock permeability. *Energies* **2019**, *12*, 478. [CrossRef]
6. Fan, Z.; Hou, J.; Ge, X.; Zhao, P.; Liu, J. Investigating influential factors of the gas absorption capacity in shale reservoirs using integrated petrophysical, mineralogical and geochemical experiments: A case study. *Energies* **2018**, *11*, 3078. [CrossRef]
7. Sha, F.; Xiao, L.; Mao, Z.; Jia, C. Petrophysical characterization and fractal analysis of carbonate reservoirs of the eastern margin of the pre-Caspian basin. *Energies* **2018**, *12*, 78. [CrossRef]

8. Lei, G.; Cao, N.; Liu, D.; Wang, H. A non-linear flow model for porous media based on conformable derivative approach. *Energies* **2018**, *11*, 2986. [CrossRef]
9. Chen, M.; Cheng, L.; Cao, R.; Lyu, C. A study to investigate fluid-solid interaction effects on fluid flow in micro scales. *Energies* **2018**, *11*, 2197. [CrossRef]
10. Zeng, F.; Peng, F.; Guo, J.; Xiang, J.; Wang, Q.; Zhen, J. A transient productivity model of fractured wells in shale reservoirs based on the succession pseudo-steady state method. *Energies* **2018**, *11*, 2335. [CrossRef]
11. Tao, H.; Zhang, L.; Liu, Q.; Deng, Q.; Luo, M.; Zhao, Y. An analytical flow model for heterogeneous multi-fractured systems in shale gas reservoirs. *Energies* **2018**, *11*, 3422. [CrossRef]
12. Wang, P.; Wang, Z.; Shen, L.; Xin, L. Lattice Boltzmann simulation of fluid flow characteristics in a rock micro-fracture based on the pseudo-potential model. *Energies* **2018**, *11*, 2576. [CrossRef]
13. Ren, L.; Wang, W.; Su, Y.; Chen, M.; Jing, C.; Zhang, N.; He, Y.; Sun, J. Multiporosity and multiscale flow characteristics of a stimulated reservoir volume (SRV)-fractured horizontal well in a tight oil reservoir. *Energies* **2018**, *11*, 2724. [CrossRef]
14. Tang, C.; Chen, X.; Du, Z.; Yue, P.; Wei, J. Numerical simulation study on seepage theory of a multi-section fractured horizontal well in shale gas reservoirs based on multi-scale flow mechanisms. *Energies* **2018**, *11*, 2329. [CrossRef]
15. Wang, D.; Shi, F.; Yu, B.; Sun, D.; Li, X.; Han, D.; Tan, Y. A numerical study on the diversion mechanisms of fracture networks in tight reservoirs with frictional natural fractures. *Energies* **2018**, *11*, 3035. [CrossRef]
16. Kamal, M.; Mohammed, M.; Mahmoud, M.; Elkatatny, S. Development of chelating agent-based polymeric gel system for hydraulic fracturing. *Energies* **2018**, *11*, 1663. [CrossRef]
17. Tang, S.; Zheng, Y.; Yang, W.; Wang, J.; Fan, Y.; Lu, J. Experimental study of sulfonate Gemini surfactants as thickeners for clean fracturing fluids. *Energies* **2018**, *11*, 3182. [CrossRef]
18. Chen, W.; Zhang, Z.; Liu, Q.; Chen, X.; Opoku Appau, P.; Wang, F. Experimental investigation of oil recovery from tight sandstone oil reservoirs by pressure depletion. *Energies* **2018**, *11*, 2667. [CrossRef]
19. Lyu, C.; Wang, Q.; Ning, Z.; Chen, M.; Li, M.; Chen, Z.; Xia, Y. Investigation on the application of NMR to spontaneous imbibition recovery of tight sandstones: An experimental study. *Energies* **2018**, *11*, 2359. [CrossRef]
20. Jiang, S.; Liang, P.; Han, Y. Effect of clay mineral composition on low-salinity water flooding. *Energies* **2018**, *11*, 3317. [CrossRef]
21. Long, X.; Zhang, K.; Yuan, R.; Zhang, L.; Liu, Z. Hydrogeochemical and isotopic constraints on the pattern of a deep circulation groundwater flow system. *Energies* **2019**, *12*, 404. [CrossRef]
22. Zhang, L.; Deng, Z.; Zhang, K.; Long, T.; Desbordes, J.; Sun, H.; Yang, Y. Well-placement optimization in an enhanced geothermal system based on the fracture continuum method and 0-1 programming. *Energies* **2019**, *12*, 709. [CrossRef]

Article

Characterization and Consecutive Prediction of Pore Structures in Tight Oil Reservoirs

Zhaohui Xu [1,*], Peiqiang Zhao [2], Zhenlin Wang [3], Mehdi Ostadhassan [4] and Zhonghua Pan [5]

1 College of Geosciences, China University of Petroleum, Beijing 102249, China
2 Institute of Geophysics and Geomatics, China University of Geosciences, Wuhan 430074, China; zhaopq@cug.edu.cn
3 Research Institute of Exploration and Development, Xinjiang Oilfield Company, PetroChina, Karamay 834000, China; wzhenl@petrochina.com.cn
4 Petroleum Engineering Department, University of North Dakota, Grand Forks, ND 58202, USA; mehdi.ostadhassan@und.edu
5 Wuhan Geomatic Institute, Wuhan 430022, China; cugpzh@outlook.com
* Correspondence: xuzhaohui@cup.edu.cn; Tel.: +86-187-0138-0799

Received: 28 September 2018; Accepted: 9 October 2018; Published: 11 October 2018

Abstract: The Lucaogou Formation in Jimuaser Sag of Junggar Basin, China is a typical tight oil reservoir with upper and lower sweet spots. However, the pore structure of this formation has not been studied thoroughly due to limited core analysis data. In this paper, the pore structures of the Lucaogou Formation were characterized, and a new method applicable to oil-wet rocks was verified and used to consecutively predict pore structures by nuclear magnetic resonance (NMR) logs. To do so, a set of experiments including X-ray diffraction (XRD), mercury intrusion capillary pressure (MICP), scanning electron microscopy (SEM) and NMR measurements were conducted. First, SEM images showed that pore types are mainly intragranular dissolution, intergranular dissolution, micro fractures and clay pores. Then, capillary pressure curves were divided into three types (I, II and III). The pores associated with type I and III are mainly dissolution and clay pores, respectively. Next, the new method was verified by "as received" and water-saturated condition T_2 distributions of two samples. Finally, consecutive prediction in fourteen wells demonstrated that the pores of this formation are dominated by nano-scale pores and the pore structure of the lower sweet spot reservoir is more complicated than that in upper sweet spot reservoir.

Keywords: Lucaogou Formation; tight oil; pore structure; prediction by NMR logs

1. Introduction

As a major unconventional resource, tight oil reservoirs have received significant attention for exploration and development all around the world [1–3]. Tight oil reservoirs are complex and highly heterogeneous, generally characterized by low porosity and ultra-low permeability [4,5]. Single wells have no natural production capacity, which requires horizontal drilling and hydraulic fracturing to obtain economic flow [5–8]. It is necessary to evaluate various properties of such reservoirs for a better exploitation of the resources. However, macroscopic petrophysical parameters such as porosity, permeability, and saturation cannot satisfy adequate evaluation of the effectiveness of tight oil reservoirs. In this regard, pore structures, in particular determine reservoir storage capacity and control rock transportation characteristics, represent microscopic properties of the rock [9–12]. Therefore, characterization and consecutive prediction of rock pore structure in wells is a key task in the study of tight oil reservoirs.

The Permian Lucaogou Formation of Jimusaer Sag, Junggar Basin, China is a typical tight oil reservoir which has been studied previously in terms of the pore structures. Kuang et al. [13],

Zhang et al. [14], Zhou [15] and Su et al. [16] used diverse imaging techniques such as CT-scanning, SEM and FIB-SEM image analysis to qualitatively characterize the pore structures. They concluded that pore types include organic matter pores, mineral pores, inter-crystalline pore, dissolved pores, and micro cracks. Zhao et al. [17] presented that the median capillary radius of this reservoir ranges from 0.0063 to 0.148 μm with an average of 0.039 μm. Zhao et al. [18] studied the complexity and heterogeneity of pore structures based on multifractal characteristics of nuclear magnetic resonance (NMR) transverse relaxation (T_2) distributions. Wang et al. [19] investigated pore size distributions and fractal characteristics of this formation by combining high pressure and constant rate mercury injection data. However, the limited number of core samples could not reflect general properties of this formation. The NMR logging which is consecutively recording the vertical variations of transverse relaxation time can reveal pore distributions and is widely used to overcome the discrete data points that core sample analysis owns.

Researchers have conducted extensive studies on the construction of mercury intrusion capillary pressure curves by NMR T_2 distributions obtained in laboratory [20–27]. The pore structure evaluation methods by NMR technique are based on the fact the rocks are water-saturated and hydrophilic. However, in oil reservoirs, it is necessary to correct the effect of hydrocarbons on T_2 spectra of NMR logging. Volokin and Looyedtijn [22,23] first studied the morphological correction of T_2 spectra of NMR logging in hydrocarbon-bearing rocks. The basic idea is that the bound water of the T_2 distribution is constant, and hydrocarbon would only affect the free fluid portion of the T_2 distribution. Therefore, when performing a hydrocarbon-containing correction on the T_2 distribution, it is only required to correct the T_2 signal of the free fluid portion and remain the bound fluid of T_2 signal intact. Xiao et al. [28] established a method for constructing capillary pressure curves based on J function and Schlumberger Doll Research (SDR) model. This method used T_2 logarithmic mean value (T_{2lm}) as an input parameter, which makes it possible for the correction of T_2 distributions regarding hydrocarbons. This is possible because T_{2lm} can be calibrated by core values. Hu et al. [29] proposed a novel method for hydrocarbon corrections where T_2 distribution measured by short echo time (T_E) was used to construct the T_2 distribution under full-water conditions with long T_E time. The difference between the measured and constructed water-saturated state T_2 distributions determines the oil signal and the water signal, thereby the correction of the hydrocarbon-containing state T_2 distribution would become achievable. Ge et al. [30] proposed a correction method through extracting oil signals from the echoes, which has been already applied to carbonate reservoirs. Xiao et al. [31] proposed a method to remove the effect of hydrocarbons on NMR T_2 response based on a point-by-point calibration method. However, the application of these methods would be challenging when the wettability of the reservoir appears to be oleophilic or neutral. This is because the bulk transversal relaxation time could not be ignored according to NMR relaxation mechanism [32–34]. Zhao [35] proposed a new method for evaluating pore structures of reservoirs with neutral wettability and oil-wetting characteristics, but the method is not firmly verified.

In this research, the major objectives are to: (a) characterize the pore structures by MICP data and SEM images; (b) further confirm the Zhao method [35] by "as-received" and water saturated state T_2 distributions; and finally (c) predict the global features of pore structures via field NMR logs.

2. Methods

2.1. Samples and Experiments

Samples were drilled from the Permian Lucaogou Formation in Jimusar Sag, Junggar Basin. The Junggar Basin is the second largest inland basin in China, which is located in north of the Xinjiang Province, Northwest China. The Jimusaer sag is structurally located in the eastern uplift of the Junggar Basin, adjacent to the Fukang Fault in the south, and the Santai Oilfield and the North Santai Oilfield in the west [36]. The Permian system is the main source rock strata in the Junggar Basin. The target Lucaogou Formation was developed in Permain System, which from bottom to top includes Jiangjunmiao, Jingjinggouzi, Lucaogou and Wutonggou Formations. The Lucaogou Formation in the Jimsar Sag is a set of stratigraphic layers deposited in an evaporitic (salt lake) environment. The formation is generally composed of dolomite dark argillaceous rocks and fine sandstones. The dolomite is mostly interbedded lacustrine deposits. The reservoir is tight, the physical properties are poor, and the dark mudstone has a high abundance of organic matter [13,37]. The Lucaogou formation consists of two "sweet spot" reservoirs and the shale source rocks is deposited between these two sweet spots [13,37]. The average porosity and permeability for "sweet spot" reservoirs are 9.93% and 0.0233 mD. The average porosity and permeability for non-sweet spot reservoirs are 7.03% and 0.0013 mD. Figure 1 depicts the depth contour of the top of Lucaogou Formation and location of the studied wells.

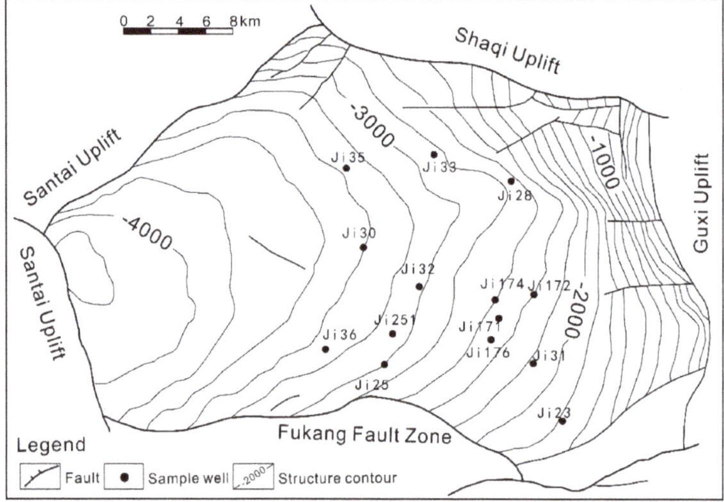

Figure 1. Depth contour in meters of the top of Lucaogou Formation and location of wells.

Mineralogical compositions of samples are determined using X-ray diffraction (XRD) analysis on non-oriented powdered samples (100 mesh) using an X-ray diffractometer equipped with a copper X-ray tube that operated at 30 kV and 40 mA [18]. The scan angle range was 5–90° at a speed of 2°/min. SEM was performed on a S4800 scanning electron microscope (Hitachi, Tokyo, Japan) with a lowest pixel resolution of 1.2 nm and accelerating voltage of 30 kV, following the standards of SY/T 5162-2014 China.

Core plugs were subjected to drying prior to porosity and permeability measurements with a helium porosimeter. A net confining pressure of 5000 psi (34.47 MPa) was carried on to simulate the formation pressure during the measurements. Mercury injection capillary pressure curves were acquired on a mercury porosimeter by following the China Standard of SY/T 5346-2005. Before the measurements, the samples were subjected to oil washing and drying at 105 °C to a constant weight. The minimum intrusion pressure was set as 0.005 MPa and the maximum intrusion pressure was as high as 163.84 MPa, corresponding to a pore-throat radius of roughly 4.5 nm.

To verify the method for predicting the pore structures, two rock samples were subjected to NMR T_2 distributions measurements at the "as received" and water saturated conditions in the lab using a Geospace2 instrument (Oxford, UK). After the measurements on "as received" state sample, core plugs were cleaned, dried, vacuumed and fully water saturated for water saturated conditions NMR measurements. The resonant frequency of a Geospace2 instrument is 2 MHZ with the polarization time or waiting time (T_w), the echo spacing, the number of echoes and the number of scans as 10,000 ms,

0.3 ms, 4096 and 128, respectively. When the echoes are recorded, the T_2 spectra are able to calculate using the Bulter-Reeds-Dawson (BRD) inversion method [38].

2.2. Prediction Method of Pore Structure by NMR Logs

According to NMR theory, for the T_2 distribution of water saturated and hydrophilic rock samples, the following equation [32,33] was deduced:

$$\frac{1}{T_2} = \rho \frac{F_s}{r} \tag{1}$$

where r is the pore radius (μm); ρ is surface relaxivitity (μm/s); F_s is the pore shape factor, equals to 2 and 3 for cylindrical and spherical pores, respectively. In this study, the pores are considered as cylindrical.

Known by reservoir physics, the relationship between injection pressure and pore throat radius is given by [39]:

$$P_c = \frac{2\sigma \cos\theta}{R_c} \tag{2}$$

where P_c is the capillary pressure (MPa); σ is the surface tension (mN/m); θ is the contact angle of mercury in air (°); and R_c is the pore throat radius (μm).

Assuming R_c to be proportional to r, both NMR and MICP would quantify similar pore size distributions. Generally, the following equation [22] is used:

$$P_c = C\frac{1}{T_2} \tag{3}$$

where C is the coefficient which can be obtained by capillary pressure curves and nuclear magnetic resonance experiments of rock samples.

The above equations can also be applied to conventional water-wet reservoirs. As mentioned earlier, the reservoirs of Lucaogou Formation in Jimusaer Basin, are either neutral or oil-wet. Zhao [35] proposed a method for evaluating pore structures of oil-wet reservoirs that has been applied to tight oil reservoirs. He realized that the bigger pores in tight oil reservoirs are highly oil saturated, while the formation water is mainly occupies smaller pores. The bigger pores are oleophilic and the smaller pores are hydrophilic. The surface relaxivity of oleophilic pores to oil is lower than hydrophilic pores to water [40,41], and the lower surface relaxivity would lead to an increase in relaxation time. Hence, the long-relaxation signal of the NMR T_2 spectra of tight oil reservoir rocks is mainly the relaxation signal of oil (referred to as oil spectrum), while the short relaxation signal of T_2 spectrum is mainly the relaxation of water signal (referred to as water spectrum).

If the water saturation at a certain depth of the reservoir is known, the $T_{2cutoff}$ value for water can be determined by the following equation [35]:

$$S_w = (\sum_{i=i}^{T_{2cutoff}} \phi_i T_{2i}) / \sum_{i=1}^{n} \phi_i T_{2i} \tag{4}$$

where S_w *is* water saturation (%); $T_{2cutoff}$ is for determining the water and oil (ms); ϕ_i and T_{2i} are porosity component (%) and T_2 corresponding to the ith component; n is the total number of T_2 distribution.

After determining the $T_{2cutoff}$ value, the water signal and the oil signal of the T_2 spectra can be respectively converted into the size distributions for pores containing water and oil by utilizing the hydrophilic pore surface relaxivity and the oleophilic pore surface relaxivity:

$$r_o = 2\rho_o T_2 \tag{5}$$

$$r_w = 2\rho_w T_2 \quad (6)$$

where r_o and r_w respectively represent the radius of pores containing oil and water (μm); ρ_o and ρ_w respectively represent surface relaxivitity of oleophilic pore and hydrophilic pore (μm/s).

By superposing the size distribution of the water-containing pores with the size distribution of the oil-bearing pores, the pore size distribution of the whole rock can be obtained. Then, the Equations (2) and (3) can be employed to construct the capillary pressure curves.

The oil and water two-phase signals are cut directly by the $T_{2cutoff}$ values, and the resulting pore size distribution would not be smooth. The weight function of the pore fluid was introduced as [35]:

$$S(T_2) = \frac{1}{1 + (T_2/T_{2cutoff})^m} \quad (7)$$

where m is the coefficient that controls the width of the transition zone for the water-containing and oil-bearing pores.

3. Results and Discussion

3.1. Mineralogical Compositions

The mineral compositions of sixteen samples obtained from the XRD analysis are listed in Table 1. As can be observed from this table, plagioclase and dolomite are the two most abundant minerals. The plagioclase contents vary from 13.7% to 44.4% with an average value of 30.9%. The dolomite content in the samples varies between 0–49.4% with an average value of 28.2%. The next most abundant mineral is quartz, ranging from 13% to 30% with an average value of 19.4%. Each sample has clay and K-feldspar minerals, with the average values of 8.9% and 4.4%, respectively. The calcite content of these samples found to vary significantly. Seven samples out of sixteen did not contain calcite, while the maximum content of calcite reaches 22.9% in the rest of the samples. In addition, a small fraction of pyrite and siderite was also detected in some samples.

Table 1. Mineralogical composition (wt.%) of the sixteen core samples of tight oil reservoirs.

No.	Clay	Quartz	K-Feldspar	Plagioclase	Calcite	Dolomite	Pyrite	Siderite
1	4.2	15.9	2.2	35.3	17.5	24.9	0.0	0.0
2	6.3	21.4	7.9	37.5	1.0	18.9	0.0	7.0
3	3.4	13.0	6.1	27.1	8.7	41.7	0.0	0.0
4	9.8	16.5	3.9	41.0	13.3	15.0	0.0	0.5
5	5.9	15.8	4.9	32.5	0.5	40.1	0.3	0.0
6	7.5	16.3	5.0	38.4	22.9	9.9	0.0	0.0
7	6.0	15.6	4.4	25.4	0.0	48.6	0.0	0.0
8	6.9	17.8	5.4	44.4	0.0	23.6	0.0	1.9
9	12.2	24.7	4.5	31.1	0.0	26.5	1.0	0.0
10	13.9	23.2	3.9	29.4	21.9	7.7	0.0	0.0
11	18.2	16.4	4.7	27.7	0.0	32.5	0.5	0.0
12	10.8	20.3	3.8	25.6	0.0	38.5	1.0	0.0
13	11.6	22.6	2.5	13.7	0.0	49.4	0.0	0.2
14	7.6	32.0	3.9	34.1	22.0	0.0	0.4	0.0
15	11.8	18.3	5.8	32.3	0.0	31.3	0.0	0.5
16	6.6	21.0	1.8	18.2	5.8	41.2	5.4	0.0
Ave.	8.9	19.4	4.4	30.9	7.1	28.2	0.5	0.6

3.2. Pore Types

According to the SEM image analysis, the primary pores in the tight oil reservoirs of the Lucaogou Formation are very rare, and the main pore types are secondary pores developed during the diagenesis stage. The pores of the studied areas can be divided into the four types: intragranular dissolution pores, intergranular dissolution pores, micro fractures and clay pores.

Intergranular dissolved pores were formed by the selective corrosion of the edge of clastic grains, early intergranular cement and matrix. This type of pore is the main reservoir porosity in the Lucaogou Formation in the studied area. These pores are mainly distributed between the dolomitic sand crumbs and belong to cement dissolved pores. Intergranular dissolved pores usually develop between albite (a type of sodium feldspar) in dolomitic siltstone. The pore sizes are commonly less than 10μm, as shown in Figure 2a–c.

Intragranular dissolved pores refer to pores formed inside the grains or grains due to selective dissolution. They are also common pore types in the reservoir understudy of the Lucaogou Formation (Figure 2c,d). The dissolved pores in the sand are mainly formed by the dissolution of albite; the dissolved pores in the debris often show the dissolution of sodium feldspar, while the dissolved pores in the dolomite are usually the result of residual dissolution of internal calcite.

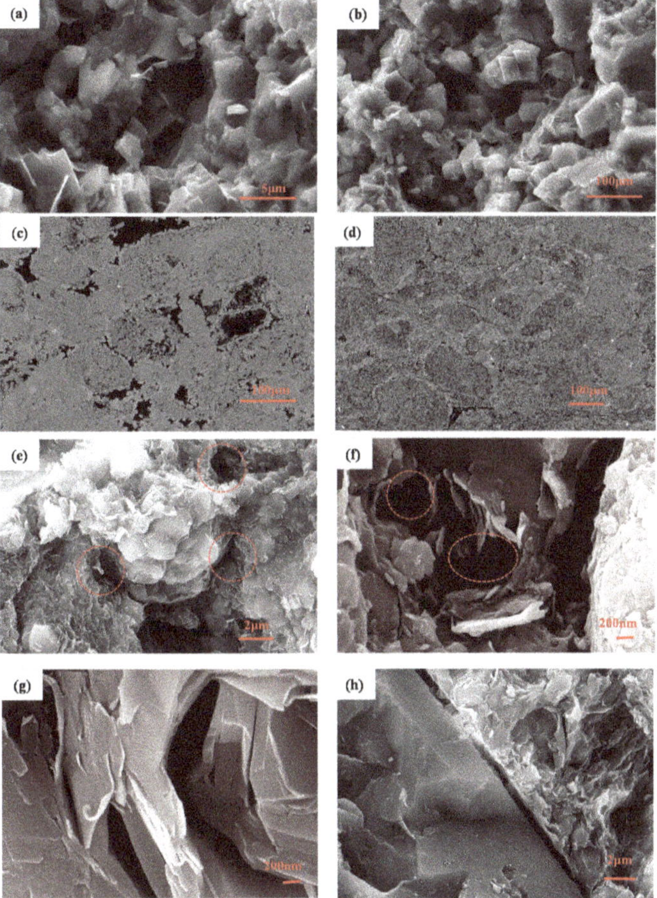

Figure 2. The pore types according to SEM analysis. (**a**) Intergranular dissolved pores; (**b**) Intergranular dissolved pores; (**c**) Intergranular and intragranular dissolved pores; (**d**) Intergranular and intragranular dissolved pores; (**e**) Illite/smectite mixed layer clay pores; (**f**) Chlorite clay pores; (**g**) Fracture pore; (**h**) Fracture pore.

Clay pores refer to pores within clay aggregates of the studied samples. The clay pores were found in the illite/smectite mixed layers (Figure 2e) and chlorite minerals (Figure 2f). The sizes of the clay pores are smaller than the dissolution pores and mainly distributed between 300 nm and 800 nm in size. Fracture pores refer to the pores that penetrate into the particles and resemble cracks. They are not structural cracks in the traditional sense, but the fluid channel formed by organic acid dissolution (Figure 2g,h).

3.3. Petrophysiccal Properties and Mercury Injection Capillary Curves

The porosity, permeability and pore structure parameters obtained from MICP experiments are listed in Table 2. The porosity ranges from 7.38% to 20.1% with an average value of 12.83%. The permeability fluctuates from 0.0023 mD to 0.1487 mD. The logarithmic average value of the permeability is 0.01 mD. Only two samples (No. 1 and 2) were measured with the permeability greater than 0.1 mD, representing the tight nature of the studied samples.

Table 2. Petrophysical parameters and types of tight oil reservoir sample.

No.	Porosity (%)	Permeability (mD)	P_d (MPa)	P_{50} (MPa)	S_{max} (%)	R_m (μm)	Type
1	14.22	0.1142	0.83	6.32	90.93	0.26	I
2	16.02	0.1487	1.19	6.51	99.25	0.19	I
3	15.19	0.0799	1.28	4.96	96.99	0.18	I
4	14.14	0.0203	1.72	11.46	94.57	0.13	I
5	15.86	0.0424	2.35	9.64	98.16	0.10	II
6	13.43	0.0128	3.19	15.09	94.38	0.07	II
7	13.63	0.0275	3.38	14.97	95.57	0.07	II
8	13.63	0.0323	3.38	16.94	93.09	0.07	II
9	14.59	0.0110	4.69	19.09	96.59	0.05	II
10	7.38	0.0034	4.69	19.18	91.71	0.05	II
11	8.26	0.0042	7.03	39.8	92.95	0.03	III
12	10.3	0.0040	6.13	60.23	89.43	0.03	III
13	8.28	0.0023	11.18	83.02	82.56	0.02	III
14	20.1	0.0168	10.42	66.27	76.68	0.02	III
15	10.23	0.0042	6.55	63.48	69.55	0.03	III
16	10.0	0.0025	13.01	66.6	76.98	0.02	III
Ave.	12.83	0.01	5.06	31.47	89.96	0.08	

Displacement pressure (P_d) represents the starting pressure of mercury entering the rock sample [42]. It is an important parameter to characterize the permeability of the rock sample. Small displacement pressure shows that the mercury is easy to be squeezed into the rock sample, attributing to a large throat radius, and higher permeability. The P_d values of the studied samples are relatively high, varying from 0.83 MPa to 13.01 MPa with an average value of 5.06 MPa. Saturation median pressure refers to the corresponding capillary pressure when the non-wetting phase saturation is 50% on the capillary pressure curve [42]. It ranges from 4.96 MPa to 83.02 MPa with an average of 31.47 MPa. The maximum mercury intrusion saturation (S_{max}) of the samples found to vary from 69.55% to 99.25% with an average of 89.96%, demonstrating that 89.96% of pores are greater than 4.5 nm (163 MPa of maximum mercury intrusion pressure). The mean capillary radius (R_m) varies from 0.02 μm to 0.26 μm with an average value measured to be 0.08 μm. In summary, the displacement pressure and median pressure are higher, and the capillary radius is smaller, revealing a poor pore structure characteristic of the samples.

MICP parameters P_d, P_{c50}, S_{max}, R_m are displacement pressure (MPa), median pressure for 50% mercury intrusion saturation (MPa), maximum mercury intrusion saturation (%), and mean pore throat radius (μm), respectively.

The MICP curves are shown in Figure 3. Based on the shape of these curves and their displacement pressure values, the rock samples were divided into three types: displacement pressure <2 MPa, 2–5 MPa and >5 MPa. Red, black and blue lines represent the types I, II and III, respectively. Type III rocks have the highest displacement pressures and the lowest maximum mercury intrusion saturation. Type I rocks have the smallest displacement pressures. Type I rocks have relatively good pore structure, whereas Type III has the worst pore structure. Unlike conventional reservoirs, the curves do not have the inflection point separating larger and smaller pores, indicating that larger pores do not exist in the tight oil reservoir samples.

The pore size distributions were calculated using Equation (2). The average pore size distributions for these three types are presented in Figure 4. These pore size distributions are found to be unimodal. The pore size distribution of Type I rocks is the widest, while Type II is the narrowest. The peaks of pore size distributions for these three types are 0.144, 0.036 and 0.009 μm, respectively. The type I rock pores are mainly dissolution pores, type III rock pores are clay pores. This can be confirmed by the cross plot of permeability and displacement pressure with clay and plagioclase contents. As it can be observed from Figure 5, the permeability is negatively correlated with clay contents and positively correlated with plagioclase contents. In Figure 6, the displacement pressure is positively correlated with clay contents and negatively correlated with plagioclase contents. The clay pores are attributed to clays, and part of dissolution pores are attributed to feldspar.

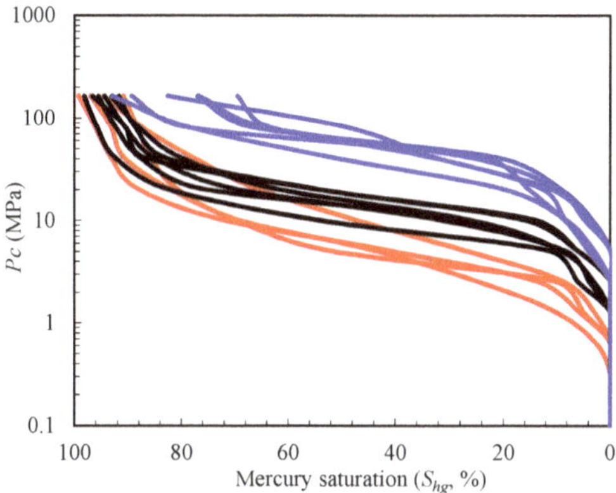

Figure 3. Classified capillary pressure curves. Red, black and blue lines represent the types I, II and III, respectively.

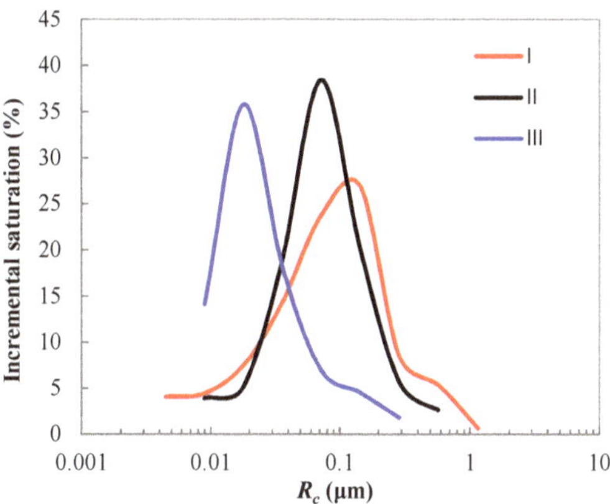

Figure 4. Average pore size distributions for the three types. Red, black and blue lines represent the types I, II and III, respectively.

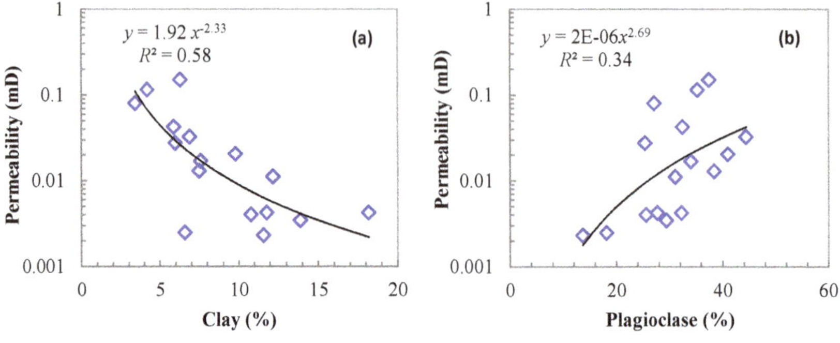

Figure 5. The cross plot of permeability with clay and plagioclase contents.

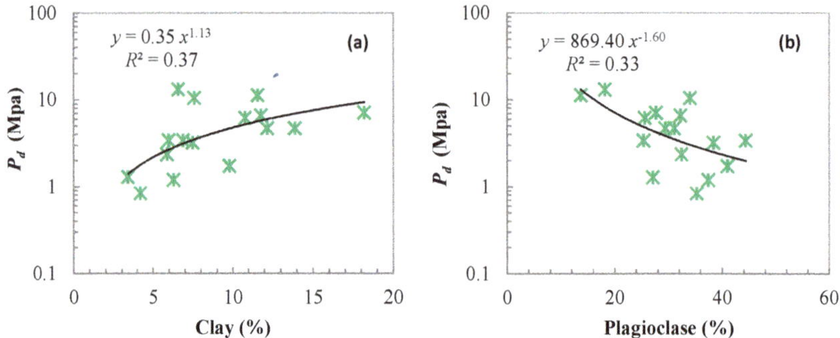

Figure 6. The cross plot of displacement pressure with clay and plagioclase contents.

3.4. Prediction by NMR Logs

3.4.1. Model Verification

Zhao [35] used several capillary pressure curves and their corresponding T_2 distributions from filed NMR logging to verify the model. However, the model was not fully verified by the NMR measurements in the laboratory. Figure 7a displays the T_2 distributions for Sample M1 at both "as received" and water-saturated conditions. The "as received" state T_2 distribution is bimodal and wider, which is similar to the T_2 characteristics of the field NMR logging, while the water saturated state T_2 distribution is narrower. The porosity and permeability for this sample is 12.7% and 0.0308 mD.

Using Equation (7), the "as received" state T_2 distribution was divided into two segments: water and oil signal distributions, as shown in Figure 7b. In this case, the $T_{2cutoff}$ was determined as 6.2 ms according to the saturation that was obtained from core analysis. The coefficient m was set as 4, equal to Zhao [35] calculations. The green dotted line represents weight function $S(T_2)$.

The different values for surface relaxivity of the hydrophilic and oleophilic pores were used to calculate the pore size distributions from water and oil signal distributions (Equations (5) and (6)). The water-containing pore, oil-bearing pore and total pore size distributions are shown in Figure 7c with the peaks for the pore size distributions at 13.8 nm, 66.6 nm and 15.9 nm, respectively.

The corrected T_2 distribution for water saturated state can be obtained using the total pore size distribution and surface relaxivity of the hydrophilic pores from Equation (5). The corrected and measured T_2 distributions for water-saturated state are shown in Figure 7d where both T_2 distributions are almost overlapping (compare with Figure 7a). The difference between the two T_2 distributions may originate from the "as-received" state T_2 distributions that does not truly represent the T_2 distribution under reservoir conditions.

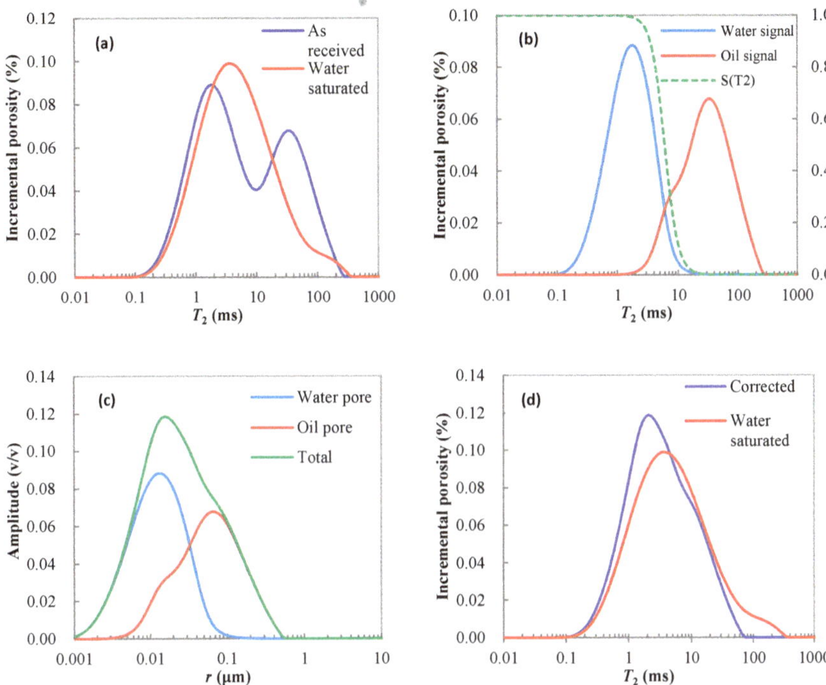

Figure 7. Sample M1: (**a**) T_2 distributions for "as received" state and water saturated state; (**b**) Water and oil signal distributions obtained from "as received" state T_2 distribution using weight function $S(T_2)$; (**c**) Water-containing pore, oil-bearing pore and total pore size distributions; (**d**) Comparison of corrected and measured T_2 distributions for water-saturated state.

Figure 8 exhibits the T_2 distributions of the sample M2. The porosity and permeability for this sample was measured 15.5% and 0.0299 mD, correspondingly. The corrected and measured T_2 distributions for water-saturated conditions are shown in Figure 8b. It can be seen that the difference between the two T_2 distributions is minor, presenting the effectiveness of the correction method.

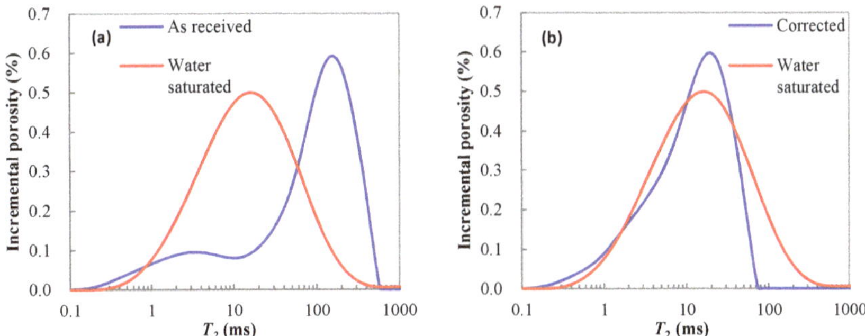

Figure 8. Sample M2: (**a**) T_2 distributions for "as received" state and water saturated state; (**b**) Comparison of corrected and measured T_2 distributions for water-saturated state.

3.4.2. Case Study

Figure 9 displays well logs from Well Ji32 from the lower sweet spot reservoir. The average hydrophilic pore surface relaxivity obtained by the capillary pressure curves and the T_2 spectra of nuclear magnetic logging is scaled as 2.5 μm/s, and the oleophilic pore surface relaxivity is 0.75 μm/s. The first track from left in the figure presents the lithology logs including GR, SP and CAL. The second track is deep and shallow lateral resistivity (LLD and LLS) logs, and the third one shows the conventional porosity logs, in terms of DEN, CNL and AC logs. Track 4 presents the total porosity obtained from NMR logging. Track 5 shows the measured NMR T_2 spectra. Track 6 presents the corrected T_2 spectra for fully water-saturated state. From this track, it is known that T_2 spectra for fully water-saturated state are narrower, revealing poor pore structure of the formation, exhibits a tight oil reservoir characteristic. Track 7 presents the capillary pressure curves constructed using the T_2 spectra of water-saturated state. The last two tracks are the comparison of the displacement pressure and the median pressure calculated by the constructed capillary pressure (red curves) with the core data (blue dots). The prediction results are in good agreement with the core analysis results (blue dots), which verifies the reliability and effectiveness of the pore structure prediction method proposed in this paper. From this figure, it can be seen that a consecutively prediction result for pore structures. The capillary pressure curves and related parameters at different depths can be seen directly. The variation in pore structure with depth cannot be observed if only core samples are used.

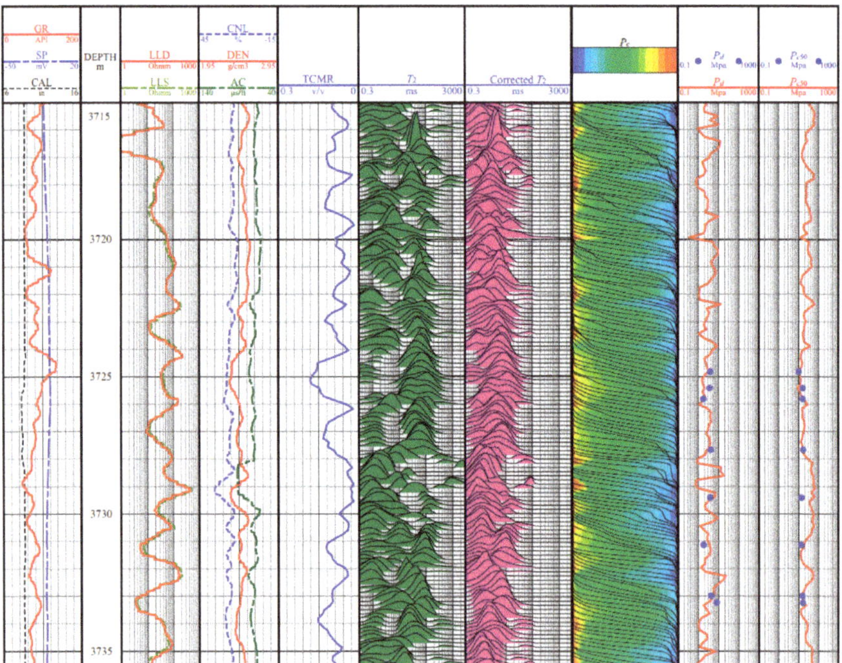

Figure 9. Pore structure prediction results for lower sweet spot reservoir in Well Ji32.

3.4.3. Overall Pore Structure Characteristics of the Studied Formation

According to classification criteria presented earlier of MICP, the constructed capillary pressure curves of the fourteen wells with NMR logging measurements in the studied area were categorized. Types I, II, and III account for 25.2%, 33.9%, and 40.9% respectively in the upper sweet spot reservoir, while Types I, II, and III make up 17.2%, 24.1%, and 58.6% in the lower sweet spot reservoir, as shown in Figure 10.

According to the constructed capillary pressure curves obtained from the fourteen wells in the studied area, the pore size distributions were further calculated for the reservoirs. Figure 11 demonstrates the average pore size distribution of the upper and lower sweet spot reservoirs in the studied area. It can be seen from Figure 11a that the main peak of the pore size is between 12 nm and 40 nm, while the pores smaller than 40 nm make up 57.4%, and the pores between 40 nm and 500 nm, 36.1% of all pores collectively. The pore size distribution of the lower sweet spot in Figure 11b is relatively dispersed, where the proportion of pores smaller than 40 nm and the pores between 40 and 500 nm are quiet the same as the upper sweet spot reservoir. However, the pores that smaller than 12 nm are more abundant in the lower sweet spot reservoir compared to the upper one. In addition, the pores smaller than 4 nm in both upper and lower sweet spots are 10.2% and 15.7%, found to be higher than similar pores calculated from capillary pressure curves.

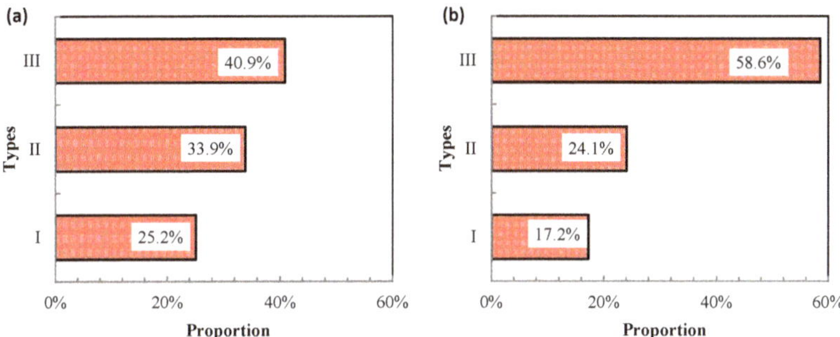

Figure 10. Proportions of reservoir types estimated by NMR logs: (**a**) Upper sweet spot reservoir; (**b**) Lower sweet spot reservoir.

Finally, from Figures 10 and 11, it is concluded that the pore structure of the upper sweet spot reservoir is relatively better than that the lower sweet spot reservoir, while the overall characteristics of the pores in the studied area is very much complex and dominated by nano-scale pores.

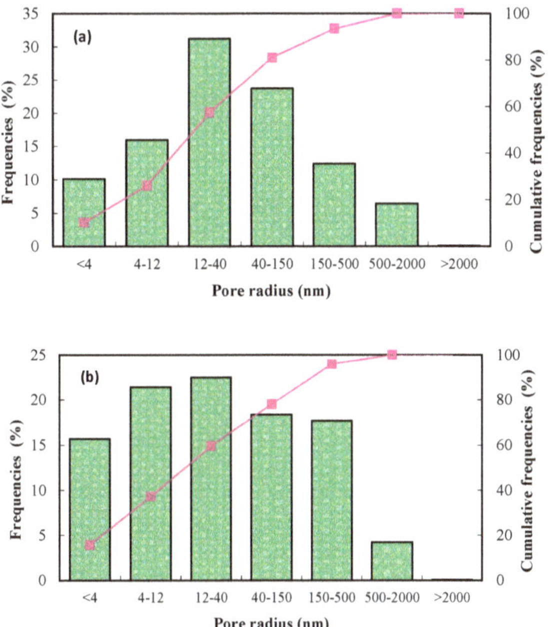

Figure 11. Average pore size distributions estimated by NMR logs: (**a**) Upper sweet spot reservoir; (**b**) Lower sweet spot reservoir.

4. Conclusions

In this paper, the pore structure of a tight oil reservoir in Permain Lucaogou formation of Jimusaer Sag was studied using SEM images and MICP data. NMR logs were used to provide a consecutive prediction of the pore structures. The following conclusions are made:

1. According to the SEM images, the main pores of the tight oil reservoirs in the Lucaogou Formation are secondary pores. These pores can be divided into four categories: intragranular dissolution, intergranular dissolution, micro fractures and clay pores.
2. The displacement pressure values of the studied samples ranges from 0.83 to 13.01 MPa with an average of 5.06 MPa. Saturation median pressure varied from 4.96 to 83.02 MPa with an average of 31.47 MPa. The mean capillary radius was measured from 0.02 to 0.26 μm.
3. The capillary pressure curves are divided into three types: displacement pressure <2 MPa, 2–5 MPa and >5 MPa. Type I rocks have the smallest displacement pressures while Type III the highest displacement pressures and lowest maximum mercury intrusion saturation. The pores of type I rocks are mainly dissolution pores, and type III are clay pores.
4. The T_2 distributions of "as-received" and water-saturated state samples were measured. The model for predicting capillary pressure curves with NMR T_2 distribution was verified by two state T_2 distributions measurements. This model was applied to well logs where the estimated pore structure parameters by NMR T_2 distribution were in a good agreement with core analysis.
5. The predicted capillary pressure curves from NMR logging data of the fourteen wells in the studied area were categorized based on the proposed model. Types I, II, and III of the upper sweet spot reservoir account for 25.2%, 33.9%, and 40.9%, while in the lower sweet spot, 17.2%, 24.1%, and 58.6% was calculated respectively. The pores smaller than 12 nm in the lower sweet spot reservoirs are more abundant than the upper sweet spot, indicating the pore structure of the lower sweet spot reservoir is more complicated than that in the upper sweet spot reservoir.

Author Contributions: Investigation, Z.X.; Methodology, P.Z.; Resources, Z.W.; Writing—review & editing, M.O. and Z.P.

Acknowledgments: This paper is supported by the China National Science and Technology Major Project (2017ZX05009-001), the National Nature Science Foundation of China (41302109) and the Foundation of State Key Laboratory of Petroleum Resources and Prospecting, China University of Petroleum, Beijing (No. PRP/open-1601).

Conflicts of Interest: The authors declare no conflict of interest.

References

1. Miller, B.A.; Paneitz, J.M.; Mullen, M.J.; Meijs, R.; Tunstall, K.M.; Garcia, M. The successful application of a compartmental completion technique used to isolate multiple hydraulic-fracture treatments in horizontal Bakken Shale wells in North Dakota. In Proceedings of the SPE Annual Technical Conference and Exhibition, Denver, CO, USA, 21–24 September 2008; SPE-116469-MS.
2. Jia, C.Z.; Zou, C.N.; Li, J.Z.; Li, D.H.; Zheng, M. Assessment criteria, main types, basic features and resource prospects of the tight oil in China. *Acta Pet. Sin.* **2012**, *33*, 343–350.
3. Li, C.; Ostadhassan, M.; Gentzis, T.; Kong, L.; Carvajal-Ortiz, H.; Bubach, B. Nanomechanical characterization of organic matter in the Bakken formation by microscopy-based method. *Mar. Pet. Geol.* **2018**, *96*, 128–138. [CrossRef]
4. Wang, L.; Zhao, N.; Sima, L.; Meng, F.; Guo, Y. Pore Structure Characterization of the Tight Reservoir: Systematic Integration of Mercury Injection and Nuclear Magnetic Resonance. *Energy Fuel* **2018**, *32*, 7471–7484. [CrossRef]
5. Zou, C.N.; Zhu, R.K.; Wu, S.T.; Yang, Z. Types, characteristics, genesis and prospects of conventional and unconventional hydrocarbon accumulations: Taking tight oil and tight gas in China as an instance. *Acta Pet. Sin.* **2012**, *33*, 173–187.

6. Ramakrishna, S.; Balliet, R.; Miller, D.; Sarvotham, S.; Merkel, D. Formation evaluation in the Bakken Complex using laboratory core data and advanced logging technologies. In Proceedings of the SPWLA 51st Annual Logging Symposium, Perth, Australia, 19–23 June 2010. SPWLA-2010-74900.
7. Wood, D.A.; Hazra, B. Characterization of organic-rich shales for petroleum exploration & exploitation: A review—Part 1: Bulk properties, multi-scale geometry and gas adsorption. *J. Earth Sci.* **2017**, *28*, 739–757. [CrossRef]
8. Wood, D.A.; Hazra, B. Characterization of organic-rich shales for petroleum exploration & exploitation: A review—Part 3: Applied geomechanics, petrophysics and reservoir modeling. *J. Earth Sci.* **2017**, *28*, 739–757. [CrossRef]
9. Ge, X.; Fan, Y.; Li, J.; Zahid, M.A. Pore structure characterization and classification using multifractal theory—An application in Santanghu basin of western China. *J. Petrol. Sci. Eng.* **2015**, *127*, 297–304. [CrossRef]
10. Kong, L.; Ostadhassan, M.; Li, C.; Tamimi, N. Pore characterization of 3D-printed gypsum rocks: A comprehensive approach. *J. Mater. Sci.* **2018**, *53*, 5063–5078. [CrossRef]
11. Lai, J.; Wang, G.; Wang, Z.; Chen, J.; Pang, X.; Wang, S.; Zhou, Z.; He, Z.; Qin, Z.; Fan, X. A review on pore structure characterization in tight sandstones. *Earth Sci. Rev.* **2018**, *177*, 436–457. [CrossRef]
12. Xia, Y.; Cai, J.; Wei, W.; Hu, X.; Wang, X.; Ge, X. A new method for calculating fractal dimensions of porous media based on pore size distribution. *Fractals* **2018**, *26*. [CrossRef]
13. Kuang, L.C.; Hu, W.X.; Wang, X.L.; Wu, H.G.; Wang, X.L. Research of the tight oil reservoir in the Lucaogou Formation in Jimusar Sag: Analysis of lithology and porosity characteristics. *Geol. J. China Univ.* **2013**, *19*, 529–535.
14. Zhang, T.; Fan, G.; Li, Y.; Yu, C. Pore-throat characterization of tight oil reservoir in the Lucaogou formation, Jimusar sag. *CT Theor. Appl.* **2016**, *25*, 425–434.
15. Zhou, P. Characterization and Evaluation of Tight Oil Reservoirs of Permian Lucaogou Formation in Jimusar Sag, Junggar Basin. Master's Thesis, Northwest University, Xi'an, China, 2016.
16. Su, Y.; Zha, M.; Ding, X. Pore type and pore size distribution of tight reservoirs in the Permian Lucaogou Formation of the Jimsar Sag, Junggar Basin, NW China. *Mar. Pet. Geol.* **2017**, *89*, 761–774. [CrossRef]
17. Zhao, P.; Sun, Z.; Luo, X.; Wang, Z.; Mao, Z.; Wu, Y.; Xia, P. Study on the response mechanisms of nuclear magnetic resonance (NMR) log in tight oil reservoirs. *Chin. J. Geophys.* **2016**, *29*, 1927–1937. [CrossRef]
18. Zhao, P.; Wang, Z.; Sun, Z.; Cai, J.; Wang, L. Investigation on the pore structure and multifractal characteristics of tight oil reservoirs using NMR measurements: Permian Lucaogou Formation in Jimusaer Sag, Junggar Basin. *Mar. Pet. Geol.* **2017**, *86*, 1067–1081. [CrossRef]
19. Wang, X.; Hou, J.; Liu, Y.; Zhao, P.; Ma, K.; Wang, D.; Ren, X.; Yan, L. Overall PSD and fractal characteristics of tight oil reservoirs: A case study of Lucaogou formation in Junggar Basin, China. *Fractals* **2019**, 27. [CrossRef]
20. Loren, J.D.; Robinson, J.D. Relations between pore size fluid and matrix properties, and NML measurements. *SPE J.* **1970**, *10*, 50–58. [CrossRef]
21. Kenyon, W.E. Petrophysical principles of applications of NMR logging. *Log Anal.* **1997**, *38*, 21–40.
22. Volokitin, Y.; Looyestijn, W. A practical approach to obtain primary drainage capillary pressure curves from NMR core and log data. *Petrophysics* **2001**, *42*, 334–343.
23. Volokitin, Y.; Looyestijn, W. Constructing capillary pressure curve from NMR log data in the presence of hydrocarbons. In Proceedings of the SPWLA 40th Annual Logging Symposium, Oslo, Norway, 30 May–3 June 1999. SPWLA-1999-KKK.
24. Liu, T.; Wang, S.; Fu, R.; Li, Y.; Luo, M. Analysis of rock pore throat structure with NMR spectra. *Oil Geophys. Prospect.* **2003**, *38*, 328–333.
25. Grattoni, C.A.; Al-Mahrooqi, S.H.; Moss, A.K.; Muggeridge, A.H.; Jing, X.D. An improved technique for deriving drainage capillary pressure from NMR T2 Distributions. In Proceedings of the International Symposium of the Society of Core Analysis, Pau, France, 21–24 September 2003; SCA2003-25.
26. He, Y.; Mao, Z.; Xiao, L.; Ren, X. An improved method of using NMR T2 distribution to evaluate pore size distribution. *Chin. J. Geophys.* **2005**, *48*, 373–378.
27. He, Y.; Mao, Z.; Xiao, L.; Zhang, Y. A new method to obtain capillary pressure curve using NMR T2 distribution. *J. Jilin Univ. (Earth Sci. Ed.)* **2005**, *35*, 177–181.
28. Xiao, L.; Mao, Z.; Wang, Z.; Jin, Y. Application of NMR logs in tight gas reservoirs for formation evaluation: A case study of Sichuan basin in China. *J. Petrol. Sci. Eng.* **2012**, *81*, 182–195. [CrossRef]

29. Hu, F.; Zhou, C.; Li, C.; Xu, H.; Zhou, F.; Si, Z. Water spectrum method of NMR logging for identifying fluids. *Petrol. Explor. Dev.* **2016**, *43*, 244–252. [CrossRef]
30. Ge, X.; Fan, Y.; Liu, J.; Zhang, L.; Han, Y.; Xing, D. An improved method for permeability estimation of the bioclastic limestone reservoir based on NMR data. *J. Magn. Reson.* **2017**, *283*, 96–109. [CrossRef] [PubMed]
31. Xiao, L.; Mao, Z.; Li, J.; Yu, H. Effect of hydrocarbon on evaluating formation pore structure using nuclear magnetic resonance (NMR) logging. *Fuel* **2018**, *216*, 199–207. [CrossRef]
32. Coates, G.R.; Xiao, L.Z.; Prammer, M.G. *NMR Logging Principles and Applications*; Gulf Publishing Company: Houston, TX, USA, 1999.
33. Dunn, K.J.; Bergman, D.J.; Latorraca, G.A. *Nuclear Magnetic Resonance Petrophysical and Logging Application*; Elservier: Amsterdam, The Netherlands, 2002.
34. Daigle, H.; Johnson, A. Combining mercury intrusion and nuclear magnetic resonance measurements using percolation theory. *Transp. Porous Med.* **2016**, *111*, 669–679. [CrossRef]
35. Zhao, P. Study on the NMR Log Responses and Models for Estimating Petrophysical Parameters from Well Logs in Tight Oil Reservoirs. Ph.D. Thesis, China University of Petroleum, Beijing, China, 2016.
36. Cao, Z.; Liu, G.; Xiang, B.; Wang, P.; Niu, G.; Niu, Z.; Li, C.; Wang, C. Geochemical Characteristics of crude oil from a tight oil reservoir in the Lucaogou Formation, Jimusar Sag, Junggar, Basin. *AAPG Bull.* **2017**, *101*, 39–72. [CrossRef]
37. Kuang, L.C.; Tang, Y.; Lei, D.W.; Chang, Q.S.; Ouyang, M.; Hou, L.H.; Liu, D.G. Formation conditions and exploration potential of tight oil in the Permian saline lacustrine dolomitic rock, Junggar Basin, NW China. *Petrol. Explor. Dev.* **2012**, *39*, 657–667. [CrossRef]
38. Butler, J.P.; Reeds, J.A.; Dawson, S.V. Estimating solutions of first kind integral equations with nonnegative constraints and optimal smoothing. *SIAM. J. Numer. Anal.* **1981**, *18*, 381–397. [CrossRef]
39. Washburn, E.D. The dynamics of capillary flow. *Phys. Rev.* **1921**, *17*, 273–283. [CrossRef]
40. Latour, L.L.; Kleinberg, R.L.; Sezginer, A. Nuclear magnetic resonance properties of rocks at elevated temperatures. *J. Colloid Interface Sci.* **1994**, *150*, 535–548. [CrossRef]
41. Looyestijn, W.J.; Hofman, J.P. Wettability determination by NMR. *SPE Reserv. Eval. Eng.* **2006**, *6*, 146–153. [CrossRef]
42. Yang, S.L.; Wei, J.Z. *Petrophysics*; Petroleum Industry Press: Beijing, China, 2004; pp. 209–233.

Article

Evolution of Coal Permeability during Gas Injection—From Initial to Ultimate Equilibrium

Xingxing Liu [1,2], Jinchang Sheng [1], Jishan Liu [3,*] and Yunjin Hu [4]

1 College of Water Conservancy and Hydropower Engineering, Hohai University, Nanjing 210098, China; xliu@hhu.edu.cn (X.L.); sh901@sina.com (J.S.)
2 Beijing Key Laboratory for Precise Mining of Intergrown Energy and Resources, China University of Mining and Technology, Beijing 100083, China
3 Department of Chemical Engineering, School of Engineering, The University of Western Australia, 35 Stirling Highway, Perth, WA 6009, Australia
4 School of Civil Engineering, Shaoxing University, Shaoxing 312000, China; huyunjin@tsinghua.org.cn
* Correspondence: jishan.liu@uwa.edu.au

Received: 4 September 2018; Accepted: 11 October 2018; Published: 17 October 2018

Abstract: The evolution of coal permeability is vitally important for the effective extraction of coal seam gas. A broad variety of permeability models have been developed under the assumption of local equilibrium, i.e., that the fracture pressure is in equilibrium with the matrix pressure. These models have so far failed to explain observations of coal permeability evolution that are available. This study explores the evolution of coal permeability as a non-equilibrium process. A displacement-based model is developed to define the evolution of permeability as a function of fracture aperture. Permeability evolution is tracked for the full spectrum of response from an initial apparent-equilibrium to an ultimate and final equilibrium. This approach is applied to explain why coal permeability changes even under a constant global effective stress, as reported in the literature. Model results clearly demonstrate that coal permeability changes even if conditions of constant effective stress are maintained for the fracture system during the non-equilibrium period, and that the duration of the transient period, from initial apparent-equilibrium to final equilibrium is primarily determined by both the fracture pressure and gas transport in the coal matrix. Based on these findings, it is concluded that the current assumption of local equilibrium in measurements of coal permeability may not be valid.

Keywords: equilibrium permeability; non-equilibrium permeability; matrix–fracture interaction; effective stress; coal deformation

1. Introduction

The permeability of coal is a key transport property in determining coalbed methane production and CO_2 storage in coal seam reservoirs. Coal permeability is often determined by regular sets of fractures called cleats, with the aperture of cleats being a key factor defining the magnitude of permeability [1]. Coal permeability may vary significantly in both space and time in response to the complex coal–gas interactions and presents complex evolutionary paths in unconventional reservoirs [2].

Significant experimental efforts have been made to investigate and interpret the evolution of coal permeability. Many factors affect coal permeability, including gas types [3,4], pore pressure, sorption-induced matrix swelling/shrinkage [5–10], effective pressure [11,12], water content [13,14] and gas exposure time [15]. Most of the above studies were performed under stress-controlled (unconstrained) boundary conditions to replicate in situ conditions. A variety of coal permeability models have been formulated to quantify permeability evolution from such laboratory experiments [16–23].

Most of these permeability models fail to explain stress-controlled results since they improperly idealize the fractured coal as a matchstick or cubic geometry, or assume local equilibrium between the matrix and fracture pressures, or ignore matrix–fracture interactions [2]. Stress-controlled conditions, applied in such models, discount matrix swelling from affecting coal permeability. This is because matrix swelling that is induced by increasing pore pressure results in an increase in matrix block size, rather than a change in fracture [24]. This analytical conclusion contradicts laboratory observations of significant change coal permeability induced by matrix swelling under constant confining stress conditions [11,25].

Previous studies suggest that the discrepancies between laboratory observations and theoretical characterizations are mainly attributed to sorption-induced swelling strain. Connell et al. [1] distinguished the sorptive strain of the coal matrix, the pores (or the cleats) and the bulk coal to relax the equilibrium strain assumption between the bulk and pore strains, and derived several different forms of permeability models for the laboratory tests. Liu et al. [24] developed a new coal-permeability model for constant confining-stress conditions, which explicitly considers fracture–matrix interactions during coal-deformation processes based on the concept of internal swelling stress. Chen et al. [26] introduced a partition factor to split the contributions of swelling strain between fracture and bulk deformation and developed relations between the partition ratio and cleat porosity change based on model fitting results. Liu et al. [27] proposed a conceptual solution to consider the matrix–fracture interaction through introduction of the concept of a switch in processes between local swelling and global swelling. Peng et al. [28] further combined the concept of local swelling and macro-swelling and matrix–fracture interactions at the micro-scale into a more rigorous dual permeability model, which was applied to generate a series of coal permeability relations exhibiting the characteristic "V" shape as observed in experiments.

While a certain degree of success has been achieved using these models to explain and match experimental observations, the interaction between coal matrix and fracture remains incompletely understood. Studies involving the partitioning of sorption-induced strain between fracture and matrix have focused on incorporating fracture–matrix interactions into permeability models, but often the necessary coefficients obtained from model fitting results are non-physics-based [24,26] or the division between global swelling and local swelling typically overestimates local fracture swelling contributions to fracture closure [28]. As the matrix and fractures have dramatically different flow characteristics and mechanical properties, this leads to non-uniform gas pressure distributions and uneven matrix strain. Thus, dynamic interactions can exert significant temporal effects on both fracture aperture changes (permeability changes) and bulk deformation [19,26,27,29]. It is clear that these studies have two deficiencies that need to be addressed. Firstly, the effects of fracture gas pressure on both fracture aperture and matrix deformation must be correctly accommodated—this is not the case in these prior studies. Prior studies have required that fracture opening is only induced by the swelling of the bridging contacts and the opening effects are less competitive than the opposite effects of the intervening free-face swelling. This leads to a decrease in permeability at the beginning of gas injection rather than an increase as demonstrated in some laboratory observations [30]. Secondly, different contributions of confining stress and matrix swelling to fracture aperture change and bulk deformation are neglected. Although many studies focused on developing governing equations in dual-continuum systems (fractured rock) when modeling coupled liquid flow and mechanical processes [31,32], the lack of consideration of the fracture–matrix interaction may cause unacceptable errors if these equations are directly used for the fractured coal. This is because many studies only noted the temporal effect on aperture change but ignored the influence on bulk deformation [28,33,34]. Fracture pressure, confining pressure and matrix swelling all cause fracture deformation, which is different from, but related to, bulk deformation [35].

The primary motivations of this study are to restore the process of fracture aperture change in coal containing discrete fractures following gas injection under unconstrained conditions and to underscore the impact of gas pressure within the fractures on aperture change. In this study, a fracture–matrix interaction model is developed to explore the dynamic interactions between fracture gas pressure, matrix swelling/shrinkage, aperture change and bulk deformation, and to illuminate mechanisms of dynamic deformation response to gas flow from fracture to matrix. Furthermore, the evolution of aperture change associated with intrinsic and extrinsic factors such as fracture properties, initial matrix permeability, injection processes and confining pressure are quantitatively evaluated. The simulated results provide a spectrum of permeability evolutions from the initial equilibrium state, through a transient state, to the final equilibrium state.

2. Conceptual Model and Governing Equations

2.1. Conceptual Model

The key to model the dynamic interactions between matrix and fractures is to recover important non-linear responses due to effective stress effects. Thus, the mechanical influence must be rigorously coupled with the gas transport system. This can be achieved through a full coupling approach. For this approach, a single set of equations (generally a large system of non-linear coupled partial differential equations) incorporating all the relevant physics is solved simultaneously. In the following section, two kinds of simulation models are presented to investigate the permeability change and bulk deformation under unconstrained conditions.

Coal is a typical dual porosity/permeability system [36] containing a porous matrix and fractures. In this study, it is assumed that cleats do not create a full separation between adjacent matrix blocks but solid rock bridges are present, as illustrated in Figure 1a. The coal bridge plays a significant role in the fracture–matrix interaction and its effect can be interpreted as follows: (a) restricting fracture opening induced by fracture pressure increase; (b) linking the spatial and temporal matrix swelling to the aperture change and bulk deformation; and (c) contributing to the final aperture increase. This assumption is also adopted in other studies [19,24]. The model examines the influence of effective stress and swelling response for a rectangular crack, similar to the matchstick model geometry, and a single component part removed from the array may be considered as a representative element. This represents the symmetry of the displacement boundary condition mid-way between flaws as shown in Figure 1b.

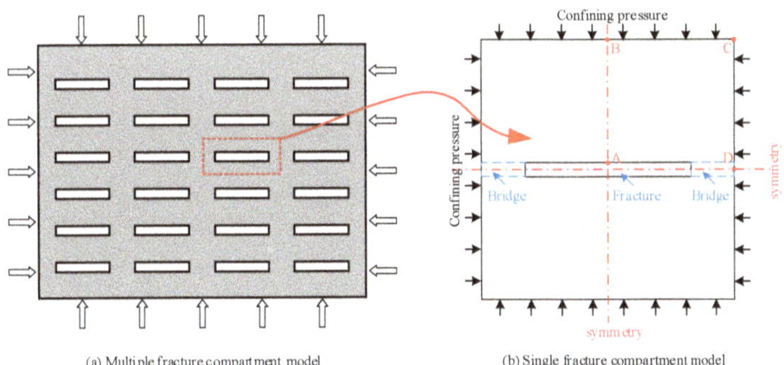

Figure 1. Numerical model under the unconstrained (constant total stress) boundary condition. (**a**) Multiple fracture compartment model; (**b**) single fracture compartment model as a representative element.

Two different methods may be used to represent the fracture:

1. The fracture may be represented as a void [27,37]. Then the porous matrix is the only object to study, leading to a stress difference between the internal and external boundaries. The external stress boundaries are controlled by the confining stress, while the internal stress boundaries are controlled by the fracture gas pressure.

2. The fracture may be represented as a softer material [19,29], with the equilibrium pressure applied on the cross section and no sorption-induced strain. Then, both the matrix and the fracture must be studied and only the external stress boundaries are relevant, and these are controlled by the constant confining stress. If the fracture pressure increases significantly faster than that in the pores in the surrounding matrix, a compressive stress due to fracture swelling will inevitably arise at the interface. This is similar to the internal swelling stress proposed by Liu et al. [24].

In the following sections, a set of field equations are defined which govern the gas transport and deformation of both the solid matrix and the fracture. Since the difference in these two approaches lies in the presence of the fractures, the governing equations are chosen accordingly.

The field equations are based on the following assumptions: (a) the matrix is homogeneous, isotropic and elastic continuum; (b) strains are infinitesimal; (c) gas contained within the pores, and its viscosity is constant under isothermal conditions; (d) gas flow through the coal matrix is assumed to be viscous flow obeying Darcy's law; (e) if the fracture is also regarded as a homogeneous, isotropic and elastic continuum, the fracture is instantly filled with gas and no sorption-induced strain arises.

2.2. Governing Equation for Mechanical Response

The strain-displacement relationship is defined as:

$$\varepsilon_{ij} = \frac{1}{2}(u_{i,j} + u_{j,i}) \tag{1}$$

where ε_{ij} denotes the component of the total strain tensor and u_i is the component of the displacement. The equilibrium equation is defined as:

$$\sigma_{ij,j} + f_i = 0 \tag{2}$$

where σ_{ij} denotes the component of the total stress tensor and f_i denotes the component of the body force.

Based on poroelasticity and by making an analogy between thermal contraction and matrix shrinkage, the constitutive relation for the coal matrix and the fracture becomes [22]:

$$\varepsilon_{ij} = \frac{1}{2G}\sigma_{ij} - \left(\frac{1}{6G} - \frac{1}{9K}\right)\sigma_{kk}\delta_{ij} + \frac{\alpha}{3K}p\delta_{ij} + \frac{\varepsilon_s}{3}\delta_{ij} \tag{3}$$

where $G = E/2(1+v)$, $K = E/3(1-2v)$, and $\sigma_{kk} = \sigma_{11} + \sigma_{22} + \sigma_{33}$, where K is the bulk modulus, G is the shear modulus, E is the Young's modulus, v is the Possion's ratio, α is the Biot coefficient, p is the gas pressure, δ_{ij} is the Kronecker delta, and ε_s is the sorption-induced volumetric strain usually expressed by a Langmuir-type equation [18]:

$$\varepsilon_s = \varepsilon_L \frac{p}{p + P_L}. \tag{4}$$

where ε_L is a constant representing the volumetric strain at infinite pore pressure and P_L is the Langmuir pressure constant representing the pore pressure at which the measured volumetric strain is equal to $\frac{\varepsilon_L}{2}$. From Equations (3) and (4), the effective stress in coal matrix, σ_{eij}, can be modified as:

$$\sigma_{eij} = \sigma_{ij} + \alpha p\delta_{ij} + \frac{\varepsilon_L}{K}\frac{p}{p + P_L}\delta_{ij} \tag{5}$$

Combining Equations (2)–(5) yields the Navier-type equation expressed as:

$$Gu_{i,kk} + \frac{G}{1-2v}u_{k,ki} - \alpha p_{,i} - \frac{K\varepsilon_L P_L}{(p + P_L)^2}p_{,i} + f_i = 0 \tag{6}$$

Equation (6) is the general form of the governing equation for the deformation of the matrix and fracture, where the gas pressure can be solved from the gas flow equation as discussed below. It should be noted that if the gradient terms of pore pressure and sorption-induced swelling are treated as a body force, the stress at boundaries should be transformed as effective stresses. The discrepancy between the matrix and the fracture is embodied in different values of mechanical parameters in the governing equations:

$$G_m u_{mi,kk} + \frac{G_m}{1-2v_m} u_{mk,ki} - \alpha_m p_{m,i} - \frac{K_m \varepsilon_L P_\varepsilon}{(p_m + P_\varepsilon)^2} p_{m,i} + f_{mi} = 0 \tag{7}$$

$$G_f u_{fi,kk} + \frac{G_f}{1-2v_f} u_{fk,ki} - \alpha_f p_{f,i} + f_{fi} = 0 \tag{8}$$

where the subscripts, m and f, denote matrix and fracture, respectively.

As the propensity of the fracture to swell is stronger than the matrix, a compressive stress arises on the interface to satisfy deformation compatibility. This restricts the fracture expansion and enhances matrix swelling as illustrated in Figure 2. The strain for the matrix and the fracture can be expressed as:

$$\Delta\varepsilon_{If} = \frac{\Delta p_f}{E_f} - \frac{\Delta\sigma_a}{E_f} \tag{9}$$

$$\Delta\varepsilon_{Im} = \Delta\varepsilon_{tr} + \zeta\frac{\Delta\sigma_a}{E_m} \tag{10}$$

where $\Delta\varepsilon_{If}$ and $\Delta\varepsilon_{Im}$ are the strain at the interface for the fracture and the matrix, respectively, Δp_f is the fracture pressure increment, $\Delta\sigma_a$ is the induced interface stress, $\Delta\varepsilon_{tr}$ is the matrix strain induced by gas transport within the matrix and ζ is the coefficient concerning the position and the geometry of the matrix and fracture. Due to equivalent strains of the fracture and the matrix at the interface, the induced interface stress can be expressed as:

$$\Delta\sigma_a = \frac{\Delta p_f E_m - E_f E_m \Delta\varepsilon_{tr}}{\zeta E_f + E_m} \tag{11}$$

The strain at the interface, $\Delta\varepsilon_I$, is then expressed as:

$$\Delta\varepsilon_I = \frac{\zeta}{\zeta E_f + E_m}\Delta p_f + \frac{E_m}{\zeta E_f + E_m}\Delta\varepsilon_{tr} \tag{12}$$

When equilibrium is achieved, the matrix strain induced by gas transport can be expressed as:

$$\Delta\varepsilon_{tr} = \frac{\Delta p_m}{E_m} + \Delta\varepsilon_s \tag{13}$$

Substituting Equation (13) into Equation (12), the strain can be obtained as:

$$\Delta\varepsilon_I = \frac{\zeta\Delta p_f}{\zeta E_f + E_m} + \frac{\Delta p_m}{\zeta E_f + E_m} + \frac{E_m\Delta\varepsilon_s}{\zeta E_f + E_m} \tag{14}$$

When the fracture is regarded as a void and the compressive stress equivalent to the fracture pressure is applied at the internal boundaries of the matrix, the strain at steady state can be expressed as:

$$\Delta\varepsilon_I = \zeta\frac{\Delta p_f}{E_m} + \frac{\Delta p_m}{E_m} + \Delta\varepsilon_s \tag{15}$$

From Equations (14) and (15), it can be seen that the two different treatments have an equivalent effect if the Young's modulus of the fracture is reduced to zero.

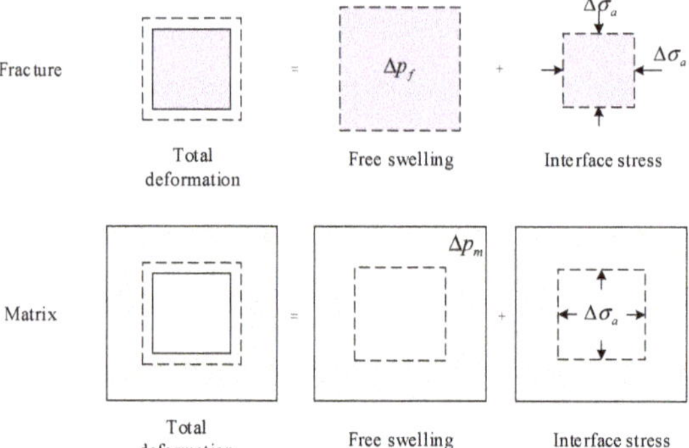

Figure 2. Schematic diagram of the compressive stress state.

2.3. Dynamic Permeability Model

Porosity, permeability and the grain-size distribution in porous media may be related via capillary models. Chilingar (1964) [38] defined this relation as:

$$k_m = \frac{d_e^2 {\phi_m}^3}{72(1-\phi_m)^2} \tag{16}$$

where k_m is the permeability, ϕ_m is porosity and d_e is the effective diameter of grains. Based on this equation, one obtains:

$$\frac{k_m}{k_{m0}} = \left(\frac{\phi_m}{\phi_{m0}}\right)^3 \left(\frac{1-\phi_{m0}}{1-\phi_m}\right)^2 \tag{17}$$

where the subscript, 0, denotes the initial value of the variable. When the porosity is much smaller than 1 (normally less than 10%), the second term of the right-hand side asymptotes to unity. This yields the cubic relationship between permeability and porosity for the coal matrix:

$$\frac{k_m}{k_{m0}} = \left(\frac{\phi_m}{\phi_{m0}}\right)^3 \tag{18}$$

Coal porosity can be defined as a function of the effective strain [2] as:

$$\frac{\phi_m}{\phi_{m0}} = 1 + \frac{\alpha}{\phi_{m0}}\Delta\varepsilon_{me} \tag{19}$$

$$\Delta\varepsilon_{me} = \Delta\varepsilon_v + \frac{\Delta p_m}{K_s} - \Delta\varepsilon_s \tag{20}$$

where $\Delta\varepsilon_{me}$ is defined as the total effective volumetric strain increment, which is responsible for permeability change, $\Delta\varepsilon_v$ is total volumetric strain increment, Δp_m is the gas pressure increment, K_s is the bulk modulus of the coal grains, $\Delta p_m/K_s$ is the compressive strain increment, and $\Delta\varepsilon_s$ is gas sorption-induced volumetric strain increment. Substituting Equation (19) into Equation (18) yields the permeability ratio as:

$$\frac{k_m}{k_{m0}} = \left(1 + \frac{\alpha}{\phi_{m0}}\Delta\varepsilon_{me}\right)^3 \tag{21}$$

Equations (19) and (21) define matrix porosity and permeability, which are derived based on the fundamental principles of poroelasticity and can be applied to the evolution of matrix porosity and permeability under variable boundary conditions.

The fracture permeability is usually defined by the well-known "cubic law" [39] and the fracture permeability ratio can be expressed as:

$$\frac{k_f}{k_{f0}} = \left(1 + \frac{\Delta b}{b_0}\right)^3 \tag{22}$$

where b_0 is initial fracture aperture and Δb is the fracture aperture change.

2.4. Governing Equation for Gas Flow within the Matrix

Conservation of mass for the gas phase is defined as:

$$\frac{\partial m}{\partial t} + \nabla \cdot \left(\rho_g \boldsymbol{q}_g\right) = Q_s \tag{23}$$

where ρ_g is the gas density, $\boldsymbol{q}_g$ is the Darcy velocity vector, Q_s is the gas source or sink, t is time, and m is the gas content including free -phase gas and adsorbed gas [40], defined as:

$$m = \phi_m \rho_{mg} + \rho_{ga} \rho_c \frac{V_L p_m}{p_m + P_L} \tag{24}$$

where ρ_{ga} is the gas density at standard conditions, ρ_c is the matrix density, ϕ_m is the matrix porosity, V_L represents the Langmuir volume constant, and P_L represents the Langmuir pressure constant. According to the ideal gas law, the relationship between gas density and pressure in the matrix is described as:

$$\rho_{mg} = \frac{M_g}{RT} p_m \tag{25}$$

where M_g is the molar mass of the gas, R is the universal gas constant, and T is the absolute gas temperature. From Equation (25), one obtains another expression for the gas density:

$$\rho_{mg} = \eta_g p_m \tag{26}$$

where,

$$\eta_g = \frac{\rho_{ga}}{p_{ga}} \frac{T_{ga}}{T} \tag{27}$$

where η_g is the coefficient between the gas density and pressure, T_{ga} and p_{ga} are the temperature and gas pressure at standard conditions. From Equation (27), the coefficient, η_g, depends on the temperature thus it is a constant for isothermal conditions.

Neglecting the effect of gravity, the Darcy velocity, $\boldsymbol{q}_g$, is defined as

$$\boldsymbol{q}_g = -\frac{k_m}{\mu} \nabla p_m \tag{28}$$

where k_m is the matrix permeability and μ is the dynamic viscosity of the gas. Substituting Equations (24) and (26)–(28) into Equation (23), yields,

$$\left(\eta_g \phi_m + \frac{\eta_{ga} p_{ga} \rho_c V_L P_L}{(p_m + P_L)^2}\right) \frac{\partial p_m}{\partial t} + \eta_g p_m \frac{\partial \phi_m}{\partial t} - \nabla \cdot \left(\eta_g p_{mg} \frac{k_m}{\mu} \nabla p_m\right) = Q_s \tag{29}$$

2.5. Governing Equation for Gas Flow within Fractures

Gas transfer through fractures is also governed by the mass conservation relation of Equation (23), but it is rarely used in models of matrix-fracture interaction, due to its rapid equilibration. Usually, a time-injection pressure is specified for the fracture [27]:

$$p_f = \begin{cases} P_{ini} + P_c\left(1 - e^{-\frac{t-t_p}{t_d}}\right) & t \geq t_p \\ P_{ini} & t < t_p \end{cases} \tag{30}$$

where P_{ini} is the initial pressure, P_c is the pressure increment due to gas injections, t_d is the characteristic time for transport, and t_p is the starting time for the gas injection.

From Equation (30), the partial differential equation of the fracture gas pressure can be expressed as:

$$\frac{\partial p_f}{\partial t} = \begin{cases} \frac{P_c}{t_d} e^{-\frac{t-t_p}{t_d}} & t \geq t_p \\ 0 & t < t_p \end{cases} \tag{31}$$

2.6. Coupled Governing Equations

From Equations (4), (19) and (20), the partial derivative of matrix porosity with respect to time is expressed as:

$$\frac{\partial \phi_m}{\partial t} = \alpha \frac{\partial \varepsilon_v}{\partial t} + \frac{\alpha}{K_s}\frac{\partial p_m}{\partial t} - \frac{\alpha \varepsilon_L P_L}{(p + P_L)^2}\frac{\partial p_m}{\partial t} \tag{32}$$

Substituting Equation (32) into Equation (29) yields the governing equation for gas flow in the coal matrix with gas sorption as:

$$\left(\phi_m + \frac{\eta_{ga}}{\eta_g}\frac{p_{ga}\rho_c V_L P_L}{(p_m + P_L)^2} + \frac{\alpha p_m}{K_s} - \frac{\alpha \varepsilon_L P_L p_m}{(p_m + P_L)^2}\right)\frac{\partial p_m}{\partial t} - \nabla \cdot \left(p_m \frac{k_m}{\mu}\nabla p_m\right) = \frac{Q_s}{\eta_g} - \alpha p_m \frac{\partial \varepsilon_v}{\partial t} \tag{33}$$

Equations (7), (21), (32) and (33) define the coupled gas flow and matrix deformation model, while Equations (8) and (30) form an uncoupled model for fracture gas pressure and fracture deformation.

The interaction between the matrix and the fracture is achieved by the stress specified on internal boundaries of the matrix induced by (a) the gas pressure in the fracture when the fracture is regarded as a void; or (b) the generated compressive stress due to fracture swelling due to the fracture gas pressure increasing when the fracture is treated as a soft inclusion.

3. Implementation and Simulation

3.1. Finite Element Implementation

The coupled processes of gas flow and coal deformation for the medium with a centrally-located void representing a fracture are defined by Equations (7), (21), (32) and (33), while those for the medium with a centrally-located soft inclusion are defined by Equations (7), (8), (21), and (30)–(33). The mathematical model comprises a fully coupled finite element approach which simultaneously solves the matrix pore pressure and the displacement of the coal matrix or fracture. COMSOL Multiphysics, a commercial partial differential equation (PDE) solver, is used as the platform for the implementation.

Exploiting the analogy between thermal contraction and matrix shrinkage, the typical example of the thermal consolidation of a column is used. The input data are given in Table 1. Both isothermal and thermoelastic consolidation are simulated and comparisons are made with the analytical solution of Biot [41] and the numerical solution of Noorishad et al. [42] (Figure 3). The excellent match establishes the validity of our modeling approach.

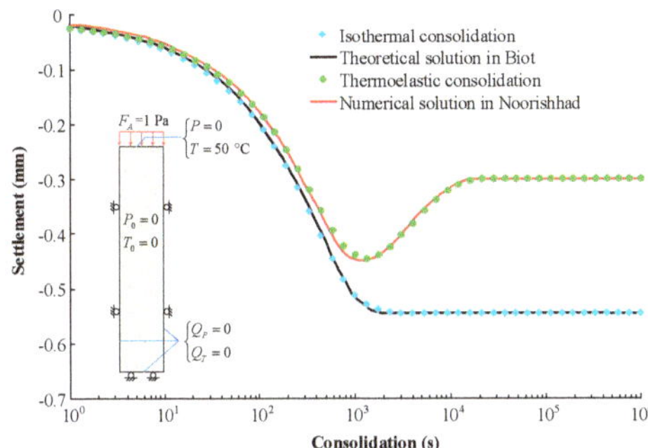

Figure 3. Comparison of simulation results with the analytical solution.

Table 1. Input parameters used for validation.

Parameter	Value
Young's modulus, MPa	6×10^{-3}
Poisson's ratio	0.4
Matrix porosity	0.2
Matrix permeability, m^2	4×10^{-6}
Biot's coefficient	1.0
Water density, kg/m^3	1000
Dynamic viscosity, Pa·s	1×10^{-3}
Thermal conductivity, kJ/(m·s·K)	0.836
Specific heat, $kJ/(m^3 \cdot K)$	167.0
Linear thermal expansion coefficient, 1/K	3×10^{-7}

3.2. Simulations

The simulation geometry is 10 mm by 10 mm with a fracture located at the center. The fracture is 5 mm in length and 0.5 mm in width. As shown in Figure 1b, all the simulation models exhibit horizontal and vertical symmetry. Because of the different treatment of the fracture, appropriate boundary conditions must be applied as shown in Figure 4:

1. The fracture is regarded as void. For the deformation model, the confining stress is applied to all the external boundaries and the fracture pressure (injection pressure) is applied to the internal boundaries. For gas flow, the injection pressure in Equation (30) is applied to the internal boundaries and no flow conditions are applied to all the external boundaries.
2. The fracture is regarded as a soft inclusion without sorption. For the deformation model, the confining stress is applied to all the external boundaries. For the gas flow model, no flow conditions are applied to all the external boundaries.

Firstly, one numerical simulation using the fracture void is conducted to investigate the evolution of fracture aperture, matrix permeability and bulk deformation and to quantify the effects of the change in (M1) the matrix pressure and (M2) the fracture pressure. With the assumption of linear elasticity, the effect of the matrix pressure change can be decomposed into that of (M1a) the body force, (M1b) the effective stress change induced by pore pressure increase on internal boundaries, and (M1c) the effective stress change induced by pore pressure increase on internal boundaries. Input parameters are listed in Table 2 and the values of these parameters are chosen from the literature [2,8]. Then, a series of numerical conditions as listed in Table 3 are simulated to investigate the impacts of factors, involving fracture properties, matrix permeability, injection processes and confining pressure, on the matrix–fracture interaction.

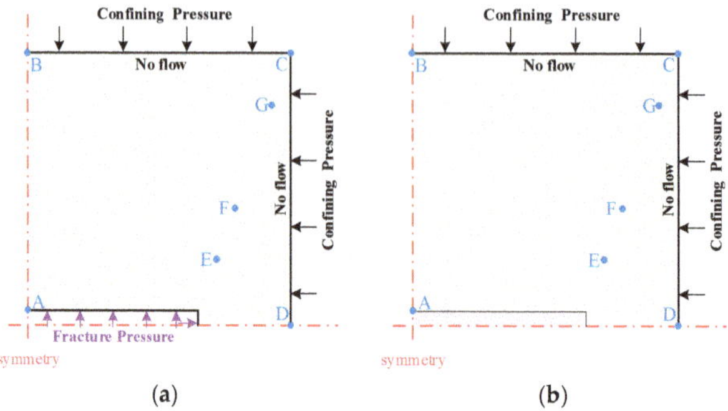

Figure 4. Boundary conditions for different treatment of the fracture. (**a**) The fracture is regarded as a void; (**b**) the fracture is regarded as a soft inclusion.

Table 2. Material properties used in simulations.

Parameter	Matrix-fracture Model CH_4	Verification Model CO_2
Matrix porosity, φ_{m0}	0.05	0.027
Matrix permeability, k_{m0} (m^2)	10^{-20}	4×10^{-23}
Matrix density, ρ_c (kg/m^3)	1500	1500
Matrix Young's modulus, E_m (GPa)	3.95	5.42
Fracture Young's modulus, E_f (GPa)	-	$E_m/2000$
Poisson ratio, v	0.1	0.34
Biot's coefficient, α	0.66	0.66
Langmuir strain constant, ε_L	0.03	0.0119
Langmuir volume constant, V_L (m^3/kg)	0.01316	0.0477
Langmuir pressure constant, P_L (MPa)	3.96	2.76
Gas density at standard condition, ρ_{ga} (kg/m^3)	0.717	1.96
Gas viscosity, μ (Pa·s)	1.2278×10^{-5}	1.84×10^{-5}
Temperature, T (K)	298.15	298.15
Confining pressure, P_{con} (MPa)	0	0
Initial reservoir pressure, P_{ini} (MPa)	0	0
Injection pressure increment, P_c (MPa)	6	-
Injection starting time, t_p (s)	5	-
Injection speed characteristic time, t_d (s)	750	-

Table 3. Simulations for the investigation of dynamic fracture–matrix interaction.

Parameter Investigated	Value
Fracture properties, E_f	Void, $Em/1000$, $Em/100$, $Em/10$
Initial matrix permeability, k_{m0} (m^2)	10^{-18}, 10^{-20}, 10^{-22}, 10^{-24}
Injection speed characteristic time, t_d (s)	5, 100, 750, 10,000
Injection pressure increment, P_c (MPa)	2, 4, 6, 8
Confining pressure, P_{con} (MPa)	0, 4, 8, 12

4. Results and Discussion

4.1. Analysis of Evolving Mechanisms

As shown in Equation (30), we use the characteristic time to define the injection process—a smaller characteristic injection time indicates faster injection. If the characteristic injection time is extremely small, then the fracture pressure reaches the maximum pressure (essentially) immediately. In this simulation, the characteristic time is set to 5 s to replicate a very rapid injection process.

As discussed in Section 2, the changes in fracture pressure, body force and effective stress at boundaries are three influencing factors and they are all related to the pressure. The pressure evolutions at four representative points within the medium are shown in Figure 5 and the evolutions of fracture aperture due to different mechanisms are shown in Figure 6a,b. Four representative points are chosen to illustrate the area of gas propagation within the matrix and to interpret the various mechanical responses. The pressure at Point A represents the fracture pressure, which acts as an internal boundary stress applied to the matrix and opens the fracture due to matrix contraction; Point B and Point D are the nearest external points in the horizontal and vertical directions, respectively, which represent the initiation of gas storage and the effective stress change at the external boundaries; Point C is the furthest external point within the matrix and represents the lowest zone to gain gas increase. From Figure 5, the fracture pressure reaches a maximum pressure at about 40 s, Point D and Point B begin to increase gas pressure at about 300 s and 700 s, respectively and the gas propagates to all external boundaries of the matrix at about 1000 s. As shown in Figure 6a,b, these four representative times are closely related to the deformation induced by different mechanisms:

(1) Gas injection with increasing pressure inflates the fracture due to an increase in the external stress applied to the internal boundaries but narrows the fracture aperture due to the increase of the pore pressure on the internal boundaries. These two effects are enhanced from 5 s to 40 s due to the continuous increase in fracture pressure and remain unchanged after the fracture pressure reaches the maximum. It should be noted that the effect of the body force is determined by the gas pressure gradient in the matrix, thus it can be influenced by the increasing rate of the fracture pressure rather than the fracture pressure itself.

(2) From 40 s to 300 s, the fracture pressure remains constant, causing no change to the opening or narrowing effects induced by effective stress on the internal boundaries, and the pore pressure on the external boundaries remains at the initial value. This induces no change in effective stress on the external boundaries and has a null effect on fracture aperture change. However, the opening effect induced by the body force is slightly weakened as the gas propagates into the matrix.

(3) From 300 s to 700 s, the pore pressure on the external vertical boundaries increases gradually and the pressure gradient on the boundary further drives gas transport inside. During this period, the horizontal body force decreases while the vertical body force continues to increase, leading to the enhanced opening of the fracture. The pore pressure increase on the external vertical boundaries results in a horizontal stress, leading to the narrowing of the fracture.

(4) From 700 s to 4000 s, the pore pressure on the external horizontal boundaries increases gradually, and the gas is transported from the center to the corner as driven by the pressure gradient. During this period, both horizontal and vertical body forces decrease, and the fracture recovers from the opening state. The pore pressure increase on the external horizontal boundaries generates vertical stress. This leads to the fracture opening after counteracting the narrowing effect of horizontal stress on the external vertical boundaries.

(5) From 4000 s, the pore pressure in the whole matrix is equalized with the fracture pressure, and an ultimate equilibrium state is achieved.

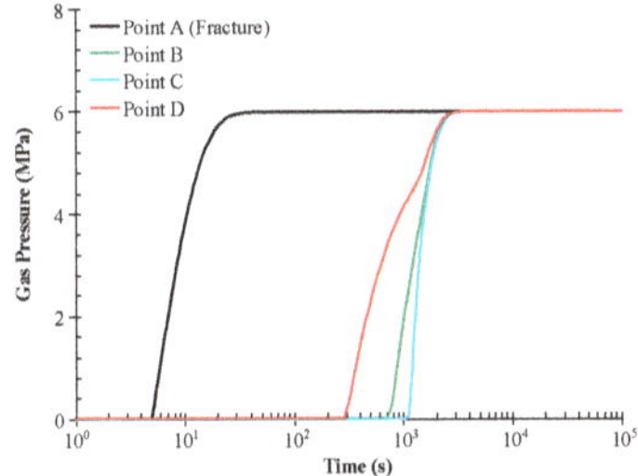

Figure 5. Temporal evolution of pore pressure at four representative points in the matrix.

If the effects of the body force and the effective stress change on the boundaries are combined and called the "effect of gas transport in the matrix" as shown in Figure 6b, the fracture would first narrow and then rebound. The maximum reduction in fracture aperture induced by gas transport in the matrix is 0.021 mm and the final fracture aperture increment is 0.003 mm. The trend of fracture aperture change is similar to our previous results [2,37] that are obtained by ignoring the mechanical effects of fracture pressure under the free swelling condition. Furthermore, it has been proposed that a switch occurs from constant volume to constant stress boundaries during the stress-controlled coal swelling process and the critical pressure or permeability should be used to define this boundary switch [27,43]. Clearly, the concept of such a switch on the external boundaries is not reasonable due to the reality of unchanged boundary conditions. For the case of rapid injection under unconstrained boundary conditions, the fracture still experiences a narrowing trend with injection time, and the critical time for the fracture aperture rebound is at about 300 s, the time at which the pressure at the nearest point on the external boundaries increases. The effects of gas transport in the matrix and fracture pressure on the bulk deformation are expressed in terms of the displacement of Point B and Point D as shown in Figure 7. It can be seen that the gas transport process causes the matrix to experience a transition from contraction to expansion in the vertical direction due to the rectangular geometry of the fracture. Although the horizontal expansion is greater than that in the vertical during gas transport, the discrepancy between horizontal and vertical expansion would vanish as a final equilibrium state results. The final values of both vertical and horizontal displacements is 0.075 mm. The equivalent effective stress change in terms of pore pressure produces a uniform volume strain in the matrix, leading to the equivalent bulk deformation in all directions and uniform matrix permeability as shown in Figure 8. At 300 s, the ratio of the vertical displacement to the final displacement is ~10%, while the ratio of the horizontal displacement to the final displacement is ~61%, meaning that the assumption of a constant volume condition before the switch is triggered may result in unacceptable errors.

If the effect of fracture pressure increase is incorporated, the fracture aperture would increase, first due to the stronger opening effect induced by the fracture pressure that overcomes the narrowing effect induced by gas transport in the vicinity of the fracture. This will then decrease due to the narrowing effect as the dominant mode with a final recovery to a stable value due to the expansion of gas invasion area. From Figure 6b, it is apparent that the fracture aperture change has a peak value of 0.014 mm and a minimum trough of 0.001 mm; the final value is obviously enlarged (from 0.004 mm to 0.026 mm). As shown in Figure 7, the bulk deformation of the matrix shows a horizontal contraction and vertical expansion with the fracture pressure exerted on the rectangular fracture surface. Consequently, the entire matrix would have a larger vertical expansion than the horizontal expansion, demonstrating the heterogeneity of the matrix imposed by the fracture geometry.

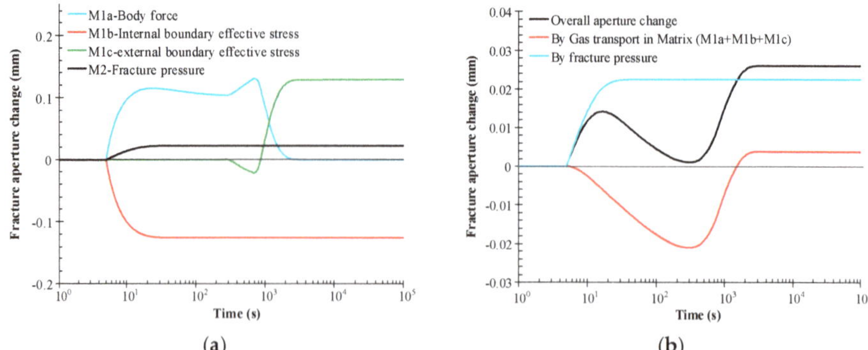

Figure 6. Temporal effects of different mechanisms on fracture aperture change. (**a**) Individual effect; (**b**) combined effect.

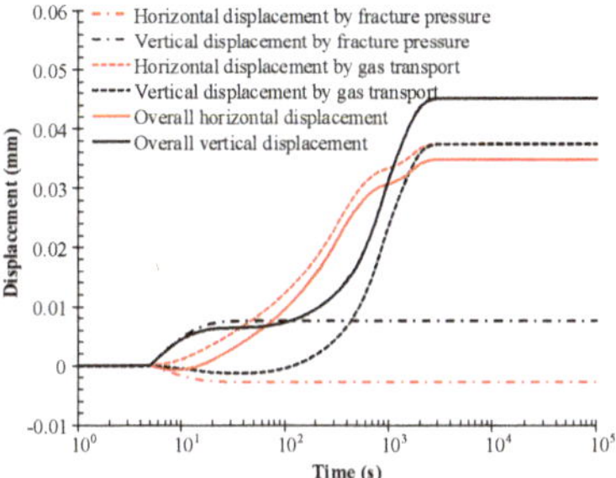

Figure 7. Temporal evolution of the bulk deformation of the representative component geometry.

The matrix permeability is also significant in controlling the coupled process and its evolution controls the gas transport behavior during gas transport in the matrix. It is clear that the matrix permeability varies with the change in volume strain, grain compression strain and sorption-induced strain (see Equations (19) and (20)). Figure 9 shows typical volumetric strain and pressure evolution in the matrix domain at Point F, and Figure 10 presents the temporal evolution of permeability ratio for three domain points. As shown in Figure 9, the gas storage increases at Point E when t = 200 s, before which the matrix volumetric strain is influenced by the fracture pressure increase and the gas transport in the propagating zone. After 200 s, the volumetric strain is influenced by the pressure-based sorption-induced strain and grain compaction strain. As a result, the effective volumetric strain for the matrix pores also experiences an increasing-then-decreasing-then-increasing period, leading to the same trend in matrix permeability as shown in Figure 10. Moreover, the permeability evolution of three fiducial points within the domain all exhibit similar trends, despite different switch times and different peak values, troughs and final values of the permeability ratio due to different positions relative to the fracture. These demonstrate the distinct spatial and temporal characteristics within fracture and matrix during gas transport from the fracture to the matrix.

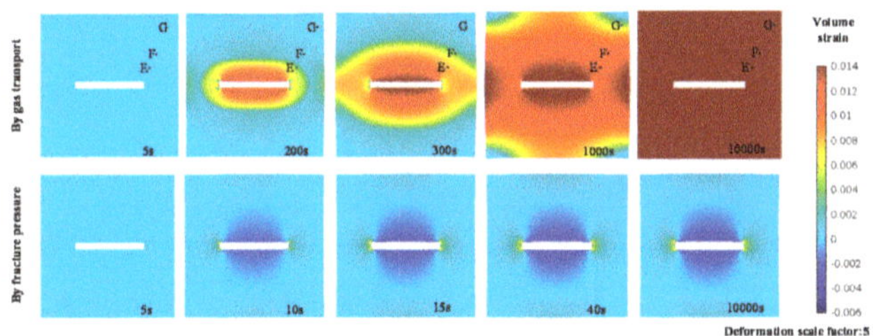

Figure 8. Matrix volume strain distribution at different times.

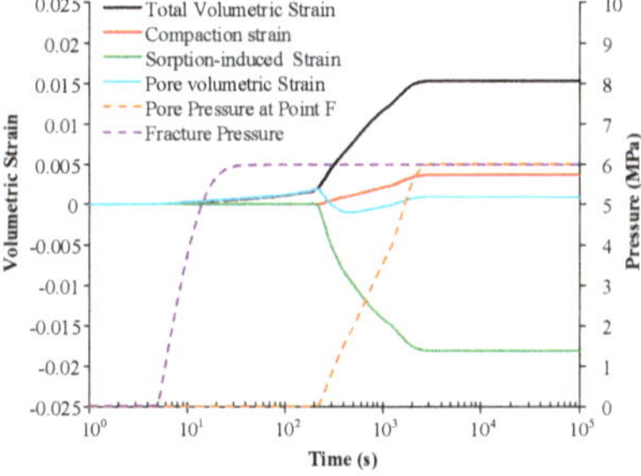

Figure 9. Evolution of volumetric strain and pore pressure at Point F, together with fracture pressure.

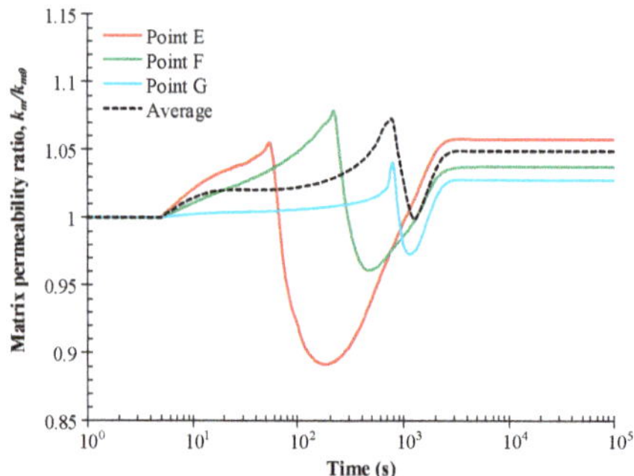

Figure 10. Temporal evolution of matrix permeability ratio.

4.2. Impacts of Fracture Properties

Two different treatments of the fracture are used in this section to investigate and compare the impacts of fracture properties on the fracture aperture change. Figure 11 illustrates the evolution of fracture aperture change due to different fracture properties. The Young's modulus of the fracture is varied from 395 MPa to 3.95 MPa when the fracture is regarded as a soft inclusion. The results show that Young's modulus has a significant effects on the evolution of fracture aperture change. Although the evolution of the fracture aperture change shows the same trend in all three cases, the magnitudes of the peak, the trough and the final fracture aperture vary in each case due to the different Young's moduli of the fracture. As the Young's modulus increases from 3.95 MPa to 395 MPa, the peak value increases from 0.0036 mm to 0.0138 mm, the value of the trough increases from -4×10^{-5} mm to 0.0011 mm and the final value increases from 0.0062 mm to 0.025 mm. It is notable that the evolution of the fracture aperture approaches that of the case with a fracture void when the Young's modulus of the fracture is reduced to one thousandth of the matrix Young's modulus. This suggests that a reduction factor for the stress induced by fracture pressure increase on the internal surfaces may be considered due to the presence of coal fragment or of proppant.

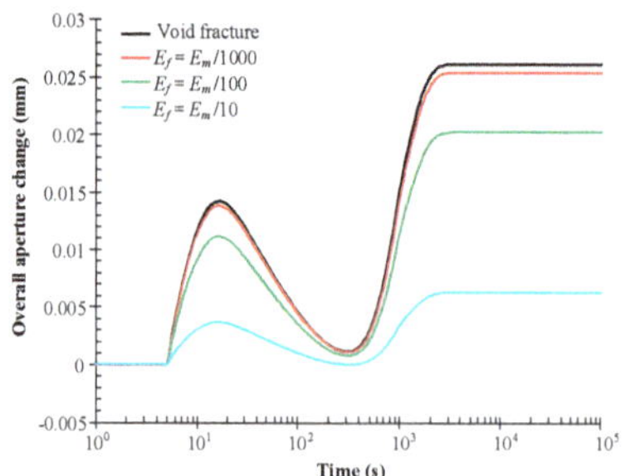

Figure 11. Evolution of fracture aperture change for different assumed fracture properties.

4.3. Impacts of Initial Matrix Permeability

Gas transport in the matrix is not only determined by the pressure gradient but also affected by initial matrix permeability. Figure 12 illustrates the evolution of fracture aperture change for different initial matrix permeabilities. The initial matrix permeability is varied from 10^{-18} m^2 to 10^{-24} m^2 when the gas is injected. The results show that the initial matrix permeability has a significant effect on the evolution of fracture aperture change. A peak and trough of fracture aperture change is observed in the first three cases with all the aperture changes finally approaching the same value (about 0.026 mm). Because the matrix permeability of 10^{-24} m^2 is very low, it requires a considerably longer time for the gas to transport into the matrix and fill the entire representative volume to the injection pressure. If the simulation time is sufficiently long, the evolution of aperture change would follow the same trend as the other three cases. As the initial matrix permeability decreases from 10^{-18} m^2 to 10^{-22} m^2, the time when the aperture change reaches the peak value varies from 7 s to 28 s while the time for aperture change to reach the trough increases from 12 s to 3×10^4 s. The trough values of aperture change are close to 0.0011 mm, while the peak values differ greatly. The peak value of aperture change increases from 0.0022 mm to 0.022 mm with the initial matrix permeability decreasing from 10^{-18} m^2 to 10^{-22} m^2.

The initial matrix permeability affects the process of gas transport and adsorption in the matrix. With an increasing initial matrix permeability, the gas flow from fracture to matrix and the gas transport within the matrix become more rapid, which may cause the gas to transport to the external boundaries before the injection pressure reaches the maximum. Then the significant effect of gas transport in the matrix advances the switch from increase to decrease in overall aperture change. If the initial matrix permeability is sufficiently high to instantly fill the matrix at the injection pressure, the switching of gas transport vanishes and the overall fracture aperture would increase to the final value with increasing time or injection pressure. Conversely, if the initial matrix permeability decreases, the narrowing effect induced by gas transport weakens in the early period, leading to a dominant effect of fracture pressure and in increasing the peak value of overall fracture aperture change. When the matrix becomes impermeable, the peak value of the aperture reaches its maximum. The reason why the trough in the fracture aperture remains almost unchanged may be since the initial matrix permeability merely prolongs the time for gas transport to the external boundaries but exerts little impact on the gas pressure distribution. It is notable that there exists a constant difference between the maximum peak value and the final value of the fracture aperture change, or between the final values of fracture aperture change in the cases of an impermeable matrix and instantly-filled matrix, which is equivalent to the aperture change induced by matrix swelling when pressure equilibrium is achieved.

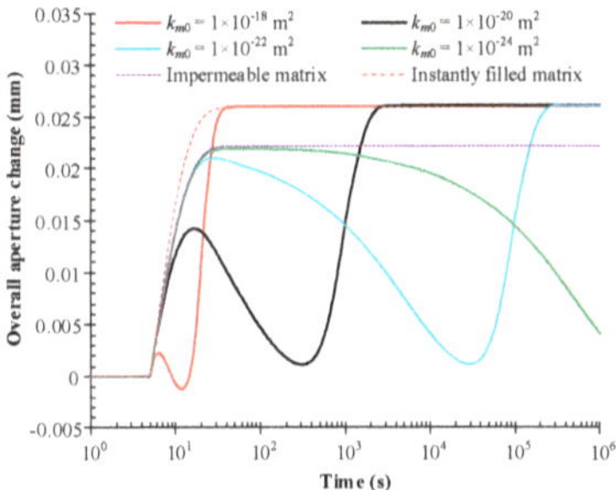

Figure 12. Evolution of fracture aperture change for different initial matrix permeabilities.

4.4. Impacts of Injection Processes

The injection process controls the fracture pressure and the rate of pressure increase on internal boundaries. This influences the fracture aperture change induced by the mechanical effects of fracture pressure and the swelling effect of gas transport in the matrix. The injection rate characteristic time and the injection pressure increment are two important factors controlling the injection processes. A higher characteristic time indicates a lower rate to get to the maximum injection pressure, while a higher increment means a larger magnitude of the maximum injection pressure.

4.4.1. Impacts of Injection Characteristic Time

Figure 13 illustrates the evolution of each component in overall fracture aperture change, fracture pressure and pore pressure at Point D for different injection rate characteristic times. The injection characteristic time is varied from 5 s to 10^4 s and the results show that the injection characteristic time has a significant effect on the evolution of fracture aperture change. A peak and trough in the fracture aperture change is observed in all four cases; however, the magnitude of the peak and trough and the time when fracture aperture change switches vary in each case, due to the different injection characteristic times.

The maximum reduction in fracture aperture induced by gas transport is ~0.02 mm when the characteristic time is 100 s. This is almost as large as that for the case with a characteristic time of 5 s, while the maximum reduction is only 0.01 mm in the case with a characteristic time of 10^4 s. A comparison in the pressure evolution suggests that whether the maximum reduction of fracture aperture induced by gas transport in the matrix and the switch time vary depends on relationships between the required times for the fracture pressure to reach the maximum (denoted by t_1) and for gas transport to Point D (denoted by t_2): if $t_1 \leq t_2$, the maximum reduction changes slightly and the switch time is equivalent to t_2; if $t_1 > t_2$, the maximum reduction decreases with increasing injection characteristic time, and the switch time then gradually lags behind due to the increasing body force induced by the injection pressure increase.

The overall aperture change is the combination of the aperture change induced by gas transport in the matrix and the fracture pressure increase. When the characteristic time is 750 s, the peak value of the overall aperture change is only 4.6×10^{-4} mm, about 3% of that in the case with characteristic time is 5 s. The trough in the overall aperture change decreases from 1.1×10^{-3} mm to -3×10^{-3} mm as the characteristic time varies from 5 s to 10^4 s. With an increase in the characteristic time, the time for the occurrence of the peak value grows from 15 s to 500 s due to the slowdown in the fracture pressure increase while that of the valley grows from 300 s to 4500 s due to the decrease of the pressure gradient within the matrix.

These results indicate that if the injection speed characteristic time approaches infinity, the whole system is always in a state of equilibrium and the fracture aperture change gradually increases with the equilibrium pressure due to the decrease in effective stress and increase in fracture pressure. This is consistent with the results obtained from the assumption of local equilibrium.

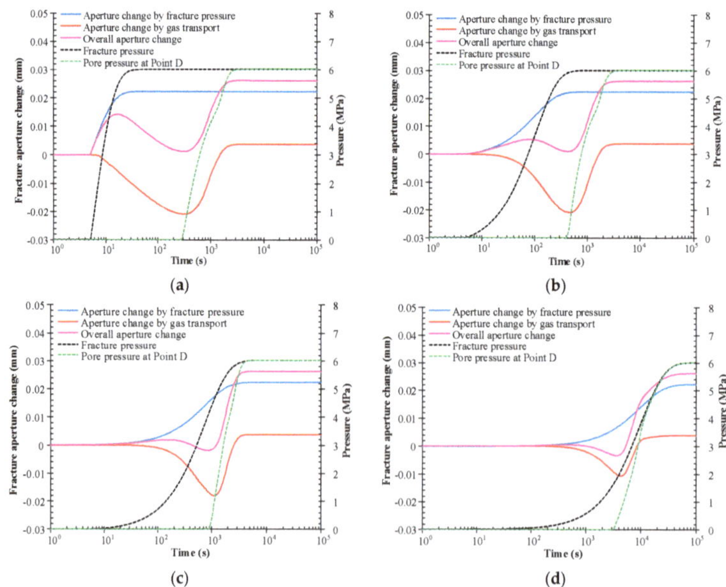

Figure 13. Evolution of each component in overall aperture change from fracture pressure and pore pressure at Point D at different injection characteristic times. (**a**) t_d = 5 s; (**b**) t_d = 100 s; (**c**) t_d = 750 s; (**d**) t_d = 10000 s.

4.4.2. Impact of Injection Pressure Increment

The evolution of fracture pressure and pore pressure at Point D at different pressure increments is shown in Figure 14. This indicates that the injection pressure increment only changes the amplification of the fracture pressure increase without impacting the time to reach the maximum. As the fracture pressure rapidly reaches the maximum, the aperture change induced by fracture pressure and gas transport in the matrix is proportional to the ratio of injection pressure increment soon after the injection initiates. The time for the peak fracture aperture change to occur remains unchanged as illustrated in Figure 15. However, the gas transport in the matrix accelerates due to the increase in pressure gradient during continuous injection, advancing the time of gas transport to Point D or the occurrence of the trough in fracture aperture change. The time for the trough is ~200 s when the pressure increment is 8 MPa, while it increases to 1500 s as the pressure increment drops to 2 MPa. Notably, the switch in aperture change is abrupt for the case with a pressure increment of 8 MPa, compared to the relatively smooth transition in the other three cases. Moreover, whether the trough of the aperture change is negative or positive depends on the relative magnitude of the maximum increase in aperture change induced by fracture pressure and the aperture maximum reduction induced by gas transport in matrix. For the simulation cases in this section, injection pressure increments of 2 MPa and 4 MPa can both cause the maximum reduction in aperture change induced by gas transport that are higher than the maximum increase of aperture change induced by fracture pressure. The maximum increase in fracture aperture change induced by fracture pressure is 0.015 mm in the case for an injection pressure increment of 4 MPa, less than the maximum reduction of aperture change induced by gas transport of 0.017 mm, leading to the trough in the fracture aperture change equivalent to −0.02 mm.

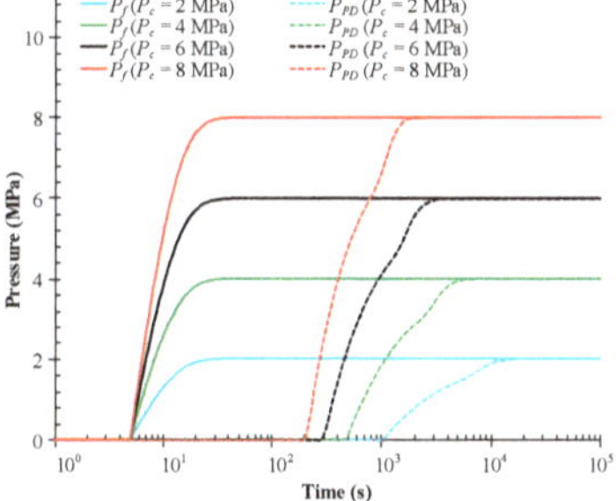

Figure 14. Evolution of fracture pressure and pore pressure at Point D for different injection pressure increments. P_f denotes the fracture pressure and P_{PD} denotes the pore pressure at Point D.

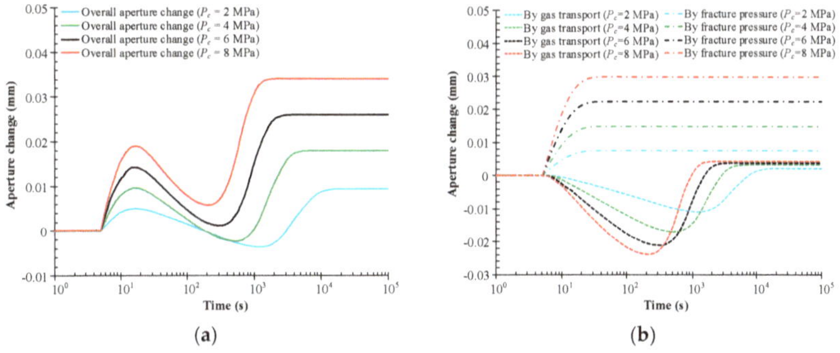

Figure 15. Evolution of fracture aperture change for different injection pressure increments. (**a**) Overall aperture change; (**b**) aperture change induced by fracture pressure and gas transport in matrix.

4.5. Impacts of Confining Pressure

Figure 16 illustrates the evolution of fracture pressure, the pore pressure at Point D and the fracture aperture change at different confining pressures. The results show that the confining pressure changes the initial and final fracture aperture and has little impact on the trend of fracture aperture change. As the confining pressure increases from 0 to 12 MPa, the initial fracture aperture reduces by 0.046 mm as well as the peak and troughs of fracture aperture. Simultaneously, the time for gas transport to Point D increases from 300 s to 400 s, slightly postponing the occurrence of the trough in the aperture. The confining pressure acts as an external stress and causes matrix shrinkage and fracture closure. However, the confining pressure is usually applied before the injection begins, which induces compressive deformation of the fracture and the matrix and thus forms a different initial fracture geometry and uneven distribution of initial matrix permeability, as illustrated in Figure 17. The permeability ratio of the transient value to the initial value is often used. However, the uneven deformation due to confining pressure presents a significant challenge in determining the initial permeability of both the matrix and fracture.

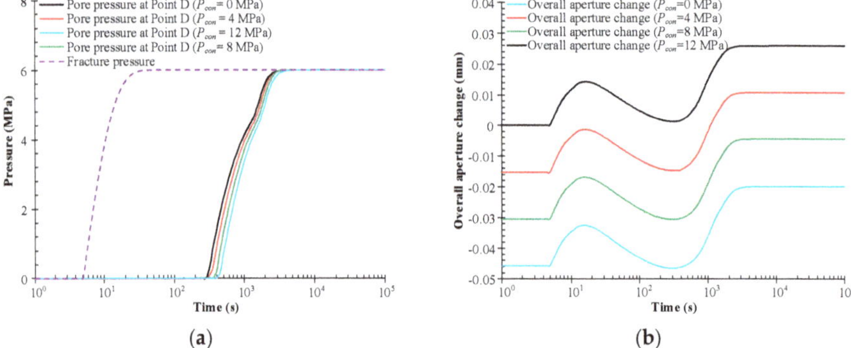

Figure 16. Evolution of pressure and aperture change for different confining pressures. (**a**) Fracture pressure and pore pressure at Point D; (**b**) overall facture aperture change.

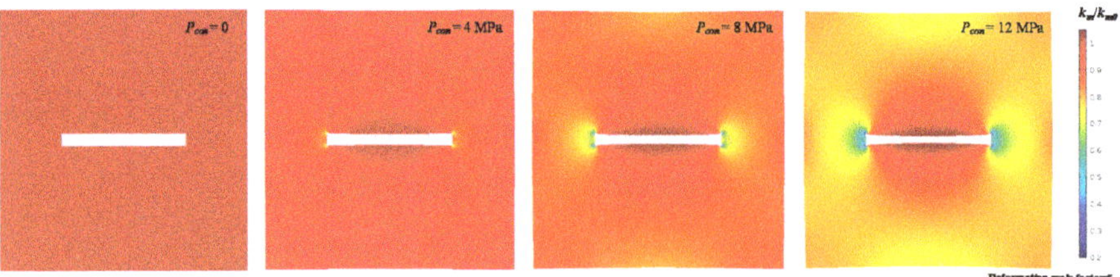

Figure 17. Uneven deformation and permeability ratio in the matrix induced by different confining pressures.

5. Verification with Experimental Data

Experimental data measured under stress-controlled conditions are now used to verify our model. The coal core sample contains an induced fracture to increase the initial permeability. This is completed by applying a compressive load to the cylindrical surface as in experiments conducted by Siriwardane et al. [15] to investigate the influence of CO_2 exposure on coal permeability. In this study, the virtual core representing the sample is cylindrical with a throughgoing fracture. The fracture is located at the center of the circular section with a diameter of 0.0375 m, as illustrated in Figure 18a. The fracture length is 0.03 m to obtain a contact ratio of 20% and the fracture aperture calculated from the initial permeability is 7×10^{-7} m, based on equivalent mass flow. Due to the contact area, roughness and tortuosity, the hydraulic aperture is smaller than the mechanical aperture [44,45]. In the simulation, the initial equivalent mechanical aperture is set to 3.5×10^{-6} m. It is noted that the initial fracture aperture is the mean value after the confining pressure is applied around the sample and a fracture pressure of 10 MPa is applied on the fracture boundaries. One quarter of the typical section is chosen as the simulation model and the corresponding boundary conditions are illustrated in Figure 18b. In the simulation, the properties of the matrix for the fractured sample is required, which are missing from the experiments. Therefore, we may only assume their values based on the literature [8,34] as listed in Table 2.

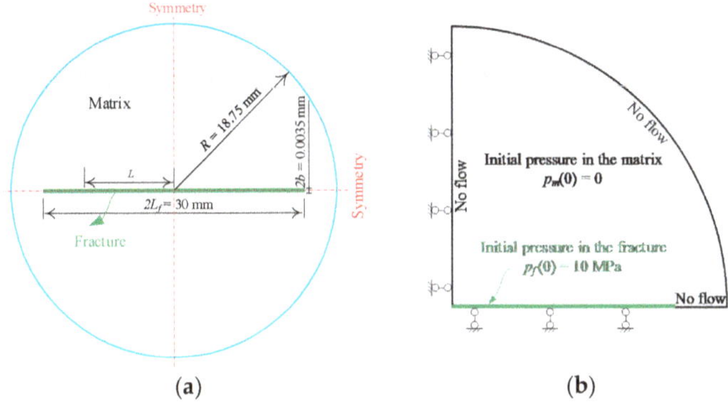

Figure 18. Verification model. (**a**) Typical section of coal sample; (**b**) Model implemented in the simulation.

5.1. Evolution of Fracture Permeability

Figure 19 illustrates the distribution of the permeability ratio along the fracture at different times for the case of a fractured core sample exposed to CO_2. The normalized distance represents the ratio of the distance from the fracture center (L) to half the fracture length (L_f). As apparent in Figure 19, fracture permeability at different locations experiences different evolutions over the simulation period. The part of the fracture with a normalized distance less than 0.8 continues to narrow, while that outside this normalized distance (greater than 0.8) rebounds after experiencing initial narrowing with a different reduction ratio. From the trend of the permeability ratio (Figure 19), the outer part of the fracture rebounds faster, indicating that the central part with a continuous permeability decrease also rebounds as the gas propagates to the distal matrix, if given sufficient time. The distance from the fracture to its perpendicular boundary increases from the end of the fracture to its center, and this may represent a constraint for fracture deformation induced by matrix swelling. At first, the CO_2 propagates in the vicinity of the fracture surface, and the induced swelling stress perpendicular to the fracture surface resulting from gas adsorption is shared by the matrix in the perpendicular direction. This may explain why the permeability of the fracture center decreases and then rebounds slowest in the case of a circular matrix.

The permeability ratio recovered from the experimental results is shown in Figure 20, together with the evolution of average fracture permeability ratio. When the coal matrix is exposed to CO_2 at a pressure of 10 MPa for 80 h, the local swelling induced by gas adsorption has an important impact on fracture permeability—the initial permeability is reduced by approximately one order of magnitude. Specifically, the average fracture permeability decreases dramatically with a reduction ratio of 70% over the initial period of 10 h. Then, the permeability decrease gradually slows with the sequential reduction ratio of 20% in the next 40 h and, finally, the permeability asymptotes to a constant magnitude. As apparent from Figure 19, all the fracture would compact rapidly. When t = 10 h, 85% of the fracture proximal to its center has a reduction ratio in the range 60%–90%. Subsequently, an increasing proportion of the outer part of the fracture begins to rebound and the central part of the fracture further narrows. The increase in the rebounding part, and its extent, slows the average permeability decrease, and the average permeability remains stable due to the equivalent effects of fracture narrowing and rebound. When $t > 80$ h, the rebounding effects may get stronger than the narrowing effect, leading to the recovery of average fracture permeability ratio.

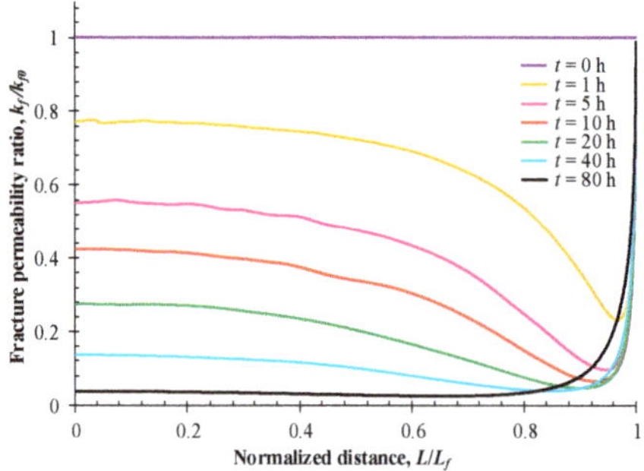

Figure 19. Distribution of permeability ratio along the fracture.

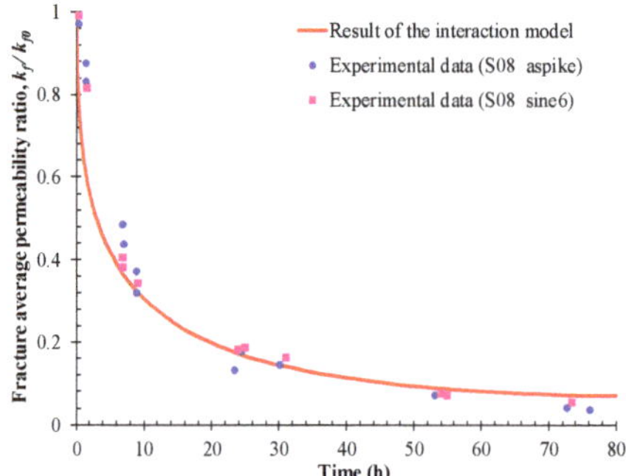

Figure 20. Evolution of coal average permeability ratio.

5.2. Sensitivity of Initial Matrix Permeability

As discussed in the previous section, the initial matrix permeability plays an important role in the permeability evolution. In the experiment conducted by Siriwardane et al. [15], the average fracture permeability continues to decrease over 80 h, by contrast with other experimental observations of permeability recovery. This monotonic decrease results because the experiments were ended while the local swelling was still the dominant mechanism—resulting in permeability reduction, alone. This stability in permeability evolution results from the close competitive effects of fracture narrowing due to local swelling and fracture opening due to global swelling at some time stage [27], instead of the disappearance of matrix swelling due to gas pressure equilibrium [34]. In order to obtain different permeability evolution in fractured coal exposed to CO_2, three different initial matrix permeabilities are adopted and the evolution of the average fracture permeability ratio is illustrated in Figure 21. When initial matrix permeability increases from 4×10^{-23} m^2 to 4×10^{-21} m^2, gas propagation into the matrix accelerates and the time for gas transport to reach the external boundaries is shortened. As a result, the rebound in the permeability the entire fracture appears in advance, leading to a shortened period for the reduction of the average fracture permeability ratio. This is consistent with the previous view that the central part of the fracture would rebound after longer than 80 h, when initial matrix permeability is assumed to be 4×10^{-23} m^2. As apparent in Figure 21, the stage of stable fracture permeability is observed during the rebounding of average fracture permeability and this stable permeability lasts for ~8 h when the initial matrix permeability is 4×10^{-21} m^2. The stability of the average fracture permeability during the rebound stage is the combined result of nearly-matched opposing effects of local swelling and global swelling. As illustrated in Figure 22, the fracture outer part, during this stage, is wider than its initial value. This is induced by the cumulative local swelling of the central part. With a gradual switch from local swelling to global swelling within the fracture, the outer part of the fracture narrows while the central part opens.

With continuous gas transport into the matrix, the matrix ultimately has a uniform equivalent pore pressure and consequently no swelling deformation induced by gas adsorption. As a consequence, the fracture permeability remains stable. The final fracture permeability is slightly greater than the initial value, and the increase in amplitude can reach a maximum of the swelling deformation of coal bridges induced by gas sorption and pore pressure increase.

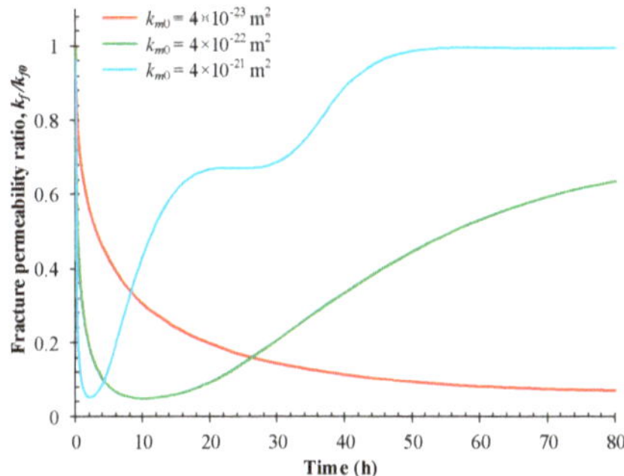

Figure 21. Evolution of fracture average permeability ratio at different initial matrix permeabilities.

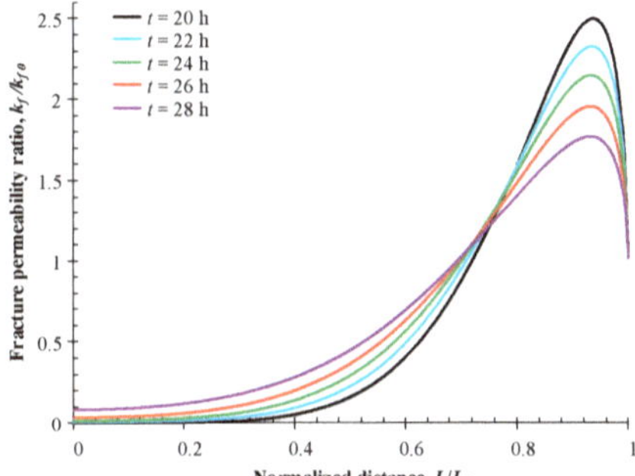

Figure 22. Distribution of permeability ratio along the fracture when $k_{m0} = 4 \times 10^{-21}$ m^2.

6. Conclusions

In this study, a matrix–fracture interaction model is applied to investigate the transient response of coal deformation and permeability to the temporal and spatial variations of effective stresses under mechanically unconstrained conditions. The effect of the increase in fracture pressure on the change in fracture aperture and on bulk deformation is incorporated into the matrix–fracture interaction and the individual effects of different mechanisms are evaluated. The impact of fracture properties, initial matrix permeability, injection processes and confining pressure are separately evaluated. Based on the model results, the following major conclusions are drawn:

- The evolution of coal permeability under unconstrained conditions is primarily controlled by the fracture pressure and gas transport in the matrix. The evolution of the non-equilibrium permeability, from the initial equilibrium permeability (when the matrix pressure is equalized with the fracture pressure prior to injection) to the final equilibrium permeability (when the matrix pressure is equalized with the matrix pressure after the gas injection) exhibits three distinct stages. In the case of gas injection, these three stages are (1) an increase in permeability due to the increased injection pressure; (2) a reduction due to the localization of matrix pressure in the vicinity of the fracture wall; and (3) recovery due to the equilibration of matrix pressure throughout the matrix block.

- The duration of the transient period, from the initial equilibrium state to the ultimate equilibrium state, is determined both by matrix transport properties and by loading processes on external boundaries and within the internal boundaries (interfaces between matrixes and fractures). For coal, this transient period may be extremely long, due to its low matrix permeability. This suggests that the permeabilities measured in many laboratory experiments may not indeed be equilibrium permeabilities, as represented in publications, and that experiments of extended duration should be developed to measure the non-equilibrium permeability in the transient state.

Author Contributions: Conceptualization and methodology, J.L., X.L. and J.S.; validation and visualization, X.L. and Y.H.; writing—original draft preparation, X.L. and J.S., writing—review and editing, X.L. and J.L.

Funding: This research is funded by the National Natural Science Foundation of China (Nos. 51474204 and 51579078), the Fundamental Research Funds for the Central Universities (No. 2015B32114) and Collaborative Innovation Center for Prevention and Control of Mountain Geological Hazards of Zhejiang Province (No. PCMGH-2017-Z-02).

Conflicts of Interest: The authors declare no conflicts of interest.

References

1. Connell, L.D.; Lu, M.; Pan, Z. An analytical coal permeability model for tri-axial strain and stress conditions. *Int. J. Coal Geol.* **2010**, *84*, 103–114. [CrossRef]
2. Liu, J.; Chen, Z.; Elsworth, D.; Qu, H.; Chen, D. Interactions of multiple processes during CBM extraction: A critical review. *Int. J. Coal Geol.* **2011**, *87*, 175–189. [CrossRef]
3. Durucan, S.; Edwards, J. The effects of stress and fracturing on permeability of coal. *Min. Sci. Technol.* **1986**, *3*, 205–216. [CrossRef]
4. Somerton, W.H.; Söylemezoḡlu, I.; Dudley, R. Effect of stress on permeability of coal. In *International Journal of Rock Mechanics and Mining Sciences & Geomechanics Abstracts*; Elsevier: Pergamon, Turkey, 1975; pp. 129–145.
5. Clarkson, C.; Bustin, R. The effect of pore structure and gas pressure upon the transport properties of coal: A laboratory and modeling study. 1. Isotherms and pore volume distributions. *Fuel* **1999**, *78*, 1333–1344. [CrossRef]
6. Gray, I. Reservoir engineering in coal seams: Part 1-The physical process of gas storage and movement in coal seams. *SPE Res. Eng.* **1987**, *2*, 28–34. [CrossRef]
7. Harpalani, S.; Schraufnagel, R.A. Shrinkage of coal matrix with release of gas and its impact on permeability of coal. *Fuel* **1990**, *69*, 551–556. [CrossRef]
8. Robertson, E.P.; Christiansen, R.L. Modeling permeability in coal using sorption-induced strain data. In Proceedings of the SPE Annual Technical Conference and Exhibition, Dallas, Texas, TX, USA, 9–12 October 2005. [CrossRef]
9. Seidle, J.R.; Huitt, L. Experimental measurement of coal matrix shrinkage due to gas desorption and implications for cleat permeability increases. In Proceedings of the International Meeting on Petroleum Engineering, Beijing, China, 14–17 November 1995. [CrossRef]
10. Tan, Y.; Ning, J.; Li, H. In situ explorations on zonal disintegration of roof strata in deep coalmines. *Int. J. Rock Mech. Min. Sci.* **2012**, *49*, 113–124. [CrossRef]
11. Pan, Z.; Connell, L.D.; Camilleri, M. Laboratory characterisation of coal reservoir permeability for primary and enhanced coalbed methane recovery. *Int. J. Coal Geol.* **2010**, *82*, 252–261. [CrossRef]
12. Pini, R.; Ottiger, S.; Burlini, L.; Storti, G.; Mazzotti, M. Role of adsorption and swelling on the dynamics of gas injection in coal. *J. Geophys. Res.-Solid Earth* **2009**, *114*. [CrossRef]
13. Han, F.; Busch, A.; van Wageningen, N.; Yang, J.; Liu, Z.; Krooss, B.M. Experimental study of gas and water transport processes in the inter-cleat (matrix) system of coal: Anthracite from Qinshui Basin, China. *Int. J. Coal Geol.* **2010**, *81*, 128–138. [CrossRef]

14. Pan, Z.; Connell, L.D.; Camilleri, M.; Connelly, L. Effects of matrix moisture on gas diffusion and flow in coal. *Fuel* **2010**, *89*, 3207–3217. [CrossRef]
15. Siriwardane, H.; Haljasmaa, I.; McLendon, R.; Irdi, G.; Soong, Y.; Bromhal, G. Influence of carbon dioxide on coal permeability determined by pressure transient methods. *Int. J. Coal Geol.* **2009**, *77*, 109–118. [CrossRef]
16. Chen, Z.; Liu, J.; Elsworth, D.; Connell, L.D.; Pan, Z. Impact of CO_2 injection and differential deformation on CO_2 injectivity under in-situ stress conditions. *Int. J. Coal Geol.* **2010**, *81*, 97–108. [CrossRef]
17. Connell, L. Coupled flow and geomechanical processes during gas production from coal seams. *Int. J. Coal Geol.* **2009**, *79*, 18–28. [CrossRef]
18. Cui, X.; Bustin, R.M. Volumetric strain associated with methane desorption and its impact on coalbed gas production from deep coal seams. *AAPG Bull.* **2005**, *89*, 1181–1202. [CrossRef]
19. Izadi, G.; Wang, S.; Elsworth, D.; Liu, J.; Wu, Y.; Pone, D. Permeability evolution of fluid-infiltrated coal containing discrete fractures. *Int. J. Coal Geol.* **2011**, *85*, 202–211. [CrossRef]
20. Palmer, I.; Mansoori, J. How permeability depends on stress and pore pressure in coalbeds: A new model. In Proceedings of the SPE Annual Technical Conference and Exhibition, Denver, CO, USA, 6–9 October 1996. [CrossRef]
21. Shi, J.Q.; Durucan, S. Drawdown Induced Changes in Permeability of Coalbeds: A New Interpretation of the Reservoir Response to Primary Recovery. *Transp. Porous Media* **2004**, *56*, 1–16. [CrossRef]
22. Zhang, H.; Liu, J.; Elsworth, D. How sorption-induced matrix deformation affects gas flow in coal seams: A new FE model. *Int. J. Rock Mech. Min. Sci.* **2008**, *45*, 1226–1236. [CrossRef]
23. Robertson, E.P.; Christiansen, R.L. A permeability model for coal and other fractured, sorptive-elastic media. In Proceedings of the SPE Eastern Regional Meeting, Canton, OH, USA, 11–13 October 2006. [CrossRef]
24. Liu, H.H.; Rutqvist, J.; Oldenburg, C.M. A new coal-permeability model: Internal swelling stress and fracture-matrix interaction. *Transp. Porous Media* **2010**, *82*, 157–171. [CrossRef]
25. Chen, Z.; Pan, Z.; Liu, J.; Connell, L.D.; Elsworth, D. Effect of the effective stress coefficient and sorption-induced strain on the evolution of coal permeability: Experimental observations. *Int. J. Greenh. Gas Control* **2011**, *5*, 1284–1293. [CrossRef]
26. Chen, Z.; Liu, J.; Pan, Z.; Connell, L.D.; Elsworth, D. Influence of the effective stress coefficient and sorption-induced strain on the evolution of coal permeability: Model development and analysis. *Int. J. Greenh. Gas Control* **2012**, *8*, 101–110. [CrossRef]
27. Liu, J.; Wang, J.; Chen, Z.; Wang, S.; Elsworth, D.; Jiang, Y. Impact of transition from local swelling to macro swelling on the evolution of coal permeability. *Int. J. Coal Geol.* **2011**, *88*, 31–40. [CrossRef]
28. Peng, Y.; Liu, J.; Wei, M.; Pan, Z.; Connell, L.D. Why coal permeability changes under free swellings: New insights. *Int. J. Coal Geol.* **2014**, *133*, 35–46. [CrossRef]
29. Peng, Y.; Liu, J.; Zhu, W.; Pan, Z.; Connell, L. Benchmark assessment of coal permeability models on the accuracy of permeability prediction. *Fuel* **2014**, *132*, 194–203. [CrossRef]
30. Robertson, E.P. Measurement and Modeling of Sorption-Induced Strain and Permeability Changes in Coal. Ph.D. Thesis, Colorado School of Mines, Arthur Lakes Library, Golden, CO, USA, 2005.
31. Bai, M.; Elsworth, D.; Roegiers, J.C. Multiporosity/multipermeability approach to the simulation of naturally fractured reservoirs. *Water Resour. Res.* **1993**, *29*, 1621–1634. [CrossRef]
32. Berryman, J.G.; Wang, H.F. The elastic coefficients of double-porosity models for fluid transport in jointed rock. *J. Geophys. Res.-Solid Earth* **1995**, *100*, 24611–24627. [CrossRef]
33. Wang, C.; Liu, J.; Feng, J.; Wei, M.; Wang, C.; Jiang, Y. Effects of gas diffusion from fractures to coal matrix on the evolution of coal strains: Experimental observations. *Int. J. Coal Geol.* **2016**, *162*, 74–84. [CrossRef]
34. Wu, Y.; Liu, J.; Elsworth, D.; Miao, X.; Mao, X. Development of anisotropic permeability during coalbed methane production. *J. Nat. Gas Sci. Eng.* **2010**, *2*, 197–210. [CrossRef]
35. Cai, M.; Horii, H. A constitutive model of highly jointed rock masses. *Mech. Mater.* **1992**, *13*, 217–246. [CrossRef]
36. Warren, J.; Root, P.J. The behavior of naturally fractured reservoirs. *Soc. Petrol. Eng. J.* **1963**, *3*, 245–255. [CrossRef]
37. Chen, Z.; Liu, J.; Elsworth, D.; Pan, Z.; Wang, S. Roles of coal heterogeneity on evolution of coal permeability under unconstrained boundary conditions. *J. Nat. Gas Sci. Eng.* **2013**, *15*, 38–52. [CrossRef]

38. Chilingar, G.V. Relationship between porosity, permeability, and grain-size distribution of sands and sandstones. In *Developments in Sedimentology*; Straaten, L.M.J.U.V., Ed.; Elsevier Science Publishers: Amsterdam, The Netherlands, 1964; Volume 1, pp. 71–75.
39. Witherspoon, P.A.; Wang, J.S.; Iwai, K.; Gale, J.E. Validity of cubic law for fluid flow in a deformable rock fracture. *Water Resour. Res.* **1980**, *16*, 1016–1024. [CrossRef]
40. Saghafi, A.; Faiz, M.; Roberts, D. CO_2 storage and gas diffusivity properties of coals from Sydney Basin, Australia. *Int. J. Coal Geol.* **2007**, *70*, 240–254. [CrossRef]
41. Biot, M.A. General Theory of Three-Dimensional Consolidation. *J. Appl. Phys.* **1941**, *12*, 155–164. [CrossRef]
42. Noorishad, J.; Tsang, C.-F. Coupled thermohydroelasticity phenomena in variably saturated fractured porous rocks–formulation and numerical solution. In *Developments in Geotechnical Engineering*; Stephansson, O., Ed.; Elsevier Science Publishers: Amsterdam, The Netherlands, 1996; Volume 79, pp. 93–134.
43. Qu, H.; Liu, J.; Pan, Z.; Connell, L. Impact of matrix swelling area propagation on the evolution of coal permeability under coupled multiple processes. *J. Nat. Gas Sci. Eng.* **2014**, *18*, 451–466. [CrossRef]
44. Renshaw, C.E. On the relationship between mechanical and hydraulic apertures in rough-walled fractures. *J. Geophys. Res.-Solid Earth* **1995**, *100*, 24629–24636. [CrossRef]
45. Zimmerman, R.W.; Bodvarsson, G.S. Hydraulic conductivity of rock fractures. *Transp. Porous Media* **1996**, *23*, 1–30. [CrossRef]

Article

Fractal Characterization of Nanopore Structure in Shale, Tight Sandstone and Mudstone from the Ordos Basin of China Using Nitrogen Adsorption

Xiaohong Li [1], Zhiyong Gao [1], Siyi Fang [2], Chao Ren [1], Kun Yang [2] and Fuyong Wang [2,*]

1 Central Laboratory of Geological Sciences, RIPED, PetroChina, Beijing 100083, China; lixiaohong5@petrochina.com.cn (X.L.); gzy@petrochina.com.cn (Z.G.); renchao2018@petrochina.com (C.R.)
2 Research Institute of Enhanced Oil Recovery, China University of Petroleum, Beijing 102249, China; 2017210530@student.cup.edu.cn (S.F.); 2017210531@student.cup.edu.cn (K.Y.)
* Correspondence: wangfuyong@cup.edu.cn

Received: 31 December 2018; Accepted: 1 February 2019; Published: 13 February 2019

Abstract: The characteristics of the nanopore structure in shale, tight sandstone and mudstone from the Ordos Basin of China were investigated by X-ray diffraction (XRD) analysis, porosity and permeability tests and low-pressure nitrogen adsorption experiments. Fractal dimensions D_1 and D_2 were determined from the low relative pressure range ($0 < P/P_0 < 0.4$) and the high relative pressure range ($0.4 < P/P_0 < 1$) of nitrogen adsorption data, respectively, using the Frenkel–Halsey–Hill (FHH) model. Relationships between pore structure parameters, mineral compositions and fractal dimensions were investigated. According to the International Union of Pure and Applied Chemistry (IUPAC) isotherm classification standard, the morphologies of the nitrogen adsorption curves of these 14 samples belong to the H2 and H3 types. Relationships among average pore diameter, Brunner-Emmet-Teller (BET) specific surface area, pore volume, porosity and permeability have been discussed. The heterogeneities of shale nanopore structures were verified, and nanopore size mainly concentrates under 30 nm. The average fractal dimension D_1 of all the samples is 2.1187, varying from 1.1755 to 2.6122, and the average fractal dimension D_2 is 2.4645, with the range from 2.2144 to 2.7362. Compared with D_1, D_2 has stronger relationships with pore structure parameters, and can be used for analyzing pore structure characteristics.

Keywords: nanopore; pore structure; shale; tight sandstone; mudstone; nitrogen adsorption; fractal

---

1. Introduction

In recent years, global energy shortages have led to more attention being paid to unconventional oil and gas sources, such as tight oil and shale gas [1,2]. The pore-size of unconventional reservoir formations such as shale and tight sandstone generally spans from micropore to mesopore and macropore. In shale a very complicated pore structure is the result of a wide pore-size distribution and abundant organic matter [3]. Therefore, it is a huge challenge to explore unconventional resources effectively. The study of nanopore structure characteristics of unconventional reservoirs is important for their effective development, as nanopores can contain huge amounts of oil and gas.

There exist various techniques to investigate the characteristics of shale and tight sandstone and their respective nanopore structures, e.g., via mercury intrusion [4,5], field emission scanning electron microscopy, transmission electron microscopy (TEM), and gas adsorption analysis [6]. For example, Ghanbarian et al. analyzed 18 tight-gas sandstones from Texas by mercury intrusion experiments, and the EMA model was used to estimate bulk electrical conductivity and permeability [7]. Low-pressure gas adsorption measurements are very important for characterization of the gas shale pore system. Based on scanning electron microscopy and nitrogen adsorption experiments, Chen et al.

found that most of the pores in shale are composed of organic pores and the pores in clay mineral layers [8]. Millán et al. proposed a truncated version of the fractal Frenkel-Halsey-Hill (FHH) model for describing H_2O-vapor adsorption and 48 H_2O-vapor adsorption isotherm data was used to verify the model [9]. Yang et al. conducted low-pressure nitrogen adsorption studies on eight core samples from upper Ordovician lower Silurian oil reservoirs in the south of Sichuan Basin to better understand the reservoir characteristics of organic-rich shale [10]. In addition, fractal theory has been used to evaluate the pore structure in porous media. Wang et al. compared six different fractal models for calculating the fractal dimensions from mercury intrusion capillary pressures, and an optimal fractal model for analyzing petrophysical properties was recommended [11]. Based on nitrogen adsorption experiments, Ming et al. found that the fractal dimensions of shale have a good positive correlation with total pore volume, micropore volume and mesopore volume, but a poor correlation with macropore volume [12]. Xiong et al. used the FHH model to calculate surface fractal dimensions and volume fractal dimensions from nitrogen adsorption data [13]. Shao et al. analyzed the pore throat structure and fractal characteristics of Longmaxi shale with a series of experiments, and found that shale pore structure is mainly determined by total organic carbon content and thermal maturity, which also affects the value of the fractal dimensions [14]. Li et al. used the FHH model to obtain the fractal dimensions of shale, and the relationships between the calculated fractal dimensions and shale composition and total organic matter content were studied [15]. In this paper, fractal theory was used to study the nanopore structure characteristics of shale, tight sandstone and mudstone from the Yanchang Formation in the Ordos Basin of China based on the nitrogen adsorption experiments. The relationships between the calculated fractal dimensions and pore structure parameters, such as pore diameter and pore volume, were investigated.

2. Core samples and Experiment Results

2.1. Core samples

A total of 14 core samples, including 10 shale core samples, three tight sandstone core samples and one mudstone core sample collected from the Ordos Basin of China were selected in this study.

All samples are taken from fresh cores of different underground depths. The parameters of the collected samples are summarized in Table 1.

Table 1. The information of collected samples in this study.

Core No.	Top Depth	Bottom Depth	Lithology
2	2069.88	2070.00	Shale
8	2070.77	2070.87	Shale
10	2071.08	2071.25	Shale
17	2071.98	2072.08	Shale
26	2073.18	2073.30	Shale
33	2074.19	2074.35	Shale
14	2000.78	2000.94	Silty mudstone
53–54	2005.19	2005.40	Sandstone
42	2028.72	2029.00	Fine sandstone
58	2049.93	2050.09	Shale
24	2054.12	2054.33	Sandstone
32	2073.93	2074.19	Shale
32–2	2073.93	2074.19	Shale
58–2	2049.93	2050.09	Shale

2.2. X-ray diffraction (XRD) measurements

XRD is an effective technique to analyze mineral composition and content. Its theoretical basis is that X-rays will diffract in different directions and the mineral composition and structure can be determined by measuring the intensities and angles of these diffracted X-ray beams. As shown in Table 2, the contents of each component in shale and tight sandstone are significantly different.

Although the four samples are both rich in clay minerals and quartz, the clay content in shale is greater than in tight sandstone. The clay content for shale samples ranged from 24.8% to 35.0%, while for tight sandstone the clay content varied from 13.3% to 19.8%. The shale quartz content is between 23.3% and 37.2%, and less than that in sandstone, which varies between 60.9% and 61.3%. The content of potash feldspar in shale and sandstone is not that much, with contents between 0.5%–2.0% and 0.3%–1.4%, respectively. Plagioclase feldspar contained in shale is not as abundant as in sandstone, which is about 6% more than shale. The shale also contains a large proportion of pyrite, and sample 32 even contains 42.5% pyrite, while no pyrite can be found in sandstone. In general, the total amount of clay minerals and non-clay minerals in shale and sandstone samples are significantly different.

Table 2. Mineralogical composition results and total clay.

Core No.	Lithology	Total Clay (%)	Mineralogical Composition Results (%)					
			Quartz	Potash Feldspar	Plagioclase Feldspar	Calcite	Dolomite	Pyrite
24	Sandstone	19.8	60.9	1.4	11.5	1.7	4.7	/
32	Shale	24.8	23.3	2.0	4.7	1.6	1.1	42.5
58	Shale	35.0	37.2	0.5	4.5	/	13.5	9.3
53	Sandstone	13.3	61.3	0.3	11.0	2.4	11.7	/

The relative contents of clay minerals of the four samples are given in Table 3. The samples do not contain S and C/S, but I and I/S are the most abundant. The content of I is 41%–80% in shale and 40%–46% in tight sandstone. In addition to sample 32 without I/S, the I/S of the other three samples is about 40%. This sample also contains certain kaolinite and chlorite. The content of kaolinite in shale is higher than that in tight sandstone, but the content of chlorite in shale is less than that in tight sandstone. The mixed-layer ratio of I/S in the sandstone is about 10%, but C/S mixed-layer is not found in tight sandstone and shale samples.

Table 3. Relative clay mineral contents and mixed-layer ratio.

Sample No.	Lithology	Relative Clay Mineral Contents (%)						Mixed-Layer Ratio (%)	
		S	I/S	I	K	C	C/S	I/S	C/S
24	Sandstone	/	39	46	/	15	/	10	/
32	Shale	/	/	80	20	/	/	/	/
58	Shale	/	40	41	7	12	/	10	/
53	Sandstone	/	34	40	6	20	/	10	/

Note: S: smectite; I/S: illite smectite mixed layer; I: illite; K: kaolinite; C: chlorite; C/S: chlorite smectite mixed-layer.

2.3. Low-Pressure Nitrogen Adsorption-Desorption Experiments

Low-pressure nitrogen adsorption-desorption experiments were conducted using an automatic specific surface area & pore size analyzer produced by Quantachrome Instruments (Boynton Beach, FL, USA). The shapes of the nitrogen adsorption-desorption isotherms can be used to analyze pore shapes. Also, nitrogen adsorption-desorption data can be used to calculate the pore structure parameters. For example, total pore volume can be calculated as the liquid molar volume of adsorbed nitrogen at the relative pressure of 0.99. Total pore volume and pore size distribution can be calculated based on the Barrett-Joyner-Halenda (BJH) model [16]. The principle of nitrogen adsorption is that the gas adsorbed on a certain surface is taken as a function of the relative pressure of the adsorbent. Under the constant temperature, the relationship between gas adsorption and gas balance relative pressure is the adsorption isotherm.

Isothermal adsorption and desorption curves were obtained by nitrogen adsorption experiment with relative pressure P/P_0 as abscissa and adsorption amount as ordinate. The nitrogen adsorption curves of 10 shale samples and one mudstone sample are shown in Figure 1, while the nitrogen adsorption curves of three sandstone samples are given in Figure 2.

Nitrogen adsorption-desorption curves can be used to characterize the characteristics of pore complexity and shape. Figures 1 and 2 show that the adsorption-desorption curves of each sample are slightly different in morphology, but the whole curve is inverted S-type. The adsorption process can be divided into three stages: The first stage ($0 < P/P_0 \leq 0.4$) is the nitrogen adsorption of low pressure stage, where the gas adsorption quantity increases slowly, and the adsorption isotherm is a gentle upward convex shape. The first stage is the single-layer adsorption of nitrogen on the pore surface, and the nitrogen adsorption curve appeared inflection point for monolayer adsorption to the transition of multilayer adsorption [17]. In the second stage ($0.4 < P/P_0 \leq 0.9$), the adsorption volume of the sample increases rapidly with the increase of relative pressure, and nitrogen adsorption isotherm rises rapidly, which leads to a hysteresis loop, and this stage is a multi-molecular layer stage. In the third stage ($P/P_0 > 0.9$), with the increase of relative pressure, the amount of gas adsorption increases dramatically.

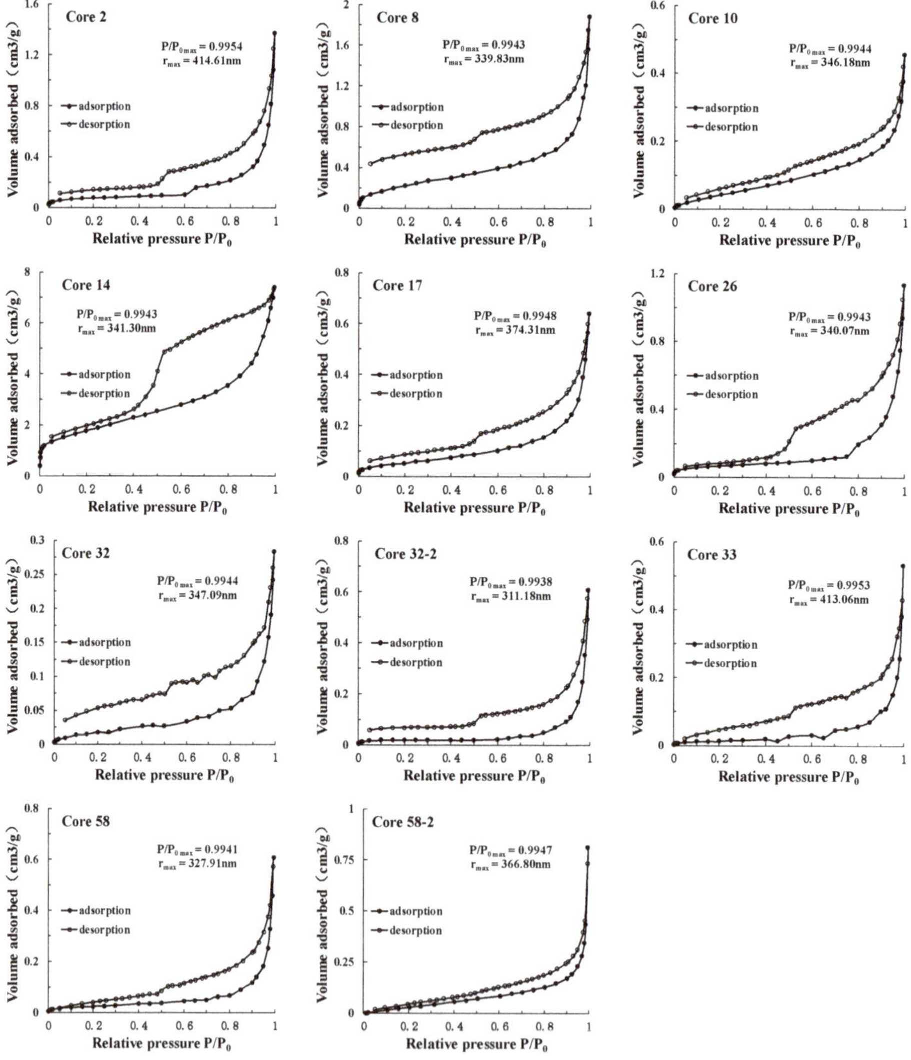

Figure 1. Nitrogen adsorption-desorption curve of samples of shale and mudstone.

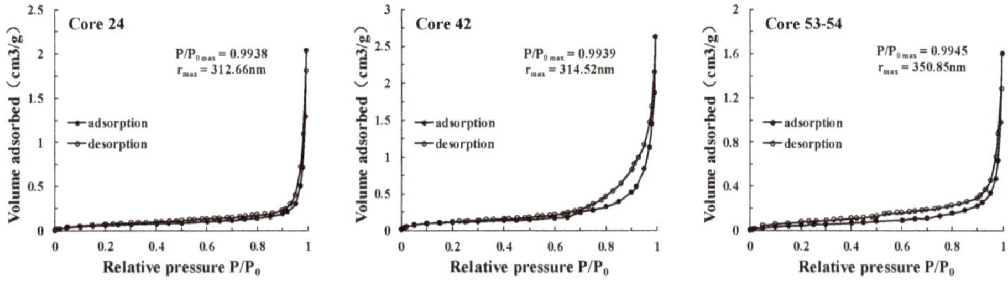

Figure 2. Nitrogen adsorption-desorption curve of three sandstone samples.

When the relative pressure was close to the saturated vapor pressure, there is no adsorption saturation phenomenon, and this is the capillary condensation stage of the sample. Due to the complex pore structure of the experimental samples, capillary condensation phenomena happen in the substrate surface, and the isothermal desorption curves of the samples show more obvious desorption hysteresis. The desorption amount is far less than the adsorption amount, and then hysteresis loop appeared. There are significant differences in the development morphology and connectivity of small pores and the adsorption of nitrogen is not fully enclosed [18].

According to IUPAC isotherm classification standard, the morphology of the nitrogen adsorption curves of the 10 shale samples and a mudstone one (sample 14) belong to the H2 and H3 type, indicating that the pore morphologies of shale are mainly similar to ink bottle holes and sheet granular matrix. As shown in Figure 2, the nitrogen adsorption curves of the sandstone samples are similar to H3 type, which indicates that the sandstone pores are mainly composed of sheet particles with non-rigid aggregate groove holes. As the pore openness is associated with increased rate of adsorption line, the larger the increasing rate is, the larger the opening of sandstone pore will be.

3. Pore Size Distribution from Nitrogen Adsorption-Desorption Isotherms

3.1. Methods

3.1.1. Specific Surface Area

The BET equation derived by Brunauer, Emmett and Teller was used to calculate the specific surface area with the range of relative pressure of 0.05–0.35 [19]. The surface area of porous media can be calculated by the amount of gas monolayer adsorption according to the Langmuir monolayer adsorption theory:

$$S_{BET} = V_m N_A A_m / 22400W \tag{1}$$

where N_A is Avogadro constant; A_m is the cross section area of adsorbed gas nitrogen molecule; W is the quality of the medium; V_m is the nitrogen saturation adsorption amount in a single layer and can be calculated by the BET equation:

$$\frac{1}{V(p_0/p-1)} = \frac{1}{V_m C} + \frac{C-1}{V_m C}\frac{p}{p_0} \tag{2}$$

where V is total volume of adsorbed gases and C is a constant relating to adsorption.

3.1.2. Pore Size

The recurrence method is usually used to calculate the pore radius of different pore size intervals based on the Kelvin principle based on the assumption of cylindrical pore [20]. Assume the thickness of film adsorbed on the pore surface is t, and the internal radius of the pore with the radius r_p is reduced to $r_p - t$, and can be calculated by [21]:

$$\ln(p/p_0) = \frac{-2\gamma V_m}{RT(r_p - t)} \quad (3)$$

where γ is the surface tension at the boiling point of nitrogen; V_m is the molar volume of liquid nitrogen; R is the gas constant; T is the boiling points (77K); p/p_0 is the relative pressure of nitrogen. The thickness of the liquid film adsorbed on the pore surface can be calculated by [22]:

$$t = [\frac{13.99}{0.034 - \log(p/p_0)}]^{1/2} \quad (4)$$

3.1.3. Pore Size Distribution

There are three different widely used methods for pore size distribution calculation based on gas adsorption isotherm, including the BJH method, HK method and DFT method [23,24]. The three method are introduced, respectively, and this paper uses the BJH model to calculate the pore size distribution.

Barrett et al. analyzed the desorption process and proposed the BJH method to calculation pore size distribution [16]. According to the desorption line of isothermal adsorption curve, the pore size distribution is obtained by calculating the nitrogen adsorption amount when the relative pressure is 0.99. The calculation formula is as follows:

$$V_{pn} = \left(\frac{r_{pn}}{r_{kn} + \Delta t_n}\right)^2 \left(\Delta V_n - \Delta t_n \sum_{j-1}^{n-1} A_{cj}\right) \quad (5)$$

where V_{pn} is the pore volume; r_{pn} is the maximum pore radius; r_{kn} is the capillary radius; V_n is capillary volume; t_n is the adsorbed nitrogen layer thickness; A_{cj} is the area after the emptying.

Horváth et al. proposed the HK method to calculation pore size distribution [25]:

$$w/w_\infty = f(l - d_a) \quad (6)$$

where w is the mass of nitrogen adsorbed on the pore surface; w_∞ is the maximum amount of nitrogen adsorbed into the pores at P/P_0 = 0.9; l is the distance between the nuclei of the two layers; d_a is the diameter of adsorbent. According to the adsorption capacity of different pore sizes, the pore size distribution f can be obtained by plotting the curve of w/w_∞ versus $(l - d_a)$.

Seaton et al. [26] calculated the pore size distribution by the adsorption isotherm using the DFT method firstly. Pore size distribution can be obtained by solving the following equation:

$$N_{\exp}(P/P_0) = \int_{D_{\min}}^{D_{\max}} N_{DFT}(P/P_0, D) f(D) dD \quad (7)$$

where $N_{\exp}(P/P_0)$ is the experimental isotherm; $N_{DFT}(P/P_0,D)$ is the theoretical isotherm; D is the pore size; $f(D)$ is the pore size distribution.

3.2. Analysis of Experimental Results

The calculated specific surface area, pore volume and pore size of shale, tight sandstone and mudstone are introduced respectively. Table 4 shows the calculation results for shale samples. It can be observed that BET specific surface area of the 10 shale samples is distributed between 0.05175 m^2/g and 0.8988 m^2/g, with an average of 0.2345 m^2/g. The pore volume of BJH is distributed between 4.359 × 10^{-4} cc/g and 27.18 × 10^{-4} cc/g, with an average of 12.7 × 10^{-4} cc/g. The weighted average pore diameter is between 12.72 nm and 63.8 nm, with an average of 30.9 nm. Shale has the characteristics of small pore size and large BET specific surface area, which is similar to the results presented in Literature [19].

Table 4. The calculated specific surface area, pore volume and pore size of shale samples.

Core No.	Specific Surface Area (10^{-2} m^2/g)			Pore Volume (10^{-4} cc/g)			Pore Diameter (nm)		
	BET	BJH	DFT	BJH	DFT	Langmuir Volume	Weighted Average Pore Diameter	BJH	DFT
2	28.30	28.60	27.67	21.58	14.21	1371.6	30.06	5.625	6.556
8	89.88	49.88	67.31	27.18	21.06	1886.5	13.02	3.414	4.887
10	22.22	16.08	16.45	6.808	5.161	455.5	12.72	3.454	6.079
17	20.22	17.31	18.84	9.686	7.641	640.7	19.66	3.414	6.079
26	24.24	20.83	24.61	17.50	12.68	1134.3	29.03	9.592	9.098
33	5.175	12.94	6.435	8.922	1.625	532.3	63.80	3.834	9.416
58	8.572	8.76	9.775	9.419	5.909	608.9	44.06	7.816	10.49
32	7.474	6.30	6.827	4.359	3.284	283.9	23.56	5.638	7.310
32-2	8.011	8.48	8.551	9.686	6.135	609.5	47.19	6.543	11.68
58-2	20.36	16.34	14.68	12.61	6.218	813	24.76	3.451	6.079
Average	23.45	18. 55	20.11	12.7	8.39	833.62	30.9	5.28	7.77

The pore structure parameters of the three tight sandstone samples obtained from the nitrogen adsorption-desorption isotherms are listed in Table 5.

Table 5. The calculated specific surface area, pore volume and pore size of tight sandstone samples.

Core No.	Specific Surface Area (10^{-2} m^2/g)			Pore Volume (10^{-4} cc/g)			Pore Diameter (nm)		
	BET	BJH	DFT	BJH	DFT	Langmuir Volume	Average Pore Diameter	BJH	DFT
24	23.70	17.23	23.54	31.25	15.36	2029.4	53.20	4.644	29.40
42	40.90	42.24	61.23	41.14	24.67	2624.1	39.79	6.547	12.55
53-54	18.60	20.17	30.10	24.84	12.04	1596.6	53.12	3.826	8.145
Average	27.73	26.55	38.29	32.41	17.36	2083.37	48.70	5.01	16.70

It is shown that the BET specific surface area of the three sandstone samples is distributed between 0.1864 m^2/g and 0.4091 m^2/g, with an average of 0.2231 m^2/g. The pore volume of BJH is distributed between 0.002484 cc/g and 0.004114 cc/g, with an average of 0.003241 cc/g. The average pore diameter ranges from 39.79 nm to 53.20 nm, and the average pore diameter is 48.70 nm. Compared with shale, BET specific surface area of the sandstone sample is smaller than that of the shale, and the pore volume and the average pore diameter are larger than those of the shale.

The experimental data of the silty mudstone sample is shown in Table 6. The BET specific surface area of the mudstone sample is 6.33 m^2/g, the BJH pore volume is 93.46×10^{-4} cc/g, and the average pore diameter is 7.272 nm. Compared with the shale and sandstone samples, the BET specific surface area of the mudstone sample is much larger, and the pore volume and average pore diameter are much smaller.

Table 6. The calculated specific surface area, pore volume and pore size of mudstone sample.

Core No.	Specific Surface Area (m^2/g)			Pore Volume (10^{-4} cc/g)			Pore Diameter (nm)		
	BET	BJH	DFT	BJH	DFT	Langmuir Volume	Average Pore Diameter	BJH	DFT
14	6.33	2.689	5.738	93.46	103	7421.1	7.272	3.819	3.78

The pore size distributions of the 10 shales, one mudstone and three sandstones obtained from the BJH method are shown in Figures 3–5, respectively. The abscissa value of the peak was marked in the figures. As Figure 3 shows, the pore size distribution of shale has a certain heterogeneity, and mainly concentrated under 30 nm, with at least one obvious peak value distributed in the range of about 3–10 nm. Figure 4 shows that the pore size distribution of mudstone sample has only one peak value which is not obvious, and the pore distribution is relatively uniform, and the pore diameter are mainly below 30 nm. As shown in Figure 5, the pore distribution of the three sandstone samples mostly had two peaks from 3 nm to 10 nm, which was similar to shale. The peak value of No. 24 is around 50 nm. Compared with shale and mudstone, the heterogeneities of tight sandstone pore are relatively large, which also indicated that large and medium pores were the main contributors to gas storage space.

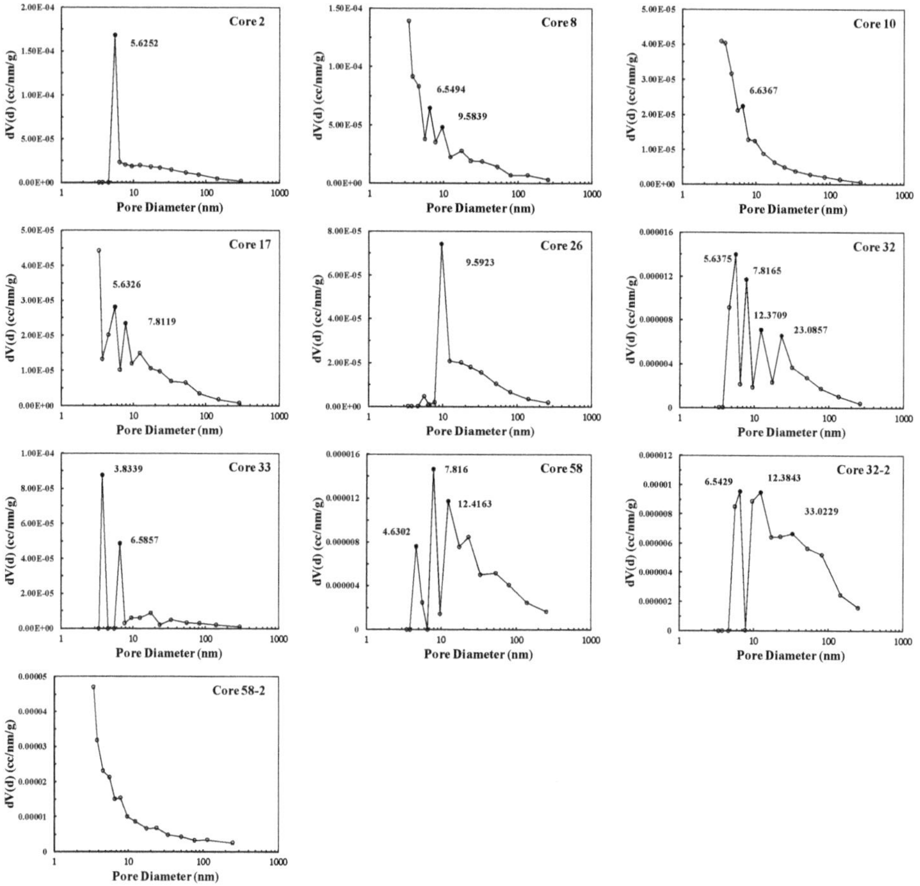

Figure 3. The pore size distribution of shale calculated by BJH method (The peak values of pore size distribution are marked by black circles, and the corresponding pore sizes are displayed).

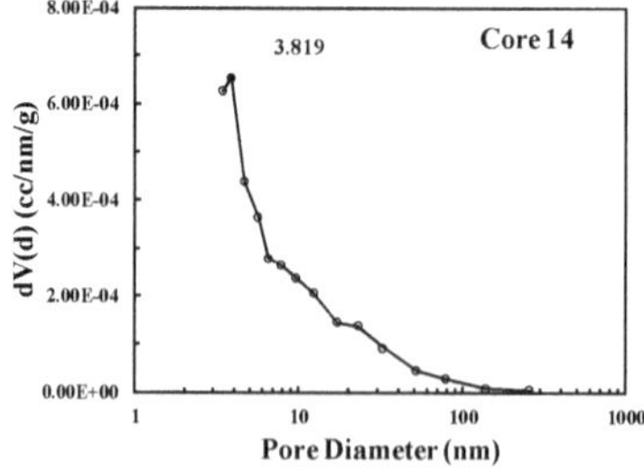

Figure 4. The pore size distribution of mudstone calculated by BJH method (The peak values of pore size distribution are marked by black circles, and the corresponding pore sizes are displayed).

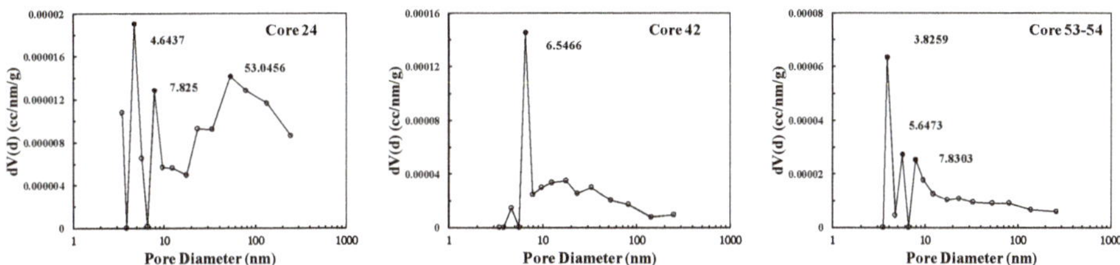

Figure 5. The pore size distribution of sandstone calculated by BJH method (the peak values of pore size distribution are marked by black circles, and the corresponding pore sizes are displayed).

3.3. Pore parameter Relationships

The relationship between average pore diameter and pore volume of the 10 shale samples is shown in Figure 6. It shows that there was no significant correlation between average pore diameter and pore volume of shale. As shown in Figure 7, the BET specific surface area of shale sample is negatively correlated with the average pore diameter, indicating that with the increase of average pore size, shale pore heterogeneity as well as the roughness decreases, and the specific surface area decreases. The relationship between shale pore volume and BET specific surface area is shown in Figure 8. The specific surface area increases with increasing total pore volume, and the correlation coefficient is nearly 0.7.

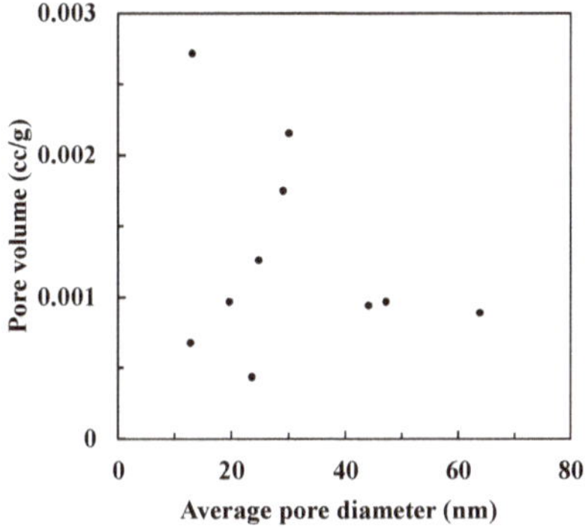

Figure 6. Relationship between average pore diameter and pore volume of shale samples.

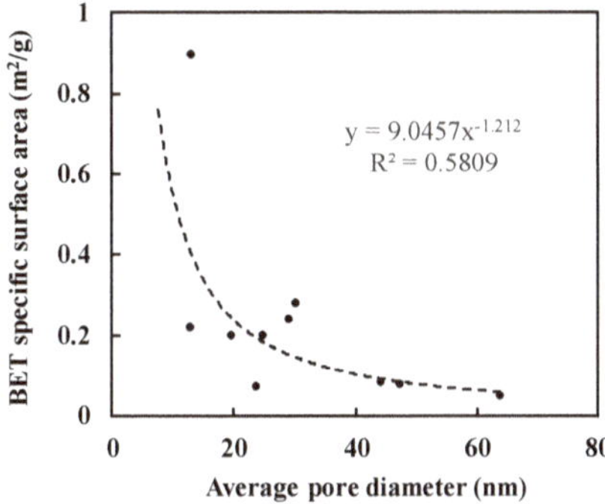

Figure 7. Relationship between average pore diameter and BET specific surface area of shale samples.

Porosity and permeability of four core samples including three tight sandstone samples and one mudstone sample were measured, as shown in Table 7. The porosity and permeability of the mudstone sample are ultra-low. The permeability of mudstone is one order of magnitude smaller that of tight sandstone, and the porosity of mudstone is only slightly greater than 1%. Figure 9 shows BET specific surface area decreases with the increase of permeability. The reason is that the larger the permeability is, the larger the corresponding average pore size will be, and it can be verified with Figure 10, which shows that the average pore size is positively correlated to the gas permeability.

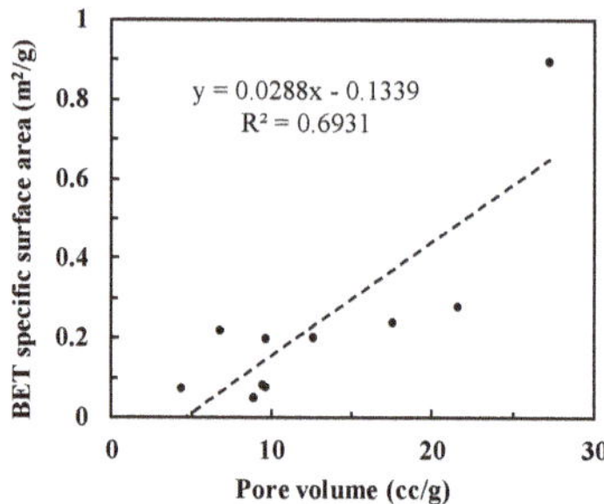

Figure 8. Relationship between pore volume and BET specific surface area of shale samples.

Table 7. Summary of pore parameters of 3 sandstone samples and 1 mudstone sample.

Core No.	Lithology	Porosity (%)	Permeability (mD)	BET Specific Surface Area (m^2/g)	Average Pore Diameter (nm)	Pore Volume (10^{-3} cc/g)
53–54	Tight sandstone	8.07	0.09	0.1864	53.12	2.48
42	Tight sandstone	6.49	0.027	0.4091	39.79	4.11
24	Tight sandstone	8.07	0.056	0.2366	53.20	3.13
14	Mudstone	1.05	0.0053	6.330	72.72	9.35

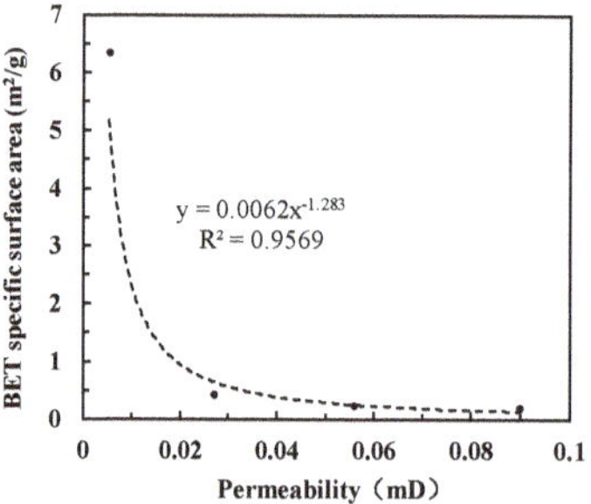

Figure 9. The relationship between gas permeability and specific surface area of 3 tight sandstone samples and 1 mudstone sample.

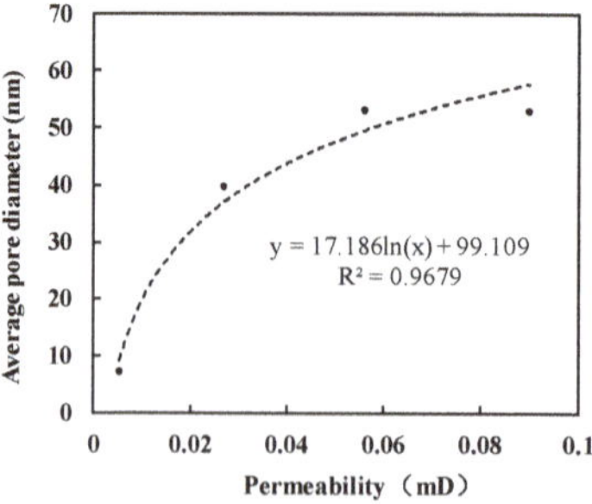

Figure 10. The relationship between gas permeability and average pore diameter of 3 tight sandstone samples and 1 mudstone sample.

4. Fractal Analysis of Nitrogen Adsorption Isotherms

4.1. Principle and Fractal Dimension Calculation Process

Fractal theory has been widely used to characterize the pore structures of unconventional reservoirs [27–30]. Fractal dimensions can quantitatively evaluate the heterogeneity of pore structure, and generally, pore heterogeneity increases with the increasing fractal dimension [31]. There are many models for calculating fractal dimension from nitrogen adsorption isotherms, including BET model, FHH model, fractal Langmuir model and thermodynamic method, and the FHH model is the one widely applied [32]. According to the FHH model, the fractal dimension D can be calculated from nitrogen adsorption experiments with the following equation:

$$LnV = KLn[Ln(P_0/P)] + C \tag{8}$$

where P is the equilibrium pressure; V is the volume of adsorbed gas corresponding to equilibrium pressure P; P_0 is the saturation pressure, K is the slope of the logarithmic curve, which is related to the adsorption mechanism and $K = D - 3$; C is a constant. If the pores has fractal characteristics, lnV and $\ln(\ln(P_0/P))$ will have a linear relationship [18].

The fractal dimension processes of the 10 shale samples are shown in Figure 11. The adsorption and desorption curves of the experimental samples produce hysteresis loops when relative pressure is about 0.4, indicating that there is a large difference in porosity before and after this relative pressure due to different adsorption behaviors. There are two fractal characteristics in the study area. In this paper, the fractal dimension calculated from the low relative pressure range of $0 < P/P_0 < 0.4$ is denoted as D_1, and the fractal dimension calculated from the high relative pressure range of $0.4 < P/P_0 < 1$ is denoted as D_2. The fractal dimension D_1 characterizes the effect of van der Waals force and reflects the surface roughness. The fractal dimension D_2 represents properties of multi-layer adsorption, which can be used to describe the spatial roughness and irregularity of pore structures [33,34]. Ghanbarian and Daigle found that the cut-off values of the upper and lower boundaries of fractal regions have a significant impact on the results of fractal dimension calculation [35]. Meanwhile, when the curve is segmented, the boundaries of relative pressure is not constant at 0.4. In order to ensure the accuracy of fractal dimension calculation, this paper divides the curve according to the change of slope. The segmentation points of the curve are shown in Table 8.

Table 8. The upper and lower boundaries of relative pressure for fractal dimension calculation.

Core No.	Upper and Lower Boundaries of Relative Pressure for Each Fractal Regime		
	Initial Point	**Segmentation Point**	**Ending Point**
2	0.0099	0.4990	0.9954
8	0.0096	0.3001	0.9896
10	0.0101	0.4035	0.9944
17	0.0100	0.4004	0.9948
26	0.0099	0.4493	0.9943
33	0.0098	0.3990	0.9953
14	0.0098	0.4003	0.9943
24	0.0099	0.3010	0.9938
32	0.0098	0.2995	0.9944
32-2	0.0095	0.3998	0.9938
42	0.0098	0.4011	0.9888
53-54	0.0096	0.4016	0.9945
58	0.0097	0.2999	0.9941
58-2	0.0545	0.3146	0.9947

In Figure 11, the curves for calculating fractal dimension D_1 and D_2 are displayed. For the core 2, core 26 and core 33, the slopes of the two curves before and after the segmentation points are close to each other can be fitted by a straight line. The reason is that the pore radii of these cores are small and their distributions are concentrated, which leads to that their surface fractal dimensions are close to the volume fractal dimension. For the core 32–2, the curve of D_2 calculation is not linear and the linear correlation is very poor. Therefore, the fractal dimension D_2 of core 32–2 was not calculated. The values of D_1 and D_2 are shown in Table 9. D_1 is distributed between 1.67 and 2.5265 with an average value of 2.1975, D_2 is distributed between 2.3076–2.6463 with an average value of 2.4791.

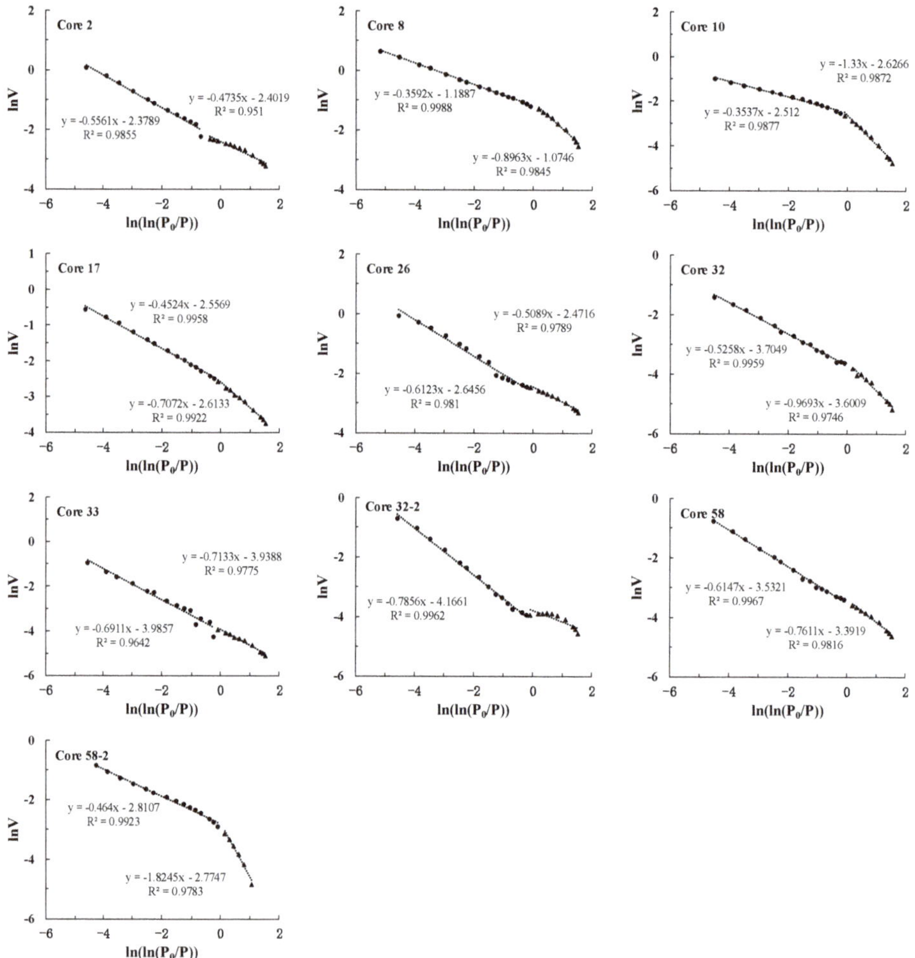

Figure 11. The curves of fractal dimension calculation for 10 shale samples.

The curves of fractal dimension calculation for the three tight sandstone samples and one mudstone sample are shown in Figures 12 and 13, respectively, and the calculated fractal dimensions D_1 and D_2 are shown in Tables 10 and 11, respectively.

Table 9. The calculated fractal dimensions of 10 shale samples.

Core No.	$0 < P/P_0 < 0.4$			$0.4 < P/P_0 < 1$		
	K_1	$D_1 = 3 + k_1$	R^2	K_2	$D_2 = 3 + K_2$	R^2
2	−0.4735	2.5265	0.951	−0.5561	2.4439	0.9855
8	−0.8963	2.1037	0.9845	−0.3592	2.6408	0.9988
10	−1.33	1.67	0.9872	−0.3537	2.6463	0.9877
17	−0.7072	2.2928	0.9922	−0.4524	2.5476	0.9958
26	−0.5364	2.4636	0.982	−0.6123	2.3877	0.981
33	−0.7461	2.2539	0.9775	−0.6924	2.3076	0.9684
58	−0.7611	2.2389	0.9816	−0.6147	2.3853	0.9967
32	−0.9693	2.0307	0.9746	-	-	-
32-2	−0.3878	2.6122	0.7652	−0.7856	2.2144	0.9962
58-2	−1.8245	1.1755	0.9783	−0.464	2.536	0.9923
Average	−0.8632	2.1368	0.9574	−0.5416	2.4584	0.9898

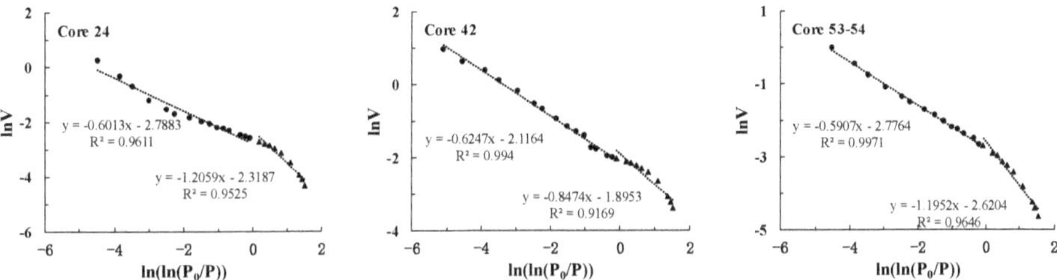

Figure 12. The curves of fractal dimension calculation for 3 tight sandstone samples.

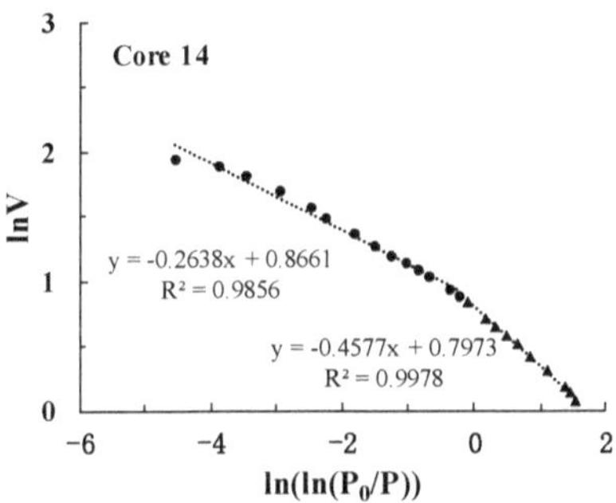

Figure 13. The curves of fractal dimension calculation for 1 mudstone sample.

Table 10. The calculated fractal dimensions of three tight sandstone samples.

Core No.	$0 < P/P_0 < 0.4$			$0.4 < P/P_0 < 1$		
	K_1	$D_1 = 3 + k_1$	R^2	K_2	$D_2 = 3 + K_2$	R^2
53–54	−1.1952	1.8048	0.9646	−0.5907	2.4093	0.9971
42	−0.8474	2.1526	0.9169	−0.6247	2.3753	0.994
24	−1.2059	1.7941	0.9525	−0.6013	2.3987	0.9611
Average	−1.0828	1.9172	0.9447	−0.6056	2.3944	0.9841

Table 11. The calculated fractal dimensions of one mudstone sample.

Core No.	$0 < P/P_0 < 0.4$			$0.4 < P/P_0 < 1$		
	K_1	$D_1 = 3 + k_1$	R^2	K_2	$D_2 = 3 + K_2$	R^2
14	−0.4577	2.5423	0.9978	−0.2638	2.7362	0.9856

For the tight sandstone samples, the calculated D_1 changes from 1.7941 to 2.1526 with an average value of 1.9172, and D_2 varies from 2.3753 to 2.4093 with an average value of 2.3944. The correlation coefficient R^2 is close to 1, indicating that the pores in tight sandstone has good fractal characteristics. For the mudstone sample, the calculated D_1 is 2.5423 and the calculated D_2 is 2.7362. It can be found that the values of D_1 and D_2 of mudstone are largest, followed by shale, and the values of D_1 and D_2 of tight sandstone are smallest.

As shown in Tables 9–11, it can be found that the surface fractal dimension D_1 is generally less than the volume fractal dimension D_2. The relationship between the fractal dimension D_1 and D_2 for shale samples is shown in Figure 14. The fractal dimension D_2 decreases with the fractal dimension D_1 increasing. The surface fractal dimensions D_1 of the cores 10, 24, 53-54, 58-2 are less than 2 which are not in the typical range: $2 < D < 3$ for three-dimensional space. Ghanbarian-Alavijeh et al. believed that the fractal dimension can be $-\infty$ when the pore size is the same, and demonstrated that theoretically fractal dimension can range between $-\infty$ and 3 [36]. Therefore, the fractal dimension D_1 less than 2 is acceptable.

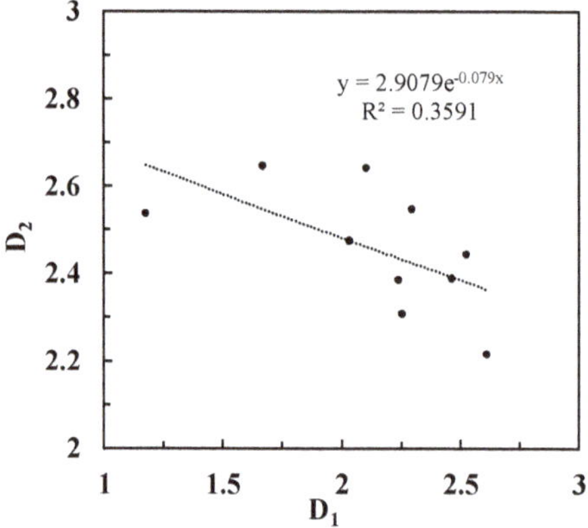

Figure 14. Relationship between the fractal dimension D_1 and D_2 of shale samples.

4.2. Relationship between Fractal Dimensions and Pore Structure Parameters

A summary of fractal dimensions and pore parameters of shale, sandstone and mudstone samples as shown in Table 12. The relationships between the calculated fractal dimensions and pore structure parameters will be studied below.

Table 12. Summary of fractal dimensions and pore parameters of shale, sandstone and mudstone.

Core No.	D_1	D_2	Specific Surface Area (m^2/g)	Average Pore Diameter (nm)	Porosity (%)	Langmuir Volume (10^{-4} cc/g)	Permeability (mD)
2	2.5265	2.4439	0.283	30.06	-	1371.6	-
8	2.1037	2.6408	0.8988	13.02	-	1886.5	-
10	1.67	2.6463	0.2222	12.72	-	455.5	-
17	2.2928	2.5476	0.2022	19.66	-	640.7	-
26	2.4636	2.3877	0.2424	29.03	-	1134.3	-
33	2.2539	2.3076	0.05175	63.8	-	532.3	-
58	2.2389	2.3853	0.08572	44.06	1.3	608.9	-
32	2.0307	2.4742	0.07474	23.56	0.8	283.9	-
32–2	2.6122	2.2144	0.08011	47.19	-	609.5	-
58–2	1.1755	2.536	0.2036	24.76	-	813	-
53–54	1.8048	2.4093	0.1864	53.12	8.07	1596.6	0.09
42	2.1526	2.3753	0.4091	39.79	6.49	2624.1	0.027
24	1.7941	2.3987	0.2366	53.2	9.8	2029.4	0.056
14	2.5423	2.7362	6.33	7.272	1.05	7421.1	0.0053
Average	2.1186	2.4645	0.679	32.95	4.59	1571.9	0.0446

4.2.1. Relationship between Fractal Dimension and Specific Surface area

The relationship between specific surface area and fractal dimension of 10 shale samples and three sandstone samples is shown in Figure 15. There is no obvious relationship between D_1 and BET specific surface area, but BET specific surface area is positively correlated with D_2, indicating that the larger the specific surface area is, the more complex pore structure will become.

4.2.2. Relationship between Fractal Dimension and Average Pore Diameter

As shown in Figure 16, average pore diameter of all samples has no obvious relationship between D_1, but is negatively correlated with D_2. With average pore diameter increasing, fractal dimension D_2 decreases exponentially, indicating properties of core samples become better.

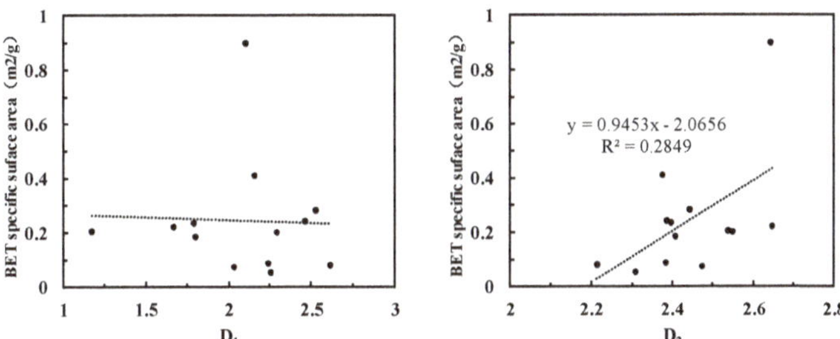

Figure 15. Relationship between fractal dimension and specific surface area.

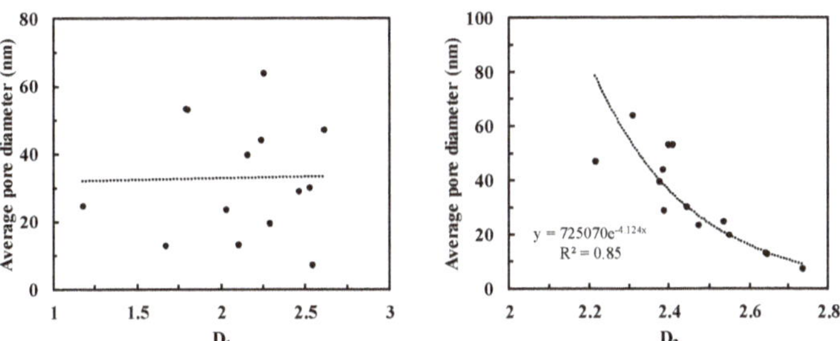

Figure 16. Relationship between fractal dimension and average pore diameter of all samples.

4.2.3. Relationship between Fractal Dimension and Porosity

As shown in Figure 17, the fractal dimension D_1 and D_2 of the three sandstone samples (Nos. 53–54, 42, 24) and one mudstone sample (No. 14) are negatively correlated with porosity.

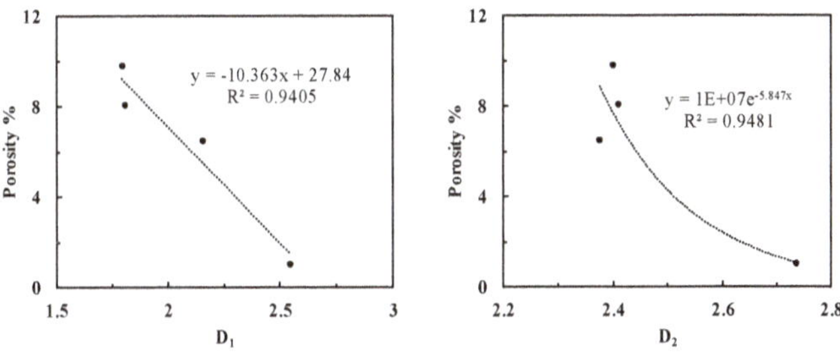

Figure 17. Relationship between porosity and fractal dimension of 3 sandstone samples and 1 mudstone sample.

The fractal dimension decreases with the increasing porosity. It is reasonable as the increasing porosity usually means the increasing pore size, and therefore properties of core samples become better.

4.2.4. Relationship between Fractal Dimension and Permeability

The relationship between permeability and fractal dimension of three sandstone samples (Nos 53–54, 42 and 24) and one mudstone (No. 14) is shown in Figure 18. Similar to porosity, permeability is both negatively correlated to D_1 and D_2. Therefore, the calculated fractal dimension D_1 and D_2 can be used for evaluating core properties.

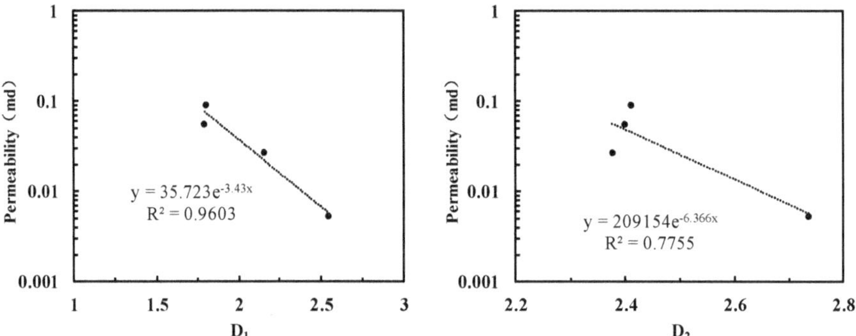

Figure 18. Relationship between fractal dimension and permeability of 3 sandstone samples and 1 mudstone sample.

5. Conclusions

Characteristics of nanopore structure in shale, tight sandstone and mudstone have been studied with nitrogen adsorption experiment and fractal theory. Several conclusions could be drawn as follows:

(1) The morphologies of nitrogen adsorption-desorption curves of 10 shale samples belong to the H2 and H3 types according to IUPAC isotherm classification standard, while the curve of nitrogen adsorption-desorption of three sandstone sample is closer to H3 type.

(2) The pore sizes of the shale have certain heterogeneity, and the pore size intervals are mainly concentrated less than 30 nm, with at least a relatively obvious peak value distributed around 3 nm–10 nm.

(3) BET specific surface area of sandstone is smaller than that of shale, and the pore volume and average pore diameter are much larger than those in shale. The BET specific surface area of mudstone is much larger than that of sandstone, and the pore volume and average diameter are much smaller than those in sandstone.

(4) The larger the average pore size of sandstone is, the smaller the BET specific surface area will be, and the BET specific surface area decreases with the increase of permeability. With the increase of porosity and permeability, BET specific surface area decreases with the increasing average pore size.

(5) Fractal dimensions calculated from the low relative pressure range D_1 are generally less than those calculated from the high relative pressure range D_2. D_1 reflects the surface roughness of pore structures and D_2 quantifies the spatial irregularity of pore spaces, and D_1 and D_2 are negatively correlated with each other. Compared with D_1, D_2 has stronger relationships with pore structure parameters and core properties. With D_2 increases, BET specific surface area increases but average pore diameter, porosity and permeability decreases.

Author Contributions: Experiment and Methodology, X.L., Z.G. and C.R.; Writing-Original Draft Preparation, S.F.; Writing-Review & Editing, K.Y.; Data Analysis & Editing, F.W.

Funding: This work was supported by the National Science and Technology Major Project (No. 2017ZX05001-001), National Natural Science Foundation of China (Nos. 51604285 and 51874320) and Scientific Research Foundation of China University of Petroleum, Beijing (No. 2462017BJB11).

Conflicts of Interest: The authors declare no conflict of interest.

References

1. Gasparik, M.; Ghanizadeh, A.; Bertier, P.; Gensterblum, Y.; Bouw, S.; Krooss, B.M. High-Pressure Methane Sorption Isotherms of Black Shales from The Netherlands. *Energy Fuels* **2012**, *26*, 4995–5004. [CrossRef]
2. Loucks, R.G.; Ruppel, S.C. Mississippian Barnett Shale: Lithofacies and depositional setting of a deep-water shale-gas succession in the Fort Worth Basin, Texas. *AAPG Bull.* **2007**, *91*, 579–601. [CrossRef]
3. Loucks, R.G.; Reed, R.M.; Ruppel, S.C.; Jarvie, D.M. Morphology, Genesis, and Distribution of Nanometer-Scale Pores in Siliceous Mudstones of the Mississippian Barnett Shale. *J. Sediment. Res.* **2009**, *79*, 848–861. [CrossRef]
4. Wang, F.; Lian, P.; Jiao, L.; Liu, Z.; Zhao, J. Fractal analysis of microscale and nanoscale pore structures in carbonates using high-pressure mercury intrusion. *Geofluids* **2018**, *2018*, 1840017. [CrossRef]
5. Wang, F.; Yang, K.; Cai, J. Fractal characterization of tight oil reservoir pore structure using nuclear magnetic resonance and mercury intrusion porosimetry. *Fractals* **2018**, *26*, 1840017. [CrossRef]
6. Ross, D.J.K.; Marc Bustin, R. The importance of shale composition and pore structure upon gas storage potential of shale gas reservoirs. *Mar. Pet. Geol.* **2009**, *26*, 916–927. [CrossRef]
7. Ghanbarian, B.; Torres-Verdín, C.; Lake, L.W.; Marder, M. Gas permeability in unconventional tight sandstones: Scaling up from pore to core. *J. Pet. Sci. Eng.* **2019**, *173*, 1163–1172. [CrossRef]
8. Chen, K.; Zhang, T.; Chen, X.; He, Y.; Liang, X. Model construction of micro-pores in shale: A case study of Silurian Longmaxi Formation shale in Dianqianbei area, SW China. *Pet. Explor. Dev.* **2018**, *45*, 412–421. [CrossRef]
9. Millán, H.; Govea-Alcaide, E.; García-Fornaris, I. Truncated fractal modeling of H2O-vapor adsorption isotherms. *Geoderma* **2013**, *206*, 14–23. [CrossRef]
10. Yang, Y.; Wu, K.; Zhang, T.; Xue, M. Characterization of the pore system in an over-mature marine shale reservoir: A case study of a successful shale gas well in Southern Sichuan Basin, China. *Petroleum* **2015**, *1*, 173–186. [CrossRef]
11. Wang, F.; Jiao, L.; Liu, Z.; Tan, X.; Wang, C.; Gao, J. Fractal Analysis of Pore Structures in Low Permeability Sandstones Using Mercury Intrusion Porosimetry. *J. Porous Media* **2018**, *21*, 1097–1119. [CrossRef]
12. Ming, M.; Guojun, C.; Yong, X.; Shijun, H.; Chengfu, L.; Lianhua, X. Fractal characteristics of pore structure of continental shale in the process of thermal evolution. *Coal Geol. Explor.* **2017**, *45*, 41–47.
13. Xiong, J.; Liu, X.; Liang, L. An Investigation of Fractal Characteristics of Marine Shales in the Southern China from Nitrogen Adsorption Data. *J. Chem.* **2015**, *2015*, 303164. [CrossRef]
14. Shao, X.; Pang, X.; Li, Q.; Wang, P.; Chen, D.; Shen, W.; Zhao, Z. Pore structure and fractal characteristics of organic-rich shales: A case study of the lower Silurian Longmaxi shales in the Sichuan Basin, SW China. *Mar. Pet. Geol.* **2017**, *80*, 192–202. [CrossRef]
15. Li, A.; Ding, W.; He, J.; Dai, P.; Yin, S.; Xie, F. Investigation of pore structure and fractal characteristics of organic-rich shale reservoirs: A case study of Lower Cambrian Qiongzhusi formation in Malong block of eastern Yunnan Province, South China. *Mar. Pet. Geol.* **2016**, *70*, 46–57. [CrossRef]
16. Barrett, E.P.; Joyner, L.G.; Halenda, P.P. The Determination of Pore Volume and Area Distributions in Porous Substances. I. Computations from Nitrogen Isotherms. *J. Am. Chem. Soc.* **1951**, *73*, 373–380. [CrossRef]
17. Jiang, F.; Chen, D.; Chen, J.; Li, Q.; Liu, Y.; Shao, X.; Hu, T.; Dai, J. Fractal Analysis of Shale Pore Structure of Continental Gas Shale Reservoir in the Ordos Basin, NW China. *Energy Fuels* **2016**, *30*, 4676–4689. [CrossRef]
18. Zhang, L.; Li, J.; Jia, D.U.; Zhao, Y.; Xie, C.; Tao, Z. Study on the adsorption phenomenon in shale with the combination of molecular dynamic simulation and fractal analysis. *Fractals* **2018**, *26*, 1840004. [CrossRef]
19. Brunauer, S.; Emmett, P.H.; Teller, E. Adsorption of Gases in Multimolecular Layers. *J. Am. Chem. Soc.* **1938**, *60*, 309–319. [CrossRef]
20. Wang, Y.; Zhu, Y.; Liu, S.; Zhang, R. Pore characterization and its impact on methane adsorption capacity for organic-rich marine shales. *Fuel* **2016**, *181*, 227–237. [CrossRef]

21. Tian, H.; Pan, L.; Xiao, X.; Wilkins, R.W.T.; Meng, Z.; Huang, B. A preliminary study on the pore characterization of Lower Silurian black shales in the Chuandong Thrust Fold Belt, southwestern China using low pressure N2 adsorption and FE-SEM methods. *Mar. Pet. Geol.* **2013**, *48*, 8–19. [CrossRef]
22. De Boer, J.H.; Lippens, B.C.; Linsen, B.G.; Broekhoff, J.; Van den Heuvel, A.; Osinga, T.J. Thet-curve of multimolecular N2-adsorption. *J. Colloid Interface Sci.* **1966**, *21*, 405–414. [CrossRef]
23. Mishra, S.; Mendhe, V.A.; Varma, A.K.; Kamble, A.D.; Sharma, S.; Bannerjee, M.; Kalpana, M.S. Influence of organic and inorganic content on fractal dimensions of Barakar and Barren Measures shale gas reservoirs of Raniganj basin, India. *J. Nat. Gas Sci. Eng.* **2018**, *49*, 393–409. [CrossRef]
24. Sun, L.; Tuo, J.; Zhang, M.; Wu, C.; Wang, Z.; Zheng, Y. Formation and development of the pore structure in Chang 7 member oil-shale from Ordos Basin during organic matter evolution induced by hydrous pyrolysis. *Fuel* **2015**, *158*, 549–557. [CrossRef]
25. Horváth, G.; Kawazo, K. Method for The Calculation of Effective Pore Size Distribution in Molecular Sieve Carbon. *J. Chem. Eng. Jpn.* **1983**, *16*, 470–475. [CrossRef]
26. Seaton, N.A.; Walton, J. A new analysis method for the determination of the pore size distribution of porous carbons from nitrogen adsorption measurements. *Carbon* **1989**, *27*, 853–861. [CrossRef]
27. Liu, K.; Ostadhassan, M.; Kong, L. Multifractal characteristics of Longmaxi Shale pore structures by N_2 adsorption: A model comparison. *J. Pet. Sci. Eng.* **2018**, *168*, 330–341. [CrossRef]
28. Cai, J.; Lin, D.; Singh, H.; Wei, W.; Zhou, S. Shale gas transport model in 3D fractal porous media with variable pore sizes. *Mar. Pet. Geol.* **2018**, *98*, 437–447. [CrossRef]
29. Xia, Y.; Cai, J.; Wei, W.; Hu, X.; Wang, X.; Ge, X. A new method for calculating fractal dimensions of porous media based on pore size distribution. *Fractals* **2018**, *26*, 1850006. [CrossRef]
30. Cai, J.; Wei, W.; Hu, X.; Liu, R.; Wang, J. Fractal characterization of dynamic fracture network extension in porous media. *Fractals* **2017**, *25*, 17500232. [CrossRef]
31. Wang, F.; Jiao, L.; Lian, P.; Zeng, J. Apparent gas permeability, intrinsic permeability and liquid permeability of fractal porous media: Carbonate rock study with experiments and mathematical modelling. *J. Pet. Sci. Eng.* **2019**, *173*, 1304–1315. [CrossRef]
32. Pfeifer, P.; Avnir, D. Chemistry in noninteger dimensions between two and three. I. Fractal theory of heterogeneous surfaces. *J. Chem. Phys.* **1983**, *79*, 3558–3565. [CrossRef]
33. Hinai, A.A.; Rezaee, R.; Esteban, L.; Labani, M. Comparisons of pore size distribution: A case from the Western Australian gas shale formations. *J. Unconv. Oil Gas Resour.* **2014**, *8*, 1–13. [CrossRef]
34. Sun, Z.; Zhang, H.; Wei, Z.; Wang, Y.; Wu, B.; Zhuo, S.; Zhao, Z.; Li, J.; Hao, L.; Yang, H. Effects of slick water fracturing fluid on pore structure and adsorption characteristics of shale reservoir rocks. *J. Nat. Gas Sci. Eng.* **2018**, *51*, 27–36. [CrossRef]
35. Ghanbarian, B.; Daigle, H. Fractal dimension of soil fragment mass-size distribution: A critical analysis. *Geoderma* **2015**, *245–246*, 98–103. [CrossRef]
36. Ghanbarian-Alavijeh, B.; Hunt, A.G. Comments on "More general capillary pressure and relative permeability models from fractal geometry" by Kewen Li. *J. Contam. Hydrol.* **2012**, *140*, 21–23. [CrossRef] [PubMed]

Article

Impact of Local Effects on the Evolution of Unconventional Rock Permeability

Xinxing Ma [1], Xianwen Li [1], Shouwen Zhang [2], Yanming Zhang [1], Xiangie Hao [3] and Jishan Liu [4,*]

1 Oil & Gas Technology Research Institute of Changqing Oilfield Company, Xi'an 710018, China; mxx1_cq@petrochina.com.cn (X.M.); lxw_cq@petrochina.com.cn (X.L.); zym_cq@petrochina.com.cn (Y.Z.)
2 Guangzhou Urban Planning & Design Survey Research Institute, Guangzhou 510060, China; shouwen.cn@gmail.com
3 Beijing Key Laboratory for Precise Mining of Intergrown Energy and Resources, China University of Mining and Technology, Beijing 100083, China; haoxianjie@cumtb.edu.cn
4 Department of Chemical Engineering, School of Engineering, The University of Western Australia, WA 6009, Australia
* Correspondence: jishan.liu@uwa.edu.au

Received: 20 December 2018; Accepted: 29 January 2019; Published: 1 February 2019

Abstract: When gas is extracted from unconventional rock, local equilibrium conditions between matrixes and fractures are destroyed and significant local effects are introduced. Although the interactions between the matrix and fracture have a strong influence on the permeability evolution, they are not understood well. This may be the reason why permeability models in commercial codes do not include the matrix-fracture interactions. In this study, we introduced the local force to define the interactions between the matrix and the fracture and derived a set of partial differential equations to define the full coupling of rock deformation and gas flow both in the matrix and in the fracture systems. The full set of cross-coupling formulations were solved to generate permeability evolution profiles during unconventional gas extraction. The results of this study demonstrate that the contrast between the matrix and fracture properties controls the processes and their evolutions. The primary reason is the gas diffusion from fractures to matrixes. The diffusion changes the force balance, mass exchange and deformation.

Keywords: shale permeability; local effect; global effect; matrix-fracture interactions

---

1. Introduction

The eastern Ordos basin of China, where shale and coal are rich in organic matter and favorable for gas accumulations, has become one of the most important gas development areas for PetroChina. Unconventional reservoirs within this area have an extremely low intrinsic permeability and low porosities. For most low permeability reservoirs, hydraulic fracturing and horizontal drilling are the key techniques to extract natural gas. Gas production in shale reservoirs is attributed to the conductivity of the matrix and fracture systems [1,2]. However, the long-term gas production from these reservoirs is known to be a function of fluid transport in the shale matrix and strongly influenced by the fluid transport process in the inorganic matrix, kerogen and fractures [3–6]. Therefore, a better understanding of effective permeability evolution in matrixes and fractures and interactions between them is important to guide the industrial production.

Gas transport in shale reservoirs is a combination of desorption and diffusion within the micropores, and Darcy flow (pressure driven volume flow) within the macro-pores, micro-fractures and fractured network system [7]. Because of the high contrasts of matrix properties and fracture ones, they have different mechanical behaviors that directly affect the permeability evolution. Many scholars established models to investigate the evolution of the permeability of different parts over the past few decades. The models were developed from a single porosity/permeability model [8] to dual porosity/permeability model [9–12]. For a single porosity model, the effects of effective stress variation and the matrix shrinkage/swelling were taken into consideration [13,14], which ignored the interactions between different range radius pores. Then researchers paid attention to the effects of adsorption-induced strain [15,16]. Based on the poroelasticity theory, Zhang et al. [15] developed a strain-based porosity model and a permeability model under variable stress conditions. These models include the coupling interaction between gas flow/diffusion and rock mechanic behavior.

In previous studies, many scholars ignored the dynamic behaviors of matrix and fracture properties, especially the interactions between them. When gas is extracted from the reservoir, the gas pressure in the inorganic matrix, kerogen and microfracture will decrease to a lower magnitude. The effective stress within them will change. The variation of effective stress will affect the pore radius of the matrix and kerogen/micro-fracture, which means that the intrinsic permeability is a variable [17,18]. Chen et al. [19] and Masoudian [20] studied the impact of effective stress and/or strain on permeability in shale fractures. However, the impact of shale matrix effective stress and/or strain on fracture permeability was not considered. Some models considered the variation of stress and just focused on gas flow principles. Cao et al. [2] and Peng et al. [21] developed a model considering the deformation induced by the changes in effective stress. In these studies, only the matrix mechanical deformation is taken into account and the mechanical interactions between the matrix and kerogen/micro-fracture induced by the differential pressure are ignored. However, matrix and micro-fracture properties are different, and their mechanical behaviors are also different under the same loading conditions. In this study, we defined mechanical equilibrium equations for the inorganic matrix and the kerogen/micro-fracture, respectively, to control the shale deformation based on our previous study [18]. Through the full coupling of two solid deformation systems and two gas flow systems in them, we studied the impact of local transient behaviors on the evolution of rock permeability.

2. Conceptual Model

In this section, an overlapping approach is introduced to analyze the full shale-gas interactions. Shale is multi-pore media including micropores, macro-pores, micro-fractures and fractured network systems. When gas is extracted from or injected into shale, local equilibrium conditions between matrixes and fractures are disturbed and significant local effects are introduced into the porous medium. Because of the high contrasts of matrix properties and fracture ones, it may take a much longer time to reach a new state of equilibrium. A schematic diagram of the conceptual overlapping approach is shown as Figure 1. For a specific domain, as shown in Figure 1a, it was meshed, generating many nodes, as shown in Figure 1b and each mathematical node can be overlapped by four physical nodes corresponding to four different physical fields representing the solid deformation in the fracture, solid deformation in the matrix, gas flow in the fracture and gas flow in the matrix, respectively, shown as Figure 1c. The physical behaviors of four physical points are not isolated but connected to each other by a set of coupling relations as shown in Figure 1d.

To explain the interactions between the matrix and fracture, we use gas injection as an example. When gas is injected into the shale as shown in Figure 1a, gas flows into micropores and fractures quickly while there is no gas in the matrix. Subsequently, the pressure in the matrix increases gradually as the gas diffuses from the fractures into the matrixes. The non-synchronization of the pressure change between the matrix and fracture generates a local force which can cause interactions. Specifically, because of the increase of the pressure in the fracture, the effective stress in the fractures firstly decrease, which results in the swelling of the fracture. At the same time, the matrix localized in the vicinity of the fracture compartment shrinks under an extra local force applied by the fracture. The swelling and shrinkage, especially induced by a local force, directly affects the permeability evolution of the matrix and fracture. When the pressure in the fracture reaches a specific magnitude, the gas diffuses into the matrix. The differential pressure starts to decline slowly to zero at a final equilibrium state.

In the same way, the matrix swells because of the decrease in effective stress. Finally, the fracture and matrix both swell and the matrix changes from local swelling to macro swelling [22]. Additionally, the permeability of the matrix and fracture will have a net increase.

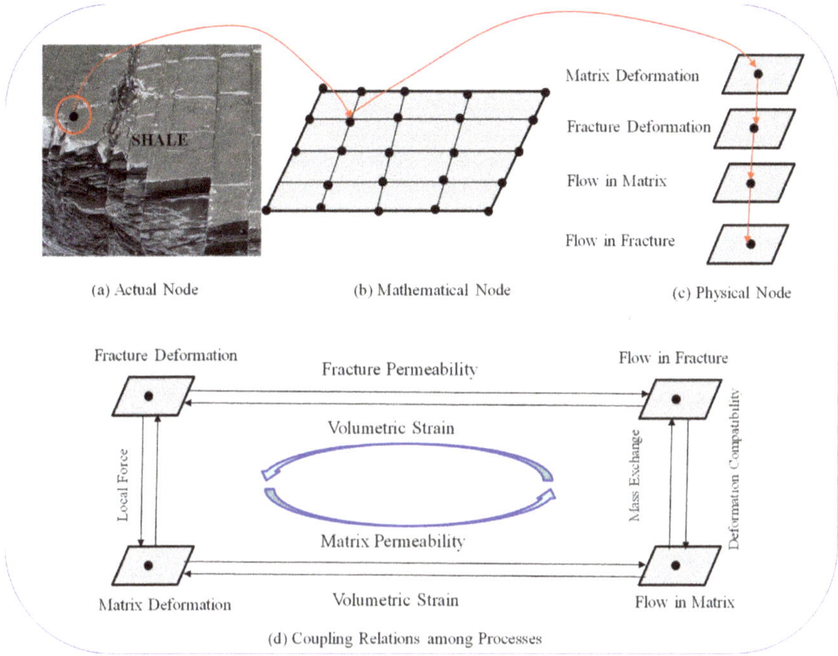

Figure 1. The schematic diagram of the conceptual overlapping approach.

3. Governing Equations

In this section, a set of partial differential equations are defined. These equations govern the deformation of the matrix and fracture deformation, and control the transport of gas flow. Originally, we develop the governing equations of single pore media based on previous studies [15] which describe the interactions in the two kinds of solid media. These derivations are based on the following assumptions:

1. Shale is a homogeneous, isotropic, dual poroelastic continuum.
2. Strains are much smaller than the length scale.
3. Gas contained within the pores is ideal, and its viscosity is constant under isothermal conditions.
4. Gas flow through the shale fracture is defined by Darcy's law and defined by Knudsen diffusion in the matrix.

Shale contains a matrix and fracture which have different mechanical properties and interact with each other. Therefore, we derived mechanical equations for the matrix and fracture, respectively. They are fully coupled with local force. Darcy's law is used for both the flow in the matrix and the flow in the fracture.

3.1. Formulation of Solid Deformation

The linear constitutive relations can be obtained by extending the known poroelasticity [23]. In previous studies, the gas sorption-induced strain is assumed to result in volumetric strain only [2,15]. However, in our new model, the volumetric strain includes both the gas sorption strain and local strain. Additionally, their effects on all three normal components of strain are the same. By making an analogy between thermal contraction and matrix shrinkage, the constitutive relations for the deformed shale matrix and fracture can be defined as

$$\varepsilon_{mij} = \frac{1}{2G_m}\sigma_{ij} - \left(\frac{1}{6G_m} - \frac{1}{9K_m}\right)\sigma_{kk}\delta_{ij} + \frac{\alpha}{3K_m}P_m\delta_{ij} - \frac{1}{3K_m}\Delta P\delta_{ij} + \frac{\varepsilon_{ms}}{3}\delta_{ij} \tag{1}$$

$$\varepsilon_{fij} = \frac{1}{2G_f}\sigma_{ij} - \left(\frac{1}{6G_f} - \frac{1}{9K_f}\right)\sigma_{kk}\delta_{ij} + \frac{\beta}{3K_f}P_f\delta_{ij} - \frac{1}{3K_f}\Delta P'\delta_{ij} + \frac{\varepsilon_{fs}}{3}\delta_{ij} \tag{2}$$

where $G_m = E_m/2(1+v_m)$ is the shear modulus of matrix, E_m and ν_m are the Young's modulus values of shale matrix and the Poisson's ratio of the matrix, respectively; $G_f = E_f/2\left(1+v_f\right)$ is the shear modulus of the fracture, E_f and ν_f are the Young's modulus values of the shale fracture and the Poisson's ratio of fracture, respectively; $K_m = E_m/3(1-2v_m)$ is the bulk modulus of matrix, $K_f = E_f/3\left(1-2v_f\right)$ is the bulk modulus of fracture; α and β are the Biot coefficients; P_m is the fluid pressure in the matrix, P_f is the fluid pressure in the fracture; ε_{ms} is the gas sorption-induced strain in the matrix, ε_{fs} is the gas sorption-induced strain in the fracture; σ_{kk} is the total stress; δ_{ij} is the Kronecker delta; ΔP is the differential pressure between the fracture and the matrix. In addition, $\Delta P = P_m - P_f$ for the constitutive relation of the matrix while $\Delta P' = P_f - P_m$ for the constitutive relation of the fracture.

Applying Langmuir isotherm, the sorption-induced volumetric strains of matrix and fracture can be defined as [15,18]

$$\varepsilon_{ms} = \varepsilon_L\frac{P_m}{P_L + P_m} \tag{3}$$

$$\varepsilon_{fs} = \varepsilon_L\frac{P_f}{P_L + P_f} \tag{4}$$

where ε_L is the Langmuir strain constant and P_L is the Langmuir pressure. When gas flows from the matrix into the fracture, the local deformations of the matrix and fracture are controlled by [18]

$$\sigma + \Delta P = \sigma_{me} + \alpha P_m \tag{5}$$

$$\sigma + \Delta P' = \sigma_{fe} + \beta P_f \tag{6}$$

where $\Delta P = P_f - P_m$, $\Delta P' = P_m - P_f$ are local forces induced by differential pressures between the fracture and matrix systems; $\sigma + \Delta P$ and $\sigma + \Delta P'$ are the dynamic effective stress values. σ_{me} and σ_{fe} are the effective stress component of the matrix and the fracture, respectively; α and β are the Biot coefficients of the matrix and the fracture. From Equations (1)–(6), the volumetric strains of matrix and fracture can be expressed as

$$\varepsilon_{mv} = -\frac{1}{K_m}(\overline{\sigma} - \alpha P_m + \Delta P) + \varepsilon_{ms} \tag{7}$$

$$\varepsilon_{fv} = -\frac{1}{K_f}\left(\overline{\sigma} - \beta P_f + \Delta P'\right) + \varepsilon_{fs} \tag{8}$$

where $\overline{\sigma} = -\sigma_{kk}/3$ is the mean compressive stress. Combining Equations (1), (2), (7) and (8) yields the general Navier-type equations for the matrix and fracture, respectively

$$G_m u_{i,kk} + \frac{2G_m}{1-2v}u_{k,ki} - \alpha P_{m,i} + \Delta P_{,i} - K_m\varepsilon_{ms,i} + F_i = 0 \tag{9}$$

$$G_f u_{i,kk} + \frac{2G_f}{1-2v}u_{k,ki} - \beta P_{f,i} + \Delta P'_{,i} - K_f\varepsilon_{fs,i} + F_i = 0 \tag{10}$$

Equation (9) and Equation (10) are the governing equations for the shale matrix and fracture deformation. They are cross-coupled by the local force, ΔP, which reflects the mechanical interaction between the matrix and the fracture.

3.2. Formulation of Gas Flow in the Fracture

The gas flow within the natural fractures obeys Darcy's law. The equation for the mass balance of the gas is defined as

$$\frac{\partial m_f}{\partial t} + \nabla \cdot \left(-\frac{k_f}{\mu} \rho_{gf} \nabla P_f \right) = -Q_{mf} \tag{11}$$

where μ is the gas dynamic viscosity, $m_f = \phi_f \rho_{gf} + \rho_g \rho_c \frac{V_L P_f}{P_f + P_L}$ is the gas content in the fracture including the free-phase gas and adsorbed gas, ϕ_f is fracture porosity, ρ_g is the gas density at standard conditions, $\rho_{gf} = \frac{M_g}{RT} P_f$ is the gas density, k_f is the permeability of the fractures and $-Q_{mf}$ is mass the transfer from the matrix to the fractures.

3.3. Formulation of Gas Flow in the Matrix

Gas flow in the matrix follows Darcy's law, so the equation for the mass transfer of the gas in the matrix is defined as

$$\frac{\partial m_m}{\partial t} + \nabla \cdot \left(-\frac{k_m}{\mu} \rho_{gm} \nabla P_m \right) = Q_{mf} \tag{12}$$

where $m_m = \phi_m \rho_{gm} + \rho_{ga} \rho_c \frac{V_L P_m}{P_m + P_L}$ is the gas content in the matrix including free-phase gas and adsorbed gas, k_m is the permeability of the matrix, μ is the dynamic viscosity of the gas, ρ_g is the gas density at standard conditions, $\rho_{gm} = \frac{M_g}{RT} P_m$ is the gas density in the matrix (M_g is the molecular mass of the gas, R is the universal gas constant, T is the absolute gas temperature), and Q_{mf} is the gas mass transfer from the fracture to the matrix.

3.4. Formulation of Cross-Couplings

The mechanisms of mass transfer for a dual porosity media are fluid expansion and viscous displacement. The final form of the transfer function for a single-phase flow from the matrix to the fracture is given as [24]

$$Q_{mf} = aV \rho_g \frac{k_m}{\mu} \left(P_m - P_f \right) \tag{13}$$

where ρ_g is the density of the gas, k_m is the permeability of the matrix, μ is the viscosity, a is called the matrix-fracture transfer shape factor and has dimensions of L^{-2} that equal 1, P_m is the matrix pressure, and P_f is the fracture pressure.

We derived the general permeability model of the shale matrix and fracture. Shale rock contains a fracture system with well-connecting macropores and a matrix system with micropores. For each system, considering it contains a solid volume V_s and pore volume V_p, the shale bulk volume can be defined as $V = V_p + V_s$ and the porosity can be defined as $\phi = V_p / V$. According to Equation (5) and (6), the volumetric evolution of the porous medium loaded by $\overline{\sigma}$, p_m or p_f and $\Delta p = p_f - p_m$ can be described in terms of $\Delta V / V$ and $\Delta V_p / V_p$, the volumetric strain of the shale matrix/fracture and the volumetric strain of the pore space, respectively [23]. The relations are

$$\frac{\Delta V_i}{V_i} = -\frac{1}{K_i} \Delta \overline{\sigma} + \left(\frac{1}{K_i} - \frac{1}{K_{is}} \right) \Delta P_i + \frac{1}{K_i} \Delta P + \Delta \varepsilon_s \tag{14}$$

$$\frac{\Delta V_{ip}}{V_{ip}} = -\frac{1}{K_{ip}} \Delta \overline{\sigma} + \left(\frac{1}{K_{ip}} - \frac{1}{K_{is}} \right) \Delta P_i + \frac{1}{K_{ip}} \Delta P + \Delta \varepsilon_s \tag{15}$$

$$\Delta P = P_f - P_m \tag{16}$$

where subscript i = 1 and 2 represent the shale fracture and matrix, respectively. If we apply $\alpha_i = 1 - K_i/K_{is}$ and $\beta_i = 1 - K_{ip}/K_{is}$, then the above equations can be expressed as

$$\frac{\Delta V_i}{V_i} = -\frac{1}{K_i}(\Delta\overline{\sigma} - \alpha_i \Delta P_i - \Delta P) + \Delta\varepsilon_s \tag{17}$$

$$\frac{\Delta V_{ip}}{V_{ip}} = -\frac{1}{K_{ip}}(\Delta\overline{\sigma} - \beta_i \Delta P_i - \Delta P) + \Delta\varepsilon_s \tag{18}$$

where K_p and K_s are the bulk moduli of the pore and the bulk modulus of the solid. We assume that the sorption-induced strain for shale is the same as for the pore space. Applying the definition of porosity, the following expressions can be defined as [15]

$$\frac{\Delta V_i}{V_i} = \frac{\Delta V_{is}}{V_i} + \frac{\Delta\phi_i}{1-\phi_i} \tag{19}$$

$$\frac{\Delta V_{ip}}{V_{ip}} = \frac{\Delta V_{is}}{V_i} + \frac{\Delta\phi_i}{\phi_i(1-\phi_i)} \tag{20}$$

By solving Equations (16)–(20), we can obtain the relationship as

$$\Delta\phi_i = \phi_i\left(\frac{1}{K_i} - \frac{1}{K_{ip}}\right)(\Delta\overline{\sigma} - \Delta P_i - \Delta P) \tag{21}$$

Substituting $K_{ip} = \phi_i K_i/\alpha_i$ [25] into the above equation yields

$$\phi_i - \phi_{i0} = \phi_i\left(1 - \frac{\alpha_i}{\phi_i}\right)\frac{\Delta\overline{\sigma} - \Delta P_i - \Delta P}{K_i} \tag{22}$$

Rearranging Equation (22) gives

$$\phi_i = \frac{\phi_{i0}}{1 - \frac{\Delta\overline{\sigma} - \Delta P_i - \Delta P}{K_i}} - \frac{\alpha_i}{1 - \frac{\Delta\overline{\sigma} - \Delta P_i - \Delta P}{K_i}}\frac{\Delta\overline{\sigma} - \Delta P_i - \Delta P}{K_i} \tag{23}$$

Because generally $(\Delta\overline{\sigma} - \Delta P_i - \Delta P)/K_i \ll 1$, the above equation can be simplified into

$$\frac{\phi_i}{\phi_{i0}} = 1 - \frac{\alpha_i}{\phi_{i0}}\frac{\Delta\overline{\sigma} - \Delta P_i - \Delta P}{K_i} = 1 + \frac{\alpha_i}{\phi_{i0}}(\Delta\varepsilon_{iet} - \Delta\varepsilon_{il}) \tag{24}$$

where $\Delta\varepsilon_{iet} = -(\Delta\overline{\sigma} - \Delta P)/K_i$ is defined as the total effective volumetric compressive strain, and $\Delta\varepsilon_{il} = -\Delta P/K_i$ is defined as the local strain induced by differential pressures between the two systems.

The typical relationship between porosity and permeability follows the cubic law [26]

$$\frac{k}{k_0} = \left(\frac{\phi}{\phi_0}\right)^3 \tag{25}$$

By substituting Equation (24) into Equation (25) we obtain shale permeability model

$$\frac{k_i}{k_{i0}} = \left(1 - \frac{\alpha_i}{\phi_{i0}}\frac{\Delta\overline{\sigma} - \Delta P_i - \Delta P}{K_i}\right)^3 = \left[1 + \frac{\alpha_i}{\phi_{i0}}(\Delta\varepsilon_{iet} - \Delta\varepsilon_{il})\right]^3 \tag{26}$$

The total effective volumetric strain can be written as

$$\Delta\varepsilon_{iv} = \Delta\varepsilon_{iet} - \Delta\varepsilon_{is} + c_l\Delta\varepsilon_{(i+1)s} \tag{27}$$

where $\Delta\varepsilon_{iv}$ is the volumetric strain, $\Delta\varepsilon_{is}$ is the volumetric strain induced by sorption, c_l is the local strain coefficient.

By substituting Equation (27) into Equation (26), we can obtain the permeability model for the shale matrix and fracture.

$$\frac{k_i}{k_{i0}} = \left[1 + \frac{\alpha_i}{\phi_{i0}}\left(\Delta\varepsilon_{iv} + \Delta\varepsilon_{is} - \Delta\varepsilon_{il} - c_l \Delta\varepsilon_{(i+1)s}\right)\right]^3 \tag{28-a}$$

$$\frac{k_m}{k_{m0}} = \left[1 + \frac{\alpha_m}{\phi_{m0}}\left(\Delta\varepsilon_{mv} + \Delta\varepsilon_{ms} - \frac{p_f - p_m}{K_m} - c_l \frac{K_f}{K_m}\Delta\varepsilon_{fs}\right)\right]^3 \tag{28-b}$$

$$\frac{k_f}{k_{f0}} = \left[1 + \frac{\alpha_f}{\phi_{f0}}\left(\Delta\varepsilon_{fv} + \Delta\varepsilon_{fs} - \frac{p_m - p_f}{K_f} - c_l \frac{K_m}{K_f}\Delta\varepsilon_{ms}\right)\right]^3 \tag{28-c}$$

$$c_l = \eta \frac{P_f - P_m}{\Delta P_{max}} \tag{28-d}$$

where subscript i = 1 and 2 represent the shale fracture and matrix, respectively. c_l is the local strain coefficient which is in proportion to differential pressures and $\left(P_f - P_m\right)/\Delta P_{max}$, η is a constant.

Shale has many nanopores. The flow regimes of the gas also strongly affect the apparent permeability [26,27]. The relation between apparent permeability, k_{mapp}, and intrinsic permeability of matrix, k_{m0}, is

$$k_{mapp} = k_{m0} g(K_n) \tag{29}$$

Kn is the Knudsen number and can be expressed as

$$g(K_n) = (1 + \zeta K_n)\left(1 + \frac{4K_n}{1 + K_n}\right) \tag{30}$$

ζ is a dimensionless rarefaction coefficient. Its value varies: $0 < \zeta < \zeta_0$ for $0 < Kn < \infty$. ζ_0 is an empirical parameter and the dimensionless rarefaction correlation is presented by Civan et al. [28]

$$\zeta = \frac{\zeta_0}{1 + \frac{A}{K_n^B}} \tag{31}$$

where A = 0.17, B = 0.4348, and ζ_0 = 1.358.

K_n is defined as the ratio of the molecular mean free path, $\lambda(nm)$ and pore radius $r(nm)$.

$$K_n = \frac{\lambda}{r} \tag{32}$$

The mean-free-path of molecules λ is given by [24]

$$\lambda = \frac{K_B T}{\sqrt{2}\pi \tilde{d}^2 p_m} \tag{33}$$

where K_B is Boltzmann constant, T is the temperature of shale reservoir, $\tilde{d}$ is the collision diameter for molecules. Based on the research of Wei et al. [29], The nanopore radius can be obtained as

$$r = r_0 - \Delta r \tag{34}$$

where r is the average nanopore radius, r_0 is the initial nanopore radius, Δr is the thickness of the adsorbed layer. The average thickness of the adsorbed layer can be expressed as [29]

$$\Delta r = t_a \exp\left(-D\left[ln\left(\frac{\rho_{van}}{\rho_g}\right)\right]^2\right) \tag{35}$$

where D is a constant and equals to 0.07 [29,30], and t_a is the thickness of the adsorbed layer at extremely high pressures, ρ_g is the density of the gas at the specific temperature and pressure, ρ_{van} is the gas density of the adsorbed phase (generally assumed to be the van der Waals density of the gas) and is 370 kg/m^3. Substituting Equation (37) into Equation (36) yields

$$r = r_0 - t_a \exp\left(-D\left[ln\left(\frac{\rho_g}{\rho_G}\right)\right]^2\right) \tag{36}$$

Therefore, the Knudsen number becomes

$$K_n = \frac{K_B T}{\sqrt{2}\pi \tilde{d}^2 p_m} \frac{1}{r_0 - t_a \exp\left(-D\left[ln\left(\frac{\rho_g}{\rho_G}\right)\right]^2\right)} \tag{37}$$

Therefore, the final formulation of apparent permeability of inorganic matrix can be expressed as

$$k_{mapp} = k_{m0}\left[1 + \frac{\alpha}{\phi_{m0}}(\Delta\varepsilon_{mv} + \Delta\varepsilon_{mI} - \Delta\varepsilon_{ms})\right]^3 (1 + \zeta K_n)\left(1 + \frac{4K_n}{1 + K_n}\right) \tag{38}$$

4. Evolution of Shale Permeability under Stress-Controlled Conditions

The above complete set of formulations, four field equations (Equations (9)–(12)) and two permeability models (Equation (28-a)–(28-c), Equation (38)), are implemented into COMSOL MULTIPHYSICS, a commercial PDE solver. The model geometry of 5 cm × 10 cm is shown in Figure 2. The new model considers mechanical deformations, sorption-induced volumetric strain, local strain induced by local force and interactions between two systems. The simulations were conducted under the constant confining stress condition, 15 MPa. Methane gas (absorbing gas) was used in the simulation and the pore pressure increased from 4 MPa to 8 MPa. The extended material properties are shown in Table 1.

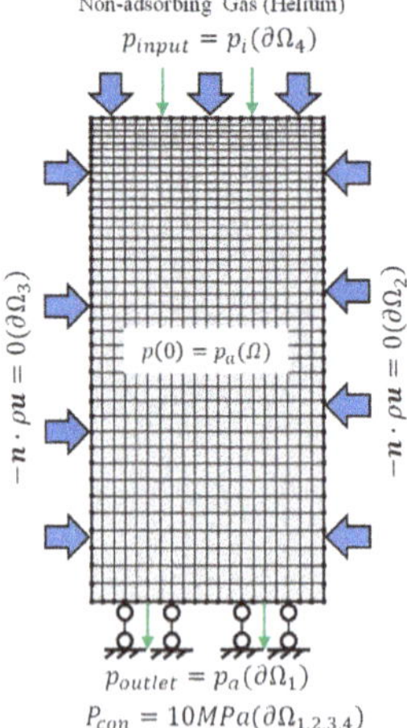

Figure 2. The simulation model for the gas transfer within the shale matrix and fracture systems under constant confining stress.

Table 1. The property parameters of the simulation sample.

Parameter	Value	Physical Meaning	Units
E_m	10	Young's modulus of the matrix	GPa
E_f	2	Young's modulus of the fracture	GPa
v_m	0.35	Poisson's ratio of the matrix	-
v_f	0.2	Poisson's ratio of the fracture	-
α	0.8	Biot coefficient of the fracture	-
β	0.4	Biot coefficient of the matrix	-
μ	1.11×10^{-5}	Viscosity of Methane	Pa·s
φ_{m0}	0.08	Initial matrix porosity	-
φ_{f0}	0.04	Initial fracture porosity	-
k_{m0}	0.5×10^{-20}	Initial matrix permeability	m^2
k_{f0}	1×10^{-19}	Initial fracture permeability	m^2
P_L	6.109	Langmuir pressure constant	MPa
ε_L	0.02	Langmuir volumetric strain constant	-
ρ_m	1250	Matrix density	kg/m^3
ρ_f	1000	Fracture density	kg/m^3
P_a	0.1	Atmosphere pressure	MPa
ρ_g	0.178	Density of gas at standard condition	kg/m^3
M	0.016	Molar mass of methane	kg/mol
R	8.314	Gas constant	J/(mol·K)
T	298.15	Temperature of the reservoir	K

From Figure 3, it can be seen that the different mechanical properties of the shale matrix and fracture play an important role in controlling the gas flow process. Pore pressure in the fracture equals that in the matrix at the initial equilibrium station. When the gas is injected into the subject, the pore pressure in the fracture increases firstly because of the higher intrinsic permeability while the pore pressure in the matrix keeps stable because of extremely low intrinsic permeability. More interestingly, the asynchronization of gas flow in the matrix and fracture generates a differential pressure which will induce local strain that has an effect on permeability. When the gas diffuses into the matrix, the pore pressure in the matrix starts to increase gradually. Until the gas diffuses into the whole shale, the subject reaches the equilibrium station again and the differential pressure becomes zero again too.

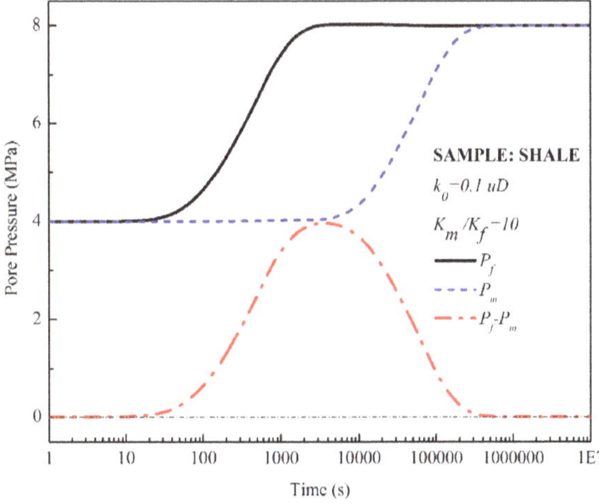

Figure 3. The pore pressure evolution in the fracture and matrix under the constant confining pressure condition.

A typical permeability evolution during gas injection is illustrated in Figure 4. It consists of two main features: the permeability ratio at point D is higher than it at the initial station; there is a wave crest B and a trough of wave C. During gas injection, the shale permeability ratio evolution can be divided into three stages. At the first stage (SI), the shale permeability ratio increases by around 20% in the short term. This is because the pore pressure in the fracture increases dramatically while that in the matrix is still the initial value. Therefore, the differential pressure will generate a local compress strain on the matrix located in the vicinity of the fracture which causes the swell of the fracture. In addition, the gas-sorption induced strain also results in the improvement of the permeability. Prior to the diffusion in the matrix, the strain induced by differential pressures and sorption both reach the maximum and the permeability reaches a crest as well. At the second stage (SII), the shale permeability ratio switches from an increase to a decrease to the initial value and then continues to decline to the minimum at point C. The pore pressure in the matrix starts to increase as the gas diffuses into the matrix. The differential pressure declines gradually. The sorption-induced swelling strain has a negative impact on the fracture and this effect will be accumulated with the expanding of the local strain region. Because the porosity of the matrix is higher than the fracture's, the accumulated sorption-induced strain of the matrix has a significant influence on the fracture permeability and this process lasts for a relatively long time. At the third stage (SIII), the permeability ratio recovered and increases and after 30 days, it finally increased by 30%. This phenomenon is contributed by two reasons: (1) the effective strain of shale changed from a local strain to a global strain when the gas spreads uniformly within the shale; (2) the effective stress decreases due to the increase of the pore pressure under the constant confining stress condition. Therefore, the permeability ratio has a net growth at the end equilibrium station.

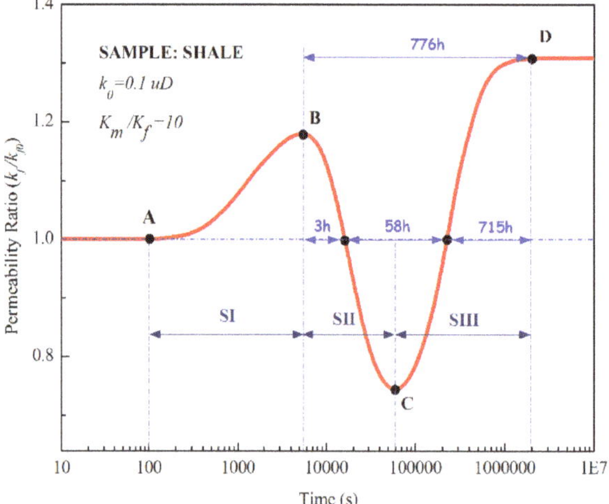

Figure 4. The permeability profile of shale sample with CH_4 gas under the constant confining pressure condition.

4.1. Impact of Local Strain on Permeability

In this study, the local strain includes two components: a local strain induced by differential pressure; the local strain induced by an interaction of the sorption strain. In order to investigate the impacts of local strain on permeability, we conducted four scenarios by considering the mechanism of local strains induced by the differential pressure and the interaction of the sorption-induced strain, both of them and none of them, respectively. The results of the resultant permeability ratio of shale are shown in Figure 5.

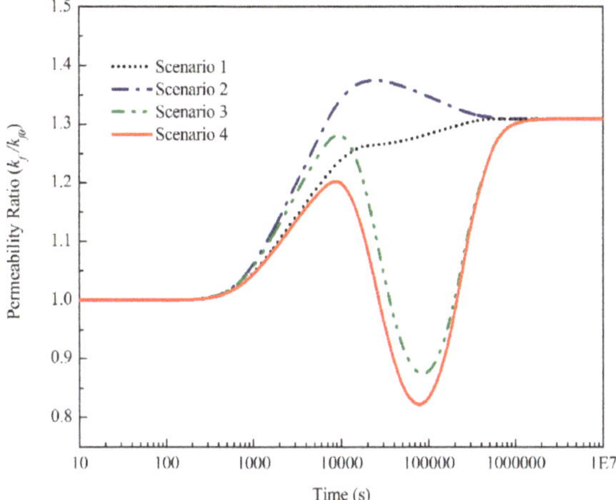

Figure 5. The impact of local strain on the evolution of permeability ratio. Scenario 1 represents the case without the local strain effects. Scenario 2 represents the case with the impact of local strain induced by the differential pressure only. Scenario 3 represents the case with the impacts of local strain induced by the sorption strain. Scenario 4 represents the case with the impact of local strain induced by the differential pressure and sorption strain.

From Figure 5, it can be seen that the local strains play an extremely important role in the shale permeability evolution. The first scenario represents the permeability ratio profile without the local strain effects. In this case, the permeability increases monotonically and is controlled by the effective strain which follows the principle of effective stress. The second scenario represents the permeability profile with the impact of local strain induced by differential pressure only. Based on the above analysis in Section 2, the local strain increases the fracture aperture. The reason is that the differential pressure generated a compress strain in the matrix located in the vicinity of the fracture. At the end equilibrium station, the local strain transforms into a global strain with the disappears of differential pressure. Therefore, the permeability declines by a small portion. The third scenario represents the permeability profile with the impacts of local strain induced by an interaction of the sorption strain. The fourth scenario represents the permeability profile with the impact of local strain induced by differential pressure and the interaction of sorption strain. From the third and fourth scenarios, we can know that with the impact of the local strain, the resultant permeability ratio increases over 1 at the initial stage and then declines dramatically under 1. With the transformation of the local strain to the global strain, the local effect finally disappears. Therefore, the resultant permeability ratio rebounds to a value over 1.

4.2. Impact of Modulus Ratios on Permeability

In order to investigate the influence of shale mechanical properties on permeability, a simulation case was conducted with different bulk modulus ratios (K_m/K_f) and the same injection pressure and confining pressure. Results corresponding to three cases ($K_m/K_f = 4$, 6 and 10) are shown in Figure 6. Permeability ratios for all the cases follow the same pattern. However, when the difference of the matrix and the fracture modulus are small like $K_m/K_f = 4$, there is a net increase of the permeability. When the modulus ratio is large enough such as $K_m/K_f = 6$ and 10, there is a significant decrease in the permeability ratio. Additionally, the higher the bulk modulus ratio is, the more the permeability ratio generated decreases. These results show that the difference of the mechanical properties between the matrix and the fracture is positively related to the impact of the local strain on permeability. These phenomena may reveal that there are more fracture or macropores in the shale with large different bulk modulus ratios and the more fractures or macropores, the more significant the local strain effects.

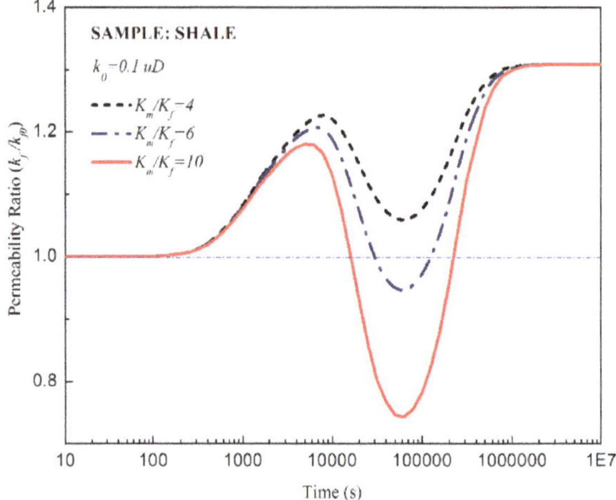

Figure 6. The shale permeability evolution with different mechanical properties (K_m/K_f).

4.3. Impact of Pore Pressure on Permeability

Figure 7 presents the evolution of shale permeability with different injection pressures. The crest and trough of the permeability ratio are higher and lower when the injection pressure is higher. Figure 7b is shale permeability ratio with pore pressure and time. It is a three dimensions graph. From this figure, we can know that the permeability ratio profile is not only related to the pore pressure but also to the time the value is obtained at.

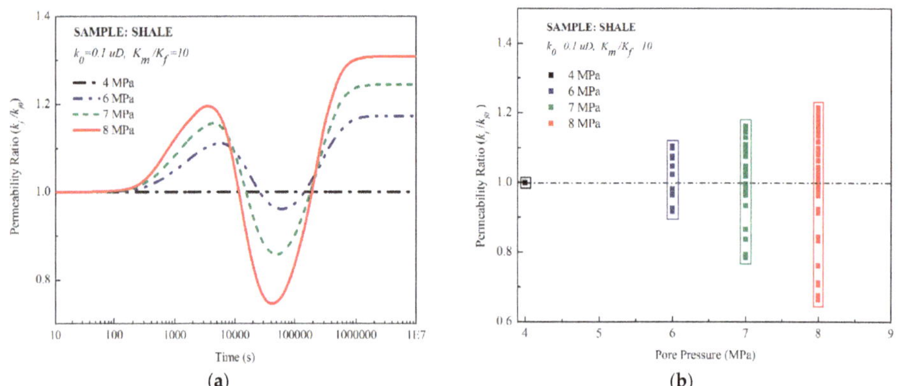

Figure 7. The evolutions of the shale permeability ratio ($k_0 = 10^{-19}$ m^2). (**a**) Shale permeability ratio evolution with time; (**b**) Shale permeability ratio evolution with pore pressure and time.

4.4. Impact of Klinkenberg Effects on Permeability

The impact of the Klinkenberg effect (Slip effect) on matrix permeability and the resultant permeability of shale is shown in Figure 8. The parameters used in this simulation case are collected from the literature [31] and are listed in Table 2. The Klinkenberg effect (slip effect) is closely related to the magnitude of the pore pressure and it has a significant influence on the matrix permeability especially under low pore pressures. From the figure, we can know that the higher the pore pressure, the lower the matrix permeability. However, compared to the impact of local strain, the Klinkenberg effect has a weak influence on the resultant permeability. In addition, the Klinkenberg effect cannot explain the net increase of the resultant permeability under the constant confining stress condition.

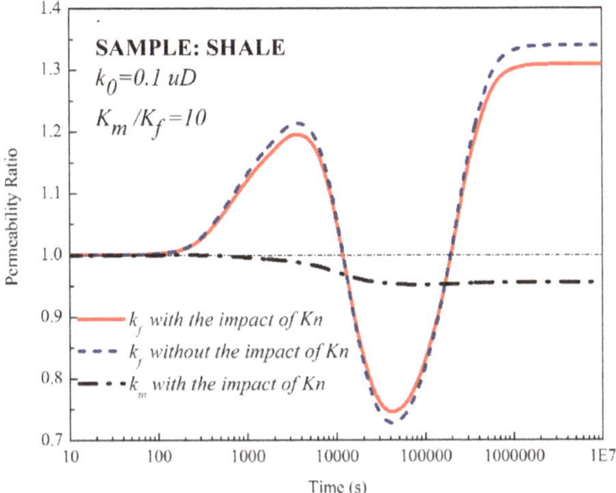

Figure 8. The evolutions of the matrix permeability, fracture permeability and resultant permeability.

Table 2. The parameters of the apparent permeability model of the shale matrix.

Symbol	Value	Physical Meanings	Units
K_B	1.38×10^{-23}	Boltzmann constant	J/K
T	298.15	Temperature	K
τ_h	1	Tortuosity of the matrix	-
A	0.178	First constant for ζ	-
B	0.4348	Second constant for ζ	-
ζ_0	0.25	Asymptotic upper limit of ζ	-

4.5. Model Evaluation and Discussions

In this section, simulations are conducted using our new model to illustrate the impact of local strains on the permeability evolution. We collect experimental data from Xiang Li at al. [32]. A series of experiments were conducted on the Green River Shale under the condition of a constant total confining stress of 20 MPa. Different gas, He, CH_4 and CO_2, were injected into the specimens with artificial fractures. The values of the mechanical parameters such as Poisson's ratio and Young's modulus are assumed based on the literature [33,34] and are listed in Table 3. The Langmuir constants of shale and dynamic viscosity of gases are listed in Table 4.

Table 3. The mechanical parameters of the shale samples.

Sample	Mechanical Properties			
	E_m (GPa)	E_f (GPa)	v_m	v_f
Green River Shale	8	6	0.25	0.25

To investigate the impact of non-sorbing gas on permeability, researchers obtained the permeability data at pore pressures of 2 MPa, 4.2 MPa, 6.16 MPa, 8.16 MPa and 10.1 MPa. We conducted simulations with the same pore pressures under the same conditions using our new model. Figure 9 shows the permeability evolution against time and pore pressure and the comparisons between the experimental data and simulation results using non-sorbing gas (He). From the experiment, we can know that the shale permeability increases with pore pressure. From the simulation results, we can obtain a three-dimensional permeability evolution not only with pore pressure but also with time. All the experimental data are located in a special zone between 2 h and 7 h.

Table 4. The Langmuir constants of shale and dynamic viscosity of gases at 300 K.

Gas	ε_L	P_L (MPa)	μ ($\mu Pa{\cdot}s$)	ρ (kg/m^3)
He	-	-	18.9	1.293
CO_2	0.0353	3.82	14.932	1.784
CH_4	0.0093	6.1	11.067	0.648

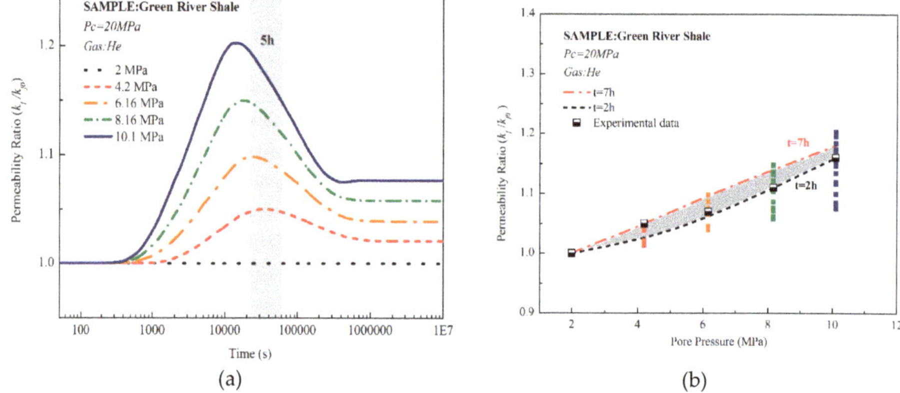

Figure 9. The permeability evolutions vary with different injection pressures with Helium gas under a constant confining stress condition and the comparison between the simulation results and experimental data. (**a**) Permeability evolutions vary with injection pressures; (**b**) Comparison of the shale permeability between the simulation results and experimental data.

To investigate the impact of the local effect of sorption-induced strain on permeability evolution. Sorbing gases (CH_4 and CO_2) were used to conduct experiments. The permeability for sorbing gases CH_4 and CO_2 all show the typical U-shaped curve, as shown in Figures 10 and 11. The permeability decreases firstly and then rebounds to an increase. Compared to the simulation results, the experimental data are in a certain zone with a diffusion time from 4.5 h to 7 h and from 9 h to 17.5 h for CH_4 and CO_2, respectively. All these experimental data are obtained in the process from the local strain to global strain.

It is obvious that the result calculated by the model with the impact of the local effects matches the experimental data well. And the model without the impact of the local effects induces significant errors that causes the apparent permeability to decrease slowly. The reason is that the local effects applied press stress on the matrix that caused an increase in the effective stress. The change of the effective stress decreased the permeability of the matrix. This phenomenon clearly illustrates the importance of local strain due to local force for the apparent permeability evolution of shale.

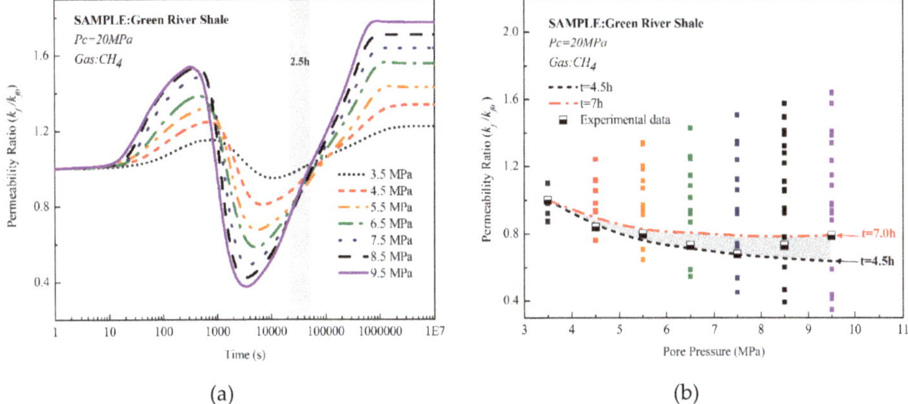

Figure 10. The permeability evolutions vary with different injection pressures with CH_4 gas under a constant confining stress condition and comparison between the simulation results and experimental data. (**a**) Permeability evolutions vary with injection pressures. (**b**) Comparison of shale permeability between the simulation results and experimental data.

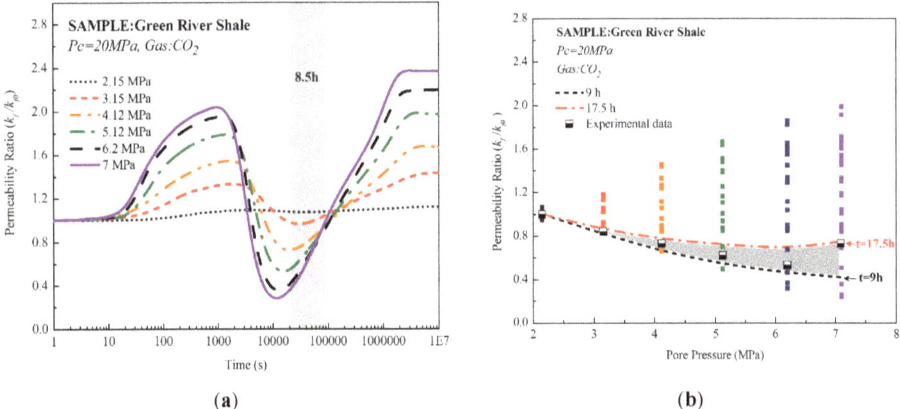

Figure 11. The permeability evolutions vary with different injection pressures with CO_2 gas under a constant confining stress condition and comparison between the simulation results and experimental data. (**a**) Permeability evolutions vary with injection pressures. (**b**) Comparison of shale permeability between the simulation results and experimental data.

5. Conclusions

The full shale matrix-fracture interactions such as mass exchange and deformation compatibility are included into a fully coupled shale deformation and gas flow model. Based on the results of this study, the following conclusions can be drawn:

For shale, gas injection induced effects can last very long because of the huge contrast between the matrix and fracture properties. The primary reason for those added effects is the gas diffusion from the fractures to the matrixes. The diffusion changes the force balance, mass exchange and deformation. Therefore, the non-equilibrium processes are much more important than the equilibrium ones.

The shale matrix and fracture's permeability experience three stages during gas injection: the initial stage of the fracture permeability increase and the matrix permeability decrease as the gas pressure in the fracture increases, the intermediate stage of the fracture permeability decrease and the matrix permeability increase as the gas diffuses from the fractures into the matrixes, and the later stage of the fracture permeability recovery as the gas desorption expands and the matrix permeability increase due to the slip effects.

Author Contributions: Conceptualization, X.M. and J.L.; Methodology, J.L., S.Z., X.L., and Y.Z.; Software, S.Z.; Validation, S.Z., X.M., and X.H.; Formal Analysis, S.Z., and J.L.; Investigation, S.Z., and X.H.; Resources, X.M., X.L.,and Y.Z. Data Curation, S.Z.; Writing-Original Draft Preparation, S.Z.; Writing-Review & Editing, J.L.; Visualization, S.Z.; Supervision, J.L.

Funding: This work was funded partially by National Key R&D Program of China (No. 2016YFC0801401), the Natural Science Foundation of China (51504235; 51474204) and Investigation and Evaluation of the geological disaster of Karst collapse, Ground subsidence and its Area Planning for Guangzhou. Those sources of support are gratefully acknowledged.

Conflicts of Interest: The authors declare no conflict of interest.

References

1. Curtis, J.B. Fractured shale-gas systems. *AAPG Bull.* **2002**, *86*, 1921–1938.
2. Cao, P.; Liu, J.; Leong, Y.K. A fully coupled multiscale shale deformation-gas transport model for the evaluation of shale gas extraction. *Fuel* **2016**, *178*, 103–117. [CrossRef]
3. Amann-Hildenbrand, A.; Ghanizadeh, A.; Krooss, B.M. Transport properties of unconventional gas systems. *Mar. Pet. Geol.* **2012**, *31*, 90–99. [CrossRef]
4. Chalmers, G.R.L.; Ross, D.J.K.; Bustin, R.M. Geological controls on matrix permeability of Devonian Gas Shales in the Horn River and Liard basins, northeastern British Columbia, Canada. *Int. J. Coal Geol.* **2012**, *103*, 120–131. [CrossRef]
5. Bustin, A.M.M.; Bustin, R.M. Importance of rock properties on the producibility of gas shales. *Int. J. Coal Geol.* **2012**, *103*, 132–147. [CrossRef]
6. Swami, V.; Settari, A. A Pore Scale Gas Flow Model for Shale Gas Reservoir. In Proceedings of the SPE Americas Unconventional Resources Conference, Pittsburgh, PA, USA, 5–7 June 2012.
7. Ghanizadeh, A.; Gasparik, M.; Amann-Hildenbrand, A.; Gensterblum, Y.; Krooss, B.M. Experimental study of fluid transport processes in the matrix system of the European organic-rich shales: I. Scandinavian Alum Shale. *Mar. Pet. Geol.* **2014**, *51*, 79–99. [CrossRef]
8. Moghaddam, R.N.; Aghabozorgi, S.; Foroozesh, J. Numerical Simulation of Gas Production from Tight, Ultratight and Shale Gas Reservoirs: Flow Regimes and Geomechanical Effects. In Proceedings of the EUROPEC 2015, Society of Petroleum Engineers, Madrid, Spain, 1–4 June 2015.
9. Warren, J.E.; Root, P.J. The Behavior of Naturally Fractured Reservoirs. *Soc. Pet. Eng. J.* **1963**, *3*, 245–255. [CrossRef]
10. Li, N.; Ran, Q.; Li, J.; Yuan, J.; Wang, C.; Wu, Y.S. A multiple-continuum model for simulation of gas production from shale gas reservoirs. In Proceedings of the SPE Reservoir Characterization and Simulation Conference and Exhibition, Abu Dhabi, UAE, 16–18 September 2013.
11. Yao, J.; Sun, H.; Fan, D.Y.; Wang, C.C.; Sun, Z.X. Numerical simulation of gas transport mechanisms in tight shale gas reservoirs. *Pet. Sci.* **2013**, *10*, 528–537. [CrossRef]
12. Sun, H.; Chawathe, A.; Hoteit, H.; Shi, X.; Li, L. Understanding shale gas production mechanisms through reservoir simulation. In Proceedings of the SPE/EAGE European Unconventional Resources Conference and Exhibition, Vienna, Austria, 25–27 February 2017.
13. Cao, P.; Liu, J.; Leong, Y.K. Combined impact of flow regimes and effective stress on the evolution of shale apparent permeability. *J. Unconv. Oil Gas Resour.* **2016**, *14*, 32–43. [CrossRef]
14. Moghaddam, R.N.; Jamiolahmady, M. Fluid transport in shale gas reservoirs: Simultaneous effects of stress and slippage on matrix permeability. *Int. J. Coal Geol.* **2016**, *163*, 87–99. [CrossRef]
15. Zhang, H.; Liu, J.; Elsworth, D. How sorption-induced matrix deformation affects gas flow in coal seams: A new FE model. *Int. J. Rock Mech. Min. Sci.* **2008**, *45*, 1226–1236. [CrossRef]
16. Pan, Z.; Connell, L.D. A theoretical model for gas adsorption-induced coal swelling. *Int. J. Coal Geol.* **2007**, *69*, 243–252. [CrossRef]
17. Masoudian, M.S.; El-Zein, A.; Airey, D.W. Modelling stress and strain in coal seams during CO2 injection incorporating the rock–fluid interactions. *Comput. Geotech.* **2016**, *76*, 51–60. [CrossRef]

18. Zhang, S.; Liu, J.; Wei, M.; Elsworth, D. Coal permeability maps under the influence of multiple coupled processes. *Int. J. Coal Geol.* **2018**, *187*, 71–82. [CrossRef]
19. Chen, D.; Pan, Z.; Ye, Z. Dependence of gas shale fracture permeability on effective stress and reservoir pressure: Model match and insights. *Fuel* **2015**, *139*, 383–392. [CrossRef]
20. Masoudian, M.S.; Hashemi, M.A.; Tasalloti, A.; Marshall, A.M. Elastic–Brittle–Plastic Behaviour of Shale Reservoirs and Its Implications on Fracture Permeability Variation: An Analytical Approach. *Rock Mech. Rock Eng.* **2018**, *51*, 1565–1582. [CrossRef]
21. Peng, Y.; Liu, J.; Pan, Z.; Connell, L.D.; Chen, Z.; Qu, H. Impact of coal matrix strains on the evolution of permeability. *Fuel* **2016**, *189*, 270–283. [CrossRef]
22. Liu, J.; Wang, J.; Chen, Z.; Wang, S.; Elsworth, D.; Jiang, Y. Impact of transition from local swelling to macro swelling on the evolution of coal permeability. *Int. J. Coal Geol.* **2011**, *88*, 31–40. [CrossRef]
23. Detournay, E.; Cheng, A.H.-D. Fundamentals of poroelasticity. In *Comprehensive Rock Engineering: Principles, Practice and Projects*; Fairhurst, C., Ed.; Pergamon Press: Oxford, UK, 1993; Volume 2, Chapter 5; pp. 113–171.
24. Ranjbar, E.; Hassanzadeh, H. Matrix–fracture transfer shape factor for modeling flow of a compressible fluid in dual-porosity media. *Adv. Water Resour.* **2011**, *34*, 627–639. [CrossRef]
25. Hudson, J.A.; Fairhurst, C. *Comprehensive Rock Engineering: Principles, Practice, and Projects, Vol. II. Analysis and Design Method*; Pergamon Press: Oxford, UK, 1993; pp. 113–121.
26. Liu, J.; Chen, Z.; Elsworth, D.; Qu, H.; Chen, D. Interactions of multiple processes during CBM extraction: A critical review. *Int. J. Coal Geol.* **2011**, *87*, 175–189. [CrossRef]
27. Beskok, A.; Karniadakis, G.E. Report: A model for flows in channels, pipes, and ducts at micro and nano scales. *Microscale Thermophys. Eng.* **1999**, *3*, 43–77.
28. Civan, F.; Rai, C.S.; Sondergeld, C.H. Shale-Gas Permeability and Diffusivity Inferred by Improved Formulation of Relevant Retention and Transport Mechanisms. *Transp. Porous Media* **2011**, *86*, 925–944. [CrossRef]
29. Wei, M.; Liu, J.; Feng, X.; Wang, C.; Zhou, F. Evolution of shale apparent permeability from stress-controlled to displacement-controlled conditions. *J. Nat. Gas Sci. Eng.* **2016**, *34*, 1453–1460. [CrossRef]
30. Sakurovs, R.; Day, S.; Weir, S. Causes and consequences of errors in determining sorption capacity of coals for carbon dioxide at high pressure. *Int. J. Coal Geol.* **2009**, *77*, 16–22. [CrossRef]
31. Wei, M.; Liu, J.; Elsworth, D.; Li, S.; Zhou, F. Influence of gas adsorption induced non-uniform deformation on the evolution of coal permeability. *Int. J. Rock Mech. Min. Sci.* **2019**, *114*, 71–78. [CrossRef]
32. Letham, E.A. Matrix Permeability Measurements of Gas Shales: Gas Slippage and Adsorption as Sources of Systematic Error. Bachelor's Thesis, The University of British Columbia, Kelowna, Canada, March 2011.
33. Wang, S.; Elsworth, D.; Liu, J. Permeability evolution in fractured coal: The roles of fracture geometry and water-content. *Int. J. Coal Geol.* **2011**, *87*, 13–25. [CrossRef]
34. Wang, J.G.; Liu, J.; Kabir, A. Combined effects of directional compaction, non-Darcy flow and anisotropic swelling on coal seam gas extraction. *Int. J. Coal Geol.* **2013**, *109*, 1–14. [CrossRef]

Article

Investigating Influential Factors of the Gas Absorption Capacity in Shale Reservoirs Using Integrated Petrophysical, Mineralogical and Geochemical Experiments: A Case Study

Zhuoying Fan [1,2], Jiagen Hou [1,2,*], Xinmin Ge [3,4], Peiqiang Zhao [5] and Jianyu Liu [3]

1 State Key Laboratory of Petroleum Resources and Prospecting, China University of Petroleum, Beijing 102249, China; fanzhuoying123@sohu.com
2 College of Geosciences, China University of Petroleum, Beijing 102249, China
3 School of Geosciences, China University of Petroleum (Huadong), Qingdao 266580, China; gexinmin2002@163.com (X.G.); liujianyu0108@163.com (J.L.)
4 Laboratory for Marine Mineral Resources, Qingdao National Laboratory for Marine Science and Technology, Qingdao 266071, China
5 Hubei Subsurface Multi-Scale Imaging Key Laboratory, Institute of Geophysics and Geomatics, China University of Geosciences, Wuhan 430074, China; zhaopq@cug.edu.cn
* Correspondence: houjg63@cup.edu.cn; Tel.: +86-532-86981315-534

Received: 17 August 2018; Accepted: 2 November 2018; Published: 8 November 2018

Abstract: Estimating in situ gas content is very important for the effective exploration of shale gas reservoirs. However, it is difficult to choose the sensitive geological and geophysical parameters during the modeling process, since the controlling factors for the abundance of gas volumes are often unknown and hard to determine. Integrated interdisciplinary experiments (involving petrophysical, mineralogical, geochemical and petrological aspects) were conducted to search for the influential factors of the adsorbed gas volume in marine gas shale reservoirs. The results showed that in shale reservoirs with high maturity and high organic content that the adsorbed gas volume increases, with an increase in the contents of organic matter and quartz, but with a decrease in clay volume. The relationship between the adsorbed gas content and the total porosity is unclear, but a strong relationship between the proportions of different pores is observed. In general, the larger the percentage of micropores, the higher the adsorbed gas content. The result is illuminating, since it may help us to choose suitable parameters for the estimation of shale gas content.

Keywords: gas adsorption capacity; shale reservoirs; influential factors; integrated methods

1. Introduction

The role of shale gas is becoming increasingly important nowadays due to the large consumption and shortage of conventional resources, and due to technological advances in oil and gas development. However, it is still challenging to estimate the gas contents of the reservoir condition since the occurrence mechanism of the shale gas reservoir is far more difficult than in conventional reservoirs. It is reported that the adsorbed gas volume accounts for more than 50% of the total gas in the pore system [1–4]. Therefore, investigating the influential and controlling factors of the adsorbed gas is significant for shale gas reservoir characterization.

Literature data have shown that gas absorption capacity is influenced by many factors, including geochemical parameters, such as the total organic matter content (TOC), kerogen types, as well as thermal maturity [1,2,5,6], pore volume and pore size distribution [1,7], petrological and mineralogical factors [8,9], and environmental factors such as the buried temperature and pressure.

Although some researchers observed that absorption capacity decreases with the increase of the TOC [10], it is generally accepted that organic matter is the primary control factor in the adsorbed gas volume and is positively correlated with the TOC [11–14]. In addition, the adsorbed gas volume increases with the confining pressure, whereas it decreases inversely to the temperature [9,15–17]. The relationship of the adsorbed gas volume to the porosity, pore size, specific surface area and mineralogical parameters are far more complex. In general, micropores represent the controlling factors for gas adsorption and storage, where the adsorption quantity increases with an increase in micro-porosity. The main reason for this is that the internal surface area and the adsorption energy of the small pores is higher than the large pores [1,6,7,18]. Nevertheless, recent reports revealed that the mesopores and macropores are also good places for methane adsorption [10,14,19], and some observed a negative correlation between the adsorbed gas volume and the porosity [13].

Various methods were proposed to investigate the influential factors of the adsorbed gas volume, most of which were independent and dispersed, and the adsorption theory and behavior are not fully understood. Hence, some relationships were established empirically, and are of local use only. Furthermore, some studies focused only on a small aspect of the influential factors, which may lead to incorrect results.

In this paper, we designed a comprehensive experiment to measure the petrophysical, petrological, mineralogical, geochemical and gas adsorption parameters. These experiments included petrophysical measurements such as porosity and permeability, pore structure measurements such as low field nuclear magnetic resonance (NMR) and carbon dioxide (CO_2)/nitrogen (N_2) adsorption experiments, geochemical measurements, such as vitrinite reflectance, pyrolysis and residual analysis, petrological analysis, such as X-ray diffraction (XRD), thin section, scanning electron microscope (SEM), and isothermal adsorption measurement. The main objective is to explore the influential and controlling factors of the gas adsorption capacity.

2. Materials and Methods

2.1. Materials

We collected 22 shale samples from the lower Cambrian Formation of Southern China. The reservoir is a typical marine shale gas reservoir in China, and most of the pore spaces are filled with methane. As seen in Figure 1, there are black shales, with ultralow porosity and permeability. Petrophysical, geochemical, mineralogical and pore structure examinations were carried out for all samples.

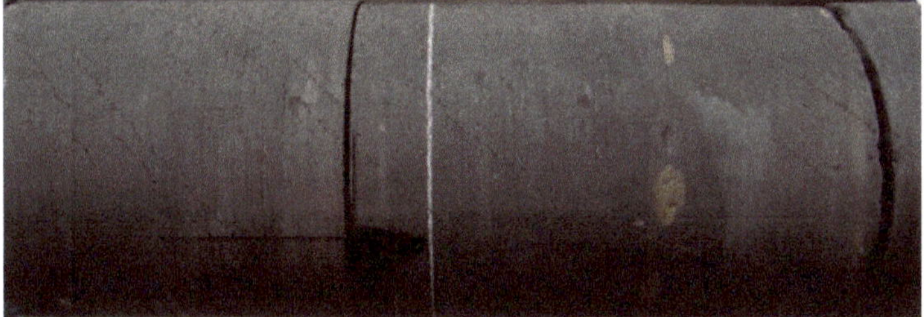

Figure 1. Typical core photo of the underground shale sample.

2.2. Petrophysical Measurements

Before the measurements, cylinder samples which were 2.5 cm in diameter and 3–5 cm in length were heated at a temperature of 100 °C, with the aim of washing out the drilling muds, light hydrocarbon, free water, and capillary bound water. In the following, the helium porosity and permeability were obtained using the AP-608 automated permeameter-porosimeter (Coretest Systems,

Inc., Reno, NV, USA). Next, the water was injected into the samples by an auto-saturator container with a confining pressure of 20 MPa to ensure that the water saturation of each sample was 100%. The low field NMR relaxation data was collected at this state using the MesoMR23-060H, with a main frequency of approximately 21.3 MHz. Compared with conventional instruments which have a main frequency of approximately 2 MHz, the dead time was reduced to 0.01 ms and the minimal echo time was reduced to 0.06 ms, through the automatic field locking and high order shimming system that was used to improve the performance of the magnetic field. Conventional Carr-Purcell-Meiboom-Gill (CPMG) pulse sequences and the Butler-Reeds-Dawson (BRD) algorithm were adopted to activate and invert the relaxation signals [20].

2.3. Low Pressure N_2 Adsorption and MIP Experiments

The low pressure nitrogen gas adsorption technique, combined with the BJH (Barret-Joyner-Halenda) model was used to obtain the pore diameter distribution and specific surface area. We used cylinder core samples, aiming to keep the original pore structure in its native state during measurement. All tests and analysis were performed with the QuadraSorb SI (Quantachrome Instruments, Boynton Beach, FL, USA) and the accessional software QuadraWin version 5.04. Prior to measurement, these samples were degassed under a vacuum at 200 °C for 12 h. Then, the degassed samples were exposed to N_2 at a temperature of −196 °C for the experiments. In addition, Mercury intrusion porosimetry (MIP) analysis was performed by the Micrometrics Autopore TM IV 9505 (Micromeritics Instruments Corporation, Norcross, GA, USA) and the maximal pressure was 200 MPa. MIP tests were performed in the last procedure. The low pressure N_2 adsorption experiments were carried out by the CNPC key well logging laboratory. To reach the pressure equilibrium during the N_2 adsorption, the time for each pressure point was more than 2 h, until the pressure variation was less than 0.003 MPa in 10 min.

2.4. Geochemical and Mineralogical Examinations

We conducted the geochemical and mineralogical experiments using the drilling cuttings at the same depth. High pressure methane adsorption isotherm experiments were carried out for samples that were crushed and dried, using the gravimetric sorption analyzer IsoSORP® that was manufactured by Rubotherm, Germany, in order to obtain the adsorbed gas volume. Before the experiments, the samples were pretreated to powders with size ranges from 20 to 40 meshes, then dried and vacuumed to remove the remaining water and unpurified gas. We used methane as the adsorbed gas. We collected 12 pressure points for every measurement. TOC and vitrinite reflectance (Ro), as well as the Backscatter electron (BSE) images were also obtained. The experimental details are elaborated in our previous publication [21].

3. Results and Discussions

3.1. Gas Adsorption Volume Correction

Figure 2a shows a typical isothermal adsorption data of one shale sample. The black dots represent the adsorbed gas content at different pressures. The data can be divided into three sections. The first section occurs at the low pressure range (with a pressure lower than 5 MPa), where the adsorbed gas content increases linearly with the pressure. The second section occurs at the medium pressure range (with pressure from 5 MPa to 15 MPa), where the adsorbed content increases smoothly and reaches the equilibrium state. The third section emerges in the high pressure range (with a pressure larger than 15 MPa), where the adsorbed content decreases with the pressure. The phenomenon mentioned above is termed as excess adsorption [22]. It is an essential characteristic of the supercritical fluid [23], which often appears in the high pressure range. The red line in Figure 2a shows the fitting results of the conventional Langmuir model. Obviously, the two-parameter Langmuir isotherm equation failed to characterize the adsorption characteristic precisely.

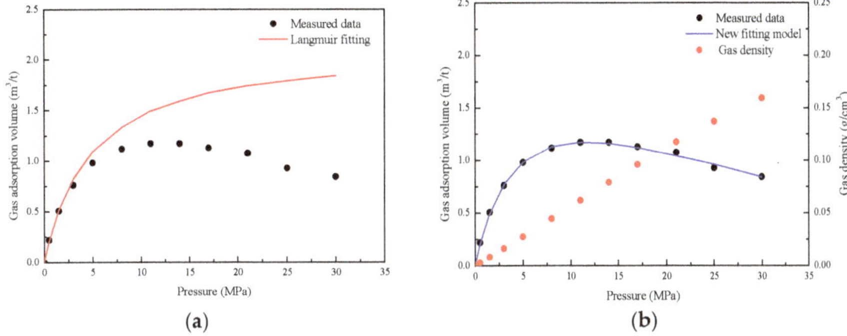

Figure 2. The typical adsorption isotherm curve and fitting results: (**a**) The fitting result of the conventional model; (**b**) the fitting result of the modified model.

In this study, we adopt the published method to correct the effect of the excess adsorption [22,24,25], which is expressed as:

$$V_i = \frac{V_L P_i}{P_L + P_i}(1 - \frac{\rho_{g,i}}{\rho_a}) \quad (1)$$

where V_i and $\rho_{g,i}$ are the adsorbed gas content in m^3/t and the gas density in g/cm^3 with the corresponding equilibrium pressure P_i; ρ_a is the density of the adsorbed phase in g/cm^3, V_L is the Langmuir volume in m^3/t, and P_L is the Langmuir pressure in MPa.

As shown in Figure 2b, the blue line represents the fitting result of the improved model, and the red dots represent gas density values at different pressures. The fitting result is improved greatly using this model. The computed Langmuir volume is required to correct to the reservoir condition, however. Based on the Langmuir equation, the adsorbed gas volume at the reservoir pressure can be expressed as:

$$V_r = \frac{V_L P_r}{P_L + P_r} \quad (2)$$

where V_r is the adsorbed gas volume.

The temperature correction equation is expressed as:

$$V_{rc} = V_r \times 10^{c(T_e - T_r)} \quad (3)$$

where V_{rc} is adsorbed gas volume after temperature correction, T_e and T_r are temperatures for experiments and reservoir condition, respectively, and c is the calibration factor. In this study, the temperature effect can be omitted since there is a slight difference between the experimental temperature and the reservoir temperature.

3.2. Pore Structure Characterization

We make full use of the pore size evaluation methods to characterize the pore structure and their distributions completely. Figure 3a,b shows the thin section and secondary electron SEM measurements of one shale sample. Due to the low resolution it is difficult to discriminate and quantify the pore information. Figure 3c gives the BSE of the same sample. In comparison, the pore morphology is cleared provided and can be characterized with image processing methods. However, the SEM cannot represent the full pore information due to the high heterogeneity of shale samples. Figure 3d–f shows the corresponding MIP, LPGA (low pressure gas adsorption) and NMR results. The pore size obtained from the MIP and LPGA results are unimodal distributed and the mainstream pore throat radius is nanoscale, which is accordance with the BSE result. However, the NMR T_2 spectrum is bimodal, revealing a higher resolution of the pore size distribution. Assuming the pore geometry is cylindrical, we can obtain the surface-relaxivity parameter as in Reference [26]:

$$\rho_2 = \frac{r_{LPGA}}{2T_2} \tag{4}$$

where ρ_2 is the surface-relaxivity parameter in μm/s, r_{LPGA} and T_2 are pore radius in μm and transversal relaxation time in ms, respectively, which can be obtained by LPGA and NMR experiments, respectively.

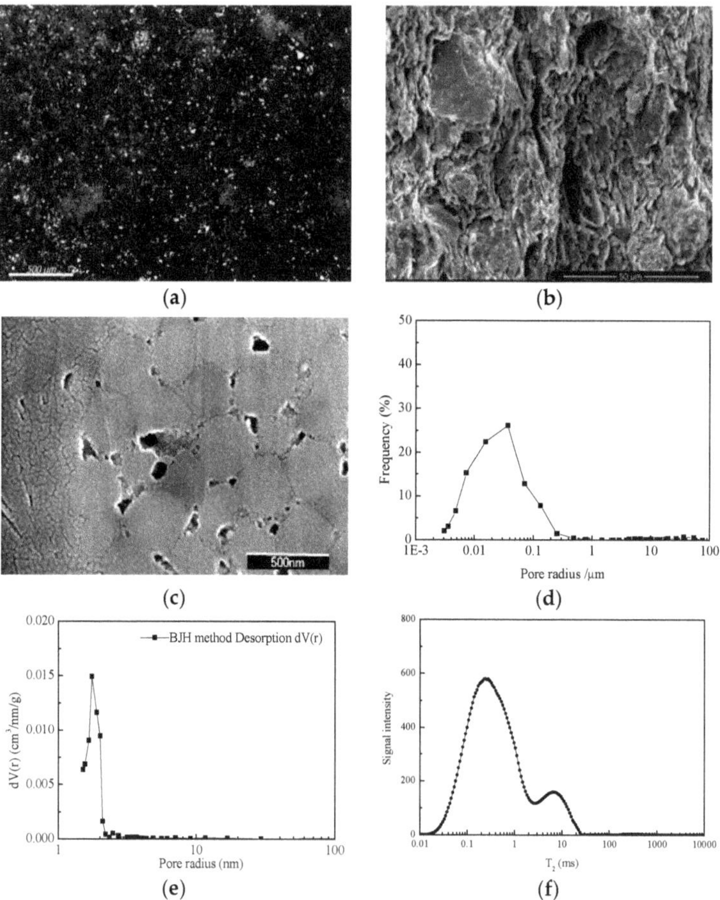

Figure 3. The typical pore morphology and pore size distributions: (**a**) Thin section; (**b**) SEM image; (**c**) BSE image; (**d**) pore radius by MIP; (**e**) pore radius by LPGA; (**f**) NMR T_2 spectrum.

The scaling factor between the pore throat radius obtained by MIP and the T_2 by NMR experiments can be can be expressed as:

$$C = \frac{r_{MIP}}{T_2} \tag{5}$$

where C is the scaling factor and r_{MIP} is the pore throat radius from the MIP experiments.

Relationships among transversal relaxation time and pore radius obtained from different methods are far more intricate, and sometimes they are difficult to convert using simple equations. In this study, ρ_2 and C can be approximately obtained using the simple peak method, where they are 3.56 μm/s and

5.42 μm/s, respectively. It was also observed that the first peak of the T_2 spectrum was symmetrical, whereas the pore radius distribution obtained by LPGA was asymmetrical. The intrinsic mechanism is unknown. This may due to the limitation of N_2, which suits the rock with mesopores (2–50 nm) and fails to characterize other pores [27]. Using the multi-Gaussian fitting technology [28] and the scaling factors, the T_2 spectrum can be transformed to the pseudo-pore radius distribution. In this study, the cutoff value for different pore types is 2.5 ms, and the corresponding pore radius is similar to the cutoff value of clay bound pore. Using ρ_2 as 3.56 μm/s, the corresponding pore radius can be computed as 17.8 nm. If we use the C as 5.42 μm/s, the pore radius can be computed as 13.55 nm. This comparison shows that both measurements and transformations include minor errors. Therefore, the T_2 cutoff value is fixed as 2.5 ms.

3.3. Effects of Geochemical Properties

As shown in Figure 4a, there is a clear positive linear relationship between the adsorbed gas content and the TOC. This was contributed to by the development of small pores in the kerogen, which are very likely to adsorb methane. The correlation of the adsorbed gas content to Ro is not obvious. However, there exists a significant negative correlation between the volume of adsorbed gas and the maximum pyrolysis temperature (T_{max}). The possible cause may be that as the pyrolysis temperature increases, the shale maturity increases and the original pores in the rock are occupied by asphaltenes or generated oil and gas, increasing the difficulty of the gas diffusing into pores, as well as reducing the adsorbed gas content. Moreover, we also observed that the volume of adsorbed gas was negatively correlated with the production index (PI). The PI is defined as $S_1/(S_1 + S_2)$, where S_1 is adsorbed free liquid hydrocarbons and S_2 represents the residual petroleum potential. This indicates that the adsorbed and free hydrocarbon bears a competitive relationship in reservoir pores.

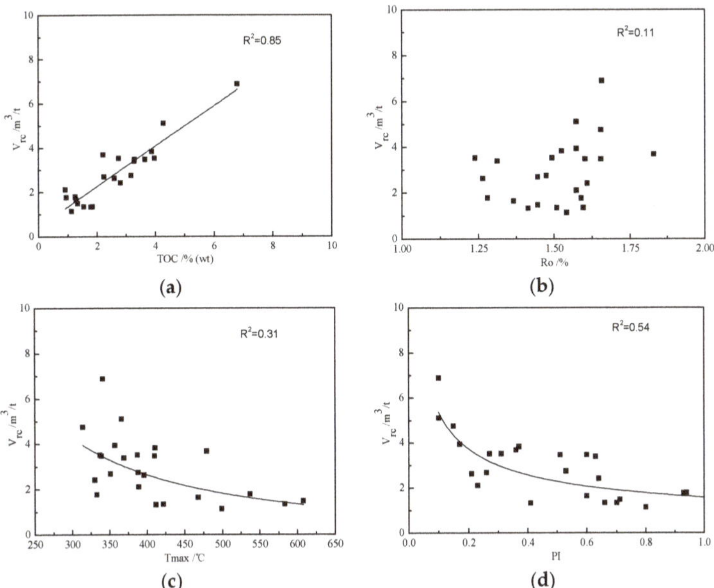

Figure 4. Relationships between the adsorbed content and geochemical parameters: (**a**) total organic matter content (TOC); (**b**) Ro; (**c**) T_{max}; (**d**) production index (PI).

3.4. Effects of Mineralogical Compositions

Figure 5 presents the influential factors of mineral compositions on adsorption capacity. It is shown that the quartz and the pyrite play a positive role on the adsorption, but the clay plays a negative role on the adsorption capacity. It is noted that no correlations between the adsorbed gas volume and the feldspar, calcite and dolomite contents were found. This may be because the quartz contains a large amount of biogenic silica, which has strong adsorption capacity. This agrees with the published results [15,29]. The target formations were deposited in the deep water shelf and contained a large number of siliceous organisms (diatoms, radiolarians, sponger, sponge bone needles, etc.). Meanwhile, we also observed a large number of fossils in the bedding of the rock samples. With the abundance of siliceous biological debris, a lot of micropores developed, which increases the specific surface of gas adsorption, leading to the positive role of the quartz. Moreover, quartz is a rigid mineral with strong compaction resistance, providing good preserving conditions in the pore space.

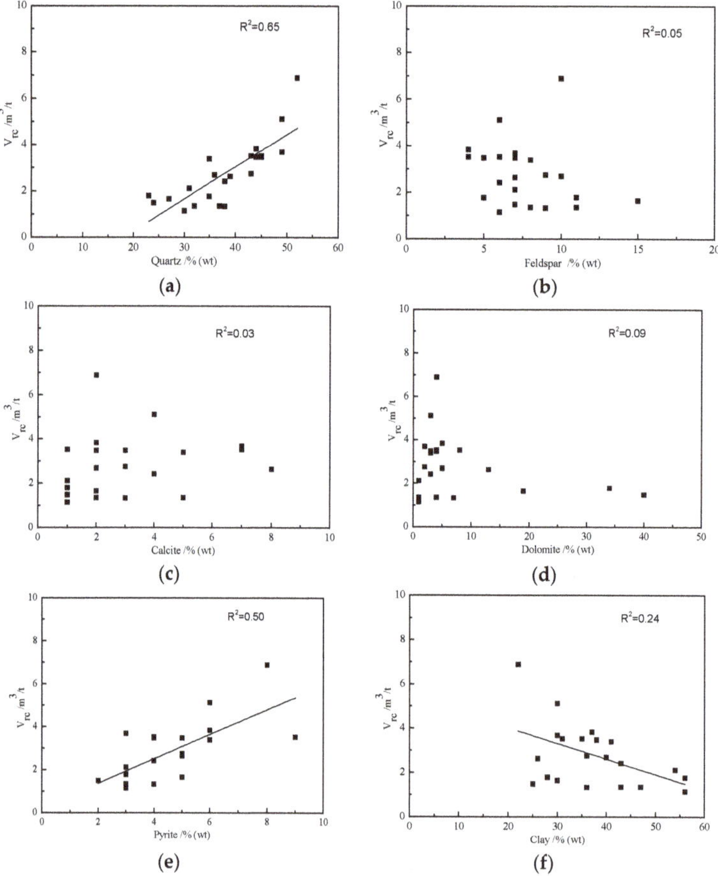

Figure 5. Relationships between the adsorbed gas content and the main compositions of mineral volumes: (**a**) Quartz volume; (**b**) feldspar volume; (**c**) calcite volume; (**d**) dolomite volume; (**e**) pyrite volume; (**f**) clay volume.

Pyrite is an indicating mineral for a strongly reducing environment, revealing the sedimentary environment is conducive to the preservation of keogen. The higher the pyrite content, the higher the degree of organic matter enrichment. Thus, its positive correlation with the adsorbed gas content was observed.

Other substances such as feldspar, calcite, and dolomite show weak relationships with the adsorbed gas content, revealing that they are not the main controlling factors of adsorption.

Additionally, an abnormal phenomenon was observed where the clay volume is inversely proportional to the adsorbed gas content. It can be interpreted that owing to the high maturity and volume of the kerogen, the contribution of the clay becomes less insignificant.

We used the XRD analysis to get the quantitative information of the clay composition. In the studied region, the clay mineral is dominated by illite, with an average proportion of 58.6%, followed by a mixed layer of illite and smectite, with an average proportion of 29.5%. Kaolinite and chlorite content was less, and no smectite was found. In order to further investigate the effect of different clay minerals on the adsorption capacity, we conducted univariate analysis on different types of clay, as shown in Figure 6. It can be seen that the illite, chlorite and mixed layers of illite and smectite positively correlated to the adsorbed gas content. The specific surface area of kaolinite is usually lower than 10 m^2/g, while smectite has a very high specific surface area of up to 900 m^2/g. The kaolinite content was too low to analyse, besides the electrification and hydrophilicity of clay minerals restrict their ability to accumulate the oil and gas. During the strong diagenesis stage, the organic acids produced by shale can dissolve the calcareous minerals and block the interlayer pores to some extent.

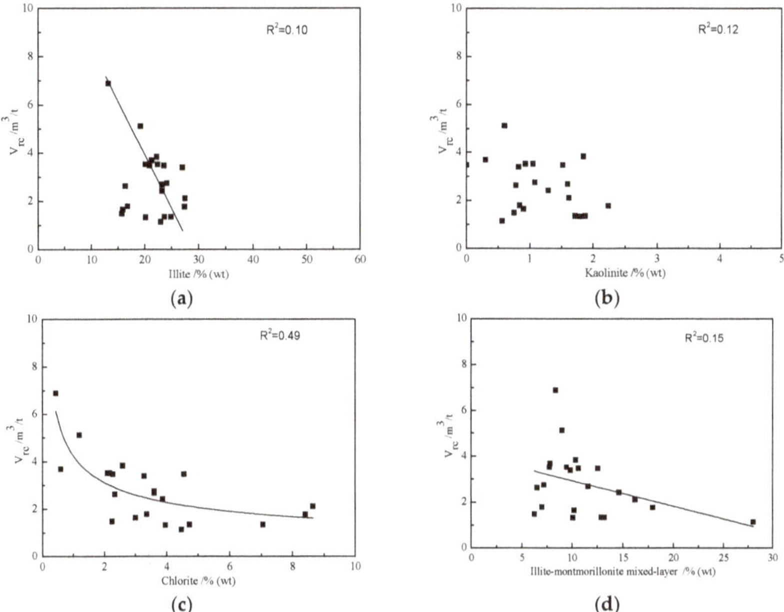

Figure 6. Relationships between the adsorbed gas content and the main compositions of clay: (**a**) Illite volume; (**b**) kaolinite volume; (**c**) chlorite volume; (**d**) illite-montmorillonite mixed-layer volume.

3.5. *Effects of Pore Size Distribution*

Figure 7 depicts the influential factors on pore volume at different pore size ranges and the pore specific surface area on the adsorption capacity. It can be seen the total porosity and the macroporosity have almost no contribution to the adsorbed gas content. Noticeably, the adsorbed gas content bears a favorable linear relationship with the number of micropores and the specific surface area, indicating that the adsorbed gas is mainly located in micropores. This relationship further supports the reasoning of the above T_2 cutoff value for the segmentation of pores. The adsorbed gas was mainly adsorbed in micropores, revealing the majority of gas adsorption was associated with the kerogen. In addition, the thermal maturation process provided favorable conditions for the development of micropores and the surface area, enlarging the adsorption space for the gas [30–32]. Moreover, much more extensive research is still required to better explain the adsorption and storage behavior of the gas in shale reservoirs.

Subsequently, we applied the criterion recommended by the International Union of Pure and Applied Chemistry (IUPAC) to classify the pore system into micropore, mesopore and macropore [33], and get their proportions by NMR. According to the surface relaxivity, this classification corresponds to the pores divided by $T_2 < 0.3$ ms, 0.3 ms < T_2 <7 ms, and $T_2 > 7$ ms. The relationship between the adsorption content and different pore proportions is shown in Figure 8. It is obvious that the content of adsorbed gas positively correlates to the microporosity, where the coefficient of correlation is lower than that which was computed using the cutoff of 2.5 ms, indicating that the criterion of IUPAC is not suitable for the studied samples.

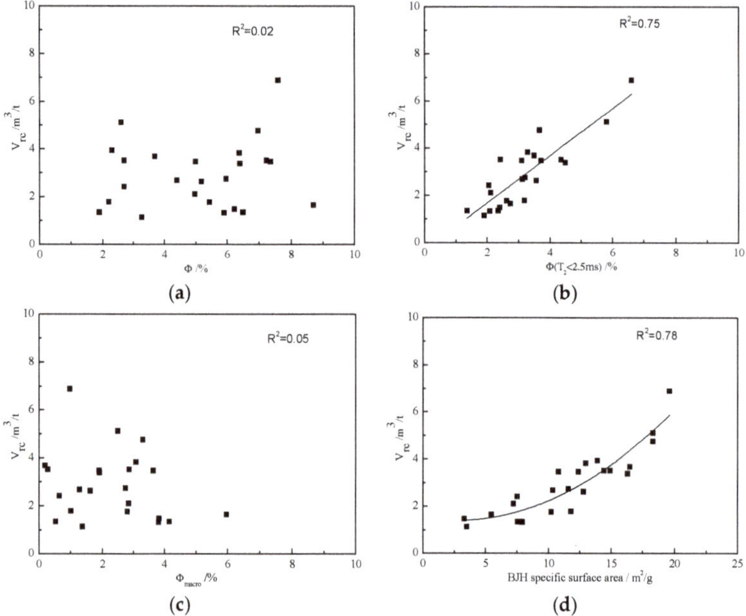

Figure 7. Relationships between the adsorbed gas content and the pore size parameters: (**a**) Total porosity; (**b**) microporosity; (**c**) cacroporosity; (**d**) BJH specific surface area.

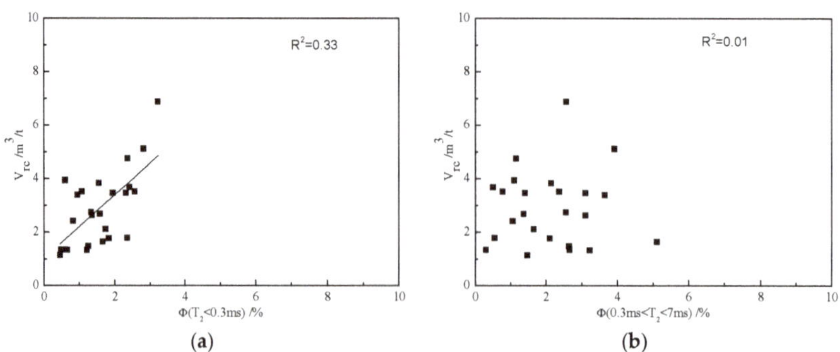

Figure 8. Relationships between the adsorbed gas content and the pore proportions by IUPAC: (**a**) Microporosity; (**b**) mesoporosity.

4. Conclusions

A series of experiments were carried out on a marine shale reservoir to investigate the factors controlling the gas adsorption capacity. Through analysis and discussion, the following conclusions were obtained: (1) The development of biogenic siliceous minerals results in an increase in the number of micropores in the rock, which in turn causes an increase in the specific surface area. This leads to an overall positive relationship between the adsorption gas content and the quartz content; (2) The adsorbed gas content is negatively correlated to clay contents the adsorption ability of clay is lower than the kerogen and the quartz; (3) The adsorbed gas is likely to store in tiny pores with smaller pore diameters; (4) The conventionally used pore classification criteria by the IUPAC may need further discussion, since the porosity of micropores, mesopores and macropores calculated by IUPAC standards cannot work well with some shale reservoirs.

However, the characteristics of petrology, source rock, and reservoir space of shale are not independent to each other. We only analyzed the influencing factors of the adsorbed gas content based on the experimental statistical relationship. Theoretical simulations were not conducted on the mechanism to study the intrinsic control factors of adsorbed gas content. In the future, we will carry out related work and enhance the reliability of the results.

Author Contributions: Conceptualization, Z.F. and X.G.; formal analysis, J.H.; investigation, Z.F. and J.H.; methodology, X.G. and P.Z., and J.L.; supervision, J.H.; validation, J.L.; Writing—original draft, X.G. and Z.F.; writing—review and editing, X.G. and Z.F.

Funding: This research was funded by the National Science and Technology Major Project of China [2017ZX05039], Fundamental Research Funds for the Central Universities [16CX05004A,18CX06025A], and National Key Foundation for Exploring Scientific Instrument of China [2013YQ170463].

Conflicts of Interest: The authors declare no conflicts of interest.

References

1. Ross, D.J.K.; Bustin, R.M. The importance of shale composition and pore structure upon gas storage potential of shale gas reservoirs. *Mar. Petrol. Geol.* **2009**, *26*, 916–927. [CrossRef]
2. Hu, H. Methane adsorption comparison of different thermal maturity kerogens in shale gas system. *Chin. J. Geochem.* **2014**, *33*, 425–430. [CrossRef]
3. Li, J.; Yan, X.; Wang, W.; Zhan, Y.; Yin, J.; Lu, S.; Chen, F.; Meng, Y.; Zhang, X.; Chen, X.; et al. Key factors controlling the gas adsorption capacity of shale: A study based on parallel experiments. *Appl. Geochem.* **2015**, *58*, 88–96. [CrossRef]
4. Montgomery, S.L.; Jarvie, D.M.; Bowker, K.A.; Pallastro, R.M. Mississippian Barnett Shale, Fort Worth basin, north-central Texas: Gas-shale play with multi-trillion cubic foot potential. *AAPG Bull.* **2005**, *89*, 155–175. [CrossRef]
5. Zhang, T.; Ellis, G.; Ruppel, S.; Milliken, K.; Yang, R. Effect of organic-matter type and thermal maturity on methane adsorption in shale-gas systems. *Org. Geochem.* **2012**, *47*, 120–131. [CrossRef]
6. Topóra, T.L.; Derkowski, A.; Ziemiański, P.; Szczurowski, J.; McCarty, D.K. The effect of organic matter maturation and porosity evolution on methane storage potential in the Baltic Basin (Poland) shale-gas reservoir. *Int. J. Coal Geol.* **2017**, *180*, 46–56. [CrossRef]
7. Wang, Y.; Zhu, Y.; Liu, S.; Zhang, R. Pore characterization and its impact on methane adsorption capacity for organic-rich marine shales. *Fuel* **2016**, *181*, 227–237. [CrossRef]
8. Aringhieri, R. Nanoporosity characteristics of some natural clay minerals and soils. *Clays Clay Miner.* **2004**, *52*, 700–704. [CrossRef]
9. Ji, W.; Song, Y.; Jiang, Z.; Wang, X.; Bai, X.; Xing, J. Geological controls and estimation algorithms of lacustrine shale gas adsorption capacity: A case study of the Triassic strata in the southeastern Ordos Basin, China. *Int. J. Coal Geol.* **2014**, *134*, 134–135. [CrossRef]
10. Zhong, J.; Chen, G.; Lv, C.; Yang, W.; Xu, Y.; Yang, S. Experimental study of the impact on methane adsorption capacity of continental shales with thermal evolution. *J. Nat. Gas Geol.* **2016**, *1*, 165–172. [CrossRef]
11. Gasparik, M.; Bertier, P.; Gensterblum, Y.; Ghanizadeh, A.; Krooss, B.M.; Littke, R. Geological controls on the methane storage capacity in organic-rich shales. *Int. J. Coal Geol.* **2014**, *123*, 34–51. [CrossRef]

12. Xiong, F.; Wang, X.; Amooiea, M.A.; Soltanian, M.R.; Jiang, Z.; Moortgat, J. RETRACTED: The shale gas sorption capacity of transitional shales in the Ordos Basin, NW China. *Fuel* **2017**, *208*, 236–246. [CrossRef]
13. Guo, S.; Lü, X.; Song, X.; Liu, Y. Methane adsorption characteristics and influence factors of Mesozoic shales in the Kuqa Depression, Tarim Basin, China. *J. Pet. Sci. Eng.* **2017**, *157*, 187–195. [CrossRef]
14. Zhou, S.; Xue, H.; Ning, Y.; Guo, W.; Zhang, Q. Experimental study of supercritical methane adsorption in Longmaxi shale: Insights into the density of adsorbed methane. *Fuel* **2018**, *211*, 140–148. [CrossRef]
15. Ji, W.; Song, Y.; Jiang, Z.; Chen, L.; Li, Z.; Yang, X.; Meng, M. Estimation of marine shale methane adsorption capacity based on experimental investigations of Lower Silurian Longmaxi formation in the Upper Yangtze Platform, south China. *Mar. Petrol. Geol.* **2015**, *68*, 94–106. [CrossRef]
16. Ma, X.; Song, Y.; Liu, S.; Jiang, L.; Hong, F. Experimental study on history of methane adsorption capacity of Carboniferous-Permian coal in Ordos Basin, China. *Fuel* **2016**, *184*, 10–17. [CrossRef]
17. Pang, Y.; Soliman, M.Y.; Deng, H.; Xie, X. Experimental and analytical investigation of adsorption effects on shale gas transport in organic nanopores. *Fuel* **2017**, *199*, 272–288. [CrossRef]
18. Han, H.; Zhong, N.; Ma, Y.; Huang, C.; Wang, Q.; Chen, S.; Lu, J. Gas storage and controlling factors in an over-mature marine shale: Acase study of the Lower Cambrian Lujiaping shale in the Dabashan arclike thrustefold belt, southwestern China. *J. Pet. Sci. Eng.* **2016**, *33*, 839–853. [CrossRef]
19. Xing, J.; Hu, S.; Jiang, Z.; Wang, X.; Wang, J.; Sun, L.; Bai, Y.; Chen, L. Classification of controlling factors and determination of a prediction model for shale gas adsorption capacity: A case study of Chang 7 shale in the Ordos Basin. *J. Pet. Sci. Eng.* **2018**, *49*, 260–274. [CrossRef]
20. Ge, X.; Liu, J.; Fan, Y.; Xing, D.; Deng, S.; Cai, J. Laboratory investigation into the formation and dissociation process of gas hydrate by low field NMR technique. *J. Geophys. Res.-Sol. Earth* **2018**, *123*, 3339–3346. [CrossRef]
21. Ge, X.; Fan, Y.; Cao, Y.; Li, J.; Cai, J.; Liu, J.; Wei, S. Investigation of organic related pores in unconventional reservoir and its quantitative evaluation. *Energy Fuel* **2016**, *30*, 4699–4709. [CrossRef]
22. Heller, R.; Zoback, M. Adsorption of methane and carbon dioxide on gas shale and pure mineral samples. *J. Uncon. Oil Gas Resour.* **2014**, *8*, 14–24. [CrossRef]
23. Zhou, S.; Wang, H.; Xue, H.; Guo, W.; Lu, B. Difference between excess and absolute adsorption capacity of shale and a new shale gas reserve calculation method. *Nat. Gas Ind.* **2016**, *36*, 12–20.
24. Gasparik, G.; Ghanizadeh, A.; Bertier, P.; Gensterblum, Y.; Bouw, S.; Krooss, B.M. High-pressure methane sorption isotherms of black shales from the Netherlands. *Energy Fuel* **2012**, *26*, 4995–5004. [CrossRef]
25. Rexer, T.F.T.; Benham, M.J.; Aplin, A.C.; Thomas, K.M. Methane adsorption on shale under simulated geological temperature and pressure conditions. *Energy Fuel* **2013**, *27*, 3099–3109. [CrossRef]
26. Singer, P.M.; Rylander, E.; Jiang, T.; McLin, R.; Lewis, R.E.; Sinclair, S.M. 1D and 2D NMR core-log intergration in organic shale. In Proceedings of the International Symposium of the Society of Core Analysts, Napa Valley, CA, USA, 16–19 September 2013.
27. Kuila, U.; Prasad, M. Specific surface area and pore-size distribution in clays and shales. *Geophys. Prospect.* **2013**, *61*, 341–362. [CrossRef]
28. Ge, X.; Fan, Y.; Cao, Y.; Xu, Y.; Liu, X.; Chen, Y. Reservoir pore structure classification technology of carbonate rock based on NMR T_2 spectrum decomposition. *Appl. Magn. Reson.* **2014**, *45*, 155–167. [CrossRef]
29. Jarvie, D.M.; Hill, R.J.; Ruble, T.E.; Pollastro, R.M. Unconventional shale-gas systems: The Mississippian Barnett Shale of north-central Texas as one model for thermogenic shale-gas assessment. *AAPG Bull.* **2007**, *91*, 475–499. [CrossRef]
30. Chalmers, G.R.; Bustin, R. Lower Cretaceous Gas Shales of Northeastern British Columbia: Geological Controls on Gas Capacity and Regional Evaluation of a Potential Resource. *Bull. Can. Pet. Geol.* **2008**, *56*, 1–21. [CrossRef]
31. Loucks, R.G.; Reed, R.M.; Ruppel, S.C.; Jarvie, D.M. Morphology, genesis, and distribution of nanometer-scale pores in siliceous mudstones of the Mississippian Barnett Shale. *J. Sediment. Res.* **2009**, *79*, 848–861. [CrossRef]

32. Milliken, K.L.; Rudnicki, M.; Awwiller, D.N.; Zhang, T. Organic matter-hosted pore system, Marcellus Formation (Devonian), Pennsylvania. *AAPG Bull.* **2013**, *97*, 177–200. [CrossRef]
33. Rouquerol, J.; Avnir, D.; Fairbridge, C.W.; Everett, D.H.; Haynes, J.M.; Pernicone, N.; Ramsay, J.D.F.; Sing, K.S.W.; Unger, K.K. Recommendations for the characterization of porous solids. *Pure Appl. Chem.* **1994**, *66*, 1739–1758. [CrossRef]

Article

Petrophysical Characterization and Fractal Analysis of Carbonate Reservoirs of the Eastern Margin of the Pre-Caspian Basin

Feng Sha [1,2,3], Lizhi Xiao [1,2,*], Zhiqiang Mao [1,2,*] and Chen Jia [4]

1 State Key Laboratory of Petroleum Resources and Prospecting, China University of Petroleum, Beijing 102249, China; shafeng.cup@gmail.com
2 Beijing Key Laboratory of Earth Prospecting and Information Technology, China University of Petroleum, Beijing 102249, China
3 CNPC Greatwall Drilling Company, Beijing 100101, China
4 Research Institute of Exploration and Development, Dagang Oilfield Company, PetroChina, Tianjin 300280, China; jiachen@petrochina.com.cn
* Correspondence: xiaolizhi@cup.edu.cn (L.X.); maozq@cup.edu.cn (Z.M.); Tel.: +86-10-8973-3305 (L.X.); +86-10-8973-3318 (Z.M.)

Received: 18 November 2018; Accepted: 24 December 2018; Published: 28 December 2018

Abstract: Petrophysical properties including pore structure and permeability are essential for successful evaluation and development of reservoirs. In this paper, we use casting thin section and mercury intrusion capillary pressure (MICP) data to investigate the pore structure characterization, permeability estimation, and fractal characteristics of Carboniferous carbonate reservoirs in the middle blocks of the eastern margin of the Pre-Caspian Basin. Rock casting thin sections show that intergranular and intragranular dissolution pores are the main storage spaces. The pore throats greater than 1 μm and lower than 0.1 μm account for 47.98% and 22.85% respectively. A permeability prediction model was proposed by incorporating the porosity, Swanson, and R_{35} parameters. The prediction result agrees well with the core sample data. Fractal dimensions based on MICP curves range from 2.29 to 2.77 with an average of 2.61. The maximum mercury intrusion saturation is weakly correlated with the fractal dimension, while the pore structure parameters such as displacement pressure and median radii have no correlation with fractal dimension, indicating that single fractal dimension could not capture the pore structure characteristics. Finally, combined with the pore types, MICP shape, and petrophysical parameters, the studied reservoirs were classified into four types. The productivity shows a good correlation with the reservoir types.

Keywords: carbonate reservoir; petrophysical characterization; pore types; pore structure; permeability; fractal dimension; reservoir classifications

1. Introduction

Carbonate reservoirs play an important role in the world's oil and gas distribution. Its oil and gas account for about 50% of the world's total oil and gas reserves and more than 60% of the world's total oil and gas production [1,2]. The reservoirs of many important oil and gas producing areas in the world are mainly carbonate rocks. The Caspian Basin located at the north of the Caspian Sea is one of the largest oil and gas-bearing basins [3,4]. The Carboniferous carbonate reservoirs of the eastern part of the Pre-Caspian Basin are favorable petroleum reservoirs [5]. Carbonate reservoirs are commonly characterized by high heterogeneity due to a variety of storage space combinations [6]. Petrophysical properties including micro pore structure, macro porosity, and permeability are essential for successful evaluation and development of reservoirs [7,8]. Therefore, it is necessary to study the petrophysical characterization of the Carboniferous carbonate reservoirs in the eastern margin of the Pre-Caspian Basin.

Previous studies on the Carboniferous carbonate reservoirs in the Pre-Caspian Basin including sequence stratigraphic and depositional setting [9,10], geochemical properties [11], oil and gas accumulation model [12,13], reservoir property including pore types [14] have been reported. However, study on the petrophysical properties has not been enough to date. He et al. [6] studied the relationship between porosity and permeability of this area and analyzed the influence factors. Miao et al. [15] reported the pore development characteristics and well logging responses of porosity, fracture, and vugs. He [16] investigated the storage space types and their evaluation and estimation using well logs. Macroscopic parameters such as porosity and permeability are usually derived from microscopic pore structure parameters.

Rock casting thin section, scanning electron microscope (SEM) [17,18], and transmission electron microscope (TEM) [19] can provide the pore types and qualitative pore space. Mercury intrusion capillary pressure (MICP) data is an important means to quantitatively study the pore structure characteristics of the reservoirs [20,21]. It can directly reflect the pore structure and performance of the reservoir, and capture the ranges of pore throat radius from 3.6 nm to a few microns in rocks [22]. Commonly used microscopic pore structure parameters include displacement pressure, median capillary pressure, irreducible water saturation, and maximum pore throat radius, etc. The MICP data are also used to estimate the permeability based on some key parameters, such as Swanson and R_{35} parameters [23–25]. Although low temperature gas adsorption curves including N_2 and CO_2 adsorption can provide smaller pore size distributions, they may be applicable in unconventional reservoirs [22]. The nuclear magnetic resonance (NMR) method is important in clastic rock and unconventional shale [26,27], but it does not always work for carbonate reservoirs as the relaxativity of carbonate minerals is too low to satisfy the theory of NMR [28].

In addition, fractal analysis conducted based on rock pore size distributions provided by MICP or gas adsorption could be used to assistant in studying pore structure of rocks [29,30]. Fractal geometry was proposed by Mandelbrot [31] to study porous media including rocks and other materials. The fractal dimension (D) is one of the key parameters in fractal geometry theory, describing the complexity and heterogeneity of pore space and particles [31–35]. Krohn [36] determined the fractal dimensions of pore–rock interface for Smackover Formation carbonates in Arkansas using SEM pictures, which range from 2.27 to 2.75. Billi [37] reported the fractal dimensions of particle size distributions in carbonate cataclastic rocks, which are from the core of a regional strike–slip fault zone in the foreland of the Southern Apennines, Italy, and are in the range of 2.09–2.93. Based on SEM images, Xie et al. [38] investigated the fractal characteristics of a Jurassic marine carbonate reservoir sample in western Hubei and eastern Sichuan region, China. The fractal dimension of pore size varies from 0.77 to 1.36. Liu et al. [39] used fractal characteristics to study the quantitative evaluation for pore structure in the carbonate reservoirs of Mishrif Formation of W oilfield in Iraq based on MICP data. It is of note that Ghanbarian-Alavijeh and Hunt [40] theoretically showed that fractal dimension can vary between minus infinity and 3. Thus, even negative D values are acceptable as reported by Ghanbarian and Sahimi [41].

In this study, taking the Carboniferous carbonate reservoirs in the middle blocks of the eastern margin of the Pre-Caspian Basin as an example, we investigated the pore structure characterization, permeability estimation, and fractal characteristics. According to the observations, we could define the reservoir types and studied their correlation with productivity. Rock casting thin section images were used to study the pore spaces and types. Mercury intrusion capillary pressure data was used to study the pore-throat size distribution, petrophysical property, and permeability estimation. Based on the box counting method, the fractal dimension of the samples was calculated. The carbonate reservoir classification was studied combining with the pore types, MICP shape, and petrophysical parameters. Thus, we could use the classification to predict the productivity.

In this paper, Section 2 includes geological setting, experimental methods and determination of fractal dimension using MICP. Section 3 presents the results and discussion of pore types and space, petrophysical characteristics, permeability estimation, fractal dimension, and reservoir classifications. Section 4 defines the main conclusions.

2. Materials and Methods

2.1. Geological Setting

The Pre-Caspian Basin, which is located in the north of the Caspian Sea, and underlies parts of Russia and Kazakhstan [12,14], is one of the world's largest oil and gas basins with an area of more than 500,000 km^2. It extends in the east–west direction, with a length of one thousand km and a maximum width of 650 km. The contour is approximately elliptical. It is a basin rich in oil and gas, but with a low degree of exploration. In the tectonic division, the Pre-Caspian Basin belongs to the southeastern part of the Eastern European platform. The northern and western parts of the basin are adjacent to the Paleozoic carbonate rock platform in the Volga–Ural Basin in the southern part of Eastern Europe. The northern and western parts of the basin are adjacent to the Hercynian fold belt (including the Southern Ural, Nanba, and Karakul etc.). The east is bordered by the Ural Haixi fold belt, the southwest is bordered by the Enba uplift and the southwest is bordered by the Karpinsky Haixi fold belt.

The Middle Block of the eastern margin of the basin is located in the Aktobe state. It is also located in the transitional zone between the Astrakhan–Akchubin central uplift and the Primm Gordgar ancient depression, which is the uplift of the Primm Gordgar Late Paleozoic. The Middle block is an important pointing zone for oil and gas migration, and the regional structural position is very favorable. Figure 1 shows the location of the study area.

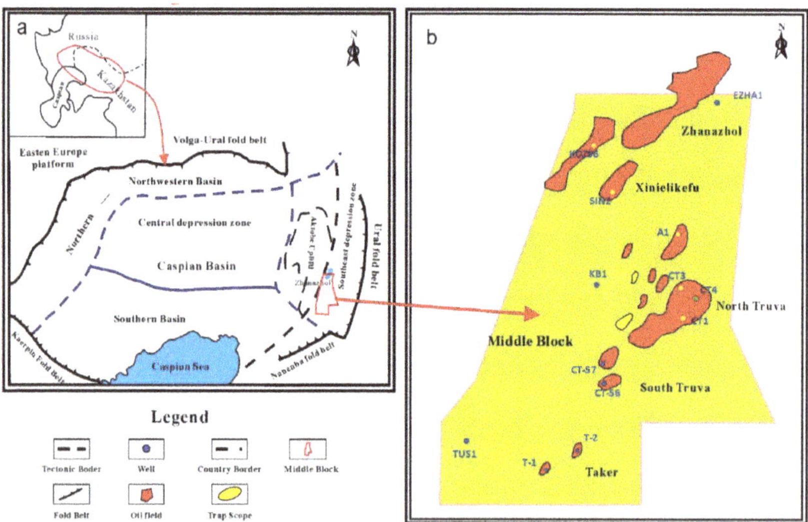

Figure 1. The location of the study area.

According to the drilling data of the block and surrounding oil and gas fields, the basin can be basically divided into the combination of carbonate and clastic rocks in the Lower Devonian-Carboniferous and Lower Permian, gypsum salt rock of Konggu Formation in the Lower Permian, and the Upper Permian-Triassic clastic rock deposit and the Jurassic-Cretaceous combination. The Mesozoic and the Upper Permian strata are clastic and dominated by sand and mudstone, and the Lower Permian pore-valley terrace is a set of salt rock strata, which was high-speed deposited. The distribution and thickness of salt rocks vary greatly. The Carboniferous in the Middle Block can be divided from top to bottom into carbonate and gypsum salt beds of KT-1 Formation, clastic rock of MKT Formation, carbonate rocks and a few mudstone of KT-2 Formation, as well as multi-Neixiqian sandstone and mudstone beds in the middle and lower part of the Uyxian Stage. The detailed information is depicted in Figure 2.

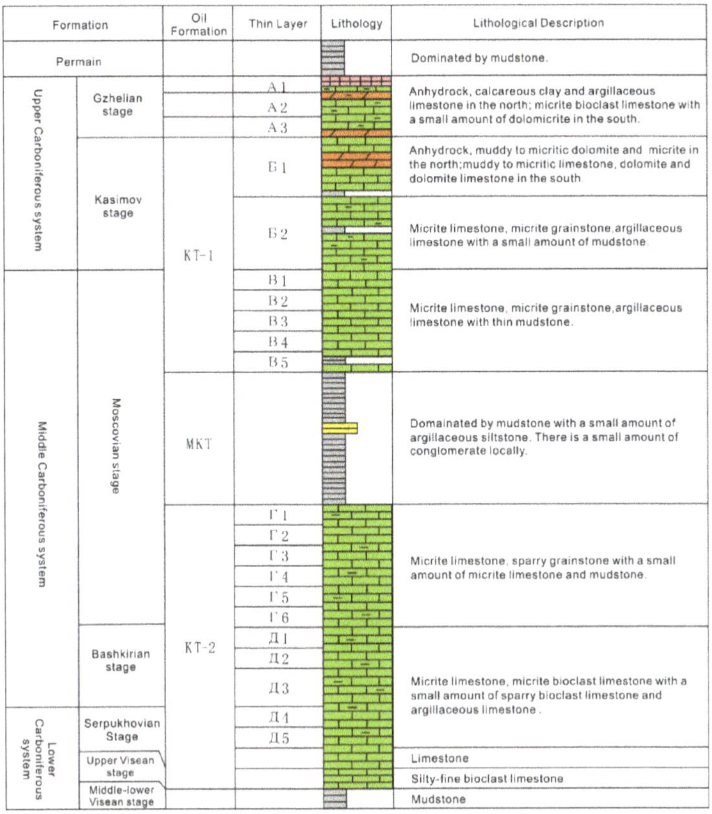

Formation		Oil Formation	Thin Layer	Lithology	Lithological Description
Permain					Dominated by mudstone.
Upper Carboniferous system	Gzhelian stage		А1, А2, А3		Anhydrock, calcareous clay and argillaceous limestone in the north; micrite bioclast limestone with a small amount of dolomicrite in the south.
	Kasimov stage	KT-1	Б1		Anhydrock, muddy to micritic dolomite and micrite in the north;muddy to micritic limestone, dolomite and dolomite limestone in the south.
			Б2		Micrite limestone, micrite grainstone,argillaceous limestone with a small amount of mudstone.
Middle Carboniferous system	Moscovian stage		В1, В2, В3, В4, В5		Micrite limestone, micrite grainstone,argillaceous limestone with thin mudstone.
		MKT			Domainated by mudstone with a small amount of argillaceous siltstone. There is a small amount of conglomerate locally.
		KT-2	Г1, Г2, Г3, Г4, Г5, Г6		Micrite limestone, sparry grainstone with a small amount of micrite limestone and mudstone.
	Bashkirian stage		Д1, Д2, Д3		Micrite limestone, micrite bioclast limestone with a small amount of sparry bioclast limestone and argillaceous limestone.
Lower Carboniferous system	Serpukhovian Stage		Д4, Д5		
	Upper Visean stage				Limestone
					Silty-fine bioclast limestone
	Middle-lower Visean stage				Mudstone

Figure 2. Characteristics of Carboniferous strata the study area.

2.2. Experiment Methods

Thirty two plunger samples were carried out for porosity, permeability, and MICP measurements. The plunger has dimensions of diameter of 2.5 cm and length of 4 cm. The porosity and permeability are measured with a helium porosimeter. Before measurements, plugs were subjected to oil and salt washing and drying. After porosity and permeability measurements, the plunger samples were subjected to drying at 100 °C until the weight remained constant. Then, MICP data were determined with a mercury porosimeter. The minimum and the maximum intrusion pressure were denoted as 0.0035 MPa and 200 MPa, respectively. The 200 MPa of intrusion pressure guarantees the mercury can enter a small pore-throat, whose radius is low at roughly 3.7 nm. In addition, many samples for rock casting thin section analysis were drilled from four wells. These samples almost cover all the depths of KT-I and KT-II formation, which make this study more accurate.

2.3. Fractal Dimension

According to fractal geometry theory, if the pore space of a rock obeys the fractal structure, the pore radius r and the number of pores with a radius larger than r would follow a power-law function [31]:

$$N(>r) = \int_{r}^{r_{\max}} P(r)dr = \alpha r^{-D} \tag{1}$$

where r and $N(>r)$ are pore radius and the number of pores with radius larger than r and $r_{\max}$ is the maximum pore radius, $P(r)$ is the distribution density function of the pore radius, α is a proportionality constant, D is the fractal dimension.

Based on some assumptions and transformation, the following equation was derived [31,42]:

$$S_v(<r) = (\frac{r}{r_{\max}})^{3-D} \tag{2}$$

where S_v is the cumulative volume fraction of pores with a radius smaller than r.

According to Washburn [20], mercury injection pressure and pore throat radius obey the following relationship:

$$P_c = \frac{2\sigma\cos\theta}{r} \tag{3}$$

where Pc is the capillary pressure, σ is the surface tension, and θ is the contact angle of mercury in air.

Combining Equations (2) and (3) and the basic principle of MICP, the follow equation is obtained:

$$1 - S_{Hg} = (\frac{P_{c\min}}{P_c})^{3-D} \tag{4}$$

where S_{Hg} is mercury saturation; $P_{c\min}$ is the minimum of the capillary pressure.

By taking the logarithm on both sides on the above equation, the following relationship was obtained [43]:

$$\log(1 - S_{Hg}) = (D-3)\log(P_c) + (3-D)\log(P_{c\min}) \tag{5}$$

For each sample, there is a series of (P_c, S_{Hg}) values. Thus, the fractal dimension can be determined by using MICP data.

3. Results and Discussion

3.1. Pore Spaces and Types

Similar to the carbonate reservoir in other study areas [44], the pore space of carbonate rocks in this study area is divided into three types: pores, fractures, and caves.

3.1.1. Pores

(1) Intergranular pores or intergranular dissolution pores

The Carboniferous carbonate granular rocks in the study area all have cements. Some of the rocks have intergranular residual pores due to insufficient cementation while some of the rocks have intergranular dissolution pores formed by later-stage dissolution of mud-crystal or columnar bright-crystal cement between the particles. The two types of pores mentioned above could be called intergranular pore. This is the main pore type in the Carboniferous reservoir, of which the visible porosity is between 0.2% and 18%. The intergranular dissolution pores are mainly distributed in the A3, Г layer, and the Д layer, which indicate good pore connectivity, and strong storage capacity of oil and gas (Figure 3). In the Г layer, granular limestone particles are coarse, the pore size is large in scale, and the connectivity is good. While particles in the Д layer are relatively fine, the size of the intergranular pores is relatively smaller, and the connectivity is relatively poorer.

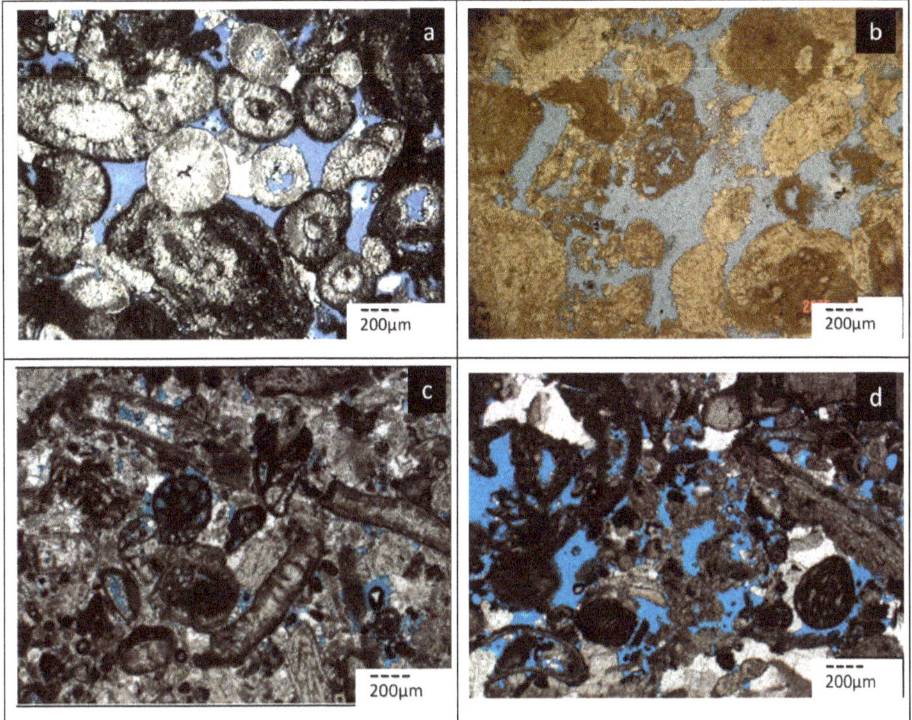

Figure 3. Characteristics of intergranular pores of Γ layer in the study area (**a**) Well CT-10, 3156.1 m, Sparry algal oolitic limeston, intergranular dissolution pores, visible porosity is 2.57%; (**b**) Well A-1, 3422.0 m, Sparry red algae foraminifera granules limestone, intergranular dissolution pores more, visible porosity is 17.2%;(**c**) Well CT-1, 3131.0 m, Sparry green algae foraminifera limestone, intergranular dissolution pores and intrafosill pores, visible porosity is 8%; (**d**)Well CT-4, 3131.0 m, Sparry foraminifera parasolitic limestone, intergranular pores, visible porosity is 15%.

(2) Intragranular dissolution pores

The intragranular dissolution pores are the pores that are formed by the later dissolution within the particles, such as ooids, biological debris, and sands (Figure 4). Pores in which the particles or grains are completely dissoluted but still retain the original particle or grain shape are called moldic pore. The pores formed in the body cavity of the biological granular which due to decay or erosion of the body are called intrafosill pores. In addition, there are a small number of intraskeletal pores. All of these pores are referred to as intragranular pores, which are also important pore types of the Carboniferous reservoirs in the North Truva.

The intrafosill, intragranular dissolution, and moldic pores are more developed in the carboniferous system of the North Truva structure. The visible porosity of the thinsection in which the intrafossil pores developed is between 0.1% and 15%, with an average of about 2%. The visible porosity of the thin section in which the intra-granular dissolution pores developed is between 0.2% and 15%, and with an average of about 2%. The visible porosity of the thin section in which moldic pores developed ranges from 0.1% to 35%, and the average value is about 5%. The larger the visible porosity is, the better the reservoir property.

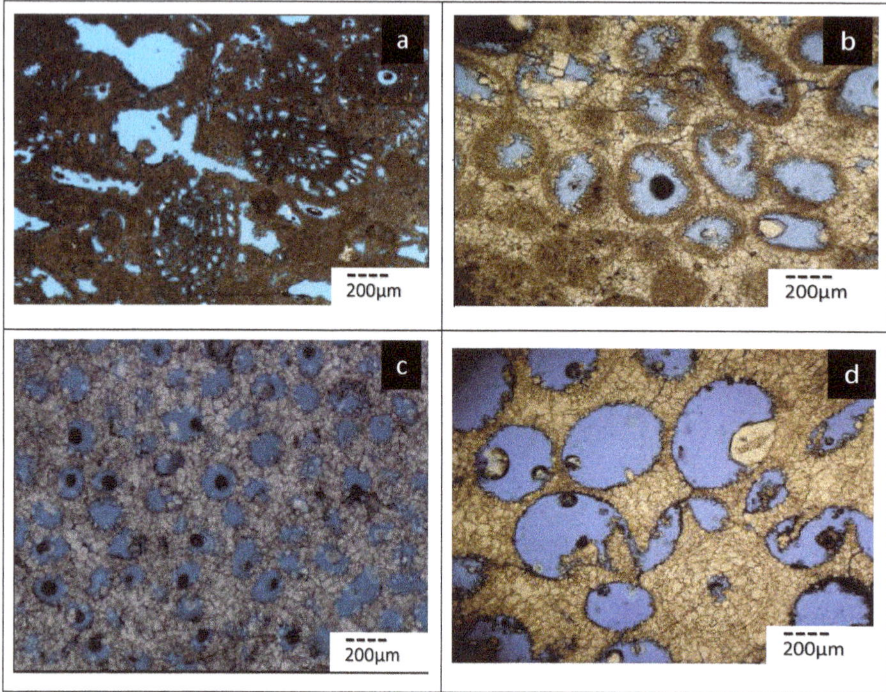

Figure 4. Intragranular dissolution pores (**a**) Well CT-4, 2341.41 m, Micrite bioclastic dolomite, more intrafosill pores and its dissolution, visible porosity is 18.3%; (**b**) Well A-1, 2846.0 m, Sparry oolitic limestone, oolites and negative oolites, ooids modic pores, visible porosity is 17.0%; (**c**) Well A-2, 3190.0 m, Sparry cast oolitic limestone, ooids modic pores, visible porosity is 23.7%; (**d**) Well A-1, 3621.0 m, Sparry oolitic limestone, intergranular pores, visible porosity is 35.0%.

(3) Intercrystalline and intercrystalline dissolution pores

The pores existing among the euhedral dolomite, subhedral dolomite or calcite grains are intercrystal (Figure 5). The void areas formed by the dissolution of the soluble components such as residual calcite or gypsum between the dolomite crystals are intercrystal dissolution pores. The visible porosity of the thin section, the intercrystal pore, is between 2% and 15%. They are mainly concentrated in the dolomite of the Carboniferous B1 layer. They can be also observed in the Д layer.

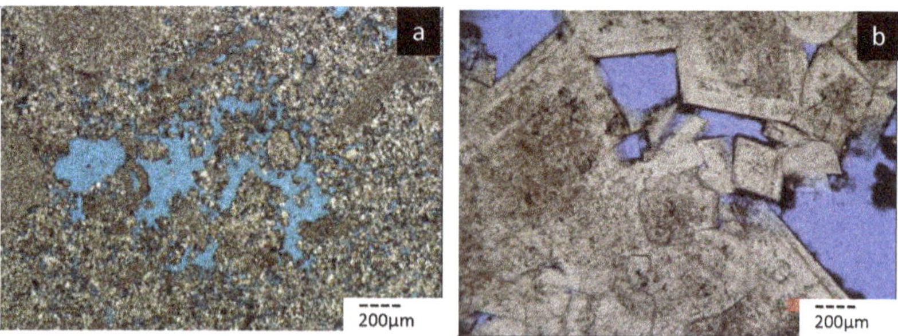

Figure 5. Intercrystal pore and intercrystal dissolution pore (**a**) Well CT-4, 2347.0 m, Residual clastic silt dolomite, intercrystal dissolution pores, visible porosity is 13.0%; (**b**) Well A-1, 3956.0 m, Sugar-like dolomite, intercrystal pores, visible porosity is 5.0%.

3.1.2. Fractures

Fractures are important percolating channels for the reservoirs. According to the core observation and the casting thin section, the North Truva Carboniferous cracks mainly include four types: dissolution fracture, tectonic fracture, stylolite fracture, and grain cracks (Figure 6). These cracks not only have a certain impact on the reservoir storage, but also have a significant effect on connecting pores and improving reservoir permeability. They are also conducive to the development of dissolution holes, thus forming a unified pore, hole, and fracture system and further improving the reservoir permeability of the reservoir. Fractures in the upper part of the KT-1 layer develop more than other locations.

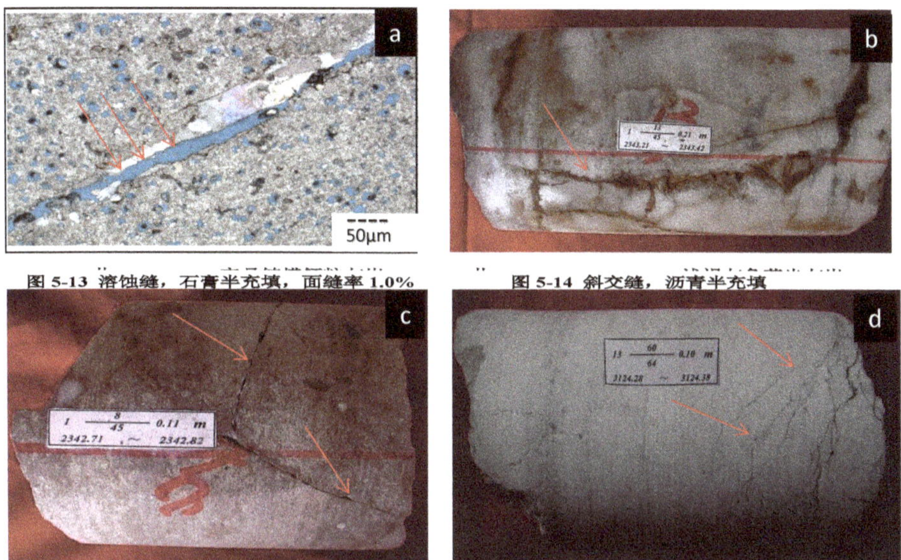

Figure 6. Fracture characteristics in the study area (**a**) Well A-1 2890.13 m, Spray oolitic limestone, dissolution fracture, semi-filled by gypsum; (**b**) Well CT-4 2343.21~2343.42 m, Vertical extension fracture; (**c**) Well CT-10 2342.71~2342.82 m, Oblique tectonic fracture; (**d**) Well CT-10 3142.28–3142.38 m, Stylolite.

3.1.3. Dissolution Cavern

Dissolution pores with diameter larger than 2 mm are called caverns. The pores with diameters between 2 and 5 mm are called small caverns, while those with diameters of 5 to 10 mm are called middle size caverns, while those with diameters larger than 10 mm are called large caverns.

The core of the CT-4 well was found to have 1742 caves with an area of 272,583 mm^2. They are mainly distributed in the dolomite section of the A_3 and B_1 layer (Figure 7). The cumulative number of caverns in this layer is 1712, indicating that the dissolution of this section is very developed.

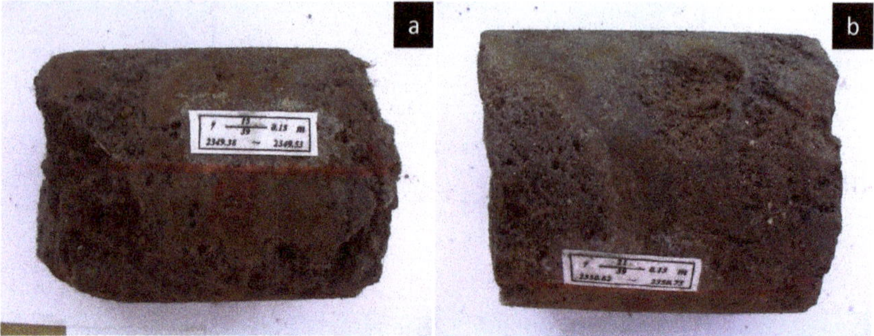

Figure 7. Cavern characteristics of the study area (**a**) Well CT-4, 2343.19~2343.29 m, Cinder-like dolomite (**b**) Well CT-4, 2344.77~2345.14 m, dissolution caverns in the dolomite rock.

3.2. Petrophysical Characteristics and MICP Data

Porosity, permeability, and related parameters derived from MICP curves of 32 samples are listed in Table 1. Porosity ranges from 4.67% to 32.4% with an average value of 13.71%. Permeability ranges from 0.002 mD to 349 mD. Among them, the permeabilities of 10 samples are lower than 1 mD. The geometric mean value of the permeability is 4.07 mD.

The MICP curves are shown in Figure 8. The red and blue curves in the lower position of this figure represent the samples with relatively good pore structure as the displacement pressure and saturation median pressure are smaller. The middle parts of these curves are concave. In contrast, the black and green curves in the upper position of this figure have much bigger displacement pressure and saturation median pressure. The middle parts of the black and green curves are straight instead of concave, demonstrating a relatively poor structure.

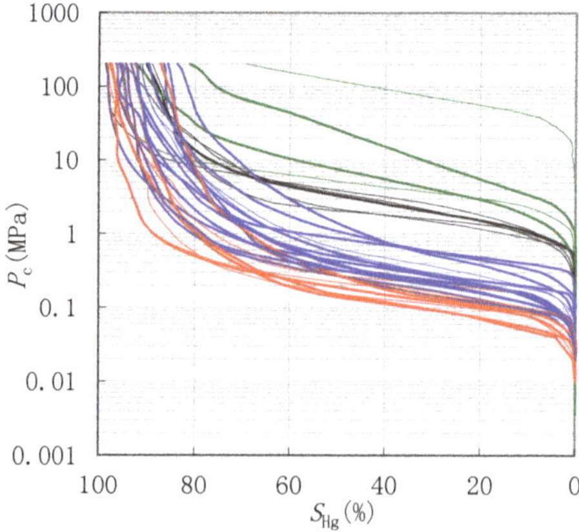

Figure 8. Mercury intrusion capillary pressure (MICP) curves of the studied samples.

The pressure at which mercury first enters the sample (after the mercury has filled any surface irregularities on the sample) is termed the displacement pressure (P_d) [2,45]. It is commonly inferred from the injection pressure at 10% saturation [46]. The P_d can be calculated for the largest pore throat radius. According to Equation (1), the smaller the P_d value, the bigger the largest pore throat radius. The P_d values of the studied samples range from 0.05 MPa to 41.39 MPa with an average of 1.75 Mpa. Saturation median pressure (P_{c50}) refers to the intrusion pressure when the non-wet phase saturation is 50% [2,45]. It varies in the range of 0.14 to 120.64 MPa with an average of 5.81 MPa. Median radii (R_{c50}) are between 0.01 μm and 5.15 μm and with an average of 1.76 μm. The maximum mercury intrusion saturation (S_{max}) of the samples ranged from 69.32% to 98.76%, with an average of 92.11%. The maximum intrusion pressure is 200 Mpa, corresponding to 3.7 nm of pore throat radius. This indicates that 92.11% of the pore radius is greater than 3.7 nm.

According to Equation (2), we calculated the pore size distribution for each sample. We defined the pore with a pore-throat radius greater than 1 μm as large pore, the pore with a pore-throat radius in the 0.1–1 μm as medium pore, and the pore with a pore-radius less than 0.1 μm as small pore. As can be seen in Table 1, the three types of pores account for 47.98%, 29.17%, and 22.85%, respectively.

Table 1. Petrophysical parameters and fractal dimensions of the studied samples.

Sample No.	Porosity	K	P_d	P_{c50}	R_{50}	Swanson	R_{35}	S_{max}	D	Large Pore	Medium Pore	Small Pore
-	%	mD	MPa	MPa	μm	v/v/MPa	μm	%	/	%	%	%
1	4.67	1.63	0.96	3.44	0.22	0.16	0.19	90.16	2.62	4.22	66.97	28.82
2	9.53	0.005	3.16	26.91	0.03	0.04	0.06	81.16	2.65	0.14	25.67	74.19
3	7.56	0.019	0.86	3.67	0.2	0.15	0.31	91.01	2.58	6.52	62.49	31.0
4	10.35	0.02	1.63	7.71	0.1	0.07	0.16	92.26	2.56	2.33	46.8	50.87
5	10.29	0.034	2.07	5.18	0.14	0.10	0.19	95.14	2.37	0.78	65.76	33.46
6	11.08	0.333	0.86	2.07	0.36	0.24	0.44	88.98	2.64	5.13	72.15	22.72
7	8.55	0.088	1.09	3.92	0.19	0.12	0.26	98.76	2.29	4.83	69.79	25.38
8	7.76	0.322	0.74	2.24	0.33	0.22	0.45	94.66	2.54	6.35	63.76	29.89
9	7.16	0.002	41.39	120.64	0.01	0.00	0.01	69.32	2.68	0.00	0.45	99.55
10	29.0	35.3	0.19	0.44	1.67	1.17	2.41	98.43	2.42	67.2	25.19	7.61
11	30.7	62.3	0.13	0.28	2.62	1.83	3.3	98.45	2.49	74.8	15.63	9.57
12	27.4	32	0.38	0.89	0.83	0.55	1.15	96.45	2.45	41.45	43.88	14.67
13	32.4	230	0.08	0.19	3.82	2.67	5.43	97.79	2.55	85.27	9.88	4.85
14	13.7	12.3	0.09	0.27	2.69	1.77	3.85	95.84	2.61	76.38	14.6.	9.02
15	12.4	8.08	0.18	0.58	1.27	1.02	2.12	91.55	2.68	55.21	24.87	19.92
16	17.3	349	0.05	0.15	4.87	3.28	6.96	97.44	2.58	76.6	14.16	9.24
17	7.8	19.7	0.09	0.28	2.65	2.07	4.33	85.74	2.77	66.46	13.98	19.56
18	16.7	15.3	0.07	0.28	2.64	2.04	4.27	96.57	2.6	68.24	19.54	12.22
19	8.5	7.54	0.15	0.4	1.85	1.41	2.82	86.64	2.74	62.4	17.47	20.13
20	9.1	24.4	0.11	0.41	1.80	1.40	3	92.31	2.67	61.52	21.08	17.40
21	11.3	77.9	0.05	0.14	5.15	3.45	7.14	92.65	2.70	77.26	10.35	12.39
22	10.2	18.4	0.14	0.56	1.32	1.14	2.39	93.42	2.65	55.69	25.30	19.00
23	7.0	0.25	0.09	0.29	2.57	2.30	4.79	87.26	2.76	65.2	14.57	20.24
24	11.4	33.5	0.19	0.64	1.15	0.79	1.76	91.57	2.68	52.63	26.75	20.63
25	13.3	3.86	0.14	0.32	2.26	1.55	2.99	94.52	2.63	74.57	14.86	10.57
26	12.2	120	0.11	0.33	2.21	1.53	3.23	95.35	2.61	67.64	19.42	12.95
27	20.5	69.5	0.1	0.27	2.73	1.8	3.94	95.47	2.61	72.14	16.41	11.45
28	12.4	9.98	0.13	0.5	1.46	1.08	2.26	92.35	2.66	60.73	24.11	15.16
29	8.8	0.931	0.23	1.41	0.52	0.56	1.18	83.76	2.75	38.33	30.32	31.35
30	9.9	13.4	0.33	1.11	0.66	0.54	1.12	89.75	2.66	38.26	38.34	23.4
31	18.7	43.7	0.09	0.23	3.25	2.11	4.38	98.26	2.5	84.63	10.52	4.85
32	21.0	232	0.07	0.16	4.71	3.22	6.1	94.36	2.67	82.34	8.40	9.27
Average	13.71	44.43	1.75	5.81	1.76	1.26	2.59	92.11	2.61	47.98	29.17	22.85

P_d is inferred from the injection pressure at 10% saturation. Swanson is the maximum of the ratio of mercury saturation to corresponding pressure. R_{35} is the calculated pore throat radius corresponding to a mercury saturation of 35%. D is fractal dimension determined using mercury intrusion capillary pressure (MICP).

3.3. Estimation of Permeability

In order to predict permeability, we calculated the R_{35} and Swanson parameters (see Table 1) as they are commonly used in the permeability prediction. The R_{35} which is the calculated pore throat radius corresponding to a mercury saturation of 35% contributes greatly to the rock permeability [23,47]. An empirical relationship between porosity, permeability, and R_{35} published by Kolodzie [23], known as the Winland model, is expressed as:

$$\log(R_{35}) = 0.732 + 0.588\log(K) - 0.864\log(\phi) \quad (6)$$

The Swanson parameter defined by Swanson [24] is the maximum of the ratio of mercury saturation to the corresponding pressure, denoted as $(\frac{S_{hg}}{P_c})_{max}$. The plot of $(\frac{S_{hg}}{P_c})_{max}$ and S_{hg} resembles a downward opening parabola. The inflection point of the curve is known as the Swanson parameter. Before the inflection point occurs, the non-wetting phase occupies an effective interconnected pore space; after the inflection point, the non-wetting phase begins to enter a finer pore space or irregular pores, the flowability of the non-wetting phase is significantly reduced. The Swanson parameter is also correlated to the permeability of sandstone and carbonate samples. The Swanson permeability model is [24]

$$K = a \times \left(\frac{S_{hg}}{P_c}\right)_{max}^{b} \tag{7}$$

where a and b are regression coefficients, varying with study area and lithology.

The relations between the permeability and porosity, Swanson, R_{35} are shown in Figure 9. The coefficients of determination are lower than 0.8, which is not high enough to predict permeability.

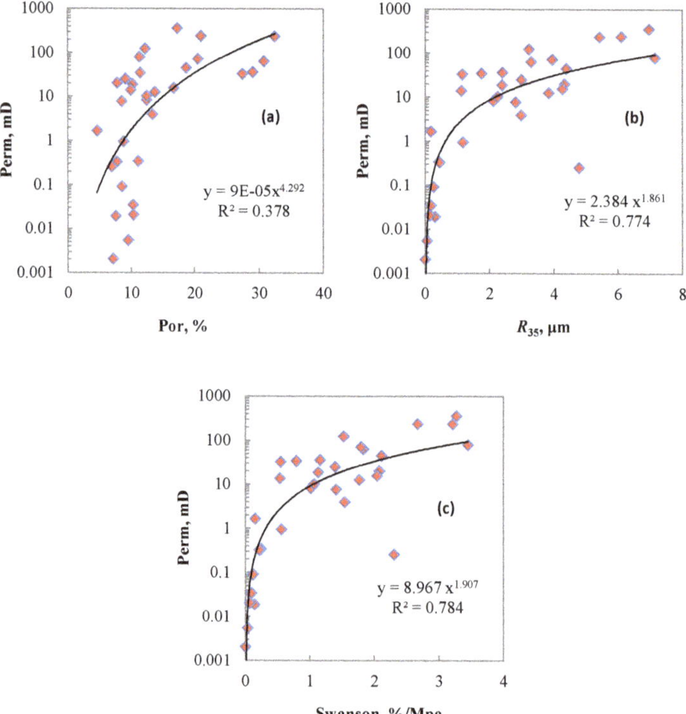

Figure 9. The relationships of permeability with parameters. (**a**) Porosity; (**b**) R_{35}; (**c**) Swanson. The Swanson parameter is the maximum of the ratio of mercury saturation to the corresponding pressure.

In this study, combined the porosity, R_{35} and Swanson parameters, we established a new model to accurately predict permeability as:

$$K = 0.704\phi^{1.760}\left(\frac{S_{hg}}{P_c}\right)_{max}^{4.463} R_{35}^{-2.779} \tag{8}$$

The cross plot of predicted and measured permeability are shown in Figure 10. As seen from this figure, the coefficient of determination is improved to 0.834 and the data dots are distributed near the 100% agreement lines.

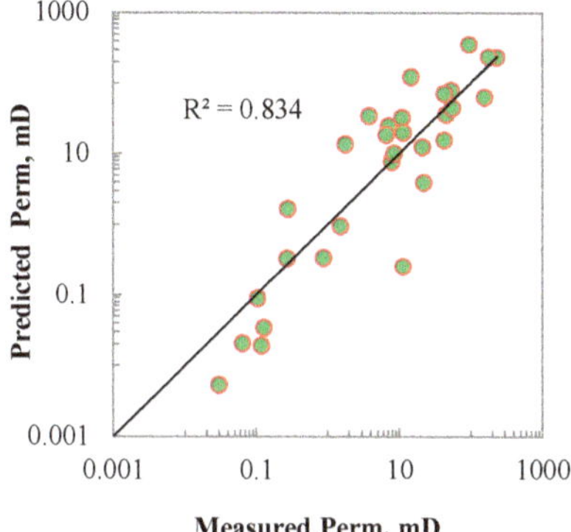

Figure 10. The comparison of predicted and measured permeability.

3.4. Fractal Dimension

Figure 11 is an example of the determination of the fractal dimension of sample 12 by using the above method. As can be observed in this figure, the slope of the regression equation is −0.51, thus, D is determined as 2.49. In addition, the coefficient of determination is high at 0.99, indicating the fractal nature of the pore space of sample 12.

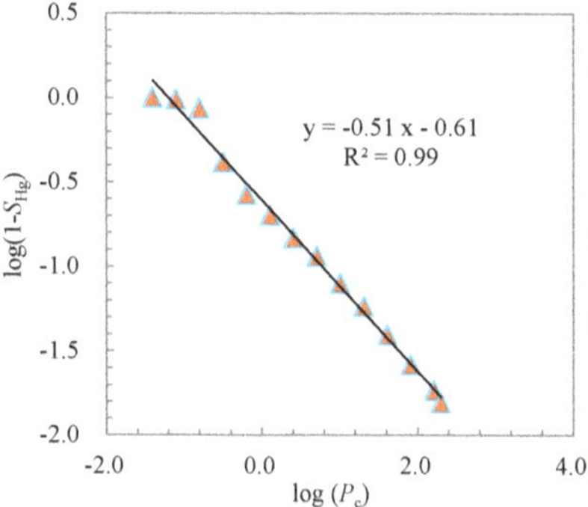

Figure 11. Determination of the fractal dimension for sample 12.

The fractal dimensions of the studied samples are listed in Table 1. D varies from 2.29 to 2.77, with an average of 2.61. Sample 17 has the largest fractal dimension, while sample 7 has the smallest value. However, the permeability of sample 17 is higher than that of sample 7. This is not consistent with previous knowledge that the fractal dimension is smaller if the pore structure is good [29,43]. In fact, maximum mercury intrusion saturation $S_{\max}$ is weakly correlated with the fractal dimension, the coefficient of determination is 0.359 (Figure 12a). Also, Figure 12b shows the logarithmic relationship of fractal dimension and wetting saturation, i.e., 1-$S_{\max}$. This is consistent with the observation by Ghanbarian-Alavijeh and Millan [48]. The other petrophysical parameters are not correlated with fractal dimension, as is seen in Table 2. This may be attributed to the fact that the single fractal dimension could not capture pore structure characteristics. We will conduct multifractal analysis for the pore structure to further investigate the carbonate reservoir property in the future.

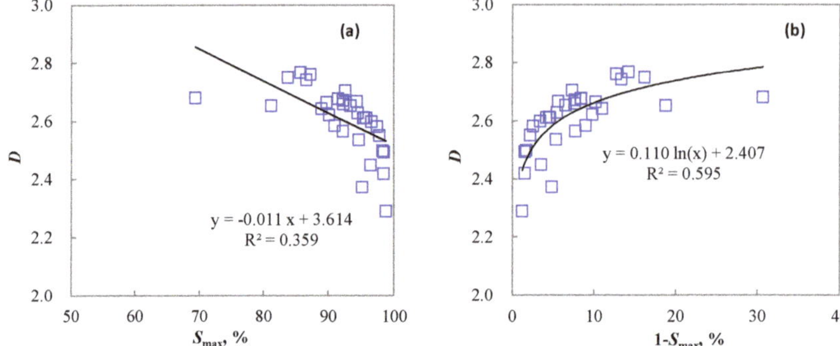

Figure 12. (**a**) Cross plot of fractal dimension with maximum mercury intrusion saturation; (**b**) Cross plot of fractal dimension with wetting saturation.

Table 2. The coefficients of determination between fractal dimension and petrophysical parameters.

Por	K	P_d	P_{c50}	R_{50}	Swanson	R_{35}	Large Pore	Medium Pore	Small Pore
0.174	0.001	0.009	0.011	0.035	0.061	0.064	0.053	0.207	0.015

3.5. Reservoir Classifications

According to the above pore space types, porosity, permeability, and MICP curves, we divided the samples into four types: Types I, II, III, and IV. We did not take the fractal characteristics into consideration as they could not effectively capture the reservoir property. In Figure 8, the red, blue, black, and green curves represent Types I, II, III, and IV, respectively. The typical MICP curves and pore throat distribution for each type are shown in Figure 13. The pore space types and related petrophysical parameters for each type are listed in Table 3. The type I reservoir has the largest porosity, permeability, median radius, and smallest displacement pressure. It has the best pore structure characteristic, while Type IV reservoir holds the worst pore structure characteristic. However, the average porosity of Type IV is larger than that of Type III. This may be attributed to the sample number of the two types being less than that of Types I and II.

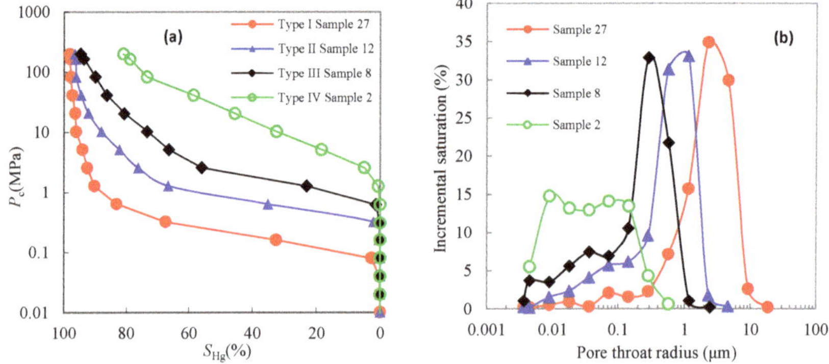

Figure 13. The MICP curves and pore throat radius distributions of different types of samples (**a**) MICP curves, (**b**) pore throat radius distributions.

Table 3. The pore space types and related petrophysical parameters for each type of reservoir.

Reservoir Types	Pore Space Types	Porosity (%)	K (mD)	R_{50} (μm)	P_d (MPa)
I	Dissolution caves; intergranular dissolution pore; fracture	7.0–32.4 19.51	0.25–349.0 42.668	2.57–5.15 3.59	0.05–0.13 0.08
II	Intergranular dissolution pore; intergranular pore	7.8–29 13.29	0.931–120.0 14.266	0.52–2.69 1.6	0.09–0.38 0.18
III	Intergranular pore; intragranular dissolution pores; intrafosill pore	4.67–11.8 7.92	0.02–1.63 0.196	0.19–0.36 0.26	0.74–0.19 0.9
IV	Intercrystal pore; or undeveloped pores	7.16–10.35 9.33	0.002–0.034 0.009	0.01–0.14 0.07	1.63–41.39 12.06

We carried out statistics on the oil production of different reservoir types. The reservoir type has a good correlation with productivity. For types I, the daily oil production is greater than 150 t. The daily oil production for types I and II are between 100 and 150 t, less than 100 t, respectively. Reservoir IV cannot produce oil. This proves the validity and reliability of reservoir classification.

4. Conclusions

In this study, taking the Carboniferous carbonate reservoirs in the middle blocks of the eastern margin of the Pre-Caspian Basin as an example, we investigated pore structure characterization, permeability estimation, and fractal characteristics. According to the observations, we made a classification for reservoirs. The following conclusions are obtained:

(1) The storage space of carbonate rocks in this study area is divided into three types: pores, fractures, and caverns. The main pore types are the intergranular pore, intergranular dissolution pore, and intragranular dissolution pore. The fractures can be divided into dissolution fracture, tectonic fracture, stylolite fracture, and grain cracks.

(2) The P_d values of the studied samples range from 0.05 MPa to 41.39 MPa, with an average of 1.75 Mpa. Median radii (R_{c50}) are between 0.01 μm and 5.15 μm, with an average of 1.76 μm. The pore throats greater than 1 μm and lower than 0.1 μm account for 47.98% and 22.85% respectively, which suggests that the pore structure in the study area is relatively good.

(3) Permeability ranges from 0.002 mD to 349 mD, and with a logarithmic mean value of 4.07 mD. A permeability prediction model was established in a power-law form which incorporated porosity, Swanson parameter, and R_{35}. The coefficient of determination between the predicted and core analysis permeability is 0.834, showing that the proposed model is effective and reliable. The proposed model could be applicable to other study areas.

(4) Fractal dimension carried out based on MICP curves ranged from 2.29 to 2.77, with an average of 2.61. The pore structure parameters were not correlated with fractal dimension, indicating that the single fractal dimension could not characterize the pore structure characteristics. Multifractal analysis of the MICP data may be more suitable for pore structure investigation.

(5) Combined with the pore types, MICP shape, and petrophysical parameters, the studied reservoirs were classified into four types: Types I, II, III, IV. Type I is the most favorable reservoir with daily oil production greater than 150 t, while Type IV is the worst reservoir and cannot produce oil. The good correlation between reservoir type and productivity demonstrates the effectiveness of the classification in this paper.

Author Contributions: Conceptualization, F.S.; Methodology, L.X. and Z.M.; Investigation, F.S. and Z.M.; Writing—Original Draft Preparation, F.S.; Writing—Review and Editing, F.S. and C.J.

Funding: This research was funded by the Major National Oil & Gas Specific Project of China (No. 2016ZX05050).

Acknowledgments: We thank anonymous reviewers and journal editor for constructive comments which improved the quality of the manuscript.

Conflicts of Interest: The authors declare no conflict of interest.

References

1. Qiang, Z. *Carbonate Reservoir Geology*; China University of Petroleum Press: Dongying, China, 2007; pp. 32–78.
2. He, G.; Tang, H. *Petrophysics*; Petroleum Industry Press: Beijing, China, 2011; pp. 1–382.
3. Volozh, Y.A.; Antipov, M.P.; Brunet, M.F.; Garagash, I.A.; Lobkovskii, L.I.; Cadet, J.P. Pre-Mesozoic geodynamics of the Precaspian basin (Kazakhstan). *Sediment. Geol.* **2003**, *156*, 35–58. [CrossRef]
4. Deng, X.; Wang, H.; Bao, Z.; Sun, N.; Zhang, X. Distribution law and exploration potential of oil and gas in Pre-Caspian Basin. *China Pet. Explor.* **2012**, *17*, 36–47, (In Chinese with English Abstract).
5. Zempolich, W.G.; Negri, A.; Leo, C.; Ojik, K.V.; Verdel, A. The Kashagan discovery: An example of the successful use of a multi-disciplined approach in reducing geologic risk. In Proceedings of the AAPG Annual Meeting, Houston, TX, USA, 10–13 March 2002; p. A197.
6. He, L.; Zhao, L.; Li, J.; Ma, J.; Liu, R.; Wang, S.; Zhao, W. Complex relationship between porosity and permeability of carbonate reservoirs and its controlling factors: A case study of platform facies in Pre-Caspian Basin. *Pet. Explor. Dev.* **2014**, *41*, 225–234. [CrossRef]
7. Zhao, H.; Ning, Z.; Wang, Q.; Zhang, R.; Zhao, T.; Niu, T.; Zeng, Y. Petrophysical characterization of tight oil reservoirs using pressure-controlled porosimetry combined with rate-controlled porosimetry. *Fuel* **2015**, *154*, 233–242. [CrossRef]
8. Rezaee, R.; Saeedi, A.; Clennell, B. Tight gas sands permeability estimation from mercury injection capillary pressure and nuclear magnetic resonance data. *J. Pet. Sci. Eng.* **2012**, *88–89*, 92–99. [CrossRef]
9. Ronchi, P.; Ortenzi, A.; Borromeo, O.; Claps, M.; Zempolich, W.G. Depositional setting and diagenetic processes and their impact on the reservoir quality in the late Visean–Bashkirian Kashagan carbonate platform (Pre-Caspian Basin, Kazakhstan). *AAPG Bull.* **2010**, *94*, 1313–1348. [CrossRef]
10. Yang, F. A Study on Carbonate Sequence Stratigraphy in Block X of the Pre-Caspian Basin. Master's Thesis, Northeast Petroleum University, Daqing, China, 2016.
11. Shi, X.; Chen, X.; Wang, J.; Li, Z.; Jin, S.; Guo, H.; Zhu, M. Geochemical characteristics of the Carboniferours KT-I interval dolostone in eastern margin of coastal Caspian Sea Basin. *J. Palaeogeogr.* **2012**, *14*, 777–785. (In Chinese with English Abstract)
12. Liu, D.; Dou, L.; Hao, Y.; Zhang, Y.; Li, J. The origin of hydrocarbon accumulation below the Lower Permian Salt Bed and the prospecting in the East Part of PreCaspian Basin. *Mar. Origin Pet. Geol.* **2004**, *9*, 53–58. (In Chinese with English Abstract)
13. Hu, Y.; Xia, B.; Wang, Y.; Wan, Z.; Cai, Z. Tectonic evolution and hydrocarbon accumulation model in eastern Precaspian Basin. *Sediment. Geol. Tethyan Geol.* **2014**, *34*, 78–81. (In Chinese with English Abstract)
14. Guo, K.; Cheng, X.; Fan, L.; Yan, S.; Ni, G.; Fu, H. Characteristics and Development Mechanism of Dolomite Reservoirs in North Truva of Eastern Pre-Caspian Basin. *Acta Sedimentol. Sin.* **2016**, *34*, 747–757. (In Chinese with English Abstract)
15. Miao, Q.; Zhu, X.; Guo, H.; Zhao, H.; Zhu, M. Log evaluation of complex carbonate reservoirs in Centre Block of the Eastern Margin of Pre-Caspian Basin. *Well Logging Technol.* **2014**, *38*, 196–200. (In Chinese with English Abstract)
16. He, L. Petrophysical Evaluation on the Microfracture-Typed Complex Carboniferous Carbonate Reservoirs in the Eastern Margin of the Caspian Sea Basin. Ph.D. Thesis, Yangtze University, Wuhan, China, 2015.
17. Sakhaee-Pour, A.; Bryant, S.L. Effect of pore structure on the producibility of tight-gas sandstone. *AAPG Bull.* **2014**, *98*, 663–694. [CrossRef]
18. Klaver, J.; Desbois, G.; Littke, R.; Urai, J.L. BIB-SEM pore characterization of mature and post mature Posidonia Shale samples from the Hills area, Germany. *Int. J. Coal Geol.* **2016**, *158*, 78–89. [CrossRef]
19. Chalmers, G.R.; Bustin, R.M.; Power, I.M. Characterization of gas shale pore systems by porosimetry, pycnometry, surface area, and field emission scanning electron microscopy/transmission electron microscopy image analyses: Examples from the Barnett, Woodford, Haynesville, Marcellus, and Doig units. *AAPG Bull.* **2012**, 1099–1119. [CrossRef]
20. Washburn, E.D. The dynamics of capillary flow. *Phys. Rev.* **1921**, *17*, 273–283. [CrossRef]

21. Feng, C.; Shi, Y.; Li, J.; Chang, L.; Li, G.; Mao, Z. A new empirical method for constructing capillary pressure curves from conventional logs in low permeability sandstones. *J. Earth Sci.* **2017**, *28*, 516–522. [CrossRef]
22. Kuila, U.; Prasad, M. Specific surface area and pore-size distribution in clays and shales. *Geophys. Prospect.* **2013**, *61*, 341–362. [CrossRef]
23. Kolodzie, S.J. Analysis of pore throat size and use of the Waxman–Smits equation to determine OOIP in spindle field. In Proceedings of the SPE Annual Technical Conference and Exhibition, Dallas, TX, USA, 21–24 September 1980. SPE-9382-MS.
24. Swanson, B.F. A simple correlation between permeabilities and mercury injection capillary pressures. *J. Pet. Technol.* **1981**, *33*, 2498–2504. [CrossRef]
25. Lafage, S. An Alternative to the Winland R35 Method for Determining Carbonate Reservoir Quality. Master's Thesis, Texas A&M University, College Station, TX, USA, 2008.
26. Xiao, L.; Wang, H.; Zou, C.; Mao, Z.; Guo, H. Improvements on "Application of NMR logs in tight gas reservoirs for formation evaluation: A case study of Sichuan basin in China". *J. Pet. Sci. Eng.* **2016**, *138*, 11–17. [CrossRef]
27. Zhao, P.; Sun, Z.; Luo, X.; Wang, Z.; Mao, Z.; Wu, Y.; Xia, P. Study on the response mechanisms of nuclear magnetic resonance (NMR) log in tight oil reservoirs. *Chin. J. Geophys.* **2016**, *29*, 1927–1937. [CrossRef]
28. Dunn, K.J.; Bergman, D.J.; Latorraca, G.A. *Nuclear Magnetic Resonance Petrophysical and Logging Application*; Elsevier: Amsterdam, The Netherlands, 2002.
29. Li, K. Analytical derivation of Brooks–Corey type capillary pressure models using fractal geometry and evaluation of rock heterogeneity. *J. Pet. Sci. Eng.* **2010**, *73*, 20–26. [CrossRef]
30. Yang, F.; Ning, Z.; Liu, H. Fractal characteristics of shales from a shale gas reservoir in the Sichuan Basin, China. *Fuel* **2014**, *115*, 378–384. [CrossRef]
31. Mandelbrot, B.B. *The Fractal Geometry of Nature*; W.H. Freeman: New York, NY, USA, 1982.
32. Cai, J.; Perfect, E.; Cheng, C.; Hu, X. Generalized modeling of spontaneous imbibition based on Hagen-Poiseuille flow in tortuous capillaries with variably shaped apertures. *Langmuir* **2014**, *30*, 5142–5151. [CrossRef] [PubMed]
33. Lyu, C.; Cheng, Q.; Zuo, R.; Wang, X. Mapping spatial distribution characteristics of lineaments extracted from remote sensing image using fractal and multifractal models. *J. Earth Sci.* **2017**, *28*, 507–515. [CrossRef]
34. Wei, W.; Cai, J.; Hu, X.; Han, Q. An electrical conductivity model for fractal porous media. *Geophy. Res. Lett.* **2015**, *42*, 4833–4840. [CrossRef]
35. Cai, J.; Wei, W.; Hu, X.; Liu, R.; Wang, J. Fractal characterization of dynamic fracture network extension in porous media. *Fractals* **2017**, *25*, 1750023. [CrossRef]
36. Krohn, C.E. Fractal measurements of sandstones, shales, and carbonates. *J. Geophys. Res. Solid Earth.* **1988**, *93*, 3297–3305. [CrossRef]
37. Billi, A.; Storti, F. Fractal distribution of particle size in carbonate cataclastic rocks from the core of a regional strike-slip fault zone. *Tectonophysics* **2004**, *384*, 115–128. [CrossRef]
38. Xie, S.; Cheng, Q.; Ling, Q.; Li, B.; Bao, Z.; Fan, P. Fractal and multifractal analysis of carbonate pore-scale digital images of petroleum reservoirs. *Mar. Pet. Geol.* **2010**, *27*, 476–485. [CrossRef]
39. Liu, H.; Tian, Z.; Xu, Z. Quantitative evaluation of carbonate reservoir pore structure based on fractal characteristics. *Lithol. Reserv.* **2017**, *29*, 97–105. [CrossRef]
40. Ghanbarian-Alavijeh, B.; Hunt, A.G. Comments on "More general capillary pressure and relative permeability models from fractal geometry" by Kewen Li. *J. Contam. Hydrol.* **2012**, *140*, 21–23. [CrossRef] [PubMed]
41. Ghanbarian, B.; Sahimi, M. Electrical conductivity of partially saturated packings of particles. *Transp. Porous Med.* **2017**, *118*, 1–16. [CrossRef]
42. Ge, X.; Fan, Y.; Deng, S.; Han, Y.; Liu, J. An improvement of the fractal theory and its application in pore structure evaluation and permeability estimation. *J. Geophys. Res. Solid Earth* **2016**, *121*, 6333–6345. [CrossRef]
43. Zhang, Z.; Weller, A. Fractal dimension of pore-space geometry of an Eocene sandstone formation. *Geophysics* **2014**, *79*, D377–D387. [CrossRef]
44. Moore, H.C. *Carbonate Reservoirs: Porosity Evolution and Diagenesis in a Sequence Stratigraphic Framework*; Elsevier: Amsterdam, The Netherlands, 2001.
45. Xu, Z.; Zhao, P.; Wang, Z.; Ostadhassan, M.; Pan, Z. Characterization and consecutive prediction of pore structures in tight oil reservoirs. *Energies* **2018**, *11*, 2705. [CrossRef]

46. Schowalter, T.T. Mechanics of secondary hydrocarbon migration and entrapment. *AAPG Bull.* **1979**, *63*, 723–760.
47. Mao, Z.; Xiao, L.; Wang, Z.; Jin, Y.; Liu, X.; Xie, B. Estimation of permeability by integrating nuclear magnetic resonance (NMR) logs with mercury injection capillary pressure (MICP) data in tight gas sands. *Appl. Magn. Reson.* **2013**, *44*, 449–468. [CrossRef]
48. Ghanbarian-Alavijeh, B.; Millán, H. The relationship between surface fractal dimension and soil water content at permanent wilting point. *Geoderma* **2009**, *151*, 224–232. [CrossRef]

MDPI

Article

A Non-Linear Flow Model for Porous Media Based on Conformable Derivative Approach

Gang Lei [1], Nai Cao [2], Di Liu [3] and Huijie Wang [4,*]

1. College Petroleum of Engineering & Geoscience, King Fahd University of Petroleum and Minerals, Dhahran 31261, Saudi Arabia; gang.lei@kfupm.edu.sa
2. College of Petroleum Engineering, China University of Petroleum, Beijing 102249, China; caonai99@gmail.com
3. School of Mechanics and Civil Engineering, China University of Mining and Technology, Beijing 100083, China; liudi@student.cumtb.edu.cn
4. College of Engineering, Peking University, Beijing 100871, China

* Correspondence: jie.wh@pku.edu.cn; Tel.: +86-010-6276-0875

Received: 11 September 2018; Accepted: 29 October 2018; Published: 1 November 2018

Abstract: Prediction of the non-linear flow in porous media is still a major scientific and engineering challenge, despite major technological advances in both theoretical and computational thermodynamics in the past two decades. Specifically, essential controls on non-linear flow in porous media are not yet definitive. The principal aim of this paper is to develop a meaningful and reasonable quantitative model that manifests the most important fundamental controls on low velocity non-linear flow. By coupling a new derivative with fractional order, referred to conformable derivative, Swartzendruber equation and modified Hertzian contact theory as well as fractal geometry theory, a flow velocity model for porous media is proposed to improve the modeling of Non-linear flow in porous media. Predictions using the proposed model agree well with available experimental data. Salient results presented here include (1) the flow velocity decreases as effective stress increases; (2) rock types of "softer" mechanical properties may exhibit lower flow velocity; (3) flow velocity increases with the rougher pore surfaces and rock elastic modulus. In general, the proposed model illustrates mechanisms that affect non-linear flow behavior in porous media.

Keywords: porous media; non-linear flow; conformable derivative; fractal

1. Introduction

Ever since Henry Darcy (1865) developed his famous linear flow model (the classical Darcy's law), based on a series of sand pack experiments, the linear flow through porous media has drawn tremendous attention in various scientific and engineering field [1,2]. However, it's a common phenomenon that experiments on low velocity flow in tight porous media, deviate from the Darcy's law and the flow velocity is lower than that predicted from Darcy's law. As stated in the literature, the existence of low velocity non-Darcy flow (or low velocity non-linear flow) in tight porous media (e.g., shale gas/oil reservoirs, coalbed, or tight gas/oil reservoirs) is due to the interaction forces between the fluid and tight pores [3,4]. Many scholars have documented that, there existed threshold Reynolds number or pressure gradient, which could be used to well describe low velocity non-linear flow [5–7]. And they concluded that there is no flow in tight porous media when the pressure gradient is beyond the certain value (i.e., threshold pressure gradient). However, Li provided contradictory evidence for the threshold pressure gradient [8]. He suggested that the threshold pressure gradient measured in labs can be probably ascribed to the difficulty in measuring lower flow velocity, and the false phenomenon of the existence of threshold pressure gradient is strengthened by the skin effect. Until now, appropriate non-linear model for fluid flow through tight porous media remains unclear, though some more formulas have been established to describe the non-linear flow, such as power function model [9], exponential function model [10], incomplete Gamma function model [11,12] and fractional derivative approach [13–15]. As stated in the literature, in some extent, these models above are suitable for the description of non-linear flow in porous media with lower permeability [9–14]. Most recently, Yang [14] and Zhou [15] also suggested that the conformable derivative approach is suitable for the describing non-linear flow in low-permeability porous media. However, these models above never took effective stress into account. It is reported that the porous media will be compressed as the effective stress increases, causing fluid flow behavior in porous media to be strongly stress-dependent [16–24].

As implied by this brief literature review, we suggest that the characteristic behavior of non-linear flow is still not definitively determined. Therefore, a major goal of this research was to develop an analytical model in a closed form for the description of low velocity non-linear flow. The specific objectives of this work were: (1) to establish a reasonable quantitative model to quantify the essential controls on non-linear flow; (2) to verify the model with available experimental data. Compared with the previous models, our model takes into account more factors, including the influence of the effective stress and the microstructural parameters of the pore space. The proposed models can reveal more mechanisms that affect the low velocity non-linear flow in porous media.

2. Mathematical Model

In this section, the analytical low velocity non-linear flow model for porous media is detailed. The conformable derivative is used to develop the Swartzendruber model for description of low velocity non-linear flow in pores, the fractal geometry theory and modified Hertzian contact theory are used to describe the complex pore structure of porous media under stress condition.

2.1. Model Assumptions

The following assumptions are made to simplify the flow system:

1. The porous media is composed by a bundle of capillary bundles and a single capillary with the equivalent radius r is made up of a packing of equivalent spherical grains.
2. The interspaces in porous media have fractal characteristics.
3. The single phase flow is under isothermal and stress condition, which is fully developed and at steady state.
4. The deformation of porous media obeys Hertzian contact theory.
5. During the flow, the fluid has constant viscosity and density.

2.2. Conformable Derivative Approach to Swartzendruber Equation

As suggested by decades of literature, the Swartzendruber equation can well describe the non-Darcian flow in tight porous media with low permeability [13,14]. Based on the Swartzendruber equation, the following equation can be written as:

$$\frac{du}{di} = \frac{K\rho g}{\mu}\left(1 - e^{-\frac{i}{I}}\right), \tag{1}$$

where u is the flow velocity in the cross section; K is the permeability of porous media; μ is fluid viscosity; ρ is fluid density, g is the gravitational acceleration, i and I represent hydraulic gradient and threshold hydraulic gradient, respectively. According to Equation (1), the flow velocity in a single capillary with radius r can be written as:

$$\frac{du}{di} = \frac{r^2\rho g}{8\mu}\left(1 - e^{-\frac{i}{I}}\right), \tag{2}$$

where r is the radius of the capillary.

As the linear operator does not inherit all the operational behaviors from the typical first derivative, the Swartzendruber equation fails to capture the full range of non-Darcy flow behavior in porous media [13–15]. Fortunately, as a well-behaved and efficient method, conformable derivative approach with real order can be applied to address this problem. The conformable derivative of the flow velocity $u(i)$: $[0, \infty) \rightarrow R$ for all $i > 0$ with order $\alpha \in (0, 1]$ can be defined by [10]:

$$T_\alpha u(i) = \lim_{\varepsilon \to 0} \frac{u(i + \varepsilon i^{1-\alpha}) - u(i)}{\varepsilon}, \tag{3a}$$

and the conformable derivative at 0 is given by $(T_\alpha u)(0) = \lim_{i \to 0} (T_\alpha u)(i)$.

As stated in the literature [10,14,25], the relationship between the conformable derivative and the first derivative can be written as:

$$T_\alpha u(i) = i^{1-\alpha} \frac{du(i)}{di}. \tag{3b}$$

Equation (3b) shows that the conformable derivative coincides with the classical first derivative with a given differential order $\alpha = 1$, which means the conformable derivative is a modification of classical derivative in direction and magnitude [25–27]. As stated in the literature [25–27], the physical interpretation of the conformable derivative is a modification of classical derivative indirection and magnitude. Replacing the first order derivative in Equation (2) with conformable derivative, the Swartzendruber equation can be rewritten as:

$$T_\alpha u(i) = \frac{r^2 \rho g}{8\mu} \left(1 - e^{-\frac{i}{I}}\right). \tag{4}$$

By solving Equation (4) with Laplace transform and inverse Laplace transform, the flow velocity in the cross section can be determined as:

$$u(i) = \frac{r^2 \rho g}{8\mu} \frac{i^\alpha}{\alpha} \left[1 - {}_1F_1\left(\alpha; \alpha + 1; -\frac{i}{I}\right)\right], \tag{5}$$

where ${}_1F_1\left(\alpha; \alpha + 1; -\frac{i}{I}\right) = \sum\limits_{j=0}^{\infty} \frac{\alpha^{(j)}}{(\alpha+1)^{(j)}} \frac{(-i/I)^j}{j!}$ is the Kummer confluent hypergeometric function [23].

Based on Equation (5), the flow rate q in the cross section can be written as:

$$q = \frac{\pi r^4 \rho g}{8\mu} \frac{i^\alpha}{\alpha} \left[1 - {}_1F_1\left(\alpha; \alpha + 1; -\frac{i}{I}\right)\right], \tag{6}$$

where q is the flow rate in the capillary with the radius r.

2.3. Non-Linear Flow Model

The fractal theory is used to develop the non-linear flow model for tight porous media. Based on fractal theory and modified Hertzian contact theory described in detail in [28,29], the total volumetric flow rate Q in the cross section under stress condition can be calculated by integrating the flow rate over the radius ranging from the minimum radius $r_{\min}$ to the maximum porous radius $r_{\max}$ [28–33]:

$$Q = N \int_{r_{\min}}^{r_{\max}} q f dr, \tag{7}$$

with:

$$\begin{cases} N = (r_{\max}/r_{\min})^{D_f}; f(r) = D_f r_{\min}^{D_f} r^{-(D_f+1)}, \\ r = r_0 \left\{ 1 - 4 \left[\frac{3\pi(1-v^2)p_{\text{eff}}}{4E} \right]^{\beta} \right\}, \\ r_{\min} = r_{\min 0} \left\{ 1 - 4 \left[\frac{3\pi(1-v^2)p_{\text{eff}}}{4E} \right]^{\beta} \right\}, \\ r_{\max} = r_{\max 0} \left\{ 1 - 4 \left[\frac{3\pi(1-v^2)p_{\text{eff}}}{4E} \right]^{\beta} \right\}. \end{cases} \tag{8}$$

In Equation (8), N is the number of pores; r_0 and r are the initial pore radius and pore radius under stress condition, respectively. f is the probability density function for pore size distribution; β is the power law index which is related to the structure of pore surface. $r_{\max 0}$ and $r_{\min 0}$ are the maximum and minimum pore radius of porous media at zero stress, respectively. $r_{\max}$ and $r_{\min}$ are the stress-dependent maximum and the stress-dependent minimum pore radius of porous media, respectively. p_{eff} is the effective stress, E is rock elastic modulus and v is rock Poisson's ratio. D_f is the fractal dimension for pore size distribution which can be determined as [32,33]:

$$\begin{cases} D_f = 2 - \frac{(2-D_{f0})r_{\max 0}}{(3-D_{f0})r_{\max} - (2-D_{f0})r_{\max 0}}, \\ D_{f0} = 2 - \frac{\ln \varphi_0}{\ln(r_{\min 0}/r_{\max 0})}, \end{cases} \tag{9}$$

where φ_0 is the initial porosity of the porous media, and D_{f0} is the fractal dimension for pore size distribution at zero stress.

Substituting Equations (6), (8) and (9) into Equation (7), the flow rate can be rewritten as:

$$Q = \frac{\pi \rho g D_f r_{\max}^{D_f}}{8\mu} \frac{i^{\alpha}}{\alpha} \left[1 - {}_1F_1\left(\alpha; \alpha+1; -\frac{i}{I}\right) \right] \frac{r_{\max}^{4-D_f} - r_{\min}^{4-D_f}}{4 - D_f}. \tag{10}$$

As stated in the literature [32–35], the cross sectional area of a unit cell A can be expressed as:

$$A = \frac{\pi D_f r_{\max}^2 \left[1 - (r_{\min}/r_{\max})^{2-D_f} \right]}{(2-D_f)(r_{\min}/r_{\max})^{2-D_f}}. \tag{11}$$

Then, based on Equations (10) and (11), the average flow velocity u_{av} can be written as:

$$u_{\text{av}} = \frac{Q}{A} = \frac{\rho g \varphi i^{\alpha}(2-D_f)(r_{\max}^2 - \varphi r_{\min}^2)}{8\alpha\mu(1-\varphi)(4-D_f)} \left[1 - {}_1F_1\left(\alpha; \alpha+1; -\frac{i}{I}\right) \right] \tag{12}$$

It is evident that the flow rate (or average flow velocity) is the function of pore structural parameters, power law index, rock elastic modulus, effective stress, and differential order α as well as hydraulic gradient and threshold hydraulic gradient.

3. Results and Discussion

This section aims at studying the novel analytical models in detail. In the following, we first compare our results with those from experimental data. Then, in order to analyze essential controls on non-linear flow in tight porous media, the effects of relevant parameters on average flow velocity are studied in detail.

The availability of the proposed model Equation (12) depends on its ability to adequately fit experimental data [15]. To verify our quantitative model, the measured average flow velocity versus hydraulic gradient relationship in [36] and that predicted by our proposed model are compared

(Figure 1). In the experiment of Prakash K. et al. [36], the flow velocity tests were conducted on soils at an effective consolidation stress of 6.25 kPa during loading process. In our proposed model, the initial porosity of rock is 15%, the rock elastic modulus is 45 GPa, the rock Poisson's ratio is 0.23 and power law index is 3/4. Furthermore, to ensure the effective consolidation stress is 6.25 kPa, the effective stress assigned is 6.25 kPa. The values of other parameters (e.g., $r_{\max 0}$, $r_{\min 0}$, differential order α and threshold hydraulic gradient I) are listed in the Figure 1. Results displayed in Figure 1 suggest that the proposed model calculated relationship between average flow velocity and hydraulic gradient is in good agreement with that determined by experimental data [36]. Results (Figure 1) also suggest a definitive positive correlation between the average flow velocity and hydraulic gradient.

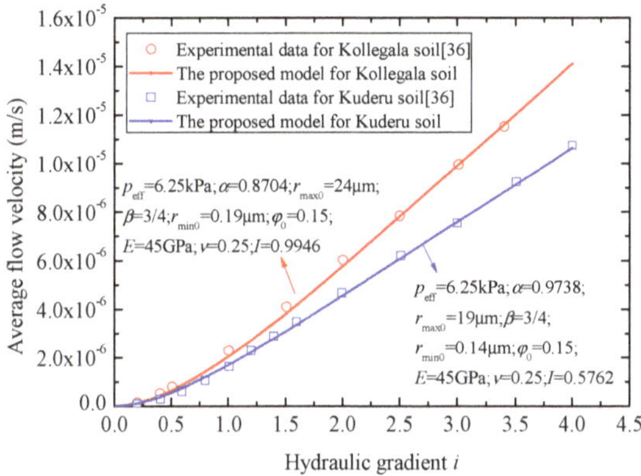

Figure 1. A comparison between the experimental data [36], and results of the proposed model.

Figure 2 provides a comparison of the relationship between average flow velocity and hydraulic gradient predicted by the proposed model with experimental data of [37]. In the experiment of Zhang et al. [37], the permeability tests were conducted on gap-graded sands (e.g., sand A, sand B and sand C) to determine the critical hydraulic gradient of piping in sands. Sand A, sand B and sand C were prepared by mixing a coarse sand with particle size of 3–5 mm and a fine sand with particle size of 0.25–0.50 mm at various ratios of 4:1, 5:1 and 6:1 respectively.

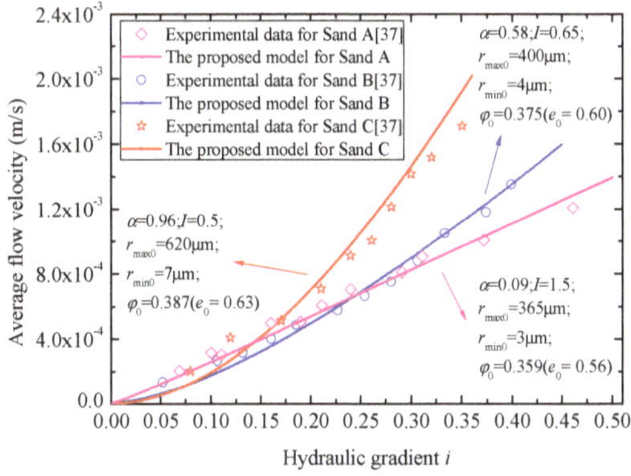

Figure 2. A comparison between the experimental data [37], and results of the proposed model.

For sand A, the hydraulic conductivity and void ratio are 2.6×10^{-3} m/s and 0.56 respectively. The hydraulic conductivity and void ratio of sand B are 3.6×10^{-3} m/s and 0.60 respectively. In addition, for sand C, the hydraulic conductivity and void ratio are 4.0×10^{-3} m/s and 0.63

respectively. To ensure the porous media simulated in the proposed model exhibits the same physical properties as the porous media tested in the experiments of [37], the initial void ratio e_0 and initial porosity φ_0 (i.e., $\varphi_0 = e_0/(1 + e_0)$) applied in the proposed model are the same with that of the sands in the experiments and the effective stress assigned is 0 MPa. The values of other parameters (e.g., $r_{\max0}$, $r_{\min0}$, differential order α and threshold hydraulic gradient I) are listed in the Figure 2. It can be seen from Figure 2 that our predicted values agree well with the corresponding experimental data [37].

As the average flow velocity is related to the threshold hydraulic gradient, differential order α, effective stress, rock elastic modulus, power law index, pore structural parameters, and the hydraulic gradient. We will then study the effects of these relevant parameters (e.g., threshold hydraulic gradient, differential order α, effective stress, rock elastic modulus, power law index and initial porosity) on average flow velocity in detail to analyze essential controls on non-linear flow in tight porous media. We plotted average flow velocity vs. hydraulic gradient with different threshold hydraulic gradient I and different differential order α (Figure 3a,b).

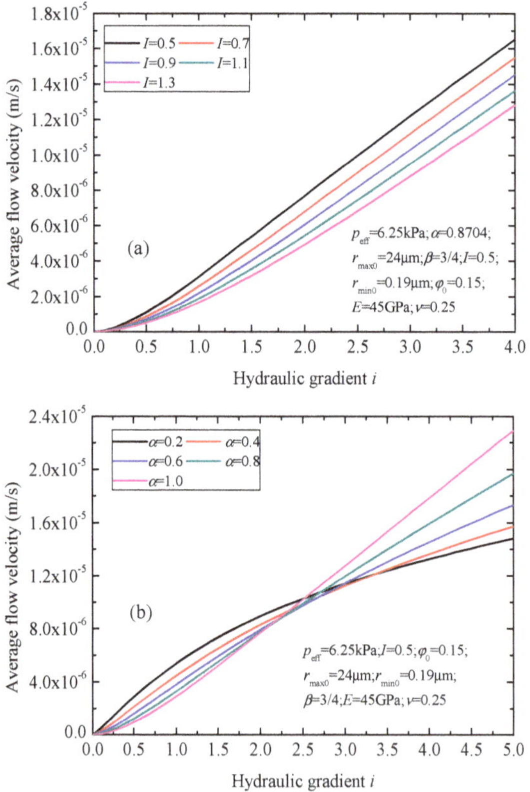

Figure 3. The average flow velocity curves: (**a**) for different threshold hydraulic gradient I; (**b**) for different differential order α.

The parameters assigned in the proposed model were listed in the Figure 3. As suggested by the results (Figure 3a), average flow velocity decreases as threshold hydraulic gradient increases. In addition, Figure 3b shows that larger different differential order leads to smaller average flow velocity with smaller hydraulic gradient, however, on the contrary, when the hydraulic gradient increases to a certain value, average flow velocity increases as differential order increases.

Figure 4 illustrates the average flow velocity vs. hydraulic gradient with different effective stress. The parameters assigned in the proposed model were listed in the Figure 4. As suggested by the results (Figure 4), average flow velocity decreases as effective stress increases. This may be attributed to the decrease of pore radius which is resulted from the pore compaction. Therefore, fluid flow behavior in tight porous media is strongly stress-dependent.

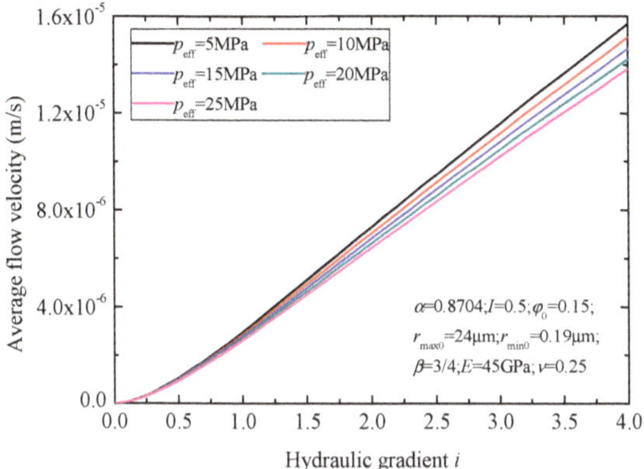

Figure 4. The average flow velocity curves versus hydraulic gradient with different effective stress.

Plotting average flow velocity vs. hydraulic gradient with different rock elastic modulus, power law index and initial porosity (Figure 5) was also useful. For the necessary calculations, the parameters assigned in the proposed model were listed in the Figure 5. As suggested by the results (Figure 5a), the average flow velocity increases as rock elastic modulus increases. We interpret this result to indicate that the larger rock elastic modulus decreases the contact surface radius of a given particle, leading to reduced pore volume compressibility and larger hydraulic conductivity. Therefore, rock types of "soft" lithology can yield lower flow velocity. As suggested by the results (Figure 5b), the average flow velocity increases as power law index increases. Correspondingly, we suggest that the larger power law index β implies rougher pore surfaces, leading to only a limited number of pores being compressed and the larger hydraulic conductivity. Figure 5c shows the average flow velocity decreases as rock initial porosity decreases, which is expected. Correspondingly, we suggest that the smaller initial porosity implies narrower pore radius, leading to the smaller hydraulic conductivity.

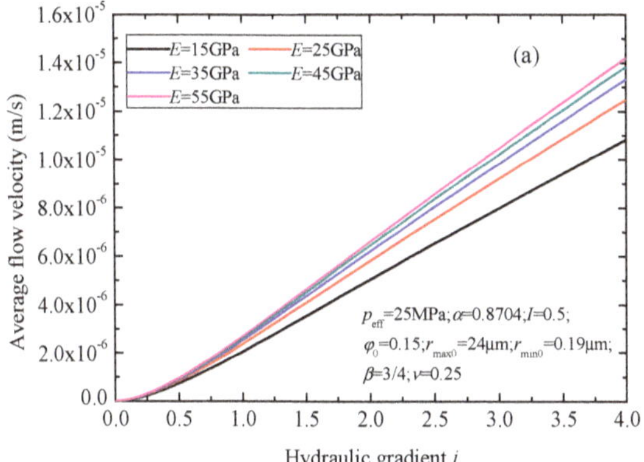

Figure 5. *Cont.*

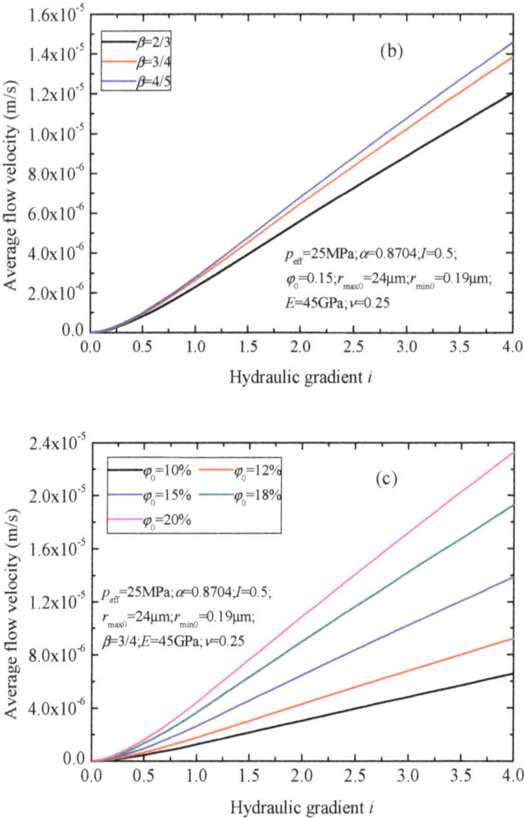

Figure 5. The average flow velocity curves: (**a**) for different rock elastic modulus; (**b**) for different power law index; (**c**) for different initial porosity.

4. Conclusions

In this study, we developed a novel non-linear flow model for tight porous media. The model is based on Swartzendruber equation and conformable derivative approach and as well as the modified Hertzian contact theory and fractal geometry, and allowed us to analyze essential controls on non-linear flow in tight porous media. An advantage of this model is that it lacks empirical constants, and, more importantly, every parameter in the model has specific physical significance. Predictions from the proposed analytical model exhibit similar variation trends as experimental data, suggesting validity of the model to predict the average flow velocity. Moreover, we analyzed resulting model predictions in detail, to confirm the model's robustness in this context. Results of the new model show the following salient conclusions:

1. The proposed models indicate that average flow velocity in tight porous media is a function of microstructural parameters of the pore space, rock lithology and differential order α as well as hydraulic gradient and threshold hydraulic gradient.
2. The parametric study reveals that average flow velocity increases with the rougher pore surfaces and rock elastic modulus, and decreases with increasing effective stress. "Softer" rock lithology may yield lower average flow velocity.
3. This non-linear model presented here considers microstructural parameters of pore space and rock lithology; we have shown that its forecasted values are robust, at least compared to experimental

data, and thus may be useful for performance predictions of non-linear flow behavior in tight porous media. Results also reveal more information about the details of specific parameters (and therefore mechanisms) that affect non-linear flow behavior in porous media. The new model presented in this work can be used to depict the non-linear flow in tight porous media, and may provide meaningful applications for design and development of tight reservoirs. In addition, as the model takes effective stress into account, it is also useful for performance predictions of the coupled flow deformation behavior (stress sensitivity) in tight porous media.

Author Contributions: Conceptualization, G.L.; Funding acquisition, H.W.; Investigation, N.C. and D.L.; Methodology, G.L.; Project administration, H.W.; Visualization, N.C.; Writing—original draft, G.L.; Writing—review & editing, H.W.

Funding: This research was funded by the State Major Science and Technology Special Project of China during the 13th Five-Year Plan (grant number: 2016ZX05025-003-007 and 2016ZX05014004-006). Appreciation and thanks are given to the organizations for their finical funding herein.

Conflicts of Interest: The authors declare no conflicts of interest.

Nomenclature

Latin symbols	
A	Cross sectional area of a unit cell, μm^2
D_{f0}	Initial pore area fractal dimension at zero stress, dimensionless
D_f	Pore area fractal dimension, dimensionless
e_0	Initial void ratio of porous media, dimensionless
E	Rock elastic modulus of porous media, GPa
f	Probability density function for pore size distribution, dimensionless
${}_1F_1$	Kummer confluent hypergeometric function
g	Gravitational acceleration, N/kg
K	Absolute permeability of porous media, μm^2
i	Hydraulic gradient, dimensionless
I	Threshold hydraulic gradient, dimensionless
N	Number of pores of a unit cell, dimensionless
p_{eff}	Effective stress, MPa
q	Flow rate in the cross section, m^3/s
Q	Total volumetric flow rate in the cross section under stress condition, m^3/s
r_0	Initial equivalent pore radius of capillary at zero stress, μm
r	Equivalent pore radius of capillary under effective stress, μm
u	Flow velocity in the cross section, m/s
u_{av}	Average flow velocity, m/s
Greek symbols	
α	Differential order
β	Power law index, dimensionless
μ	Fluid viscosity, mPa·s
ρ	Fluid density, kg/m^3
φ_0	Initial porosity of porous media, dimensionless
φ	Porosity under effective stress, dimensionless
ν	Poisson's ratio, dimensionless
Subscript	
av	Average
eff	Effective
max	maximum values
max0	Initial maximum values at zero stress
min	minimum values
min0	Initial minimum values at zero stress

References

1. Freeze, R.A.; Cherry, J.A. *Groundwater*; Prentice-Hall: Upper Saddle River, NJ, USA, 1979.
2. Zhang, J.; Xu, Q.; Chen, Z. Seepage analysis based on the unified unsaturated soil theory. *Mech. Res. Commun.* **2001**, *28*, 107–112. [CrossRef]
3. Xu, C.; Kang, Y.; Chen, F.; You, Z. Analytical model of plugging zone strength for drill-in fluid loss control and formation damage prevention in fractured tight reservoir. *J. Pet. Sci. Eng.* **2017**, *149*, 686–700. [CrossRef]
4. Cai, J.; Lin, D.; Singh, H.; Wei, W.; Zhou, S. Shale gas transport model in 3D fractal porous media with variable pore sizes. *Mar. Geol.* **2018**, *98*, 437–447. [CrossRef]
5. Thomas, L.K.; Katz, D.L.; Tek, M.R. Threshold pressure phenomena in porous media. *Soc. Petrol. Eng. J.* **1968**, *8*, 174–184. [CrossRef]
6. Tian, W.; Li, A.; Ren, X.; Josephine, Y. The threshold pressure gradient effect in the tight sandstone gas reservoirs with high water saturation. *Fuel* **2018**, *226*, 221–229. [CrossRef]
7. Morozov, P.; Abdullin, A.; Khairullin, M. An analytical model of SAGD process considering the effect of threshold pressure gradient. *IOP Conf. Ser. Earth Environ. Sci.* **2018**, *155*, 012001. [CrossRef]
8. Li, C. Is a starting pressure gradient necessary for flow in porous media? *Acta Petrol. Sin.* **2010**, *31*, 867–870.
9. Hansbo, S. Consolidation equation valid for both Darcian and non-Darcian flow. *Geotech* **2001**, *51*, 51–54. [CrossRef]
10. Swartzendruber, D. Modification of Darcy's law for the flow of water in soils. *Soil Sci.* **1962**, *93*, 22–29. [CrossRef]
11. Liu, H.H.; Birkholzer, J. On the relationship between water flux and hydraulic gradient for unsaturated and saturated clay. *J. Hydrol.* **2012**, *475*, 242–247. [CrossRef]
12. Liu, H.H. Non-Darcian flow in low-permeability media: Key issues related to geological disposal of high-level nuclear waste in shale formations. *Hydrogeol. J.* **2014**, *22*, 1525–1534. [CrossRef]
13. Wang, R.; Zhou, H.W.; Zhong, J.C.; Yi, H.; Chen, C.; Zhao, Y. The study on non-Darcy seepage equation of low velocity flow. *Sci. Sin. Phys.* **2017**, *47*, 064702. (In Chinese) [CrossRef]
14. Yang, S.; Wang, L.; Zhang, S. Conformable derivative: Application to non-Darcian flow in low-permeability porous media. *Appl. Math. Lett.* **2018**, *79*, 105–110. [CrossRef]
15. Zhou, H.W.; Yang, S.; Zhang, S.Q. Conformable derivative approach to anomalous diffusion. *Physica A* **2018**, *491*, 1001–1013. [CrossRef]
16. Zhang, H.J.; Jeng, D.S.; Barry, D.A.; Seymour, B.R.; Li, L. Solute transport in nearly saturated porous media under landfill clay liners: A finite deformation approach. *J. Hydrol.* **2013**, *479*, 189–199. [CrossRef]
17. Srinivasacharya, D.; Srinivasacharyulu, N.; Odelu, O. Flow and heat transfer of couple stress fluid in a porous channel with expanding and contracting walls. *Int. Commun. Heat Mass Transf.* **2009**, *36*, 180–185. [CrossRef]
18. Neto, L.B.; Kotousov, A.; Bedrikovetsky, P. Elastic properties of porous media in the vicinity of the percolation limit. *J. Pet. Sci. Eng.* **2011**, *78*, 328–333. [CrossRef]
19. Mokni, N.; Olivella, S.; Li, X.; Smets, S.; Valcke, E. Deformation induced by dissolution of salts in porous media. *Phys. Chem. Earth Parts A/B/C* **2008**, *33*, S436–S443. [CrossRef]
20. Gangi, A.F. Variation of whole and fractured porous rock permeability with confining pressure. *Int. J. Rock Mech. Min. Sci. Geomech. Abstr.* **1978**, *15*, 249–257. [CrossRef]
21. Archer, R.A. Impact of stress sensitive permeability on production data analysis. In Proceedings of the SPE Unconventional Reservoirs Conference, Keystone, CO, USA, 10–12 January 2008. SPE-114166-MS.
22. Schreyer-Bennethum, L. Theory of flow and deformation of swelling porous materials at the macroscale. *Comput. Geotech.* **2007**, *34*, 267–278. [CrossRef]
23. Jennings, J.B.; Carroll, H.B.; Raible, C.J. The relationship of permeability to confining pressure in low permeability rock. In Proceedings of the SPE/DOE Low Permeability Gas Reservoirs Symposium, Denver, CO, USA, 27–29 May 1981; p. SPE-9870-MS.
24. Tan, X.H.; Li, X.P.; Liu, J.Y.; Zhang, L.H.; Fan, Z. Study of the effects of stress sensitivity on the permeability and porosity of fractal porous media. *Phys. Lett. A* **2015**, *379*, 2458–2465. [CrossRef]
25. Khalil, R.; Al Horani, M.; Yousef, A.; Sababheh, M. A new definition of fractional derivative. *J. Comput. Appl. Math.* **2014**, *264*, 65–70. [CrossRef]
26. Abdeljawad, T. On conformable fractional calculus. *J. Comput. Appl. Math.* **2015**, *279*, 57–66. [CrossRef]

27. Zhao, D.; Luo, M. General conformable fractional derivative and its physical interpretation. *Calcolo* **2017**, *54*, 903–917. [CrossRef]
28. Lei, G.; Dong, Z.; Li, W.; Wen, Q.; Wang, C. Theoretical study on stress sensitivity of fractal porous media with irreducible water. *Fractals* **2018**, *26*, 1850004. [CrossRef]
29. Lei, G.; Mo, S.; Dong, Z.; Wang, C.; Li, W. Theoretical and experimental study on stress-dependency of oil-water relative permeability in fractal porous media. *Fractals* **2018**, *26*, 1840010. [CrossRef]
30. Yu, B.; Li, J. Some fractal characters of porous media. *Fractals* **2001**, *9*, 365–372. [CrossRef]
31. Cai, J.; Yu, B.; Zou, M.; Luo, L. Fractal characterization of spontaneous co-current imbibition in porous media. *Energy Fuels* **2010**, *24*, 1860–1867. [CrossRef]
32. Lei, G.; Dong, P.; Wu, Z.; Mo, S.; Gai, S.; Zhao, C.; Liu, Z.K. A fractal model for the stress-dependent permeability and relative permeability in tight sandstones. *J. Can. Pet. Technol.* **2015**, *54*, 36–48. [CrossRef]
33. Cai, J.; Wei, W.; Hu, X.; Wood, D.A. Electrical conductivity models in saturated porous media: A review. *Earth Sci. Rev.* **2017**, *171*, 419–433. [CrossRef]
34. Sheng, M.; Li, G.; Tian, S.; Huang, Z.; Chen, L. A fractal permeability model for shale matrix with multi-scale porous structure. *Fractals* **2016**, *24*, 1650002. [CrossRef]
35. Lu, T.; Duan, Y.; Fang, Q.; Dai, X.; Wu, J. Analysis of fractional flow for transient two-phase flow in fractal porous medium. *Fractals* **2016**, *24*, 1650013. [CrossRef]
36. Prakash, K.; Sridharan, A.; Prasanna, H.S. Dominant parameters controlling the permeability of compacted fine-grained soils. *Indian Geotech. J.* **2016**, *46*, 408–414. [CrossRef]
37. Zhang, J.; Jiang, S.; Wang, Q.; Hou, Y.; Chen, Z. Critical hydraulic gradient of piping in sand. In Proceedings of the Twentieth International Offshore and Polar Engineering Conference, Beijing, China, 20–25 June 2010.

MDPI

Article

A Study to Investigate Fluid-Solid Interaction Effects on Fluid Flow in Micro Scales

Mingqiang Chen *, Linsong Cheng, Renyi Cao and Chaohui Lyu

State Key Laboratory of Petroleum Resources and Prospecting, China University of Petroleum, Beijing 102249, China; lscheng@cup.edu.cn (L.C.); caorenyi@126.com (R.C.); 2016312050@student.cup.edu.cn (C.L.)

* Correspondence: 2016312031@student.cup.edu.cn; Tel.: +86-178-0115-8991

Received: 17 July 2018; Accepted: 20 August 2018; Published: 22 August 2018

Abstract: Due to micro-nanopores in tight formation, fluid-solid interaction effects on fluid flow in porous media cannot be ignored. In this paper, a novel model which can characterize micro-fluid flow in micro scales is proposed. This novel model has a more definite physical meaning compared with other empirical models. And it is validated by micro tube experiments. In addition, the application range of the model is rigorously analyzed from a mathematical view, which indicates a wider application scope. Based on the novel model, the velocity profile, the average flow velocity and flow resistance in consideration of fluid-solid interaction are obtained. Furthermore, the novel model is incorporated into a representative pore scale network model to study fluid-solid interactions on fluid flow in porous media. Results show that due to fluid-solid interaction in micro scales, the change rules of the velocity profile, the average flow velocity and flow resistance generate obvious deviations from traditional Hagen-Poiseuille's law. The smaller the radius and the lower the displacement pressure gradient (∇P), the more obvious the deviations will be. Moreover, the apparent permeability in consideration of fluid-solid interaction is no longer a constant, it increases with the increase of ∇P and non-linear flow appears at low ∇P. This study lays a good foundation for studying fluid flow in tight formation.

Keywords: fluid-solid interaction; velocity profile; the average flow velocity; flow resistance; pore network model

1. Introduction

With the development of petroleum industry, tight oil is gradually becoming one of the main fields to improve oil recovery [1–3]. However, there are large numbers of micro-nanopores in tight formation [4–7]. The large specific surface area and surface effect exhibiting in micro-nanoscales cause micro scale flow different from fluid flow in macro scales [8,9]. Therefore, figuring out microscopic flow law in consideration of micro scale effect is of great importance to the development of tight oil reservoirs. The research of micro-machining technology and micro-electro-mechanical system triggers a new field for the study of micro scale flow, which provides a new insight for studying fluid flow in tight reservoirs [10–13].

Recently, many micro flow experiments have been carried out and results show obvious deviations from traditionally theoretical prediction, which indicates that fluid flow in micro tubes no longer abides by traditional N-S equation [14–18]. Pfaler et al. [19] found that the experimental result is consistent with theoretical prediction when the micro channel size is large enough. However, an obvious deviation occurs when the size is reduced to 0.8 μm. Makihara et al. [20] conducted a water flow experiment in micro tubes with Silica and stainless steel and found that the relationship of the Reynolds number versus displacement pressure gradient did not obey theoretical values when the micro-tube diameter was smaller than 150 μm. Qu et al. [21] performed a water flow experiment in Trapezoidal microtubules and concluded that fluid flow deviated from the theoretical value of N-S equation. Wu [22] carried out a deionized water flow experiment in micro tubes with radii ranging from 1.38–10.03 μm at low ∇P and found that water flow did not agree with classical Hagen-Poiseuille law; the boundary layer was formed in the near wall area.

Fluid-solid interactions are used to account for these deviations. With the decrease of flow scale, microscopic forces acting on fluid flow become dominated, which eventually results in micro flow characteristics different from macro flow [23,24]. The Coulomb force generated by the wall molecules on the liquid, the Van der Waals force by the molecular polarization and the space configuration force affect micro fluid flow greatly [25]. Due to strong interaction between fluid and solid, the fluid near the solid wall is adsorbed on the wall surface and cannot move. Researchers define the immovable layer as boundary layer [26]. Under the influence of fluid-solid interaction, the effective flow space is compressed and the flow resistance becomes larger. Pertsin et al. [27] theoretically proved that there exists a density profile in a cylindrical pore. The closer the distance is away from the solid wall, the larger the density will be. Rene et al. [28] studied the meniscus thickness of pure water extended on clean quartz surface by image analysis interferometer and found that the thickness near the solid wall is larger than 0.1 μm, indicating the great effect of boundary layer on fluid flow in micro scales. The boundary layer thickness is not a constant [29–31], it is a function of hydrodynamics. The boundary layer thickness is large due to strong fluid-solid interaction at low ∇P. With the increase of ∇P, the shear force of the wall fluid increases, more and more fluid begins to flow, the boundary layer becomes thinner and the flow curve is closer to classical Hagen-Poiseuille's law [32–35].

Boundary layer effect on micro scale flow cannot be neglected owing to strong fluid-solid interaction. Its effect on fluid flow will become more and more significant with the decrease of flow scale. In addition, the wettability of the fluid on the solid surface will also affect micro fluid flow. When fluid has a strong wettability on the solid surface, the microscopic forces such as electrostatic forces and Van der Waals forces play a dominant role in micro fluid flow. The fluid in micro tubes is strongly affected by the fluid-solid interaction. There will be a large portion of fluid absorbed on the solid surface, which results in large boundary layer thickness. In order to characterize fluid flow in micro scales, boundary layer thickness must be described quantitatively in advance. While molecular dynamics simulation offers a good way to characterize fluid-solid interaction [36], it is very time-consuming and powerless for large-scale flow simulation, especially for porous media with thousands of pores and throats. In addition, the theoretical model which can characterize boundary layer thickness is hard to propose as fluid- solid interaction is very complex. Therefore, many empirical models combined with micro tube experiments have been developed so as to characterize boundary layer thickness [26,33,37–39]. At present, many empirical models are obtained in a traditional way. In addition, these models either have limited application range or lack physical meaning. Therefore, it is necessary to establish a novel model which not only offers definite physical meaning but also has a wider application range so as to study fluid flow in porous media with various pores.

In this work, a novel model which can characterize boundary layer thickness and fluid flow in micro scales is developed from a new prospective. Different micro tube experiments are used to validate the novel model. Furthermore, its application range is strictly proved mathematically, indicating a wide application scope. Based on the novel model, fluid-solid interaction effects on micro flow in micro tubes are studied, mainly including three parts: velocity profile, the average flow velocity and flow resistance. Finally, the novel model is incorporated into pore scale network model to study fluid-solid effects on fluid flow in tight formation.

2. Establishment of a Novel Model Considering Fluid-Solid Interaction

2.1. Construction of Modified Hagen-Poiseuille's Formula

So as to obtain the modified model in consideration of boundary layer effect which is caused by fluid-solid interaction, we assume that fluid flow belongs to steady flow in lateral micro tubes.

Since the radius is small and the micro-tube is laterally placed (see Figure 1), gravity can be ignored. X-axis is set along the flow direction in micro tubes, while r-axis is set vertically to the flow direction with its origin located in the center of micro tube. The radial and circumferential velocity component is zero. The velocity component parallel to the micro tube axis is u_x (only depends on r) and the pressure gradient along X-axis is a constant. The fluid viscosity is μ. The radius and the length of the micro tube is R and l respectively. The boundary layer thickness is h (caused by fluid-solid interaction).

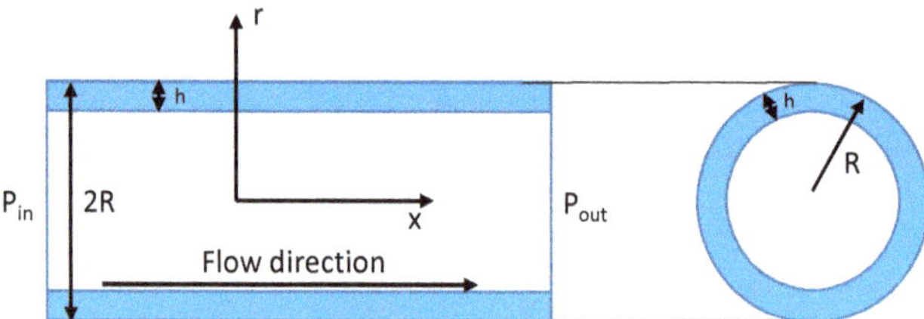

Figure 1. Scheme of fluid flow in lateral placed micro tube.

The modified Hagen-Poiseuille's formation is derived in terms of Newton's law of viscosity. The detailed derivation process is shown as follows.

Based on the element of cylindrical fluid, the pressure difference in the horizontal direction is,

$$\Delta F = \Delta P \pi r^2 \tag{1}$$

The viscous force of the surrounding fluid acting on the surface of the cylindrical fluid is,

$$f = \mu 2\pi r l \frac{dv}{dr} \tag{2}$$

With the increase of micro tube radius, the velocity decreases. Therefore, the velocity gradient $\frac{dv}{dr} < 0$.

As fluid flow in micro tube belongs to steady flow, the resultant force above is zero. That is,

$$\Delta F + f = 0 \tag{3}$$

Equations (1) and (2) are then incorporated into (3). After simplification, the expression is,

$$dv = -\frac{r}{2\mu} \nabla P dr \tag{4}$$

Integrate the Equation (4) from r to $R-h$,

$$\int_{v}^{0} dv = -\int_{r}^{R-h} \frac{r}{2\mu} \nabla P dr \tag{5}$$

The velocity distribution in micro tube in consideration of boundary layer effect is obtained,

$$v = \left[\frac{(R-h)^2 - r^2}{4\mu} \right] \nabla P \tag{6}$$

Integrate (6) along the flow section and the flow flux expression is acquired,

$$Q = \int_{0}^{R-h} v(r) \times 2\pi r dr = \frac{\pi (R-h)^4}{8\mu} \nabla P \quad (7)$$

The average flow velocity can be calculated as follow,

$$\overline{v} = \frac{Q}{A} = \frac{(R-h)^4}{8\mu R^2} \nabla P \quad (8)$$

If boundary layer effect is ignored, the Equations (7) and (8) degenerate into classical Hagen-Poiseuille's law.

Boundary layer thickness should be known in advance so as to predict the velocity and flow flux in micro tubes.

2.2. Establishment of Boundary Layer Thickness Expression

In this section, the expression of boundary layer thickness is developed from a new perspective based on Li 's micro tube experiment [31].

2.2.1. Micro Tube Experiment

The system of micro tube experiment and micro-flow parameters are satisfied with the assumption of micro fluid flow in Section 2.1. The micro tube system mainly consists of three parts: pressure supply unit, micro flow unit and measurement unit (See Figure 2). Every unit is specially designed to guarantee the accuracy of the experimental results.

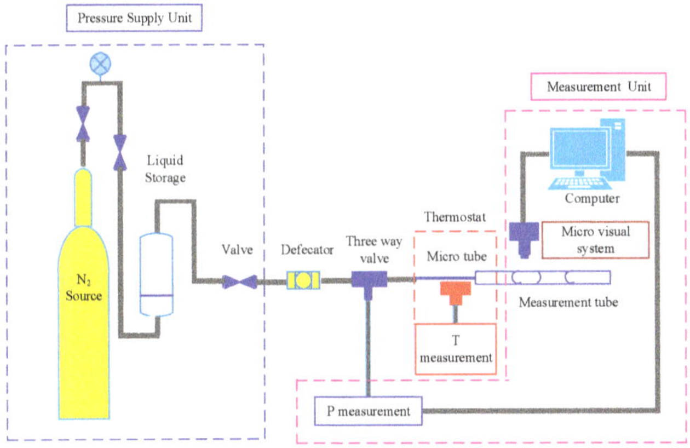

Figure 2. Schematic diagram of micro-tube experiment.

1. Pressure Supply Unit

In core scale displacement experiment, high precision displacement pump is usually chosen as the driving source. However, the actual pressure presents periodic fluctuations within a certain range in micro tube experiment. Constant pressure nitrogen is selected as the driving source after many experiments and screening, which guarantees the constant pressure boundary conditions at both ends of the micro tube.

2. Micro Flow Unit

Micro tube is the most important part in the experimental system. A fused silica micro tube made in the world's most advanced micro capillary manufacturing company—Polymicro Technologies, Inc. (Phoenix, AZ, USA)—is used. The micro tube is coated with polymide on the outer wall, which guarantees its flexibility and intensity. Quanta200 environmental scanning electron microscopy (ESEM) made in the company of FEI (Eindhoven, The Netherlands) is used to measure the radius of micro tube with its measurement accuracy 0.05 μm. And the measured radius is used for further calculation rather than nominal size. Fluid flow in micro tubes cannot be considered as Hagen-Poiseuille flow when roughness is large enough. After measurement through atomic force microscope, the relative roughness is much lower than 5%, which can be considered as hydraulic smooth pipe.

In micro tube experiment, deionized water is used as the flow medium. As the flow scale is extremely small, a small amount of impurity may lead to pipeline jam. Therefore, the deionized water must experience the process of sterilization, filtration and degassing before the experiment.

3. Measurement Unit

In order to reduce the error of measurement, capillary glass tube made in Sutter instrument Company is used as the measuring tube. The tube is treated with quenching and polishing. And its radius is uniform and the character is stable. As the flow rate in micro tube is rather small, photoelectric sensor is used to measure the process of displacement.

The process of the experiment is as follow.

N_2 is expelled out from high pressure nitrogen cylinder. Through the pressure relief valve, the pressure reduces to required value. Then, N_2 flows into liquid storage device to displace deionized water to micro tube and measurement tube. Photoelectric sensor is used to record the time that is elapsed after a period of distance in measurement tube.

Fluid flow in micro tubes with the nominal radius of 10 μm, 7.5 μm, 5 μm and 2.5 μm (the measured radius is actually 10.03 μm, 6.79 μm, 5.62 μm and 2.62 μm) is respectively carried out. The experimental velocity can be calculated by Equation (9),

$$v_{\exp} = \frac{l}{\Delta t} \tag{9}$$

where l is the distance of deionized water in measurement tube; Δt represents the time that deionized water travels through l.

The Hagen-Poiseuille's velocity can be calculated by Equation (10),

$$v_{HP} = \frac{Q_{HP}}{A} = \frac{R^2}{8\mu}\nabla P \tag{10}$$

The deviation is defined as follow,

$$S_v = \frac{v_{HP} - v_{\exp}}{v_{HP}} \tag{11}$$

Through data processing, the relationship of flow velocity versus displacement pressure gradient is obtained (See Figure 3).

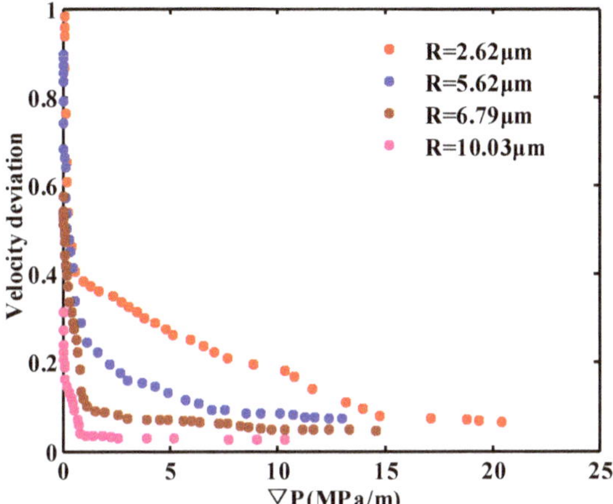

Figure 3. Relationship of velocity deviation versus displacement pressure gradient in different radial micro-tubes.

As can be seen from Figure 3, the deviation between experimental velocity and traditional Hagen-Poiseuille's velocity becomes larger with the decrease of micro-tube radius, which indicates that boundary layer effect (caused by fluid-solid interaction) on micro fluid flow cannot be ignored.

2.2.2. Representation of Boundary Layer Thickness

In this part, we will develop a representative boundary layer thickness model from the perspective of deviation between the experimental and traditional Hagen-Poiseuille's velocity.

As can be known, boundary layer forms near the wall surface due to fluid-solid interaction. When the displacement pressure gradient is zero, it can be reckoned that the boundary layer thickness is equivalent to the radius of the micro tube since there is no fluid flow in micro tubes. With the increase of the displacement pressure gradient, the shear force of the wall fluid becomes larger and the proportion of movable fluid grows, indicating a thinner boundary layer [22] (See Figure 4).

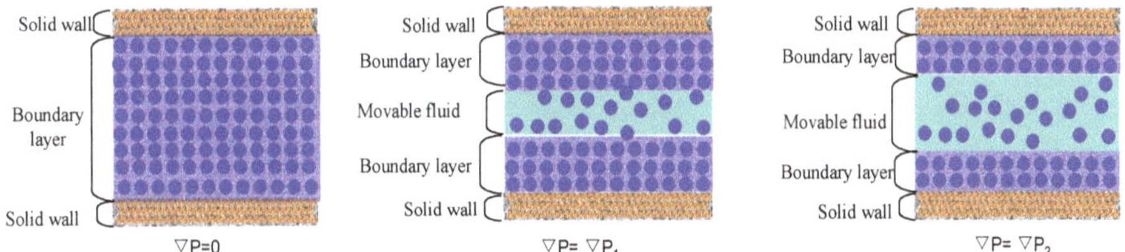

Figure 4. Schematic diagram of boundary layer thickness versus displacement pressure gradient ($\nabla P_1 < \nabla P_2$).

From Equations (8), (10) and (11), the expression of boundary layer thickness can be obtained.

$$h = R - R\sqrt[4]{1 - S_v} \tag{12}$$

From Equation (12), we can know that boundary layer thickness can be obtained through determining the expression of deviation between the Hagen-Poiseuille's velocity and experimental velocity.

According to the above analysis of boundary layer thickness and Equation (12), the deviation (S_v) reaches the maximal value 1 when the displacement pressure gradient is zero. And the deviation declines in the form of exponential function with the increase of ∇P (See Figure 3). So as to reflect the physical meaning when ∇P is zero and changing trend of deviation. The deviation model can be expressed as follow,

$$S_v = e^{-b\nabla P^c} \tag{13}$$

Parameters (b and c) in Equation (13) need to be determined so as to obtain S_v. Here, single variable method is used to obtain the parameters. Through fitting the experimental result of S_v by Equation (13) (See Figure 5), the parameters versus radius of micro tubes are obtained (See Table 1).

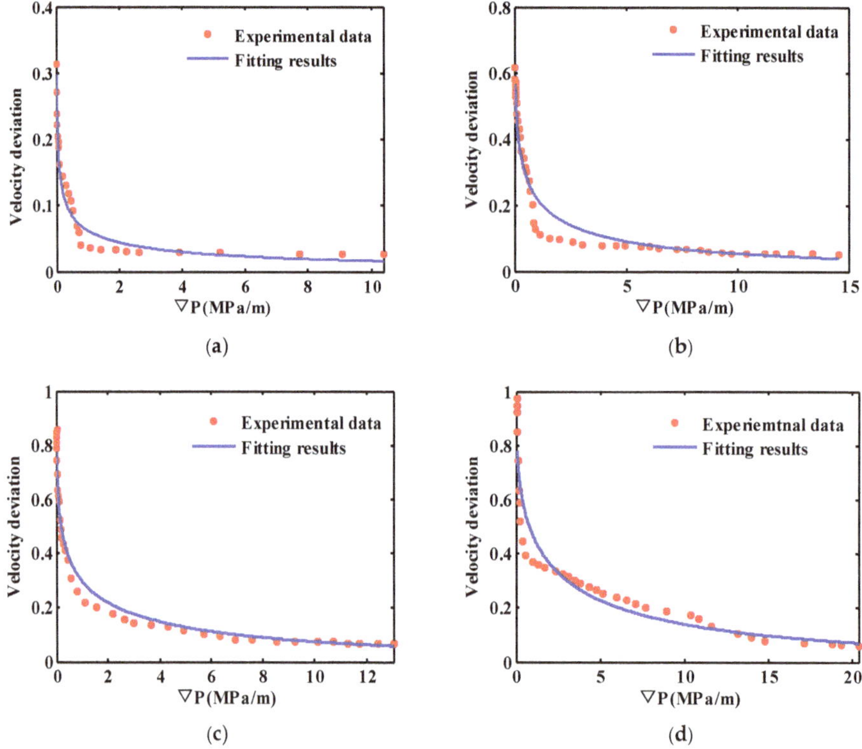

Figure 5. Fitting results with the novel deviation model: (**a**) R = 10.03 μm; (**b**) R = 6.79 μm; (**c**) R = 5.62 μm; (**d**) R = 2.62 μm.

Table 1. Parameters (b and c) versus radius of micro tubes.

Radius (μm)	b	c
2.62	0.7187	0.486
5.62	1.199	0.3372
6.79	1.527	0.284
10.03	2.77	0.1712

In order to make the model appropriate to wilder range, the exponential form is used to represent the relationship between the parameters and radius. The fitting result is shown in Figure 6.

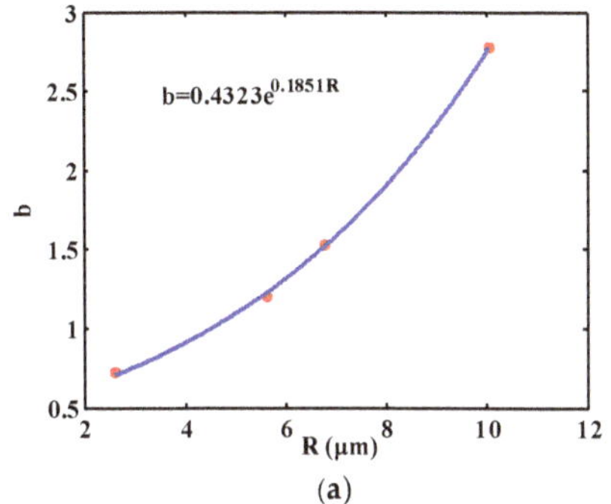

(a)

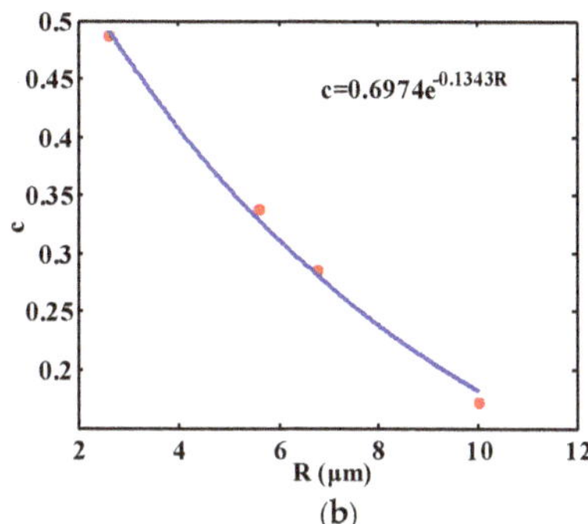

(b)

Figure 6. Parameters (b and c) versus radius: (**a**) b; (**b**) c.

The exponential expression is shown as follow,

$$\begin{cases} b = 0.4323e^{0.1851R} \\ c = 0.6974e^{-0.1343R} \end{cases} \tag{14}$$

Through substituting Equations (13) and (14) into (12), the boundary layer expression is eventually obtained,

$$h = R - R\sqrt[4]{1 - e^{-0.4323e^{0.1851R}\nabla P^{0.6974e^{-0.1343R}}}} \tag{15}$$

Finally, the modified Hagen-Poiseuille's Formula can be expressed as,

$$\begin{array}{l} Q = \frac{\pi(R-h)^4}{8\mu}\nabla P \\ h = R - R\sqrt[4]{1 - e^{-0.4323e^{0.1851R}\nabla P^{0.6974e^{-0.1343R}}}} \end{array} \tag{16}$$

3. Validation and Application Range Analysis of the Model

3.1. Validation of the Model

As there exists some deviation in fitting the relationship of S_v versus ∇P and the parameters (b and c) versus radius, we will firstly use the modified model to predict Li's micro-tube experimental results in turn. The predictive results in contrast with Li's experiments are shown in Figure 7. We can see that the predictive results are consistent with the experimental ones, indicating the accuracy of the modified model.

Furthermore, micro-tube experimental experiments at low ∇P conducted by Wu [22,40] are used to validate the accuracy and reliability of the novel model (See Figure 8). The predictive values by the novel model are still in agreement with the experimental results.

Through validation by the experimental results of Li [31] and Wu [22,40], it can be seen that the novel model can accurately characterize fluid flow in micro tubes. In addition, it has a definite physical meaning, which provides a good foundation for studying fluid-solid interaction effects on fluid flow in porous media.

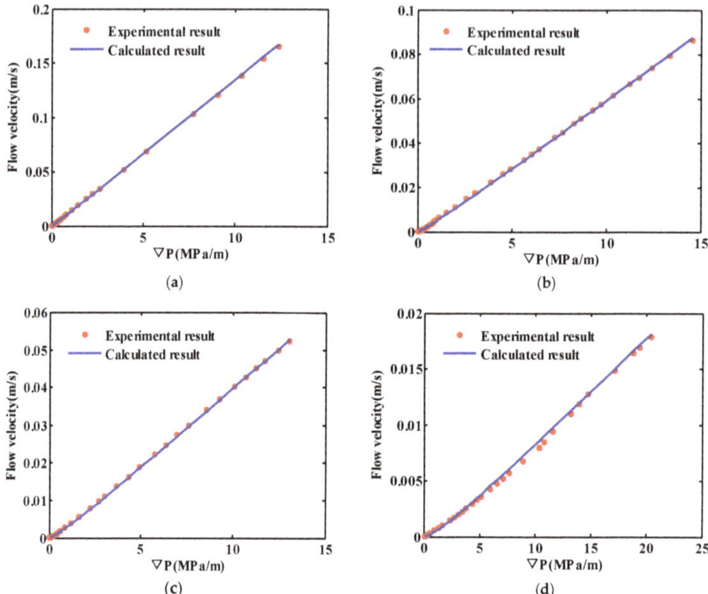

Figure 7. Comparison of predictive velocity versus ∇P and Li's experimental results [31]: (**a**) R = 10.03 µm; (**b**) R = 6.79 µm; (**c**) R = 5.62 µm; (**d**) R = 2.62 µm.

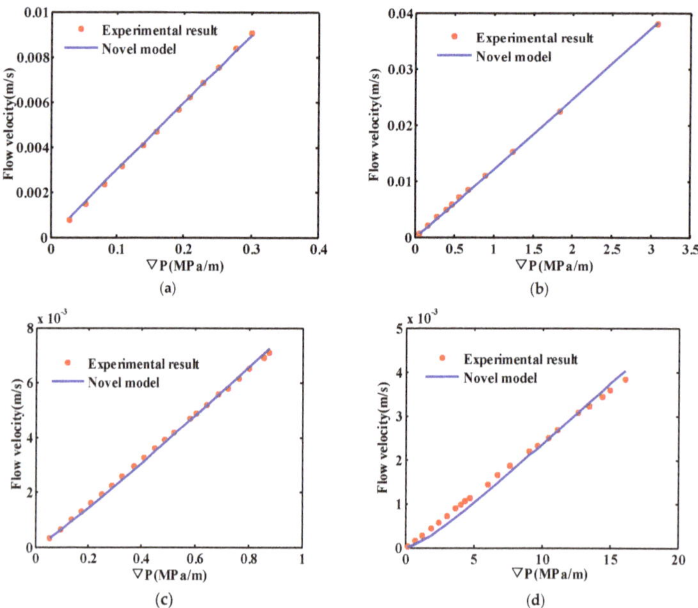

Figure 8. Comparison of calculated velocity versus ∇P and Wu's experimental results [22,40]: (**a**) R = 15.36 µm; (**b**) R = 10.03 µm; (**c**) R = 7.61 µm; (**d**) R = 1.38 µm.

3.2. Application Range Analysis of the Model

As there exist different sizes of pores and throats in tight formation, the model's application range and change rule must be discussed in advance when applied to microscopic flow in porous media. Otherwise, there may appear some singular values in some pores and throats, which may lead to inaccurate flow law in porous media. In this part, the application range and change rule of this novel model will be analyzed mathematically.

From the novel model (16) and Equation (12), we can obtain its range and change rule through analyzing the range of the deviation between experimental and traditional Hagen-Poiseuille's velocity. By taking the partial derivative of Equation (13) with respect to ∇P, the following equation can be obtained,

$$\frac{\partial S_v}{\partial \nabla P} = -bc\nabla P^{c-1} e^{-b\nabla P^{c-1}} \tag{17}$$

As $b > 0$, $c > 0$ and $\nabla P \geq 0$, then $\frac{\partial S_v}{\partial \nabla P} \leq 0$. That is to say S_v declines with the increase of ∇P. When ∇P equals to zero, both S_v and the ratio of the boundary layer thickness arrive at the maximal value 1, which obeys to the physical meaning. When ∇P tends to infinite, almost all the fluids in micro tubes start to flow, S_v and the ratio of boundary layer thickness tend to zero, which is also consistent with common sense. The detailed changing rule of the ratio of boundary layer thickness versus ∇P is shown in Figure 9.

As can be seen from Figure 9, the ratio of boundary layer thickness declines sharply with the increase of ∇P at first and then goes down as ∇P increases further. The larger the radius is, the quicker the decline rate will be. In addition, the boundary layer thickness will not be out of range at any ∇P as long as the radius is given and the flow pattern belongs to laminar flow.

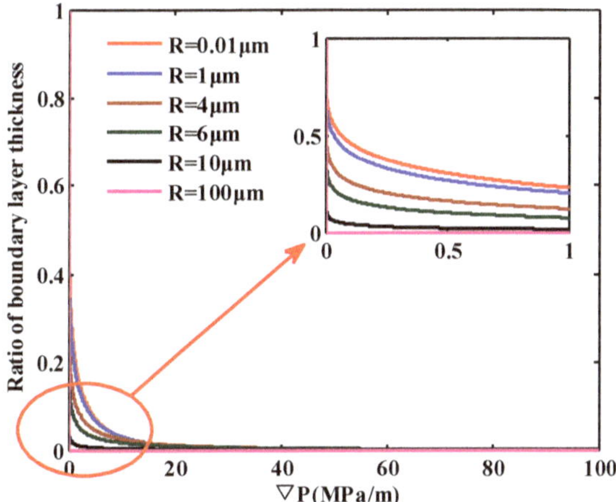

Figure 9. Ratio of boundary layer thickness versus ∇P under different radial micro tubes.

From the above analysis, the novel model has a broader application range than traditional empirical models. And singular values can be avoided in flow simulation in porous media with various sizes of pores and throats.

4. Fluid-Solid Interaction Effects on Microscopic Flow

In this section, the novel model will be applied to study fluid-solid interaction effects on microscopic flow from three aspects: the velocity profile, the average flow velocity and flow resistance.

Furthermore, we will incorporate the novel model into porous media to study fluid-solid interaction in tight formation.

4.1. Velocity Profile

Micro-tube radii of 2.62 μm, 5.62 μm, 6.79 μm and 10.03 μm are selected to study fluid-solid interaction effects on velocity profile. The viscosity and displacement pressure gradients are respectively set to be 0.92 mPa·s and 0.1 MPa/m.

As can be inferred from Figure 10, the velocity in the micro tube is smaller than that predicted by the traditional Hagen-Poiseullie's formula due to fluid-solid interaction while the velocity profile is still parabolic. The closer fluid is away from the wall surface, the stronger the fluid-solid interaction will be, which results in an immovable layer (boundary layer) near the surface. The velocity profile in consideration of fluid-solid interaction is consistent with the simulation result of dissipative particle dynamics (DPD) [41], which further confirms the accuracy of the model.

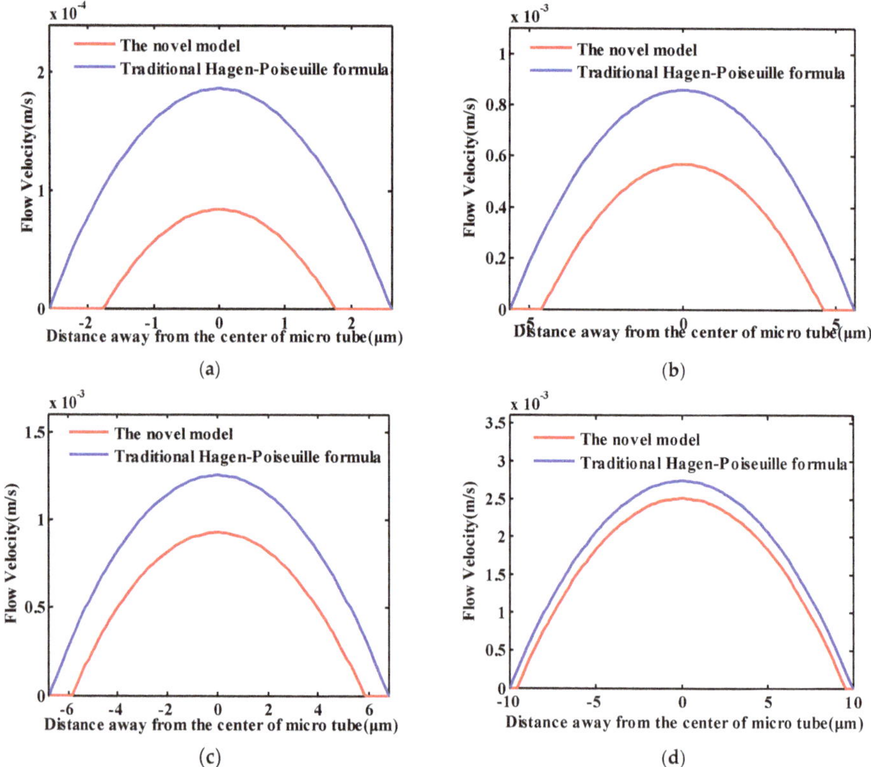

Figure 10. Velocity profile in different radial micro tubes: (**a**) R = 2.62 μm; (**b**) R = 5.62 μm; (**c**) R = 6.79 μm; (**d**) R = 10.03 μm.

In order to see the influence degree of fluid-solid interaction on fluid flow at different positions in micro tubes quantitatively, a velocity deviation which is described by Equation (18) is defined,

$$D_v = \frac{V'_{HP} - V'_{model}}{V'_{HP}} \tag{18}$$

where V'_{HP} represents Hagen-Poiseullie's velocity in the micro tubes; V'_{model} represents the velocity, in consideration of the fluid-solid interaction in the micro tubes.

The calculated velocity deviations in different radial micro tubes are shown in Figure 11.

As can be seen from Figure 11, the velocity deviation at different positions in every radial micro tube varies a lot. When the fluid is close to the wall surface, the fluid-solid interaction is strong enough to adsorb the boundary fluid to the wall and generates an immovable layer, which results in the maximal deviation 1. As the distance away from the wall surface increases, the decline rate of velocity deviation goes down quickly at first and then slows down, which indicates that the forces of the fluid-solid interaction belong to a short-range force. With the increase of the distance away from the wall surface, fluid-solid interaction effects on fluid flow decreases significantly. The fluid-solid interaction is the weakest in the center of the micro tube, which leads to minimal deviation. The smaller the radius of micro tube, the stronger the fluid-solid interaction, the larger the deviation will be.

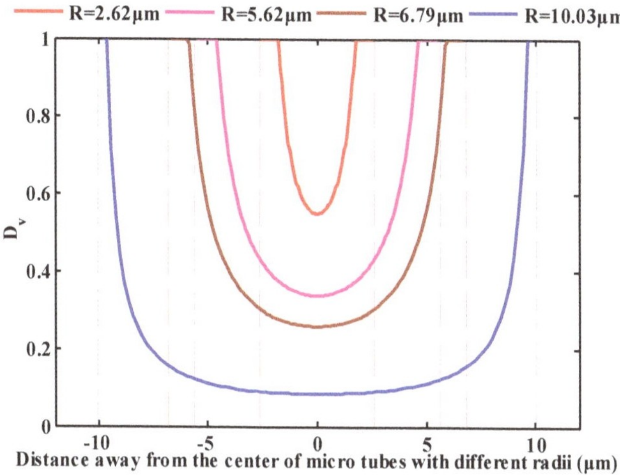

Figure 11. Deviations of velocity profile in different radial micro tubes.

4.2. The Average Flow Velocity

In this part, fluid-solid interaction on average flow velocity will be studied under the same micro tube radius and the same displacement pressure gradient respectively. The parameters used are shown in Table 2.

Table 2. Parameters used to study fluid-solid interaction on average flow velocity.

Constant Radius of Micro Tube		Constant Displacement Pressure Gradient	
Fluid viscosity (mPa·s)	0.92	Fluid viscosity (mPa·s)	0.92
Radius (μm)	5	∇P (MPa/m)	0.5

The calculation results are shown in Figure 12.

As shown in Figure 12a, the average flow velocity in consideration of fluid-solid interaction ($V_{new\ model}$) is smaller than traditional Hagen-Poiseuille's velocity (V_{HP}) at the same micro tube radius. With the increase of displacement pressure gradient, fluid-solid interaction effects on fluid flow decline and the boundary layer thickness is reduced, resulting in larger effective flow space and smaller deviation between $V_{new\ \mathrm{model}}$ and V_{HP}. It can be known from Figure 12b that $V_{new\ model}$ is smaller than V_{HP} and decreases with the increase of radius at the same displacement pressure gradient, which indicates that the effect of fluid-solid interaction on fluid flow declines as micro-tube radius increases.

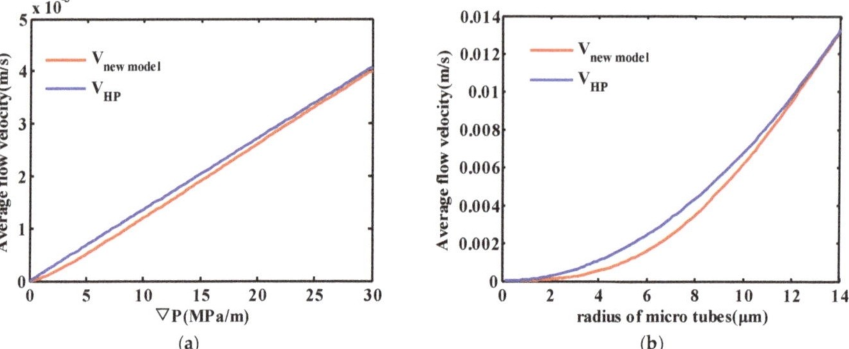

Figure 12. Fluid-solid interaction on average flow velocity under two conditions: (**a**) Average flow velocity versus ∇P; (**b**) Average flow velocity versus radius.

4.3. Flow Resistance

As can be seen from the above analysis, the flow law in micro scale is different from traditional Hagen-Poiseuille's law due to fluid-solid interaction. In this part, the flow resistance in micro tubes in consideration of fluid-solid interaction will be further analyzed. The Reynolds number (R_e), Resistance coefficient (f) and Poiseuille number (P_o) after considering fluid-solid interaction can be respectively calculated as,

$$R_e = \frac{\rho v D}{\mu} = \frac{\rho D (R-h)^4 \nabla P}{8\mu^2 R^2} \tag{19}$$

$$f = \frac{2D}{\rho v^2} \nabla P = \frac{2D}{\rho\left(\frac{(R-h)^4}{8\mu R^2} \nabla P\right)^2} \nabla P \tag{20}$$

$$P_o = f R_e = \frac{16 R^2 D^2}{(R-h)^4} \tag{21}$$

where D is diameter of micro tubes; ρ is fluid density.

The relationships of R_e and f versus ∇P are shown in Figure 13. With the increase of displacement pressure gradient, the fluid initially adsorbed on solid wall begins to flow reducing the boundary layer thickness. The average flow velocity increases which eventually leads to the increase of R_e and decrease of resistance coefficient. Since the fluid-solid interaction effects on micro fluid flow weaken with the decrease of flow scale at the same displacement pressure gradient, the deviations of R_e and f between the novel model and traditional Hagen-Poiseuille's Formula becomes smaller and smaller as flow scale increases. In addition, we notice that the Reynolds number in micro fluid flow is far less than 2300 which suggests that the fluid flow in micro tubes belongs to laminar flow. This phenomenon further declares the reality of the novel model's assumption. It can be seen that the resistance coefficient is always larger than 1, which demonstrates the non-negligible effects of fluid-solid interaction on micro scale flow.

In terms of classical laminar flow, the Poiseuille number is a constant with the value of 64 when fluid flow in horizontal circular tube is fully developed. However, some researchers hold the idea that the Poiseuille number is no longer a constant as the flow scale becomes smaller. Here, the change rule of Poiseuille number in micro tubes is analyzed based on the novel model. And the calculated results are shown in Figure 14.

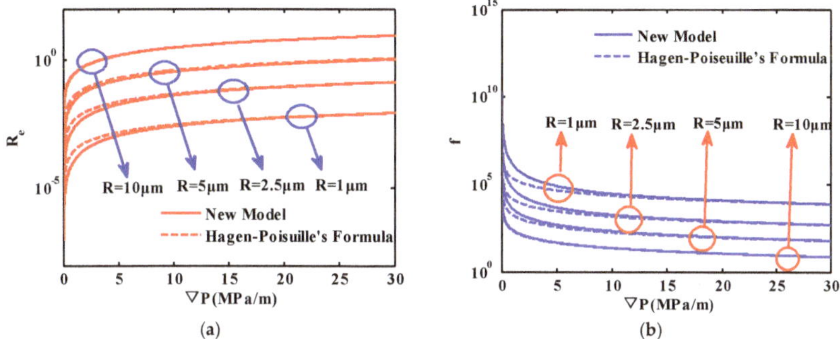

Figure 13. Relationships of R_e and f versus ∇P: (**a**) R_e versus ∇P; (**b**) f versus ∇P.

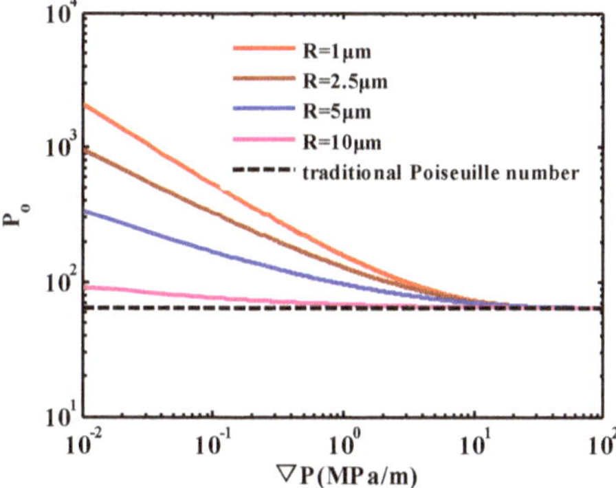

Figure 14. The relationship of P_o versus ∇P.

As can be seen from Figure 14, the Poiseuille number is indeed no longer a constant when considering fluid-solid interaction. It declines with the increase of ∇P. When ∇P is large enough, the effect of fluid-solid interaction on fluid flow is reduced and the boundary layer thickness becomes thinner. As a result, the Poiseuille number in consideration of fluid-solid interaction is more and more close to traditional Poiseuille number. The deviation between the Poiseuille number considering fluid-solid interaction and traditional one increases with the decrease of micro-tube radius at the same displacement pressure gradient.

4.4. Pore Scale Network Model

The above investigation of fluid-solid interaction effects on microscopic flow mainly focuses on micro tubes. In this part, we will apply this novel model to pore scale network model which can represent the complex structure of tight formation to study fluid-solid interactions on fluid flow in porous media. Pore-throat radii and throat lengths are assumed to obey the truncated Weibull distribution in the pore network model.

$$R = (R_{\max} - R_{\min})\left(-\delta \ln\left[x\left(1 - e^{\frac{-1}{\delta}}\right) + e^{\frac{-1}{\delta}}\right]\right)^{\frac{1}{\eta}} + R_{\min} \tag{22}$$

where R represents pore and throat radii; $R_{\max}$ and $R_{\min}$ represent respectively the maximal throat and the minimal throat radius; δ represents scale distribution parameters; η represents shape distribution parameters; $x \in [0,1]$, which is a random number.

Detailed parameters in the pore network model are shown in Table 3. As the aspect ratio (the value of pore radius divided by throat radius) is large, the fluid-solid interaction effects on fluid flow will be only considered into throats.

Table 3. Parameters of pore scale network model.

Parameters	Values	Parameters	Values
Model size	10 × 10 × 10	Throat radius (μm)	1–4
Pore radius (μm)	10–100	Throat length (μm)	30–40
Fluid viscosity (mPa s)	0.92		

The conductance in consideration of fluid-solid interaction in pore network model can be modified as,

$$g = \frac{\pi r_{eff}^2}{8\mu} = \frac{\pi (R-h)^2}{8\mu} = \frac{\pi \left(R\sqrt[4]{1 - e^{-0.4323 e^{0.1851R} \nabla p^{0.6974 e^{-0.1343R}}}} \right)^2}{8\mu} \tag{23}$$

Fluid flow through every pore satisfies mass conservation law at every displacement pressure gradient (See Figure 15a),

$$\sum_k q_{jk} = 0 \tag{24}$$

where q_{jk} represents the flow flux between pore i and a neighboring pore j. The flow flux between two neighboring pores can be calculated as follow,

$$q_{jk} = \frac{g_{jk}}{L_{jk}} (P_j - P_k) \tag{25}$$

where g_{jk} represents the conductance between pore j and k. It can be calculated by the harmonic mean of the conductance of the throat and two neighboring pores (See Figure 15b).

$$g_{jk} = \frac{L_{jk}}{\frac{L_j}{g_j} + \frac{L_t}{g_t} + \frac{L_k}{g_k}} \tag{26}$$

where L_{jk} is the distance between pore j and k; P_j and P_k represent fluid pressure in pore j and k respectively.

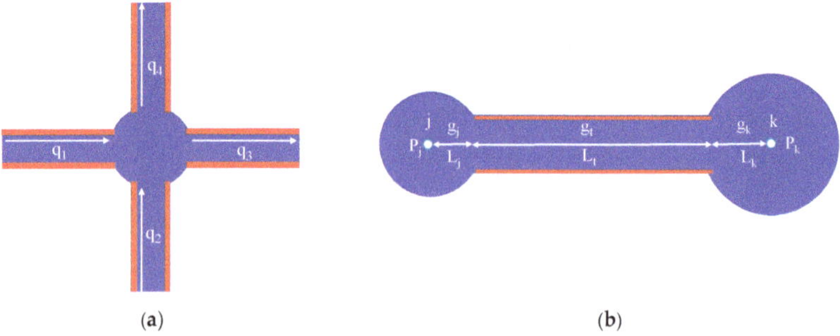

Figure 15. Scheme of pore network model (red part represents boundary layer while blue part represents the effective flow space): (**a**) Flow through every pore; (**b**) Conductance between two neighboring pores.

The apparent permeability in the pore network model in consideration of fluid-solid interaction can be calculated by Equation (27),

$$K_a = \frac{\mu Q L}{A(P_{in} - P_{out})} \tag{27}$$

The detailed calculation flow chart considering fluid-solid interaction is shown in Figure 16.

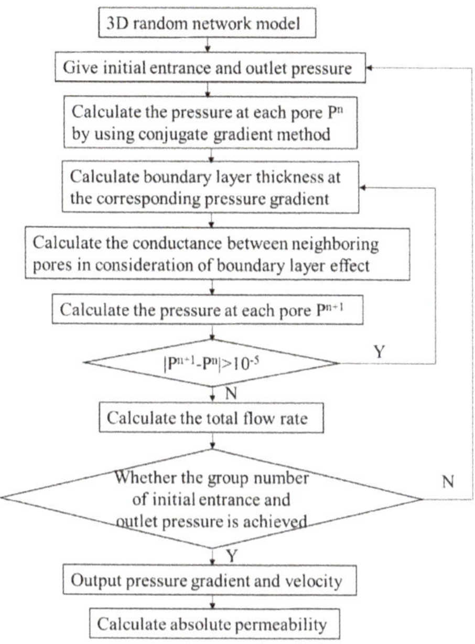

Figure 16. Flow chart of calculation in consideration of fluid-solid interaction.

The calculation results are shown in Figure 17.

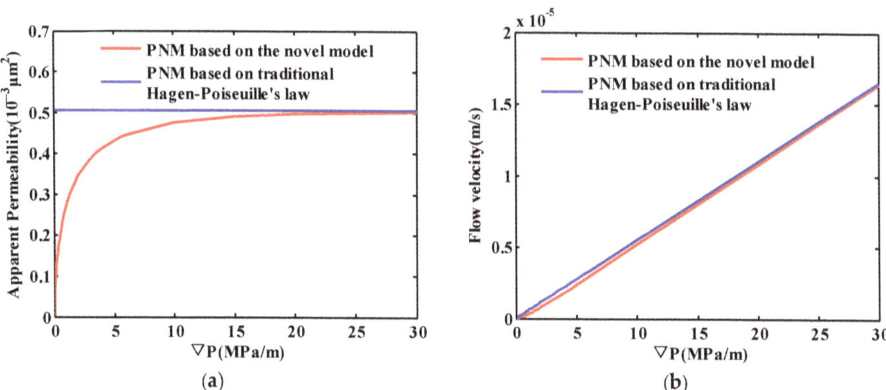

Figure 17. Relationship between apparent permeability and flow rate versus ∇P: (**a**) Apparent permeability versus ∇P; (**b**) Flow velocity versus ∇P.

As is known from Figure 17a, the apparent permeability is no longer a constant when considering fluid-solid interaction. The effect of fluid-solid interaction on fluid flow weakens with the increase of displacement pressure gradient, the boundary layer thickness declines and the effective flow space is enlarged. As a result, the deviations of the apparent permeability grow smaller. When the displacement pressure gradient is large enough, the apparent permeability is basically consistent with the one ignoring fluid-solid interaction. From Figure 17b, we can see that nonlinear flow occurs at low displacement pressure gradient due to fluid-solid interaction. The velocity deviation also decreases with the increase of displacement pressure gradient.

In order to further investigate fluid-solid interactions on fluid flow in porous media, the throat radii are modified to change the flow scale in pore network model. The average aspect ratio is used to reflect the flow scale. The deviation of apparent permeability is defined in Equation (28).

$$D_k = \frac{K_{hp} - K_{f-s}}{K_{hp}} \tag{28}$$

where K_{hp} represents the apparent permeability, ignoring the fluid-solid interaction, and K_{f-s} in consideration of fluid-solid interaction.

For calculation, we respectively set the displacement pressure gradient as 0.1 MPa/m, 0.3 MPa/m and 0.5 MPa/m. The fluid viscosity is 0.92 mPa·s. The results are respectively shown in Figure 18.

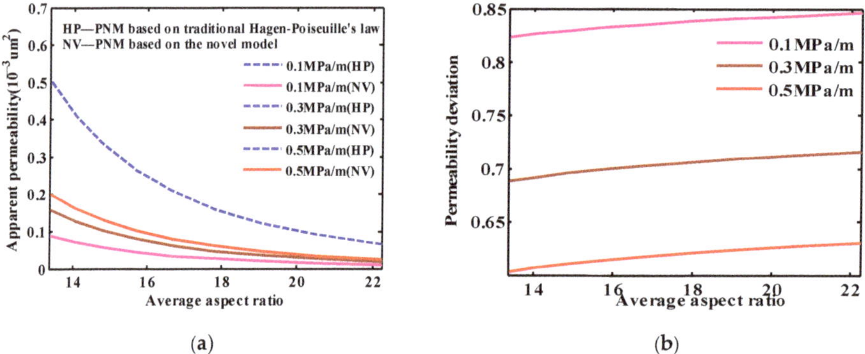

Figure 18. Relationship between apparent permeability versus average aspect ratio at different ∇P: (**a**) Apparent permeability; (**b**) Permeability deviation.

It can be seen from Figure 18 that the curves of apparent permeability versus the average aspect ratio without considering the fluid-solid interaction overlap, which indicates that apparent permeability has nothing to do with ∇P at different average aspect ratios. The apparent permeability decreases with increased average aspect ratio which suggests that the apparent permeability is the function of pore structures. When taking fluid-solid interaction into consideration, the apparent permeability at different ∇P is smaller than that ignoring its effects. Meanwhile, the curves of apparent permeability versus average aspect ratio no longer overlap. The effect of fluid-solid interaction on fluid flow is weakened and the boundary layer thickness is reduced with the increase of ∇P, which results in a larger effective flow space and smaller deviation of apparent permeability from that ignoring fluid-solid interaction. The decline trend of the apparent permeability also exhibits great difference with the increase of average aspect ratio due to fluid-solid interaction.

5. Conclusions

In this study, a novel model which can characterize fluid flow in micro scales is developed from a new perspective. Micro tube experiments are used to verify the novel model. Its application range is further analyzed mathematically. After the analysis of the novel model, fluid-solid interaction effects on the velocity profile, the average flow velocity and flow resistance in micro tubes are respectively studied. Finally, the novel model is incorporated into pore scale network model to study fluid-solid interaction effects on fluid flow in porous media. The following conclusions are arrived at:

1. When fluid-solid interaction is taken into consideration, the velocity in micro tube is smaller than that predicted by traditional Hagen-Poiseuille's law. The fluid-solid interaction declines significantly as the distance away from the solid wall grows larger, which results in largest velocity deviation near the wall surface and the smallest one in the center of micro tube.
2. Non-linear flow occurs both in micro tubes and porous media at low displacement pressure gradient due to fluid-solid interaction. Moreover, the effect of fluid-solid interaction on micro flow declines with increased displacement pressure gradient, which leads to smaller and smaller velocity deviation.
3. The changing rules of the Reynolds number, the Resistance coefficient and Poiseuille number do not obey traditional Hagen-Poiseuille's law owing to fluid-solid interaction. The smaller the radius and the lower the displacement pressure gradient, the stronger the fluid-solid interaction, the larger the deviation will be.
4. The apparent permeability in porous media is no longer a constant when incorporating the novel model into pore scale network model. The apparent permeability increases and the permeability deviation declines with increased displacement pressure gradient.

Author Contributions: M.C.; Methodology & Writing—Original Draft Preparation, L.C. and R.C.; Supervision & Editing; C.L.; Model analysis & Editing.

Acknowledgments: This work is supported by the National Basic Research Program of China (Grant No. 2015CB250902), the National Natural Science Foundation of China (Grant Nos. 51574258 and 51674273) and University Doctoral Special Research Fund (Grant No. 20130007120014).

Conflicts of Interest: The authors declare no conflict of interest.

References

1. Clarkson, C.R.; Pedersen, P.K. Tight oil production analysis: Adaptation of existing rate-transient analysis techniques. In Proceedings of the Canadian Unconventional Resources and International Petroleum Conference, Calgary, AB, Canada, 19–21 October 2010.
2. Du, J.H.; He, H.Q.; Yang, T.; Huang, F.X.; Guo, B.C.; Yan, W.P. Progress in China's tight oil exploration and challenges. *China Pet. Exp.* **2014**, *19*, 1–9.
3. Lin, S.H.; Zou, C.N.; Yuan, X.J.; Yang, Z. Status quo of tight oil exploitation in the United States and its implication. *Lithol. Reserv.* **2011**, *23*, 25–30.
4. Cui, X.; Bustin, A.M.M.; Bustin, R.M. Measurements of gas permeability and diffusivity of tight reservoir rocks: Different approaches and their applications. *Geofluids* **2009**, *9*, 208–223. [CrossRef]
5. Liu, Y.F.; Hu, W.X.; Cao, J.; Wang, X.L.; Tang, Q.S.; Wu, H.G.; Kang, X. Diagenetic constraints on the heterogeneity of tight sandstone reservoirs: A case study on the Upper Triassic Xujiahe Formation in the Sichuan Basin, southwest China. *Mar. Pet. Geol.* **2018**, *92*, 650–669. [CrossRef]
6. Lyu, C.H.; Ning, Z.F.; Wang, Q.; Chen, M.Q. Application of NMR T2 to pore size distribution and movable fluid distribution in Tight Sandstones. *Energy Fuels* **2018**, *32*, 1395–1405. [CrossRef]
7. Zhao, H.W.; Ning, Z.F.; Wang, Q.; Zhang, R.; Zhao, T.Y.; Niu, T.F.; Zeng, Y. Petrophysical characterization of tight oil reservoirs using pressure-controlled porosimetry combined with rate-controlled porosimetry. *Fuel* **2015**, *154*, 233–242. [CrossRef]
8. Wang, F.; Liu, Z.; Jiao, L.; Wang, C.; Guo, H.U. A fractal permeability model coupling boundary-layer effect for tight oil reservoirs. *Fractals* **2017**, *25*, 1750042. [CrossRef]
9. Wu, J.S.; Hu, D.Z.; Li, W.J.; Cai, X. A review on non-Darcy flow u2014 forchheimer equation, hydraulic radius model, Fractal model and experiment. *Fractals-Complex Geom. Patterns Scaling Nat. Soc.* **2016**, *24*, 1782–1788.
10. Soroori, S.; Rodriguez-Delgado, J.M.; Kido, H.; Dieck-Assad, G.; Madou, M.; Kulinsky, L. The use of polybutene for controlling the flow of liquids in centrifugal microfluidic systems. *Microfluid. Nanofluid.* **2016**, *20*, 26. [CrossRef]

11. Guo, J.S.; Huang, B.C. Hyperbolic quenching problem with damping in the micro-electro mechanical system device. *Discret. Contin. Dyn. Syst.* **2014**, *2*. [CrossRef]
12. Campo-Deaño, L. Fluid-Flow Characterization in Microfluidics. In *Complex Fluid-Flows in Microfluidics*; Galindo-Rosales, F.J., Ed.; Springer International Publishing: New York, NY, USA, 2018; pp. 53–71.
13. Kavallaris, N.I.; Suzuki, T. Micro-electro-mechanical-systems (MEMS). In *Non-Local Partial Differential Equations for Engineering and Biology: Mathematical Modeling and Analysis*; Springer International Publishing: New York, NY, USA, 2018; pp. 3–63.
14. Ho, C.M.; Tai, Y.C. Micro-electro-mechanical-systems (MEMS) and fluid flows. *Annu. Rev. Fluid Mech.* **1996**, *30*, 579–612. [CrossRef]
15. Wang, F.; Yue, X.; Xu, S.; Zhang, L.; Zhao, R.; Hou, J. Influence of wettability on flow characteristics of water through microtubes and cores. *Chin. Sci. Bull.* **2009**, *54*, 2256–2262. [CrossRef]
16. Yue, X.A.; Wang, N.; Zhang, L.J.; Wang, F. Flow experiments of HPAM solution in quartz micro-tubes. *Mech. Eng.* **2010**, *32*, 81–84.
17. Ling, Z.Y.; Ding, J.N.; Yang, J.C.; Fan, Z.; Li, C.S. Research Advance in Microfluid and Its Influencing Factors. *J. Jiangsu Univ. Sci. Technol.* **2002**, *6*, 1–5.
18. Li, Z.H.; Cui, H.H. Characteristics of Micro Scale Flow. *J. Mech. Strength* **2001**, *4*, 476–480.
19. Pfahler, J.; Harley, J.; Bau, H.; Zemel, J. Liquid transport in micron and submicron channels. *Sens. Actuators A Phys.* **1990**, *22*, 431–434. [CrossRef]
20. Makihara, M.; Sasakura, K.; Nagayama, A. The Flow of Liquids in Micro-Capillary Tubes: Consideration to Application of the Navier-Stokes Equations. *J. Jpn. Soc. Precis. Eng.* **1993**, *59*, 399–404. [CrossRef]
21. Qu, M.L.; Mala, M.; Li, D.Q. Pressure-driven water flows in trapezoidal silicon microchannels. *Int. J. Heat Mass Transf.* **2000**, *43*, 353–364. [CrossRef]
22. Wu, J.Z.; Cheng, L.S.; Li, C.L.; Cao, R.Y.; Chen, C.C.; Cao, M.; Xu, Z.Y. Experimental study of nonlinear flow in micropores under low pressure gradient. *Transp. Porous Media* **2017**, *119*, 247–265. [CrossRef]
23. Sandeep, A.; Saleem, K.; Akhil, V.; Harneet, K.; Parveen, L. Microfluidic mechanics and applications: A review. *J. Nano-Electron. Phys.* **2013**, *5*, 04047-1.
24. Gad-El-Hak, M. The Fluid Mechanics of Micro-devices—The Freeman Scholar Lecture. *ASME J. Fluids Eng.* **1999**, *121*, 5–33. [CrossRef]
25. Zhang, X.L.; Zhu, W.Y.; Cai, Q.; Liu, Q.P.; Wang, X.F.; Lou, Y. Analysis of weakly compressible fluid flow in nano/micro-size circular tubes considering solid wall force. *Acta Sci. Nat. Univ. Pekin.* **2014**, *36*, 569–575.
26. Мархасин, И.Л. *Physical and Chemical Mechanism of Reservoir*; Petroleum Industry Press: Beijing, China, 1987.
27. Pertsin, A.; Grunze, M. Water-graphite interaction and behavior of water near the graphite surface. *J. Phys. Chem. B* **2004**, *108*, 1357–1364. [CrossRef]
28. Mazzoco, R.R.; Wayner, P.C. Aqueous Wetting Films on Fused Quartz. *J. Colloid Interface Sci.* **1999**, *214*, 156–169. [CrossRef] [PubMed]
29. Liu, D.X.; Yue, X.A.; Hou, J.R.; Wang, L.M. Experimental Study of Adsorbed Water Layer on solid particle surface. *Acta Miner. Sin.* **2005**, *1*, 15–19.
30. Huang, Y.Z. *Infiltration Mechanism of Low-Permeability Reservoir*; Petroleum Industry Press: Beijing, China, 1999.
31. Li, Y. Study of Microscale Nonlinear Flow Characteristics and Flow Resistance Reducing Methods. Ph.D. Thesis, Institute of Porous Flow and Fluid Mechanics, Lanfang, China, 2010.
32. Wu, J.Z.; Cheng, L.S.; Li, C.L.; Cao, R.Y.; Chen, C.C.; Xu, Z.Y. Flow of Newtonian fluids with different polarity in micro scale. *Chin. Sci. Bull.* **2017**, *62*, 2988–2996.
33. Xu, S.; Yue, X.; Hou, J. Experimental study on the flow characteristics of deionized water in microcircular tubes. *Chin. Sci. Bull.* **2007**, *6*, 120–124.
34. Zhu, C.J.; Zhang, J.; Zhang, P. Study on Influence of Boundary Layer on the Non-Darcy Seepage Law. *J. Converg. Inf. Technol.* **2013**, *8*, 960–968.
35. Zhang, P.; Zhang, L.Z.; Li, W.Y.; Wang, Y.F. Experiment on the influence of boundary layer on the Non-Darcy seepage law. *J. Hebei Univ. Eng.* **2008**, *25*, 70–72. (In Chinese)
36. Thomas, J.A.; McGaughey, A.J.H. Reassessing fast water transport through carbon nanotubes. *Nano Lett.* **2008**, *8*, 2788–2793. [CrossRef] [PubMed]
37. Li, Z.F.; He, S.L. Influence of boundary layers upon filtration law in low-permeability oil reservoirs. *Pet. Geol. Oilfield Dev. Daqing* **2005**, *24*, 57–59, 77–107.

38. Liu, W.D.; Liu, J.; Sun, L.H.; Li, Y.; Lan, X.Y. Influence of Fluid Boundary Layer on Fluid Flow in Low Permeability Oilfields. *Sci. Technol. Rev.* **2011**, *29*, 42–44.
39. Cao, R.Y.; Wang, Y.; Cheng, L.S.; Ma, Y.Z.; Tian, X.F.; An, N. A New Model for Determining the Effective Permeability of Tight Formation. *Transp. Porous Media* **2016**, *112*, 21–37. [CrossRef]
40. Wu, J.Z.; Cheng, L.S.; Cao, R.Y.; Xu, Z.Y.; Ding, G.Y.; Ding, G.Y. A Novel Characterization of Effective Permeability of Tight Reservoir—Based on the Flow Experiments in Microtubes. In Proceedings of the 19th European Symposium on Improved Oil Recovery, Stavanger, Norway, 23–28 April 2017.
41. Tian, X.F. Non-Linear Flow Behavior and Mathematical Model in Ultra-Low Permeability Reservoirs. Ph.D. Thesis, China University of Petroleum (Beijing), Beijing, China, 2015.

Article

A Transient Productivity Model of Fractured Wells in Shale Reservoirs Based on the Succession Pseudo-Steady State Method

Fanhui Zeng [1,*], Fan Peng [1], Jianchun Guo [1,*], Jianhua Xiang [2], Qingrong Wang [2] and Jiangang Zhen [3]

1 State Key Laboratory of Oil and Gas Reservoir Geology and Exploitation, Southwest Petroleum University, Chengdu 610500, China; 201721000711@stu.swpu.edu.cn
2 PetroChina Limited Company Southwest Oil and Gas Field Branch, Chengdu 610017, China; xiangjh@petrochina.com.cn (J.X.); wang_qr@petrochina.com.cn (Q.W.)
3 PetroChina Changqing Oil Field Branch Twelfth Oil Production Plant, Xi'an 710200, China; zjig_cq@petrochina.com.cn
* Correspondence: zengfanhui@tch.swpu.edu.cn (F.Z.); guojianchun@tch.swpu.edu.cn (J.G.)

Received: 11 August 2018; Accepted: 25 August 2018; Published: 1 September 2018

Abstract: After volume fracturing, shale reservoirs can be divided into nonlinear seepage areas controlled by micro- or nanoporous media and Darcy seepage areas controlled by complex fracture networks. In this paper, firstly, on the basis of calculating complex fracture network permeability in a stimulated zone, the steady-state productivity model is established by comprehensively considering the multi-scale flowing states, shale gas desorption and diffusion after shale fracturing coupling flows in matrix and stimulated region. Then, according to the principle of material balance, a transient productivity calculation model is established with the succession pseudo-steady state (SPSS) method, which considers the unstable propagation of pressure waves, and the factors affecting the transient productivity of fractured wells in shale gas areas are analyzed. The numerical model simulation results verify the reliability of the transient productivity model. The results show that: (1) the productivity prediction model based on the SPSS method provides a theoretical basis for the transient productivity calculation of shale fractured horizontal well, and it has the characteristics of simple solution process, fast computation speed and good agreement with numerical simulation results; (2) the pressure wave propagates from the bottom of the well to the outer boundary of the volume fracturing zone, and then propagates from the outer boundary of the fracturing zone to the reservoir boundary; (3) with the increase of fracturing zone radius, the initial average aperture of fractures, maximum fracture length, the productivity of shale gas increases, and the increase rate gradually decreases. When the fracturing zone radius is 150 m, the daily output is approximately twice as much as that of 75 m. If the initial average aperture of fractures is 50 μm, the daily output is about half of that when the initial average aperture is 100 μm. When the maximum fracture length increases from 50 m to 100 m, the daily output only increases about by 25%. (4) When the Langmuir volume is relatively large, the daily outputs of different Langmuir volumes are almost identical, and the effect of Langmuir volume on the desorption output can almost be ignored.

Keywords: shale gas reservoir; fractured well transient productivity; succession pseudo-steady state (SPSS) method; complex fracture network; multi-scale flow; analysis of influencing factors

1. Introduction

Shale gas reservoirs have the characteristics of tiny pores and throats, extremely low permeability, abundant natural fractures and diverse gas storage modes [1–9]. After the reservoir is fractured, a complex fracture network is generated, composed of activated natural fractures and induced hydraulic fractures, and the gas flow channels are converted from nanoscale pores to microscale pores so that the gas flow state in the stimulated zone is changed [10–18]. The Darcy flow in the stimulated region near the wellbore is coupled with the nano-/microscale flow in the matrix, resulting in the complex multiscale flow in shale gas reservoirs [19]. Because interactions between desorption, diffusion and seepage exist in reservoir shale gas, the original linear seepage theory is no longer applicable as a result of the change of flow patterns [20]. Therefore, it is necessary to establish a new transient productivity calculation model of fractured wells considering of multi-scale flow and nonlinear seepage.

Currently, there are few studies on the transient productivity of fractured wells in shale gas reservoirs, and the characteristics of multi-scale flow and non-linear flow are neglected in the majority of productivity models [21–23]. The current analytical models are based on Laplace transform, superposition principle, source function, complex function theory and other mathematical methods to analyze the transient pressure response for shale reservoir, while the Laplace transform and source function are too complicated to solve. Meanwhile, the assumption that the pressure has already reached the boundary of the reservoir under initial condition is inconsistent with the actual situation. The new model established in this paper can avoid these disadvantages.

For the complex fracture network intermingled with natural fractures and artificial fractures near the wellbore [24–26], Min et al. [27] used the cubic law to calculate the permeability of complex fracture networks, and fracture closure and shear failure of complex fracture networks under different stresses are considered. In a study of the unsteady productivity prediction of shale gas, Stalgorova et al. [28] divided the seepage area of shale reservoir into stimulated region and un-stimulated region, and a pressure transient model based on tri-line flow model was established. However, this model neglected shale gas desorption and diffusion effects so that it was inaccurate for shale gas reservoir productivity evaluation. On the basis of the tri-line flow model, Swami et al. [29] established a dual-porosity seepage model of shale gas reservoir considering adsorption and desorption in matrix. However, this model still assumed that shale gas is linear seepage and the characteristic of multi-scale flow is neglected. Based on the previous studies, Deng [12] corrected the B-K model with different slip coefficients, and established a multi-scale flow model for shale gas reservoir considering diffusion, slippage, desorption and adsorption, while the assumption of homogeneous medium does not match with volume fracturing and not take into account the unsteady seepage of shale reservoir. Zhang et al. [30] and Su et al. [31] extended the dual-region composite flow model proposed by Zhao et al. [32] to the shale gas productivity model. The transient pressure response model of horizontal well was established by point-source function and Laplace transform, while the inappropriate assumption that the initial pressure extends to the reservoir boundary at the beginning was made. In addition, some researchers have studied other methods for calculating fracturing well properties and productivity [33–37]. With the production of shale gas, the effective stress of the reservoir increases, and a stress-sensitive phenomena occur. Meanwhile, adsorbed shale gas will undergo surface diffusion in addition to desorption [38–40].

In order to overcome the deficiencies of the source function and other mathematical methods, Shahamat et al. [41] analyzed the pressure response under the condition of transition flow and boundary flow with the succession pseudo-steady state (SPSS) method, which considers the correlation of pressure wave propagation with reservoir properties, fluid properties, and production time. The SPSS method is simple and easy to implement programming, which avoids the unreasonable assumption that the pressure has already reached the boundary of the reservoir under the initial condition. However, Shahamat did not take into consideration the multi-scale flow and non-linear flow after fracturing in the shale gas reservoir.

In this paper, first, based on the Beskok-Karniadakis apparent permeability model in regard of different slip coefficients, the seepage region in shale gas reservoir is divided into a stimulated zone and a matrix zone. The radial Darcy flow is considered in the stimulated zone and non-linear flow characterized by Knusen Number is considered in the matrix zone. Second, the mathematical model of steady-state productivity of shale horizontal wells with volume fracturing is derived. Then, a transient productivity calculation model of fractured wells in shale combined the material balance equation is established with the SPSS method, and the numerical model simulation results verify the reliability of the transient productivity model. Finally, the horizontal well productivity prediction and the analysis of influencing factors are carried out.

2. Physical Model and Basic Assumptions

As is shown in Figure 1, based on the fracture network after the volume fracturing in shale reservoir, a composite flow model for fractured horizontal well is established, and the seepage regions in shale reservoir are divided into inside flow region and outside flow region [41].

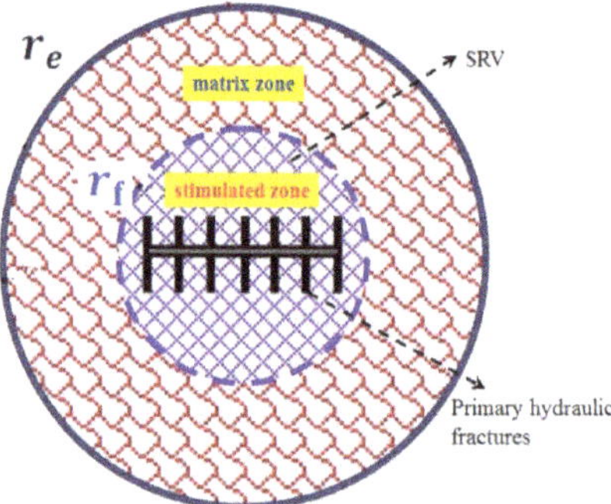

Figure 1. Physical model of fractured horizontal well in shale reservoir.

The basic assumptions are made for multi-stage fractured horizontal wells:

(1) The model is for isothermal single-phase shale gas flow, and vertical flow is neglected.
(2) The reservoir is composite with the matrix zone and stimulated zone, and the reservoir has a constant and uniformed thickness with the upper and lower boundaries closed.
(3) The gas seepage is characterized by Kundsen number in matrix zone with the radius of r_e while the gas flow is consistent with Darcy law in stimulated zone with the radius of r_f.

3. Mathematical Model

3.1. Steady-State Productivity Model

The shale reservoir after volume fracturing is divided into stimulated zone and matrix seepage zone, and the gas flow equation at different zone is established respectively. On this basis, the mathematical model of steady-state productivity of shale horizontal well with volume fracturing is derived.

3.1.1. Shale Matrix Gas Seepage Model

Knudsen dimensionless number K_n is defined as:

$$K_n = \frac{\overline{\lambda}}{r_p} \quad (1)$$

where $\overline{\lambda}$ is the gas molecules' mean free path, nm; r_p is the pore radius, nm.

With regard to the shale matrix gas seepage model, Beskok and Karniadakis [42] established an ideal gas flow equation that is universally applicable to continuous flow, slip flow, transition flow, and molecular flow:

$$v = -\frac{K_{\mathrm{m}}}{\mu}(1+\alpha K_{\mathrm{n}})(1+\frac{4K_{\mathrm{n}}}{1-\mathrm{b}K_{\mathrm{n}}})\frac{\mathrm{d}p}{\mathrm{d}x} \tag{2}$$

where v is the gas seepage velocity, m/s; K_{m} is the permeability of matrix, $10^{-3}\mu\mathrm{m}^2$; μ is the viscosity of shale gas, mPa·s; α is the rarefaction effect factor, dimensionless; K_{n} is the Knudsen number, dimensionless; b is the gas slippage constant, dimensionless.

Knudsen diffusion coefficient is given by Faruk [43]:

$$D_{\mathrm{k}} = \frac{4r_{\mathrm{p}}}{3}\sqrt{\frac{2Z\mathrm{R_g}T}{\pi M_{\mathrm{w}}}} \times 10^{-3} \tag{3}$$

where D_{k} is the diffusion coefficient, mm^2/s; Z is the gas deviation factor, dimensionless; $\mathrm{R_g}$ is the universal gas constant, $\mathrm{J \cdot mol^{-1} \cdot K^{-1}}$; T is the shale formation temperature, K; M_{w} is the molar mass of the gas, $\mathrm{kg \cdot mol^{-1}}$.

The permeability in shale matrix is defined as [44]:

$$K_{\mathrm{m}} = \frac{{r_{\mathrm{p}}}^2\phi_{\mathrm{m}}}{8\tau^2} \times 10^{-3} \tag{4}$$

where ϕ_{m} is the matrix porosity of shale, dimensionless; τ is the tortuosity, dimensionless (The value is 1).

The gas mean free path of a molecule is defined as [28]:

$$\overline{\lambda} = \sqrt{\frac{\pi Z\mathrm{R_g}T}{2M_{\mathrm{w}}}}\frac{\mu}{p} \tag{5}$$

where p is the formation pressure, MPa.

By combining Equation (1), Equation (3), Equation (4), and Equation (5), K_{n} can be expressed as follows:

$$K_{\mathrm{n}} = \frac{3\pi\mu\phi_{\mathrm{m}}D_{\mathrm{k}}}{64K_{\mathrm{m}}p} \tag{6}$$

If $\alpha = 0$, $b = 1$, combining Equations (2) and (6), differential equation of gas seepage in shale matrix pores can be expressed as follows:

$$v = -\frac{K_{\mathrm{m}}}{\mu}(1+\frac{3\pi\mu\phi_{\mathrm{m}}D_{\mathrm{k}}}{16K_{\mathrm{m}}p})\frac{\mathrm{d}p}{\mathrm{d}x} \tag{7}$$

The seepage mathematical model in the external matrix area can be expressed as follows:

$$v = -\frac{K_{\mathrm{m}}}{\mu}(1+\frac{3\pi\mu\phi_{\mathrm{m}}D_{\mathrm{k}}}{16K_{\mathrm{m}}p})\frac{\mathrm{d}p}{\mathrm{d}r} \tag{8}$$

3.1.2. Stimulated Region Gas Seepage Model

After volume fracturing, the inner area will form a complex fracture network containing natural fractures and artificial fractures [45]. Assuming the fracture fluid flows in a smooth plane, the permeability is calculated by cubic law according to the projection length of each fracture on the x and y axis. By superimposing the permeability of each fracture, the permeability of the whole fracture network of x and y directions without any stress is calculated.

$$k_x = \sum_{i=1}^{N}\frac{{b_i}^3 l_i}{12A} \cdot \cos\theta_i \tag{9}$$

$$k_y = \sum_{i=1}^{N} \frac{b_i{}^3 l_i}{12A} \cdot \sin\theta_i \quad (10)$$

where k_x and k_y are the permeability in x direction and y direction respectively, 10^{-3} μm^2; N is the number of fractures, dimensionless; A is the area of fractured region, mm^2; b_i is the aperture of different fracture, μm; l_i is the length of different fracture, mm; θ_i is the angle of the fracture and the direction of X, °.

The permeability of fractures under different stress is composed of normal open fracture and shear type fracture. In the calculation of permeability, the normal stress and shear stress are calculated, respectively, and the ultimate permeability is the superposition of the permeability caused by the normal stress and shear stress [27]:

$$k_x = k_{nx} + k_{dx} \quad (11)$$

$$k_y = k_{ny} + k_{dy} \quad (12)$$

where k_{nx} and k_{ny} are the permeability in the x and y direction due to normal closure of fractures respectively, 10^{-3} μm^2; k_{dx} and k_{dy} are the permeability in the x and y direction due to shear failure of fractures, respectively, in 10^{-3} μm^2.

In the fracture network, single fractures do not reflect fractures with different lengths and apertures. Therefore, the concept of equivalent fracture frequency (f_x, f_y) and equivalent aperture (b_x, b_y) are used to express the complex fracture deformation in fracture network. The permeability of the open or closed fractures produced by the positive stress in the direction of X and Y (k_{nx}, k_{ny}) can be calculated by the following equations [46]:

$$k_{nx} = \frac{f_x}{12} b_x^3 \quad (13)$$

$$k_{ny} = \frac{f_y}{12} b_y^3 \quad (14)$$

where f_x and f_y are the equivalent frequency caused by the positive stress in the x direction and y direction, respectively, 1/m; b_x and b_y are the equivalent aperture caused by the positive stress in the x direction and y direction respectively, in μm.

The equivalent frequency of fractures can be obtained by numerical experiments and initial aperture inversion.

$$f_x = \frac{12k_{nx}}{b_{ei}^3} \quad (15)$$

$$f_y = \frac{12k_{ny}}{b_{ei}^3} \quad (16)$$

where b_{ei} is the initial average aperture of fractures, μm.

The size of the fracture aperture in the x and y direction can be expressed as follows respectively [47]:

$$b_x = b_r + b_{max} = b_r + b_{max}\exp\{-(\alpha_x\sigma_x + \alpha_y\sigma_y)\} \quad (17)$$

$$b_y = b_r + b_{max} = b_r + b_{max}\exp\{-(\beta_x\sigma_x + \beta_y\sigma_y)\} \quad (18)$$

where b_r is the residual aperture, μm; b_{max} is the maximum deformable mechanical aperture caused by positive stress, μm; α_x and α_y are the x aperture stress coefficient in the x and y direction, respectively, dimensionless; β_x and β_y are the y aperture stress coefficient in the x and y direction, respectively, dimensionless; σ_x and σ_y are the normal stress in the x direction and y direction, respectively, MPa.

Similar equations to Equation (13) and Equation (14), fracture permeability under shear stress can be expressed as follows:

$$k_{dx} = \frac{f_{dx}}{12} d_x^3 \quad (19)$$

$$k_{dy} = \frac{f_{dy}}{12} d_y^3 \quad (20)$$

where f_{dx} and f_{dy} are the equivalent frequency caused by the shear stress in the x direction and y direction, respectively, 1/m; d_x and d_y are the equivalent apertures caused by the shear stress in the x direction and y direction respectively, μm.

Only when the pressure reaches the critical pressure, the fracture will be shear failure, so only some fractures will appear shear deformation, and other fractures will not appear shear deformation. The aperture produced by shear deformation can be calculated by the following formulae:

$$k \prec k_c, d_x = 0, d_y = 0 \quad (21)$$

$$k \succ k_c, \begin{cases} d_x = d_{\max}\{1 - \exp[-\gamma_x(k - k_c)]\} \\ d_y = d_{\max}\{1 - \exp[-\gamma_y(k - k_c)]\} \end{cases} \quad (22)$$

where k is the stress ratio of x and y, dimensionless; k_c is the critical stress ratio, dimensionless; $d_{\max}$ is the maximum aperture of fracture after shear failure, μm; γ_x and γ_y are the shear stress coefficient in x direction and y direction respectively, dimensionless;

Assuming that the fracture is not cohesive, the critical pressure ratio and critical direction of fracture failure can be calculated by the following Coulomp failure criterion [48]:

$$k_c = \frac{1 + \sin\varphi}{1 - \sin\varphi} \quad (23)$$

$$\varphi_{\mathrm{f}} = 90 - \left(45 + \frac{\varphi}{2}\right) \quad (24)$$

where φ is the internal friction angle of rock, °; φ_{f} is the critical failure angle of rock, °.

Finally, the permeability (k_x, k_y) of the x and y directions in the stimulated area can be obtained by superimposing the fracture permeability of normal closure and shear deformation [27]:

$$k_x = \frac{f_x}{12}\left[b_r + b_{\max}\exp\{-(\alpha_x\sigma_x + \alpha_y\sigma_y)\}\right]^3 + \frac{f_{dx}}{12}[d_{\max}[1 - \exp\{-\gamma_x(k - k_c)\}]]^3 \quad (25)$$

$$k_y = \frac{f_y}{12}\left[b_r + b_{\max}\exp\{-(\beta_x\sigma_x + \beta_y\sigma_y)\}\right]^3 + \frac{f_{dy}}{12}\left[d_{\max}\left[1 - \exp\{-\gamma_y(k - k_c)\}\right]\right]^3 \quad (26)$$

The comprehensive permeability of stimulated fracture network area is obtained through the root mean square of x and y permeability [49]:

$$K_{\mathrm{f}} = \frac{\sqrt{k_x^2 + k_y^2}}{2} \quad (27)$$

where K_{f} is the permeability in volume fracturing zone, 10^{-3} μm^2.

According to the hypothesis, the Darcy seepage equation in the internal volume fracturing area can be expressed as follows:

$$\frac{\mathrm{d}p}{\mathrm{d}r} = -\frac{\mu}{K_{\mathrm{f}}} v \quad (28)$$

3.1.3. Steady-State Productivity Model

Considering the flow is the product of gas seepage velocity and gas seepage area, the flow of volume fracturing area can be expressed as follows:

$$q_{1\mathrm{sc}} = \frac{\pi K_{\mathrm{f}} h T_{\mathrm{sc}}({p_{\mathrm{f}}}^2 - {p_{\mathrm{wf}}}^2)}{\overline{\mu}\overline{Z} p_{\mathrm{sc}} T \ln\frac{r_{\mathrm{f}}}{r_{\mathrm{w}}}} \quad (29)$$

where q_{1sc} is the volume fracturing zone flow, m^3/d; h is the reservoir thickness, m; T_{sc} is the temperature under standard condition, K; p_f is the outer boundary pressure of volume fracturing zone, MPa; p_{wf} is the bottom hole flow pressure, MPa; r_f is the radius of volume fracturing zone, m; r_w is the borehole radius, m; p_{sc} is the pressure under standard conditions, MPa; $\overline{\mu}$ is the gas viscosity under average formation pressure, m Pa·s; $\overline{Z}$ is the compression factor under average formation pressure, dimensionless.

The flow of external matrix seepage area is obtained from Equation (8):

$$q_{2sc} = \frac{\pi K_m h T_{sc}(p_e{}^2 - p_f{}^2)}{\overline{\mu}\overline{Z}p_{sc}T\ln\frac{r_e}{r_f}} + \frac{\pi K_m T_{sc} h(3\pi\overline{\mu}\phi_m D_k)(p_e - p_f)}{8K_m p_{sc}T\overline{\mu}\overline{Z}\ln\frac{r_e}{r_f}} \tag{30}$$

where q_{2sc} is the matrix seepage area flow, m^3/d; r_e is the supply radius, m.

According to the law of mass conservation [50], the gas volume flow of the matrix zone is equal to that of volume fracturing zone under the standard conditions:

$$q_{sc} = q_{2sc} = q_{1sc} \tag{31}$$

where q_{sc} is the production of shale gas under the standard condition, m^3/d.

By combining Equation (29), Equation (30), and Equation (31), the volume flow for steady seepage in fractured horizontal well in shale gas reservoir can be expressed as follows:

$$q_{sc} = \frac{p_e{}^2 - p_{wf}{}^2}{A + BC} \tag{32}$$

where $A = \frac{\overline{\mu}\overline{Z}p_{sc}T\ln\frac{r_f}{r_w}}{\pi K_f h T_{sc}}$; $B = \frac{\overline{\mu}\overline{Z}p_{sc}T\ln\frac{r_e}{r_f}}{\pi K_m h T_{sc}}$; $C = \frac{p_e{}^2 - p_f{}^2}{p_e{}^2 - p_f{}^2 + \frac{3\pi\mu\phi_m D_k(p_e - p_f)}{8K_m}}$.

3.2. Unsteady-State Productivity Model

The seepage of shale gas is an unstable process, and the initial production decline is very obvious. Only when the pressure wave reaches to the boundary in the later stage of production, the production tends to be stable. In shale reservoir, the amount of dissolved gas is fairly small, so the effect of dissolved gas is not considered in the process of deriving the unstable mathematical model, and the expansion of the adsorbed gas volume with pressure drop is ignored.

Considering the effects of shale gas adsorption and desorption, the Langmuir isotherm equation [51] represents the desorbed volume of shale gas from the matrix as pressure changes:

$$Q = V_L\left(\frac{p_i}{p_L + p_i} - \frac{p}{p_L + p}\right) \tag{33}$$

where Q is the desorption gas volume of per kilogram mass shale matrix, m^3/kg; V_L is the Langmuir volume, m^3/kg; p_i is the original formation pressure, MPa; p_L is the Langmuir pressure, MPa.

From the initial stage of production to the end of production, the propagation of pressure waves in the reservoir can be divided into two stages. In the first stage, the pressure wave propagates from the bottom of the well to the boundary of the reservoir, and the pressure wave radius gradually increases to r_e. In the second stage, when the pressure wave reaches the boundary, the pressure wave radius no longer changes, and the boundary pressure gradually decreases. Meanwhile, the gas production decreases, and finally, the pressure on the boundary tends to the bottom hole flow pressure. In this paper, the first stage of pressure wave propagation is divided into two periods, that is, the pressure wave firstly propagates from the bottom of the well to the outer boundary of the volume fracturing zone, and then it propagates from the outer boundary of the fracturing zone to the reservoir boundary. The distance of pressure wave propagation is only related to time, the physical properties of reservoir and the fluid.

In order to derive a transient productivity model for shale reservoir fractured well, this paper adopts a succession pseudo-steady state method (SPSS) proposed by Shahamat [41]. Specifically, on the basis of the pressure wave radius formula, steady state shale gas production formula, and material balance equation, assuming a time step, it is considered that the seepage within this time step is a steady flow, and the cumulative output is calculated. The material balance equation can calculate the average formation pressure in the area affected by the pressure wave during this period of time. The average formation pressure is used as the boundary pressure for the next time step to calculate the output at the next time step. By analogy, the relationship between output and time is obtained, that is, the output under non-steady state. The specific steps are as follows:

3.2.1. The Solution of Initial Production

The whole production stages of fractured horizontal well in shale reservoir are divided into several time steps. For the first time step Δt, Hsieh et al. [52] proposed the formula for calculating the propagation radius of pressure wave at the time of Δt. The value of the Δt is as small as possible in order to avoid R_1 from exceeding the outer boundary of the volume fracturing area:

$$R_1 = 0.5879\sqrt{\frac{K_f \Delta t}{\phi_f \mu C_{tf}}} \tag{34}$$

where R_1 is the propagation radius of pressure wave at the time of Δt, m; Δt is the time of a production time step, d; ϕ_f is the fracture porosity in volume fracturing area, dimensionless; C_{tf} is the comprehensive compression coefficient of fractured zone, MPa^{-1}.

According to the formula of steady-state productivity, the production at the time of Δt can be expressed as follows:

$$q_1 = \frac{\pi K_f h T_{sc}\left(p_i{}^2 - p_{wf}{}^2\right)}{\overline{\mu}\overline{Z} p_{sc} T \ln\frac{R_1}{r_w}} \tag{35}$$

where q_1 is the production at the time of Δt, m^3.

3.2.2. The Solution of Production at the Next Production Time Step

Using q_1 as initial output, the pressure wave propagation distance R_2 at the time of Δt + Δt_f is obtained:

$$R_2 = 0.5879\sqrt{\frac{K_f(\Delta t + \Delta t_f)}{\phi_f \mu C_{tf}}} \tag{36}$$

where R_2 is the propagation radius of pressure wave at the time of Δt + Δt_f, m; Δt_f is the propagation time of pressure waves in fractured zone, d.

In the time of the Δt_f, the seepage of shale gas is stable seepage, and the cumulative output Gp_2 can be expressed as follows:

$$G_{p2} = q_1 \times \Delta t_f \tag{37}$$

where Gp_2 is the cumulative yield in the time of the Δt_f, m^3.

The free gas geological reserves in the pressure wave propagation radius is calculated by the volume method:

$$G_{m2} = \pi R_2{}^2 h \phi_m (1 - S_w)/B_{gi} \tag{38}$$

$$G_{f2} = \pi R_2{}^2 h \phi_f (1 - S_w)/B_{gi} \tag{39}$$

where G_{m2} is the free gas volume in the shale matrix in the pressure wave radius at the time of Δt + Δt_f, m^3; S_w is the irreducible water saturation, dimensionless; B_{gi} is the volume coefficient of shale gas under the condition of original formation, dimensionless; G_{f2} is the free gas volume in the fractures within pressure wave radius at the time of Δt + Δt_f, m^3.

According to the principle of material balance, under the ground standard condition, material balance equation can be expressed as follows [53]:

$$G_{\mathrm{m}} + G_{\mathrm{f}} + G_{\mathrm{a}} = G_{\mathrm{p}} + G'_{\mathrm{m}} + G'_{\mathrm{f}} + G'_{\mathrm{a}} \tag{40}$$

where G_{m} is the free gas volume in the matrix under the original formation pressure, m^3; G_{f} is the free gas volume in the fractures under the original formation pressure, m^3; G_{a} is the adsorptive gas volume under the original formation pressure, m^3; G_{p} is the produced gas volume, m^3; G'_{m} is the free gas volume in the matrix under the current formation pressure, m^3; G'_{f} is the free gas volume in the fractures under the current formation pressure, m^3; G'_{a} is the adsorptive gas volume under the current formation pressure, m^3.

Equation (40) can be rewritten as follows:

$$\begin{aligned} & G_{\mathrm{m}} + G_{\mathrm{f}} + \frac{G_{\mathrm{m}} B_{\mathrm{gi}}}{\phi_{\mathrm{m}}(1-S_{\mathrm{w}})} \rho_{\mathrm{s}} \frac{V_{\mathrm{L}} p_{\mathrm{i}}}{p_{\mathrm{L}}+p_{\mathrm{i}}} \\ & = G_{\mathrm{p}} + G_{\mathrm{m}} \frac{B_{\mathrm{gi}}}{B_{\mathrm{g}}} \left(1 - \frac{C_{\mathrm{m}}+C_{\mathrm{w}} S_{\mathrm{w}}}{1-S_{\mathrm{w}}} \Delta p\right) + G_{\mathrm{f}} \frac{B_{\mathrm{gi}}}{B_{\mathrm{g}}} + \frac{G_{\mathrm{m}} B_{\mathrm{gi}}}{\phi_{\mathrm{m}}(1-S_{\mathrm{w}})} \rho_{\mathrm{s}} \frac{V_{\mathrm{L}} p}{p_{\mathrm{L}}+p} \end{aligned} \tag{41}$$

The equation about the average formation pressure p_2 in the range of pressure propagation radius (R_2) can be expressed as follows:

$$\begin{aligned} & \frac{p_2}{Z}\left[G_{\mathrm{m2}}\left(1 - \frac{C_{\mathrm{m}}+C_{\mathrm{w}} S_{\mathrm{w}}}{1-S_{\mathrm{w}}}(p_{\mathrm{i}} - p_2)\right) + G_{\mathrm{f2}}\right] = \\ & \frac{p_{\mathrm{i}}}{Z_{\mathrm{i}}}\left[G_{\mathrm{m2}} + G_{\mathrm{f2}} - G_{\mathrm{p2}} + \frac{G_{\mathrm{m2}} B_{\mathrm{gi}} \rho_{\mathrm{s}}}{\phi_{\mathrm{m}}(1-S_{\mathrm{w}})}\left(\frac{V_{\mathrm{L}} p_{\mathrm{i}}}{p_{\mathrm{L}}+p_{\mathrm{i}}} - \frac{V_{\mathrm{L}} p}{p_{\mathrm{L}}+p}\right)\right] \end{aligned} \tag{42}$$

where p_2 is the average formation pressure in the pressure wave radius at the time of $\Delta t + \Delta t_{\mathrm{f}}$, MPa; C_{m} is the compression coefficient of shale matrix, MPa^{-1}; C_{w} is the compression coefficient of water, MPa^{-1}; Z_{i} is the gas compression factor in the original state, dimensionless; ρ_{s} is the shale matrix density, kg/m^3.

According to the formula of steady-state productivity, the output at the time of $\Delta t + \Delta t_{\mathrm{f}}$ can be rewritten as follows:

$$\begin{cases} q_2 = \dfrac{{p_2}^2 - {p_{\mathrm{wf}}}^2}{A} \\ A = \dfrac{\overline{\mu} \overline{Z} p_{\mathrm{sc}} T \ln \frac{R_2}{r_{\mathrm{w}}}}{\pi K_{\mathrm{f}} h T_{\mathrm{sc}}} \end{cases} \tag{43}$$

Then we can repeat step (2) to solve for the production at the next production time, and calculate the average formation pressure and the sweep radius under different production times. Furthermore, we can combine the steady-state output calculation formula to obtain the output at different production times. Repeating the above steps, we can obtain the production of shale gas reservoir fracturing wells throughout the production phase. It can be seen that using the SPSS method to calculate the non-steady state production is very convenient, and it is also extremely easy to program. At the same time, it avoids the unreasonable assumption that the pressure has already reached to the boundary of the reservoir under the initial conditions when other mathematical methods solve the productivity model. The different seepage law of gas in volume fracturing zone and shale matrix zone, the adsorption and desorption effect of shale gas are also considered, which is more consistent with the actual situation.

3.3. Model Validation

This model divides the seepage area into volume fracturing area and shale matrix area, taking into account the multi-scale flow and non-linear seepage characteristics of shale gas in horizontal fracturing well. In order to verify the reliability of the SPSS method for solving multi-scale flow problems, a two-region radial seepage numerical model considering shale gas desorption and diffusion is established by using Eclipse [54]. The outer zone of the model is the matrix seepage zone based on the dual media model, and the inner zone is the volume fracturing zone.

By inputting the same model parameters, the cumulative gas production of model in this paper and the Eclipse numerical model are obtained. The two curves have a high degree of conformity, and the output change trend is consistent (Figure 2). The cumulative output of volume fracturing horizontal well composite flow model is slightly higher than that of the Eclipse numerical model. The reason for this is that the SPSS method adopted in this paper avoids the false assumption that the pressure wave spreads to the reservoir boundary under the initial condition. Therefore, it has more desorption gas and describes the actual productivity more accurately.

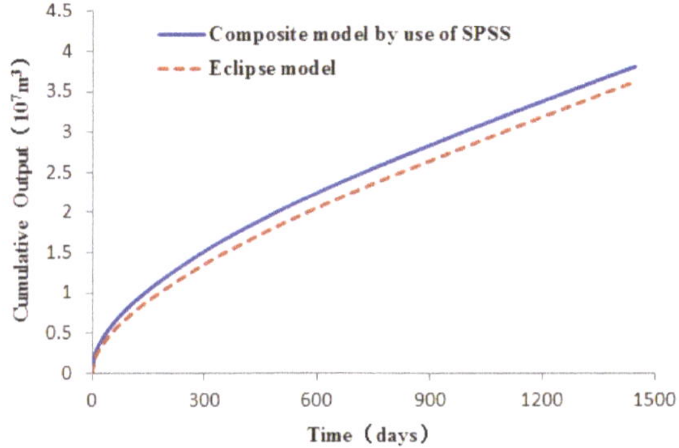

Figure 2. Comparison of cumulative gas production of model in this paper and the Eclipse numerical model.

Figure 3 shows the reservoir pressure distribution of the Eclipse model at 300 days of production. It can be seen that in the shale gas production process, the reservoir pressure gradually decreases from near the wellbore to the reservoir boundary. It reflects that the pressure wave propagates from the bottom of the well to the outer boundary of the volume fracturing zone, and then propagates from the outer boundary of the fracturing zone to the reservoir boundary.

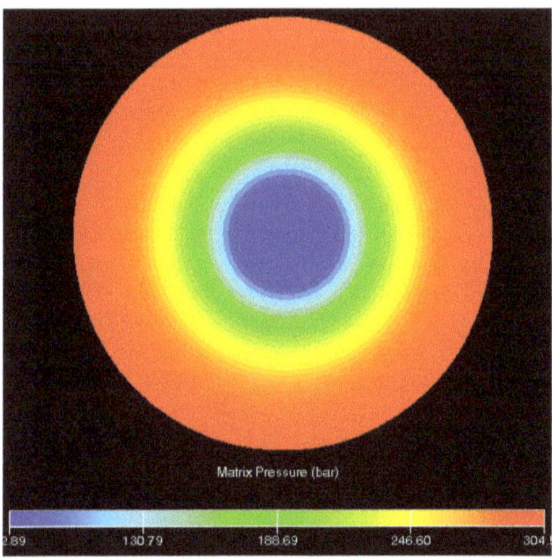

Figure 3. Reservoir pressure simulation of the Eclipse model.

4. Results and Discussion

Due to the complexity of geological features and seepage laws, shale gas productivity is affected by many factors after horizontal well volume fracturing. Based on the combined flow model of fracturing horizontal wells and the parameters of fracturing well in a certain shale gas reservoir (Table 1), according to the SPSS method, the effect of different factors such as fracturing zone radius (r_f), Langmuir volume (V_L), the initial average aperture of fractures (b_{ei}), maximum fracture length (l_{max}) on gas productivity are studied, and the results are shown in Figures 4–7.

Table 1. Model calculation basic parameters table.

Parameters (Unit)	Value
Radius of volume fracturing zone (m)	110
Permeability in volume fracturing zone (10^{-3}μm^2)	200
Comprehensive compression coefficient of fractured zone (MPa^{-1})	0.035
Fracture porosity in volume fracturing area (-)	0.1
Supply radius (m)	400
Diffusion coefficient (mm^2/s)	281
Comprehensive compression coefficient of shale matrix (MPa^{-1})	0.019
Compression coefficient of shale matrix (MPa^{-1})	0.0001
Matrix porosity (-)	0.044
Shale matrix density (kg/m^3)	2500
Bottom hole flow pressure (MPa)	5
Viscosity of shale gas (mPa·s)	0.02
Langmuir pressure (MPa)	10
Langmuir volume (m^3/kg)	0.05
Original formation pressure (MPa)	30
Formation temperature (K)	360
Borehole radius (m)	0.1
Compressibility of water(MPa^{-1})	0.0004
Irreducible water saturation (-)	0.1
Reservoir thickness (m)	30
Permeability of matrix (10^{-3} μm^2)	0.005
Pore radius of matrix (nm)	500
Gas slippage constant (-)	−1
Rarefaction effect factor (-)	-

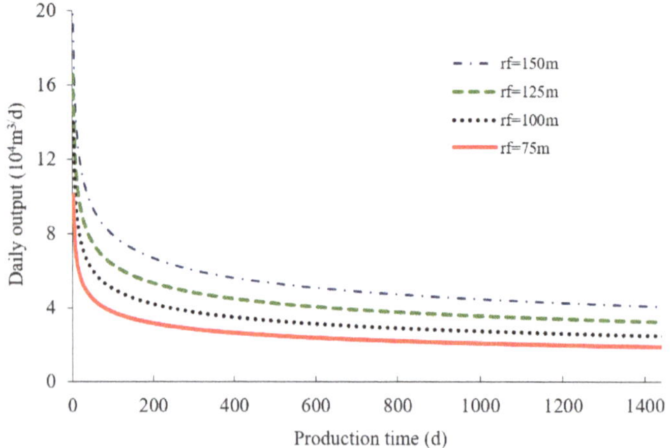

Figure 4. The influence of the radius of volume fracturing area on the daily gas production.

Figure 4 shows that with the increase of the radius of complex fracture network in the stimulated area, the daily output will also increase significantly. However, with the increase of production time, the difference of daily output under each fracturing radius is also decreasing.

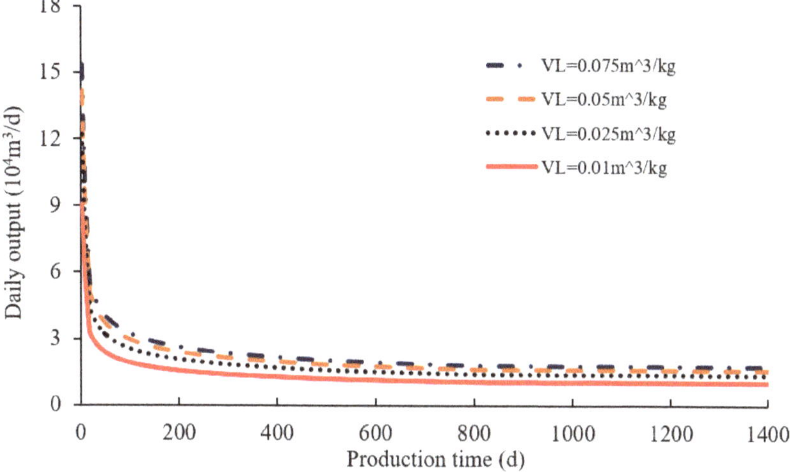

Figure 5. The influence of Langmuir volume on the daily gas production.

Figure 5 shows that with the increase of Langmuir volume, daily gas production gradually increased, but the growth rate gradually decreased. This is because the larger Langmuir volume indicates that the reservoir has more adsorbed gas, so in the process of production, more adsorbed gas is desorbed and the free gas is taken out when the pressure is reduced. However, when the Langmuir volume reaches 0.075 m^3/kg, the daily output of different Langmuir volumes are almost identical, at this time, the effect of Langmuir volume on the desorption output can almost be ignored.

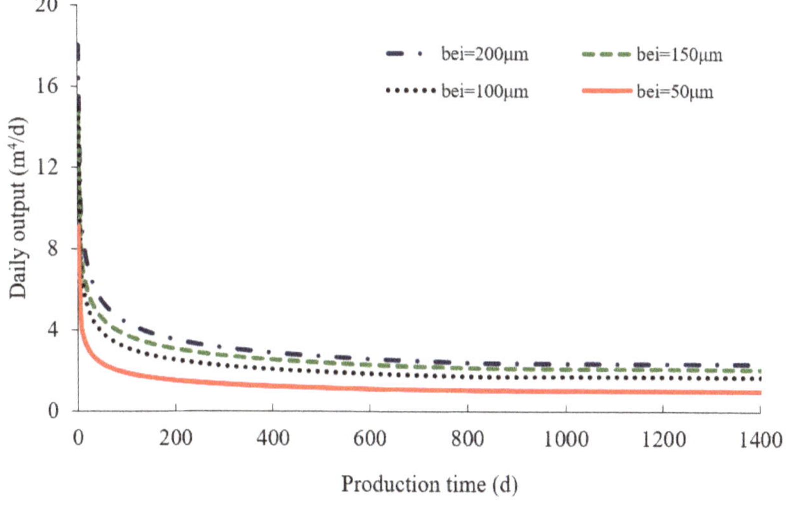

Figure 6. The influence of initial average aperture of fractures on the daily gas production.

Figure 6 shows that under the condition of small initial fracture aperture, the influence of initial fracture aperture on daily gas production is significant. The flow of gas in open fractures is the main channel for shale gas transport. The larger the aperture of fractures, the more the flow of gas provided. With the increase of initial fracture aperture, daily gas production gradually increased, but the growth rate gradually decreased.

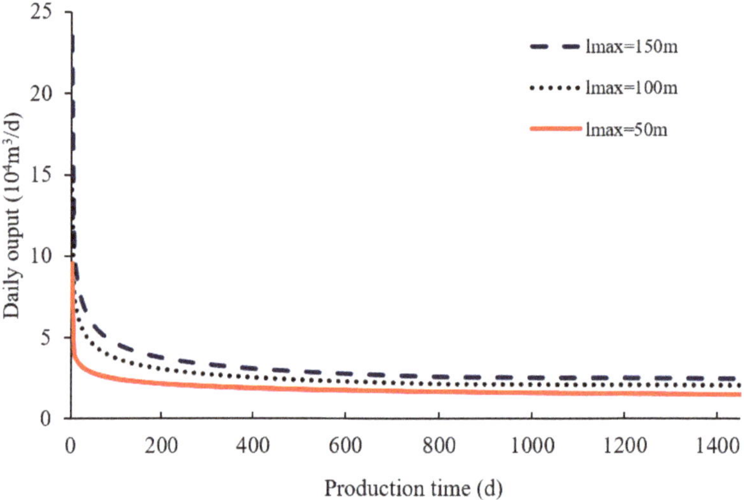

Figure 7. The influence of maximum fracture length on the daily gas production.

The fracture length of shale gas well after fracturing is an important index affecting shale gas production. The longer the fracture length, the wider the area affected by volume fracturing. Figure 7 shows that maximum fracture length has the most obvious effect on the initial production of shale gas well. As production continues, the daily output is getting closer.

5. Conclusions

(1) The productivity prediction model based on the SPSS method provides a theoretical basis for the transient productivity calculation of shale fractured horizontal wells, and it has the characteristics of simple solution process, fast computation speed and high agreement with numerical simulation results.

(2) The pressure wave propagates from the bottom of the well to the outer boundary of the volume fracturing zone, and then propagates from the outer boundary of the fracturing zone to the reservoir boundary.

(3) With the increase of fracturing zone radius, the initial average aperture of fractures, maximum fracture length, the productivity of shale gas increases, and the increase rate gradually decreases. When the fracturing zone radius is 150 m, the daily output is approximately twice as much as that of 75 m. If the initial average aperture of fractures is 50 μm, the daily output is about half of that when the initial average aperture is 100 μm. When the maximum fracture length increases from 50 m to 100 m, the daily output only increases about by 25%.

(4) When the Langmuir volume is relatively large, the daily outputs of different Langmuir volumes are almost identical, and the effect of Langmuir volume on the desorption output can almost be ignored.

Author Contributions: Conceptualization, F.Z. and J.G.; Methodology, F.Z. and F.P.; Validation, F.Z. and F.P.; Formal Analysis, F.P.; Writing-Original Draft Preparation, F.Z. and F.P.; Writing-Review & Editing, J.G. and J.X.; Supervision, J.G.; Project Administration, Q.W.; Funding Acquisition, Q.W. and J.Z.

Funding: The Natural Science Foundation of China: 51525404; 51504203. National Key Research and Development Program of China: 2017ZX05037-004.

Acknowledgments: This research was funded by the Natural Science Foundation of China (Grant Nos. 51525404; 51504203) and National Key Research and Development Program of China (Grant No. 2017ZX05037-004).

Suggestion: The proposed model for the transient productivity of fractured well in shale reservoir includes the effect of Knudsen number effect and the adsorption/desorption. We excluded the increasing effective stress for simplicity in this model. Including the effect of the increasing effective stress should be carried out in the next study stage.

Conflicts of Interest: We declare that we have no financial and personal relationships with other people or organizations that can inappropriately influence our work. There is no professional or other personal interest of any nature or kind in any product, service and/or company that could be construed as influencing the position presented in the manuscript entitled, "A Transient Productivity Model of Fractured Wells in Shale Reservoirs Based on the Succession Pseudo-Steady State Method".

References

1. Li, J.; Chen, Z.; Wu, K.; Li, R.; Xu, J.; Liu, Q.; Qu, S.; Li, X. Effect of water saturation on gas slippage in tight rocks. *Fuel* **2018**, *225*, 519–532. [CrossRef]
2. Dejam, M.; Hassanzadeh, H.; Chen, Z. Pre-Darcy Flow in Porous Media. *Water Resour. Res.* **2017**, *53*, 8187–8210. [CrossRef]
3. Ziarani, A.S.; Aguilera, R. Knudsen's permeability correction for tight porous media. *Tran. Porous Media* **2012**, *91*, 239–260.
4. Zou, C.; Dong, D.; Wang, S.; Li, J.; Li, X.; Wang, Y.; Li, D.; Cheng, K. Geological characteristics and resource potential of shale gas in China. *Petroleum Explor. Dev. Online* **2010**, *37*, 641–653. [CrossRef]
5. Zuo, C.; Chengjin, X.; Tingxue, J.; Yuming, Q. Proposals for the application of fracturing by stimulated reservoir volume (SRV) in shale gas wells in China. *Nat. Gas Ind.* **2010**, *30*, 30–32.
6. Rui, Z.; Wang, X.; Zhang, Z.; Lu, J.; Chen, G.; Zhou, X.; Patil, S. A realistic and integrated model for evaluating oil sands development with steam assisted gravity drainage technology in Canada. *Appl. Energy* **2018**, *213*, 76–91.
7. Wu, K.; Chen, Z.; Li, X.; Dong, X. Methane storage in nanoporous material at supercritical temperature over a wide range of pressures. *Sci. Rep.* **2016**, *6*, 33–46. [CrossRef] [PubMed]
8. Dejam, M.; Hassanzadeh, H. Diffusive leakage of brine from aquifers during CO_2 geological storage. *Adv. Water Resour.* **2018**, *111*, 36–57. [CrossRef]
9. Rui, Z.; Cui, K.; Wang, X.; Chun, J.; Li, Y.; Zhang, Z.; Lu, J.; Chen, G.; Zhou, X.; Patil, S. A comprehensive investigation on performance of oil and gas development in Nigeria: Technical and non-technical analyses. *Energy* **2018**, *158*, 666–680. [CrossRef]
10. Clarkson, C.R. Production data analysis of unconventional gas wells: review of theory and best practices. *Int. J. Coal Geol.* **2013**, *109*, 101–146. [CrossRef]
11. Zeng, F.; Cheng, X.; Guo, J.; Chen, Z.; Xiang, J. Investigation of the initiation pressure and fracture geometry of fractured deviated wells. *J. Petroleum Sci. Eng.* **2018**, *165*, 412–427. [CrossRef]
12. Deng, J.; Zhu, W.; Ma, Q. A new seepage model for shale gas reservoir and productivity analysis of fractured well. *Fuel* **2014**, *124*, 232–240. [CrossRef]
13. Zeng, F.; Guo, J.; Ma, S.; Chen, Z. 3D observations of the hydraulic fracturing process for a model non-cemented horizontal well under true triaxial conditions using an X-ray CT imaging technique. *J. Petroleum Sci. Eng.* **2018**, *52*, 128–140. [CrossRef]
14. Gao, S.; Liu, H.; Ye, L.; Hu, Z.; Chang, J.; An, W. A coupling model for gas diffusion and seepage in SRV section of shale gas reservoirs. *Nat. Gas Ind. B* **2017**, *4*, 120–126. [CrossRef]
15. Zeng, F.H.; Guo, J.C. Optimized design and use of induced complex fractures in horizontal wellbores of tight gas reservoirs. *Rock Mech. Rock Eng.* **2016**, *49*, 1411–1423. [CrossRef]
16. Loucks, R.G.; Reed, R.M.; Ruppel, S.C.; Jarvie, D.M. Morphology, genesis, and distribution of nanometer-scale pores in siliceous mudstones of the Mississippian Barnett Shale. *J. Sediment. Res.* **2009**, *79*, 848–861. [CrossRef]

17. Wendong, W.; Yuliang, S.; Qi, Z.; Gang, X.; Shiming, C. Performance-based Fractal Fracture Model for Complex Fracture Network Simulation. *Petroleum Sci.* **2018**, *1*, 1–9.
18. Rui, Z.; Cui, K.; Wang, X.; Lu, J.; Chen, G.; Ling, K.; Patil, S. A quantitative framework for evaluating unconventional well development. *J. Petroleum Sci. Eng.* **2018**, *166*, 900–905. [CrossRef]
19. Zhu, W.Y.; Qian, Q.I. Study on the multi-scale nonlinear flow mechanism and model of shale gas. *Sci. Sin.* **2016**, *46*, 111–119. [CrossRef]
20. Duan, Y.G.; Wei, M.Q.; Li, J.Q.; Tang, Y. Shale gas seepage mechanism and fractured wells' production evaluation. *J. Chongqing Univ.* **2011**, *4*, 11–17.
21. Wang, W.; Shahvali, M.; Su, Y. A semi-analytical model for production from tight oil reservoirs with hydraulically fractured horizontal wells. *Fuel* **2015**, *158*, 612–618. [CrossRef]
22. Zeng, F.H.; Yubiao, K.E.; Guo, J.C. An optimal fracture geometry design method of fractured horizontal wells in heterogeneous tight gas reservoirs. *Sci. China Technol. Sci.* **2016**, *59*, 241–251. [CrossRef]
23. Li, L.; Jiang, H.; Li, J.; Wu, K.; Meng, F.; Xu, Q.; Chen, Z. An analysis of stochastic discrete fracture networks on shale gas recovery. *J. Petroleum Sci. Eng.* **2018**, *167*, 78–87. [CrossRef]
24. Yuan, Y.; Yan, W.; Chen, F.; Li, J.; Xiao, Q.; Huang, X. Numerical Simulation for Shale Gas Flow in Complex Fracture System of Fractured Horizontal Well. *Int. J. Nonlinear Sci. Numerical Simul.* **2018**, *19*, 367–377. [CrossRef]
25. Wang, W.; Su, Y.; Yuan, B.; Wang, K.; Cao, X. Numerical Simulation of Fluid Flow through Fractal-Based Discrete Fractured Network. *Energies* **2018**, *11*, 286. [CrossRef]
26. Wang, W.; Su, Y.; Zhang, X.; Sheng, G.; Ren, L. Analysis of the complex fracture flow in multiple fractured horizontal wells with the fractal tree-like network models. *Fractals* **2015**, *23*, 155–164. [CrossRef]
27. Min, K.B.; Rutqvist, J.; Tsang, C.F.; Jing, L. Stress-dependent permeability of fractured rock masses: A numerical study. *Int. J. Rock Mech. Min. Sci.* **2004**, *41*, 1191–1210. [CrossRef]
28. Stalgorova, E.; Mattar, L. Practical analytical model to simulate production of horizontal wells with branch fractures. In Proceedings of the SPE Canada Unconventional Resources Conference, Society of Petroleum Engineers, Calgary, AB, Canada, 13–14 March 2018; p. 162515.
29. Swami, V. Shale gas reservoir modeling: from nanopores to laboratory. In Proceedings of the SPE Annual Technical Conference and Exhibition, San Antonio, TX, USA, 8–10 October 2012; p. 163065.
30. Zhang, D.; Zhang, L.; Zhao, Y.; Guo, J. A composite model to analyze the decline performance of a multiple fractured horizontal well in shale reservoirs. *J. Nat. Gas Sci. Eng.* **2015**, *26*, 999–1010. [CrossRef]
31. Su, Y.; Zhang, Q.; Wang, W.; Sheng, G. Performance analysis of a composite dual-porosity model in multi-scale fractured shale reservoir. *J. Nat. Gas Sci. Eng.* **2015**, *26*, 1107–1118. [CrossRef]
32. Zhao, Y.L.; Zhang, L.H.; Luo, J.X.; Zhang, B.N. Performance of fractured horizontal well with stimulated reservoir volume in unconventional gas r3eservoir. *J. Hydrol.* **2014**, *512*, 447–456. [CrossRef]
33. Wang, W.; Shahvali, M.; Su, Y. Analytical solutions for a quad-linear flow model derived for multistage fractured horizontal wells in tight oil reservoirs. *J. Energy Resour. Technol.* **2017**, *139*, 77–85. [CrossRef]
34. Lu, C.; Wang, J.; Zhang, C.; Cheng, M.; Wang, X.; Dong, W.; Zhou, Y. Transient pressure analysis of a volume fracturing well in fractured tight oil reservoirs. *J. Geophys. Eng.* **2017**, *14*, 15–23. [CrossRef]
35. Daryasafar, A.; Joukar, M.; Fathinasab, M.; Da Prat, G.; Kharrat, R. Estimating the properties of naturally fractured reservoirs using rate transient decline curve analysis. *J. Earth Sci.* **2017**, *28*, 848–856. [CrossRef]
36. Luo, W.; Tang, C.; Zhou, Y.; Ning, B.; Cai, J. A new semi-analytical method for calculating well productivity near discrete fractures. *J. Nat. Gas Sci. Eng.* **2018**, *57*, 216–223. [CrossRef]
37. Zeng, F.H.; Cheng, X.Z.; Guo, J.C.; Long, C.; Ke, Y.B. A new model to predict the unsteady production of fractured horizontal wells. *Sains Malaysiana* **2016**, *45*, 1579–1587.
38. Dong, J.J.; Hsu, J.Y.; Wu, W.J.; Shimamoto, T.; Hung, J.H.; Yeh, E.C.; Wu, Y.H.; Sone, H. Stress-dependence of the permeability and porosity of sandstone and shale from TCDP Hole-A. *Int. J. Rock Mech. Min. Sci.* **2010**, *47*, 1141–1157. [CrossRef]
39. Jia, B.; Tsau, J.; Barati, R. A workflow to estimate shale gas permeability variations during the production process. *Fuel* **2018**, *220*, 879–889. [CrossRef]
40. Singh, H.; Cai, J. A mechanistic model for multi-scale sorption dynamics in shale. *Fuel* **2018**, *234*, 996–1014. [CrossRef]

41. Shahamat, M.S.; Mattar, L.; Aguilera, R. A physics-based method to forecast production from tight and shale petroleum reservoirs by use of succession of pseudosteady states. *SPE Reserv. Eval. Eng.* **2015**, *18*, 45–53. [CrossRef]
42. Ali Beskok, G.E.K. Report: A model for flows in channels, pipes, and ducts at micro and nano scales. *Microsc. Thermophys. Eng.* **1999**, *3*, 43–77. [CrossRef]
43. Faruk, C. A triple-mechanism fractal model with hydraulic dispersion for gas permeation in tight reservoirs. In Proceedings of the SPE International Petroleum Conference and Exhibition, Villahermosa, Mexico, 10–12 February 2002; p. 74368.
44. Chapuis, R.P.; Aubertin, M. Predicting the coefficient of permeability of soils using the Kozeny-Carman equation. *Can. Geotech.* **2003**, *40*, 616–628. [CrossRef]
45. Rui, Z.; Guo, T.; Feng, Q.; Qu, Z.; Qi, N.; Gong, F. Influence of Gravel on the propagation pattern of hydraulic fracture in the Glutenite Reservoir. *J. Petroleum Sci. Eng.* **2018**, *165*, 627–639. [CrossRef]
46. Snow, D.T. Anisotropic permeability of fractured media. *Water Resour. Res.* **1969**, *5*, 1273–1289. [CrossRef]
47. Rutqvist, J.; Tsang, C.F.; Tsang, Y. Analysis of stress and moisture induced changes in fractured rock permeability at the yucca mountain drift scale test. *Elsevier Geo-Eng. B Ser.* **2004**, *2*, 161–166.
48. Brady, B.H.G.; Brown, E.T. *Rock Mechanics: For Underground Mining*; Springer Science & Business Media: Berlin, Germany, 2013.
49. Rahman, M.K.; Hossain, M.M.; Rahman, S.S. A shear-dilation-based model for evaluation of hydraulically stimulated naturally fractured reservoirs. *Int. J. Numer. Anal. Methods Geomech.* **2002**, *26*, 469–497. [CrossRef]
50. Zhu, W.; Deng, J.; Yang, B.; Hongying, Q.I. Seepage model of shale gas reservoir and productivity analysis of fractured vertical wells. *Mech. Eng.* **2014**, *15*, 178–183.
51. Civan, F.; Rai, C.S.; Sondergeld, C.H. Shale-gas permeability and diffusivity inferred by improved formulation of relevant retention and transport mechanisms. *Trans. Porous Media* **2011**, *86*, 925–944. [CrossRef]
52. Hsieh, B.Z.; Chilingar, G.V.; Lin, Z.S. Propagation of radius of investigation from producing well. *Energy Sour.* **2007**, *29*, 403–417. [CrossRef]
53. Cheng, Y. Pressure transient characteristics of hydraulically fractured horizontal shale gas wells. In Proceedings of the SPE Eastern Regional Meeting, Columbus, OH, USA, 17–19 August 2011; p. 149311.
54. Sang, Y.; Chen, H.; Yang, S.; Guo, X.; Zhou, C.; Fang, B.; Zhou, F.; Yang, J.K. A new mathematical model considering adsorption and desorption process for productivity prediction of volume fractured horizontal wells in shale gas reservoirs. *J. Nat. Gas Sci. Eng.* **2014**, *19*, 228–236. [CrossRef]

Article

An Analytical Flow Model for Heterogeneous Multi-Fractured Systems in Shale Gas Reservoirs

Honghua Tao [1,2,3,*], Liehui Zhang [1], Qiguo Liu [1,*], Qi Deng [1], Man Luo [4] and Yulong Zhao [1,*]

1 State Key Laboratory of Oil and Gas Reservoir Geology and Exploitation, Southwest Petroleum University, Chengdu 610500, China; zhangliehui@swpu.edu.cn (L.Z.); 201411000031@stu.swpu.edu.cn (Q.D.)
2 John and Willie Leone Family Department of Energy and Mineral Engineering, The Pennsylvania State University, University Park, PA 16802, USA
3 Key Laboratory of Shale Gas Exploration, Ministry of Land and Resources Engineering, Chongqing 400042, China
4 Petro China West Pipeline Company, Urumqi 830013, China; luoluo-wo@163.com
* Correspondence: hut91@psu.edu (H.T.); LiuQiguo@swpu.edu.cn (Q.L.); swpuzhao@swpu.edu.cn (Y.Z.); Tel.: +1-814-852-8648 (H.T.); +86-28-8303-2052 (Q.L.); +86-159-8232-4747 (Y.Z.)

Received: 14 October 2018; Accepted: 4 December 2018; Published: 6 December 2018

Abstract: The use of multiple hydraulically fractured horizontal wells has been proven to be an efficient and effective way to enable shale gas production. Meanwhile, analytical models represent a rapid evaluation method that has been developed to investigate the pressure-transient behaviors in shale gas reservoirs. Furthermore, fractal-anomalous diffusion, which describes a sub-diffusion process by a non-linear relationship with time and cannot be represented by Darcy's law, has been noticed in heterogeneous porous media. In order to describe the pressure-transient behaviors in shale gas reservoirs more accurately, an improved analytical model based on the fractal-anomalous diffusion is established. Various diffusions in the shale matrix, pressure-dependent permeability, fractal geometry features, and anomalous diffusion in the stimulated reservoir volume region are considered. Type curves of pressure and pressure derivatives are plotted, and the effects of anomalous diffusion and mass fractal dimension are investigated in a sensitivity analysis. The impact of anomalous diffusion is recognized as two opposite aspects in the early linear flow regime and after that period, when it changes from 1 to 0.75. The smaller mass fractal dimension, which changes from 2 to 1.8, results in more pressure and a drop in the pressure derivative.

Keywords: fractional diffusion; fractal geometry; analytical model; shale gas reservoir

1. Introduction

The development of shale gas in North America has achieved large-scale commercial success [1–3], which has set off a shale gas revolution worldwide. As a key technology in shale gas exploration and development, well testing plays an irreplaceable role. The characteristics of shale gas reservoirs can be obtained through the transient pressure analysis of multiple fractured horizontal wells (MFHWs) in shale gas reservoirs.

In order to describe the random and complex fractures, some works [4,5] have investigated discrete fracture networks through numerical simulation approaches. Tang et al. [4] established a three-dimensional numerical model based on the construction of spatial discretization by the finite volume method. Wang [5] proposed a unified model for shale gas reservoirs based on discrete fracture networks to investigate shale gas production by rate transient analysis. However, this requires numerical simulation, and the process is time-consuming and occupies a large amount of computing resources.

Fortunately, the analytical approach is a convenient and rapid method for the evaluation of dynamic characteristics of the shale gas reservoir, which takes less time and needs less reservoir data compared with numerical simulation approaches. Thus, the analytical approach has been used in more applications in recent years.

Two types of analytical model are used to analyze transient pressure behaviors. One type is the detailed analytical model, which is based on the source function and superposition principle [6–8]. This characterizes the stimulated reservoir volume (SRV) region in a shale gas reservoir as a circular or rectangular zone and extends the one-region model to a dual-region composite model. Similarly, the shortcomings of the detailed analytical model also cause a large increase in the amount of calculation required. In order to describe the SRV region more concisely and conveniently, the other type, which is linear models, such as the tri-linear flow model [9] and the five-region flow model [10], was developed. The five-region flow model was established based on the tri-linear flow model and takes into account not only the stimulated region, but also the nearby unstimulated region. These two models represent a rapid way to capture key characteristics in shale gas reservoirs.

Based on these two analytical models (the detailed analytical model and linear model), other improved models were developed, e.g., models considering the effects of fractures in the SRV region [11], non-equal spacing fractures [12], fracture networks in the shale matrix [13,14], the non-Darcy high-speed flow inside the hydraulic fracture [15], the shale matrix diffusion and dual porosity model [16], a transient flow approach [17], and non-Darcy flow with a threshold pressure gradient in tight gas reservoirs [18]. Recently, Zeng et al. [19], Zeng [20], and Zeng et al. [21] proposed a seven-region flow model, which takes into account the spatial heterogeneity and typical seepage features, such as ad-desorption and diffusion in shale gas reservoirs. Unfortunately, all of the models described above only consider the linear flow in all regions, and thereby neglect the fractal features and sub-diffusive flow in the SRV region.

In order to capture the features of fractal geometry and sub-diffusive flow in highly heterogeneous porous media, an analytical flow model that considers anomalous diffusion and other significant features to describe the flow characteristics in the SRV region has been proposed [22–26].

Chen and Raghavan [22] utilized fractional derivatives to characterize the process of anomalous diffusion in the complex fractures and took into account the continuous-time random walk in hydraulically fractured reservoirs with single porosity. Subsequently, Ren and Guo [23] presented a dual porosity and anomalous diffusion model for shale gas reservoirs. Unfortunately, they did not consider the heterogeneity of multi-fractured systems by applying a three-region or five-region model. Later, Albinali and Ozkan [24] proposed a tri-linear anomalous diffusion and dual-porosity model that uses fractional calculus to account for non-uniform velocity in porous media. However, the fractal geometry features of the induced fractures in the SRV region are not considered in the model. Wang et al. [25] considered the fractal characteristics in the complex system by coupling fractal relations to account for the heterogeneity in the SRV region. Fan and Ettehadtavakkol [26] applied micro-seismic data to verify the fractal flow model and proposed a semi-analytical model for rate transient analysis in shale gas reservoirs.

All the models described above do not fully consider the various diffusion of shale gas in the shale matrix, the dual porosity in the SRV region, or the stress sensitivity of permeability and fractal-anomalous diffusion in complex fractures. Table 1 demonstrates the differences by comparing previous analytical flow models with the present model. Previous models only considered homogeneous properties and simple transport mechanisms in shale gas reservoirs.

Based on the above, this work proposes a new analytical model based on fractal-anomalous diffusion. Firstly, the present model is coupled with anomalous diffusion and other significant features, such as ad-desorption, slip flow, surface flow, pressure-dependent permeability, and fractal geology. Using the Laplace transformation method and Duhamel's theorem [27], the analytical solution of the present model is obtained. Then, the flow regimes are identified, and the effects of relevant parameters are analyzed.

Therefore, the present model can effectively describe the complex fracture networks in the SRV region and more accurately account for the various transport mechanisms of MFHWs in shale gas reservoirs. Due to the lack of well-testing data in shale gas reservoirs, the present model has only been applied to one case, but more cases will be studied in the future.

Table 1. Feature comparisons of analytical models for multiple fractured horizontal wells (MFHWs). SRV: stimulated reservoir volume.

Serial Number	Features	Models				
		Stalgorova and Mattar [10]	Albinali and Ozkan [24]	Wang et al. [25]	Fan and Ettehadtavakkol [26]	Present Model
1	Fractal permeability in SRV	-	-	Fractal	Tortuosity- dependent	Fractal
2	Dual porous media in SRV	Cubic geometry	Spherical geometry	Cubic geometry	Slab geometry	Spherical geometry
3	Diffusion in fractures	Normal	Anomalous	Normal	Normal	Anomalous
4	Pressure-dependence of permeability	-	-	-	-	Exponential
5	Slip flow in shale matrix	-	-	-	Klinkenberg	Klinkenberg
6	Diffusion in shale matrix	-	-	-	Knudsen	Composite
7	Ad-desorption	-	-	-	Langmuir	Langmuir
8	Flow types	Five regions	Three regions	Five regions	Three regions	Five regions

2. Physical Model

Figure 1 is a schematic of the typical five-region flow model and the improved five-region flow model (new model) in a shale gas reservoir. Higher fractal permeability, dual-porosity, and anomalous diffusion in the SRV region are taken into account around each fracture. The other three regions occupy the remaining space between adjacent fractures. One-quarter of each hydraulic fracture is taken into account due to the assumption of symmetry in the reservoir.

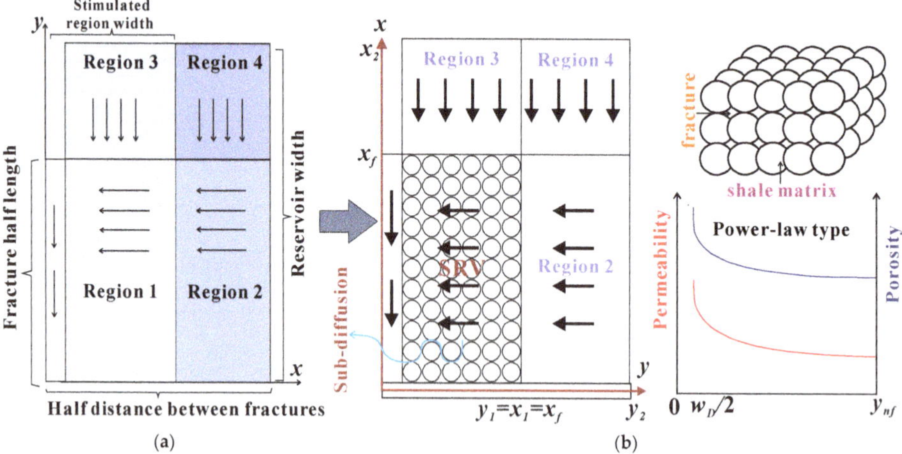

Figure 1. Schematic of physical models for hydraulically fractured horizontal wells. (**a**) The typical five-region flow model proposed by Stalgorova and Mattar [10]. (**b**) The improved five-region flow model (new model). Fracture half-length: x_f; width of the hydraulic fracture: w_D; distance from the hydraulic fracture to stimulated reservoir volume (SRV): y_1; no flow bound: x_2, y_2.

As shown in Figure 1, the reservoir between two adjacent fractures is subdivided into five regions in the improved five-region flow model. There is vertical linear flow from region 4 to region 2 and from region 3 to region 1 (SRV). Similarly, horizontal linear flow exists from region 2 to region 1 and from region 1 to each hydraulic fracture. Compared with the typical five-region flow model, ad-desorption and various diffusion in the shale matrix, dual-porosity (shown as spherical matrix in Figure 1), the fractal geometry (shown as a power-law type in Figure 1) and anomalous diffusion (sub-diffusion) in the SRV region, and stress-sensitive permeability in each region are considered in this work. The main assumptions of this new model are as follows:

(1) A hydraulically fractured horizontal well is at the center of a closed shale gas reservoir;
(2) Each hydraulic fracture is perpendicular to the horizontal well, spaced uniformly along the horizontal wellbore, and has the same length;
(3) Fluid flow in each region is a one-dimensional single-phase flow;
(4) Desorption in shale matrix yields to the Langmuir isotherm adsorption law;
(5) The continuity of flux and pressure at interfaces is used to couple the adjacent regions.

3. Mathematical Model

3.1. Mechanisms and Properties

3.1.1. Adsorption/Desorption and Apparent Permeability

Shale gas adsorption in the shale matrix typically yields to the Langmuir isotherm adsorption law, and pseudo-pressure can be written as follows [28,29]:

$$V_E = V_L \frac{P}{P_L + P} \tag{1}$$

where V_E is defined as the adsorption equilibrium concentration (sm^3/m^3), the Langmuir adsorption concentration is represented by V_L (sm^3/m^3), the Langmuir pressure is represented by P_L (MPa), and P means the pressure in the reservoir (MPa).

$$\sigma_m = 1 + \frac{\rho_{gsc} V_L p_L}{c_g \rho_m \phi_m (p_L + p)^2} \tag{2}$$

where σ_m is the adsorption factor.

$$m(p) = 2 \int_{p_0}^{p} \frac{p}{\mu} dp \tag{3}$$

where $m(p)$ is the pseudo-pressure (MPa^2/(mPa·s)), the gas viscosity is represented by μ (mPa·s), and the real gas deviation factor is represented by z.

The main transport mechanisms in the shale matrix are surface diffusion, Knudsen diffusion, and slip flow. Based on the results of previous research, the expression of total equivalent permeability (apparent permeability) is as follows [30]:

$$k_{ma} = k_e + k_d + k_s = \beta_t k_{ins} \tag{4}$$

where k_{ma} is defined as an apparent permeability which is related to surface diffusion, Knudsen diffusion, and slip flow (m^2); k_e is the equivalent slip rate of slip flow (m^2); the Knudsen diffusion equivalent permeability is represented by k_d (m^2); the surface diffusion equivalent permeability is represented by k_s (m^2); and β_t is the matrix comprehensive diffusion factor that considers the slip flow, Knudsen, and surface diffusion.

3.1.2. Fractal Permeability and Porosity in Induced Fractures

The distribution of induced fractures is extremely complex and irregular, and therefore, it is not accurate enough to describe the porosity of induced fractures in Euclidean geometry. Fractal geometry has been verified as an effective method to describe the complex pore structure of fibrous porous media [31–34]. Based on fractal geometry, fractal permeability and fractal porosity in induced fractures comply with a power-law type as follows [35–38]:

$$K_f(r) = K_{fr}\left(\frac{r}{L_{ref}}\right)^{d_f-d_e-\theta} \tag{5}$$

where K_{fr} is the permeability at the reference length, L_{ref} is the reference length; the mass fractal dimension of the inducec fractures is represented by d_f, the Euclidean dimension is represented by d_e, the radial coordinate value at any point is represented by r, and the tortuosity index is represented by θ.

$$\varnothing_f(r) = \varnothing_{fr}\left(\frac{r}{L_{ref}}\right)^{d_f-d_e} \tag{6}$$

where $\varnothing_{fr}$ is the porosity at the reference length.

3.1.3. Anomalous Diffusion in Induced Fractures

In induced fractures, the disorder, non-local, and memory features should be considered in the SRV region. This complex transport process is anomalous diffusion, which is described by fractional calculus. The modified Darcy flow velocity is given by the following form [22]:

$$v(r,t) = -\frac{k_\alpha}{\mu}\frac{\partial_{1-\alpha}}{\partial t}\nabla p(r,t). \tag{7}$$

The fractional derivative $\frac{\partial^\alpha f(t)}{\partial t^\alpha}$ is defined as follows [39]:

$$\frac{\partial^\alpha f(t)}{\partial t^\alpha} = \frac{1}{\Gamma(1-\alpha)}\int_0^t (t-t')^{-\alpha}\frac{\partial f(t')}{\partial t'}dt' \tag{8}$$

where the Gamma function is represented by $\Gamma(x)$. The Laplace transform of the fractional derivative $\frac{\partial^\alpha f(t)}{\partial t^\alpha}$ is

$$\int_0^\infty e^{-st}\frac{\partial^\alpha f(t)}{\partial t^\alpha}dt = s^\alpha f(s) - s^{\alpha-1}f(0). \tag{9}$$

when $\alpha = 1$, Equation (7) is reduced to the classical Darcy's law as follows [23]:

$$v(r,t) = -\frac{k_\alpha}{\mu}\nabla p(r,t). \tag{10}$$

3.1.4. Pressure-Dependent Permeability

The permeability in hydraulically fracturing shale gas reservoirs is sensitive to pore pressure, according to previous experiments [3,40]. Given the relationship with pore pressure, fractal permeability is introduced by permeability modulus as follows:

$$k = k_i e^{-\gamma(m_i - m)} \tag{11}$$

where k_i is the permeability under the initial pseudo-pressure (m_i), the corresponding pseudo-pressure in the reservoir is represented by m, and γ is the stress sensitivity factor.

3.2. Governing Flow Equations and Solutions

In order to obtain the final solution, the governing diffusivity equations for each region are written with the relevant initial and boundary conditions. Definitions of all dimensionless terms are given in Appendix A.

3.2.1. Unstimulated Regions (Region 4 + Region 3 + Region 2)

Starting with the fourth region, the diffusivity equation that considers the ad-desorption and various diffusion can be written in a dimensionless form:

$$e^{-\gamma_D^* m_{4D}}[\frac{\partial^2 m_{4D}}{\partial x_D^2} - \gamma_D^* (\frac{\partial m_{4D}}{\partial x_D^2})^2] = \frac{\sigma_m}{\beta_t \eta_{4D}} \frac{\partial m_{4D}}{\partial t_D} \tag{12}$$

where γ_D^* is the dimensionless stress-sensitive factor.

The perturbation inversion proposed by Pedrosa Jr. [41] is applied to pseudo-pressure, as presented in Equation (13).

$$m_D = -\frac{1}{\gamma_D^*} ln(1 - \gamma_D^* \varphi_D(r_D, t_D)) \tag{13}$$

Additionally, a zero-order approximation is performed to linearize the diffusivity equation. Then, the diffusion Equation (12) can be approximately written in a Laplace form, as follows:

$$\frac{\partial^2 \overline{\varphi}_{4D}}{\partial x_D^2} = \frac{\sigma_m s}{\beta_t \eta_{4D}} \overline{\varphi}_{4D}. \tag{14}$$

The outer boundary condition (no-flow) is

$$\frac{\partial \overline{\varphi}_{4D}}{\partial x_D}|_{x_D = x_{eD}} = 0. \tag{15}$$

The inner boundary condition (pressure continuity) is

$$\overline{\varphi}_{4D}|_{x_D = x_{nfD} = 1} = \overline{\varphi}_{2D}|_{x_D = x_{nfD} = 1}. \tag{16}$$

Therefore, the general form of the solution in the fourth region can be given as follows:

$$\overline{\varphi}_{4D} = \overline{\varphi}_{2D}|_{x_D = x_{nfD} = 1} \frac{cosh\left[\sqrt{f_4(s)}(x_D - x_{eD})\right]}{cosh\left[\sqrt{f_4(s)}\left(x_{nfD} - x_{eD}\right)\right]}|_{x_{nfD} = 1} \tag{17}$$

where

$$f_4(s) = \frac{\sigma_m s}{\beta_t \eta_{4D}} \tag{18}$$

and η_{4D} is the dimensionless conductivity in region 4.

Region 3, which has low permeability, can only flow vertically to region 1. Similarly, a general form of the solution for the third region can be given as follows:

$$\overline{\varphi}_{3D} = \overline{\varphi}_{1D}|_{x_D = x_{nfD} = 1} \frac{cosh\left[\sqrt{f_3(s)}(x_D - x_{eD})\right]}{cosh\left[\sqrt{f_3(s)}\left(x_{nfD} - x_{eD}\right)\right]}|_{x_{nfD} = 1} \tag{19}$$

where

$$f_3(s) = \frac{\sigma_m s}{\beta_t \eta_{3D}}. \tag{20}$$

Also, the governing equation of region 2 becomes

$$\frac{\partial^2 \overline{\varphi}_{2D}}{\partial y_D^2} = \frac{\sigma_m s}{\beta_t \eta_{2D}} \overline{\varphi}_{2D} - \frac{k_{4a}}{k_{2a} x_{nfD}} \frac{\partial \overline{\varphi}_{4D}}{\partial x_D} |_{x_D = x_{nfD} = 1}. \tag{21}$$

The outer boundary condition (no-flow) is

$$\frac{\partial \overline{\varphi}_{2D}}{\partial y_D} |_{y = y_{eD}} = 0. \tag{22}$$

The inner boundary condition (pressure continuity) is

$$\overline{\varphi}_{2D} |_{y_D = y_{nfD}} = \overline{\varphi}_{nfD} |_{y_D = y_{nfD}}. \tag{23}$$

Therefore, the solution for region 2 becomes

$$\overline{\varphi}_{2D} = \overline{\varphi}_{nfD} |_{y_D = y_{nfD}} \frac{cosh\left[\sqrt{f_2(s)}(y_D - y_{eD})\right]}{cosh\left[\sqrt{f_2(s)}\left(y_{nfD} - y_{eD}\right)\right]} \tag{24}$$

where

$$f_2(s) = \frac{\sigma_m s}{\beta_t \eta_{2D}} - \frac{k_{4a}}{k_{2a} x_{nfD}} \sqrt{f_4(s)} tanh\left[\sqrt{f_4(s)}\left(x_{nfD} - x_{eD}\right)\right] |_{x_{nfD} = 1}. \tag{25}$$

3.2.2. Region 1 (SRV)

Region 1 represents the SRV region in which the transient inter-porosity flow from the matrix to fracture subsystem is applied. Moreover, the anomalous diffusion, fractal permeability, and porosity in induced fractures are also considered.

- Matrix subsystem:

Similarly, the pressure solution in the matrix subsystem of region 1 can be obtained:

$$\overline{\varphi}_{1mD} = \frac{sinh\left(\sqrt{u_{1m}(s)} r_{mD}\right)}{r_{mD} sinh\left(\sqrt{u_{1m}(s)}\right)} \overline{\varphi}_{nfD} \tag{26}$$

where

$$u_{1m}(s) = \frac{15(1 - \omega_1)\sigma_m s}{\beta_t \lambda_1 \eta_{1D}}. \tag{27}$$

- Induced fractures subsystem:

The diffusivity equation of the complex fractures networks can be derived in the following dimensionless form. More detailed derivations are given in Appendix B.

$$\frac{\partial^2 \overline{\varphi}_{nfD}}{\partial y_D^2} + \frac{d_f - \theta - 2}{y_D} \frac{\partial \overline{\varphi}_{nfD}}{\partial y_D} = f_1(s) {y_D}^{\theta} \overline{\varphi}_{nfD} \tag{28}$$

where

$$f_1(s) = \frac{\omega_1}{\eta_{1D}} s^{\alpha} + \left\{ \frac{\beta_t \lambda_1}{5} [\sqrt{u_{1m}(s)} coth(\sqrt{u_{1m}(s)} - 1)] - \left(\frac{\eta_{nf}}{x_f^2}\right)^{\alpha - 1} \frac{k_{3a}}{k_{nf}} \sqrt{f_3(s)} tanh[\sqrt{f_3(s)}(x_{nfD} - x_{eD})] \right\} s^{\alpha - 1} \tag{29}$$

The outer boundary condition (flow continuity) is

$$k_{2a} \frac{\partial \overline{\varphi}_{2fD}}{\partial y_D} |_{y_D = y_{nfD}} = \left(s \frac{\eta_{nf}}{x_f^2}\right)^{1 - \alpha} k_{nf} {y_D}^{d_f - \theta - 2} \frac{\partial \overline{\varphi}_{nfD}}{\partial y_D} |_{y_D = y_{nfD}}. \tag{30}$$

The inner boundary condition (pressure continuity) is

$$\overline{\varphi}_{nfD}|_{y_D=w_D/2} = \overline{\varphi}_{FD}|_{y_D=w_D/2}. \tag{31}$$

Therefore, the general form of the pressure solution in the SRV is

$$\overline{\varphi}_{nfD} = y_D{}^a\left\{AI_n\left[\alpha y_D{}^{\frac{1}{\alpha}}\sqrt{f_1(s)}\right] + BK_n\left[\alpha y_D{}^{\frac{1}{\alpha}}\sqrt{f_1(s)}\right]\right\} \tag{32}$$

where

$$\begin{cases} a = \frac{1-b}{2},\, b = d_f - \theta - 2,\, n = \frac{1-b}{2+\theta},\, \alpha = \frac{2}{2+\theta},\, c = \alpha\sqrt{f_1(s)} \\ A = \frac{h_{22}\overline{\varphi}_{FD}|_{y_D=w_D/2}}{h_{11}h_{22}-h_{12}h_{21}},\, B = \frac{-h_{21}\overline{\varphi}_{FD}|_{y_D=w_D/2}}{h_{11}h_{22}-h_{12}h_{21}} \\ h_{11} = (\frac{w_D}{2})^a I_n[c(\frac{w_D}{2})^{\frac{1}{\alpha}}] \\ h_{12} = (\frac{w_D}{2})^a K_n[c(\frac{w_D}{2})^{\frac{1}{\alpha}}] \\ h_{21} = g(y_{nfD})^a\sqrt{f_2(s)}tanh\left[(y_{nfD} - y_{eD})\sqrt{f_2(s)}\right]I_n\left[c(y_{nfD})^{\frac{1}{\alpha}}\right] - \frac{c}{\alpha}(y_{nfD})^{a+\frac{1}{\alpha}-1}I_{n-1}[c(y_{nfD})^{\frac{1}{\alpha}}] \\ h_{22} = g(y_{nfD})^a\sqrt{f_2(s)}tanh\left[(y_{nfD} - y_{eD})\sqrt{f_2(s)}\right]K_n\left[c(y_{nfD})^{\frac{1}{\alpha}}\right] + \frac{c}{\alpha}(y_{nfD})^{a+\frac{1}{\alpha}-1}K_{n-1}[c(y_{nfD})^{\frac{1}{\alpha}}] \end{cases} \tag{33}$$

3.2.3. Hydraulic Fracture Region

Considering that the stress sensitivity of permeability and flow exchange is directly related to the quality dimension, the diffusivity equation in hydraulic fractures becomes

$$e^{-\gamma_D^* m_{FD}}\left[\frac{\partial^2 m_{FD}}{\partial x_D^2} - \gamma_D^*\left(\frac{\partial m_{FD}}{\partial x_D^2}\right)^2\right] = \frac{1}{\eta_{FD}}\frac{\partial m_{FD}}{\partial t_D} - \frac{2}{F_{CD}}(\frac{w_D}{2})^{-\theta}\left(s\frac{\eta_{nf}}{x_f^2}\right)^{1-\alpha}\frac{\partial m_{nfD}}{\partial y_D}|_{y_D=\frac{w_D}{2}} \tag{34}$$

where

$$F_{CD} = \frac{k_F w_D}{k_{nf}}. \tag{35}$$

The perturbation inversion [41] and zero order approximation in the Laplace form are applied, and the diffusivity equation then becomes

$$\frac{\partial^2\overline{\varphi}_{FD}}{\partial x_D^2} = \frac{s}{\eta_{FD}}\overline{\varphi}_{FD} - \frac{2}{F_{CD}}(\frac{w_D}{2})^{-\theta}\left(s\frac{\eta_{nf}}{x_f^2}\right)^{1-\alpha}\frac{\partial\overline{\varphi}_{nfD}}{\partial y_D}|_{y_D=\frac{w_D}{2}}. \tag{36}$$

Equation (35) can be written as follows:

$$\frac{\partial^2\overline{\varphi}_{FD}}{\partial x_D^2} = F(s)\overline{\varphi}_{FD} \tag{37}$$

where

$$\begin{cases} F(s) = \frac{s}{\eta_{FD}} - \frac{2}{F_{CD}}(\frac{w_D}{2})^{-\theta}\left(s\frac{\eta_{nf}}{x_f^2}\right)^{1-\alpha}\frac{\partial\overline{\varphi}_{nfD}}{\partial y_D}|_{y_D=\frac{w_D}{2}} \\ \frac{\partial\overline{\varphi}_{nfD}}{\partial y_D}|_{y_D=\frac{w_D}{2}} = \frac{c}{\alpha}(\frac{w_D}{2})^{a+\frac{1}{\alpha}-1}\left\{AI_{n-1}\left[c(\frac{w_D}{2})^{\frac{1}{\alpha}}\right] - BK_{n-1}\left[c(\frac{w_D}{2})^{\frac{1}{\alpha}}\right]\right\} \end{cases}. \tag{38}$$

Boundary condition 1 is

$$\frac{\partial\overline{\varphi}_{FD}}{\partial x_D}|_{x_D=1} = 0. \tag{39}$$

Boundary condition 2 is

$$\frac{\partial\overline{\varphi}_{FD}}{\partial x_D}|_{x_D=0} = -\frac{\pi}{F_{CD}s}. \tag{40}$$

The pressure solution for the hydraulic fracture region is

$$\overline{\varphi}_{FD} = \frac{\pi}{F_{CD}s}\frac{1}{\sqrt{F(s)}}\frac{cosh\left[\sqrt{F(s)}\left(x_D - x_{nfD}\right)\right]}{sinh\left[\sqrt{F(s)}x_{nfD}\right]}|_{x_{nfD}=1}. \tag{41}$$

Thus, the pressure solution at the wellbore can be given as follows:

$$\overline{\varphi}_{wD} = \overline{\varphi}_{FD}(0) = \frac{\pi}{F_{CD}s\sqrt{F(s)}tanh\left[x_{nfD}\sqrt{F(s)}\right]}|_{x_{nfD}=1}. \tag{42}$$

However, by applying the superposition principle and Duhamel's principle [27], the final solution for wellbore pressure considering convergence and storage is written as follows:

$$\overline{\varphi}_{wD}(s_c, c_D) = \frac{s\overline{\varphi}_{wD} + s_c}{s[1 + c_D s(s\overline{\varphi}_{wD} + s_c)]}. \tag{43}$$

Then, the perturbation inversion [41] and Stehfest numerical inversion [42] are applied. Finally, the pressure solution at the downhole can be written with the real-time data as

$$m_{wD} = -\frac{ln[1 - \gamma_D^*]L^{-1}(\overline{\varphi}_{wD})}{\gamma_D^*}. \tag{44}$$

4. Discussion and Analysis

4.1. Flow Regimes

In order to obtain the main flow regimes of the improved five-region flow model, the type curves of the pressure-transient response were plotted by employing pseudo-steady inter-porosity flow in the SRV region.

The related parameters are listed in Table 1. Figure 2 shows the pressure-transient response of MFHWs in shale gas reservoirs. There are five flow stages on the type curves: (1) bilinear flow in each hydraulic fracture and in the SRV region (region 1), where the pressure derivative curve's slope is 1/4 (α = 1, and d_f = 2); (2) first linear flow in the SRV region, where the pressure derivative curve shows a straight line with a slope of 1/2 (α = 1, and d_f = 2); (3) inter-porosity and fractal-anomalous diffusion in the SRV region; (4) second linear flow from the USRV to SRV region, where the pseudo-pressure derivative curve presents a straight line with a slope of 1/2 (α = 1, and d_f = 2); and (5) pseudo-steady flow (boundary control flow), where the pseudo-pressure and pseudo-pressure derivative curves are all represented by straight lines with a unit slope.

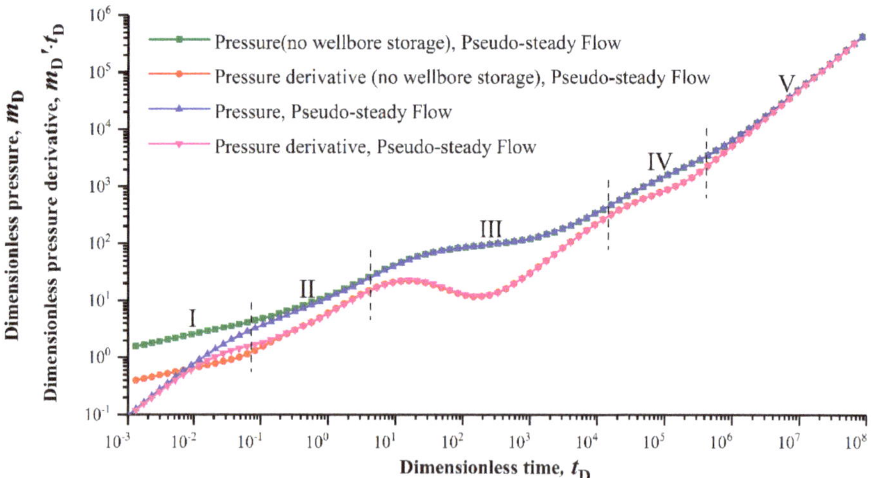

Figure 2. Transient pressure type curves of multiple fractured horizontal wells (MFHWs) in a shale gas reservoir.

4.2. Sensitivity Analysis

In the corresponding sensitivity analysis, firstly, one relevant parameter was changed while keeping the other parameters at their original values. Then, all the relevant parameters were changed at the same time. Model parameters were given values in the simulation by referring to relevant literature [6,12,16,17,23,25,26], and they are listed in Table 2.

Table 2. Model parameters.

Parameter Name	Parameter Value
Dimensionless half fracture length, x_{fD}	1
Dimensionless fracture conductivity, F_{CD}	2
Inter-porosity flow coefficient, λ	$\lambda = 0.2$
Storage capacity coefficient, ω	$\omega = 0.2$
Dimensionless distance in x direction, x_{nfD}/x_{eD}	$x_{nfD} = 1$, $x_{eD} = 50$
Dimensionless distance in y direction, y_{nfD}/y_{eD}	$y_{nfD} = 1$, $y_{eD} = 50$
Ratio of permeability, k_i/k_j	$k_{3a}/k_{nf} = 0.0005$, $k_{2a}/k_{nf} = 0.1$, $k_{4a}/k_{2a} = 0.02$
Absorption factor, σ_m	5
Diffusion factor (apparent permeability coefficient), β_t	1.1
Dimensionless stress sensitivity factor, γ_D^*	0.00009
Anomalous diffusion exponent, α	0.85
Tortuosity index, θ	0.35
Mass fractal dimension, d_f	1.9
Number of fractures, n	10

Figure 3 shows that the fracture conductivity mainly affects the early flow stages. The greater the fracture conductivity is, the smaller the gas flow resistance is, and the smaller pressure consumption is with the same production. It is not difficult to see that the fracture conductivity mainly influences the pressure and pressure derivative curves in the bilinear flow and first linear flow stages. With an increase in the fracture conductivity, the duration of the bilinear flow stage decreases and the duration of the first linear flow stage increases. As seen in Figure 3, when $F_{CD} = 25$, only the first linear flow regime can be observed.

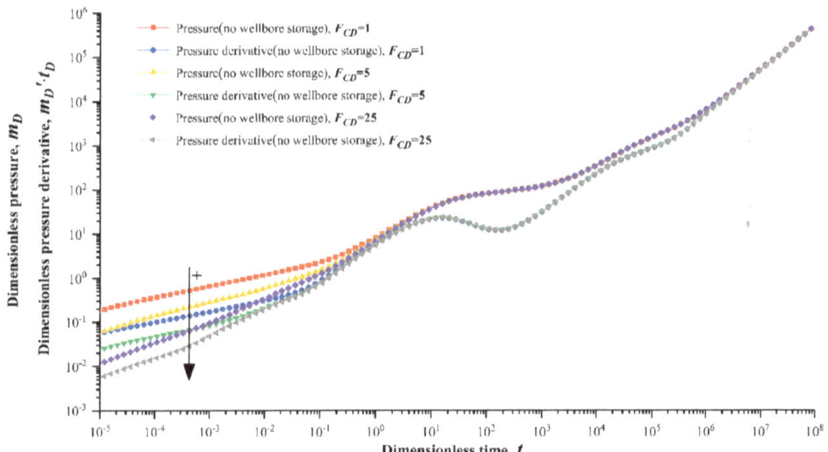

Figure 3. Effect of fracture conductivity on type curves.

Figure 4 demonstrates the type curves of the pressure and pressure derivative for MFHWs in a shale gas reservoir with various anomalous diffusion exponent (α) and tortuosity index (θ) values. As can be seen, one intersection point exists between the anomalous diffusion and classical diffusion pressure derivative curves. At the early bilinear and linear flow stages, the pressure and pressure derivative for $\alpha < 1$ or $\theta > 0$ (anomalous diffusion) are smaller than those for $\alpha = 1$ or $\theta = 0$ (classical diffusion). When the value of α increases (θ decreases), the pressure and its derivative will also increase. The reason for this is that anomalous diffusion delays the performance of pressure derivative behaviors. However, after the inter-porosity flow stage, with different α values, the difference will be more obvious, and the trend is the opposite. In other words, a decrease in α (θ increasing) causes the pressure and its derivative to increase over time. This accounts for the characteristic of sub-diffusion (slower flow) when $\alpha < 1$ or $\theta > 0$ (anomalous diffusion).

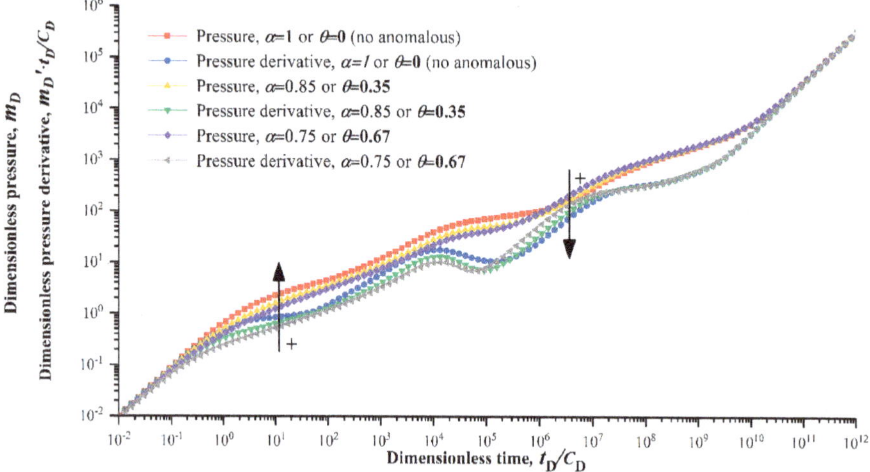

Figure 4. Effect of the anomalous diffusion exponent on type curves.

Figure 5 shows that the mass fractal dimension of induced fractures (Hausdorff index) has a significant effect on the pressure behavior at almost all the stages, except for the wellbore storage stage. Overall, the smaller the mass fractal dimension is, the larger the gas flow resistance is and the greater the pressure consumption is with the same production. As can be seen, the locations of the type curves are higher with a smaller d_f. The reason for this is that a smaller d_f value represents more resistance in the complex induced fractures.

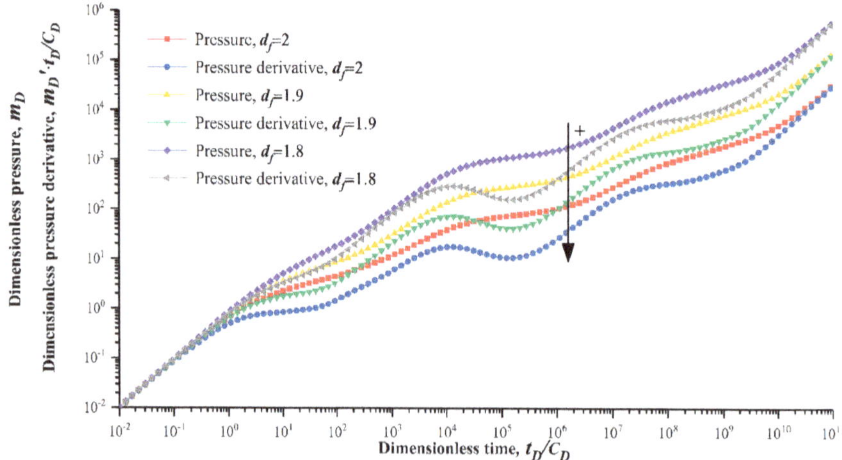

Figure 5. Effect of mass fractal dimension on type curves.

Figures 6–8 demonstrate the influences of the adsorption factor, apparent permeability coefficient, and inter-porosity flow coefficient on the type curves of MFHWs. As shown in Figure 6, the adsorption factor mainly influences the position of the type curves at the inter-porosity flow stage. A larger adsorption factor represents a stronger adsorption and production capacity and therefore makes the "concave" appear wider and deeper on the type curves. Figure 7 shows the effect of the apparent permeability coefficient on the transient pressure response. The apparent permeability has a similar effect to that of the inter-porosity coefficient in Figure 8. The total seepage and diffusion ability of the shale matrix is represented by the apparent permeability coefficient. The smaller the apparent permeability coefficient or inter-porosity coefficient is, the later the "depression" appears on the type curves.

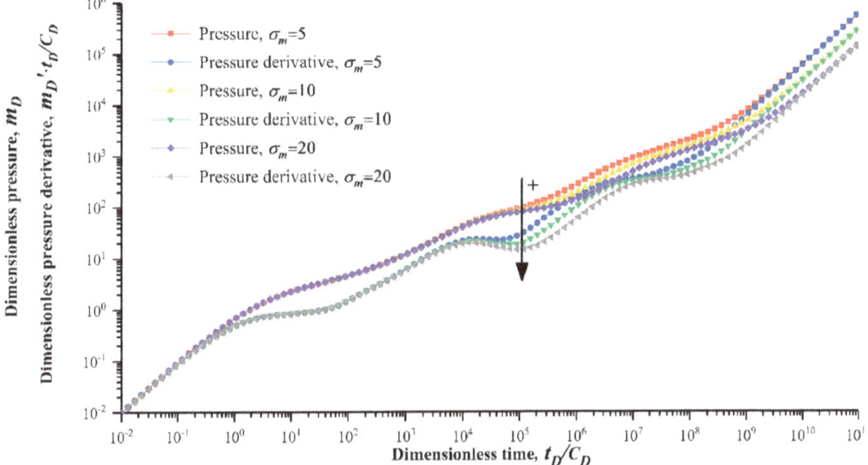

Figure 6. Effect of the adsorption factor on type curves.

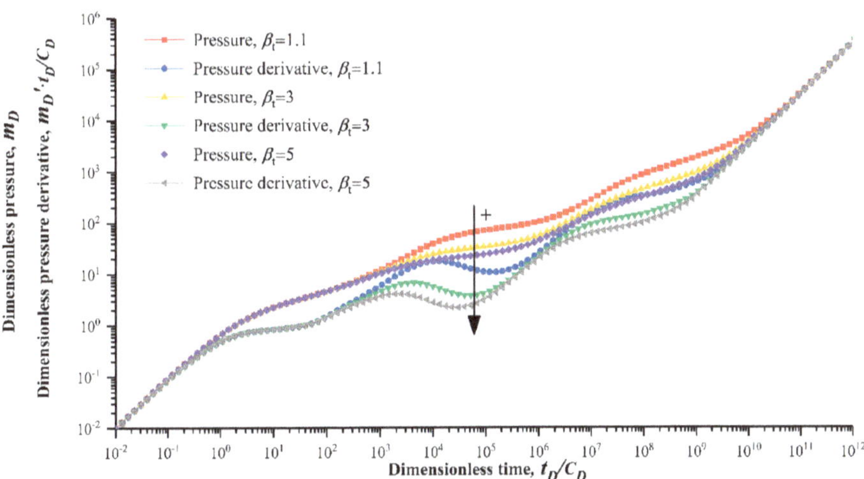

Figure 7. Effect of the apparent permeability coefficient on type curves.

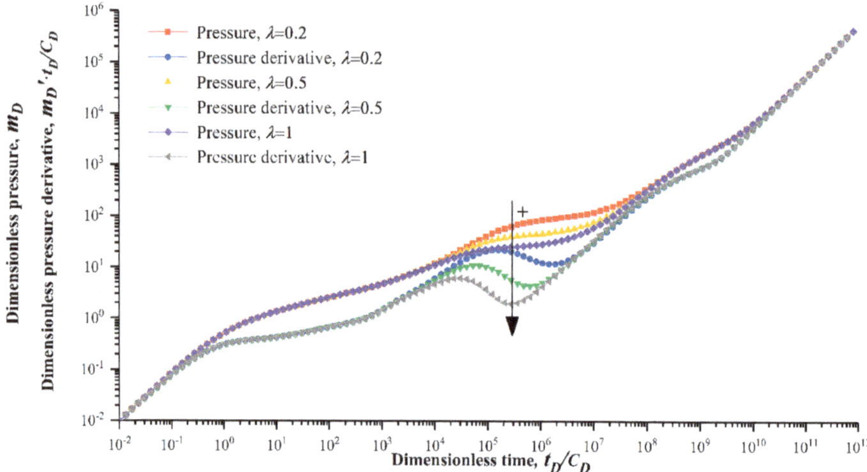

Figure 8. Effect of the inter-porosity flow coefficient on type curves.

Figure 9 shows the impact of the stress sensitivity factor on the pressure-transient response of MFHWs. It can be seen that stress sensitivity affects the whole flow stage, and it has a greater impact in the late time period. The reason for this is that the pressure drop becomes greater in the late time period. The greater the stress sensitivity is, the higher the positions of the pressure and pressure derivative curves are. This depicts the weaker seepage capacity.

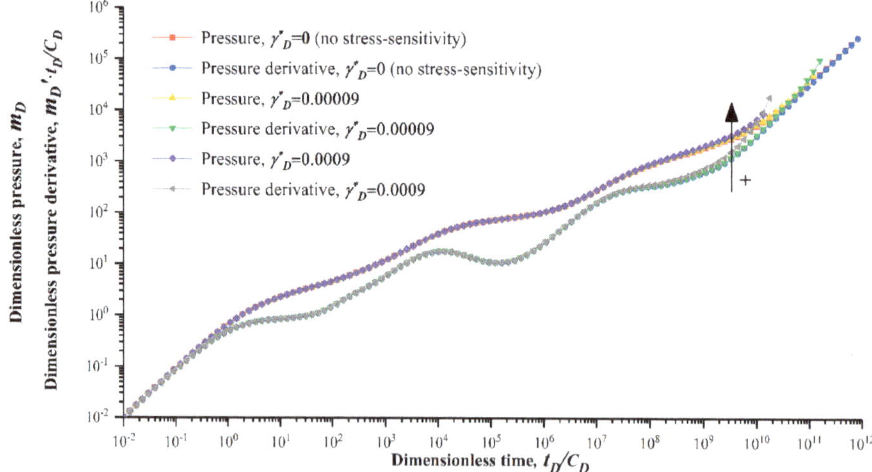

Figure 9. Effect of the stress sensitivity factor on type curves.

As shown in Figure 10, when all the factors are changed at the same time from a smaller parameter group ① to a larger parameter group ②, the positions of type curves for parameter group ② are obviously lower than the positions of type curves for parameter group ①. This indicates that when all the factors become larger, the final pressure drop becomes smaller. The reason for this is that most factors with greater values, such as F_{CD}, α, d_f σ_m, β_t, and λ, can have positive effects by making the pressure consumption smaller, and only γ^*_D has the opposite influence on pressure and pressure derivatives.

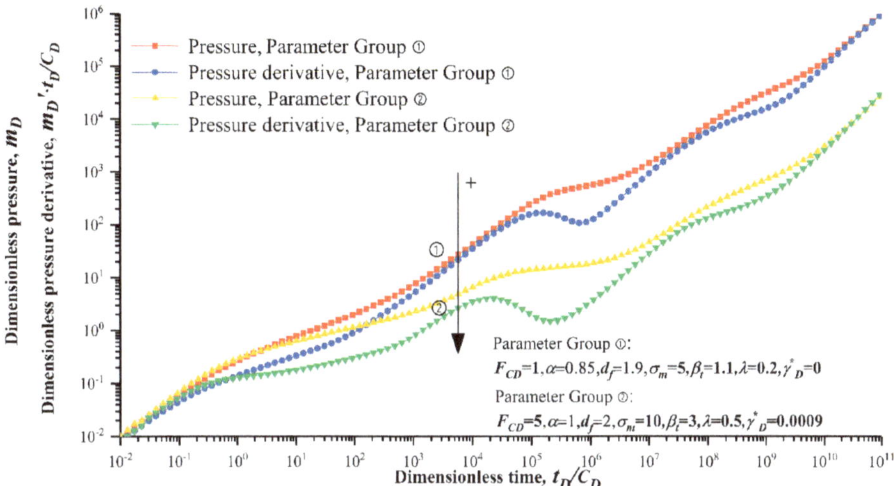

Figure 10. Effect of characteristic factors on type curves (F_{CD}, α, d_f, σ_m, β_t, λ, and γ_D^*).

5. Case Study

This section shows an application of the presented model in a fractured horizontal well (A1) of an actual shale gas field in the Sichuan basin, which has 12 fractures evenly distributed along its horizontal wellbore. The depth of well A1 is 880 m and the thickness of the shale layer is 76 m. The production was 2400 cubic meters per day for 16 h, and then it was shut down for 73 h during the pressure build-up test. For more details, refer to the related literature [16]. After transferring the build-up testing data to dimensionless forms, the actual log-log curves were plotted.

As shown in Figure 11, the improved five-region flow model proposed in this work was applied to match the build-up testing data and was able to perfectly match the real testing data by adjusting the relevant parameters. The results of the interpretation are listed in Table 3. The results reveal that hydraulic fracturing greatly increases the permeability of the fractured zone and produces complex induced fractures with fractal features.

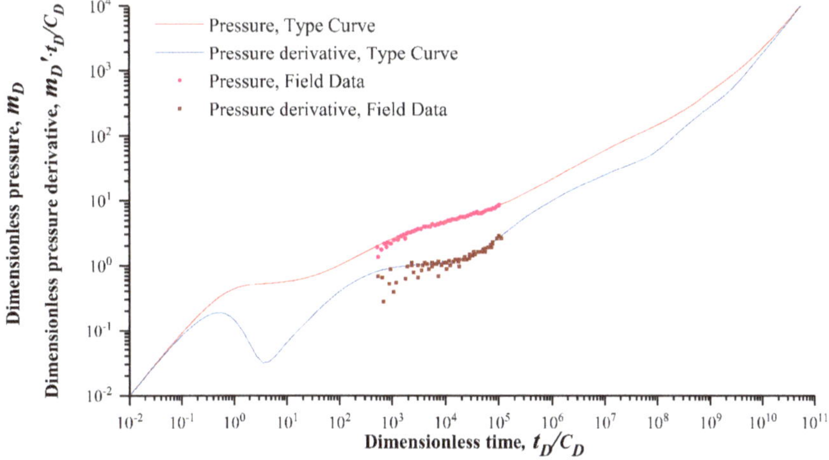

Figure 11. Type curve matching for well A1.

Table 3. Interpretation results for the build-up test of well A1.

Parameter	Parameter Value
Half fracture length, x_f	35 m
Inter-porosity flow coefficient, λ	$\lambda = 0.1$
Storage capacity coefficient, ω	$\omega = 0.05$
Permeability of hydraulic fracture, k_F	4000 mD
Fracture permeability in SRV, k_{nf}	0.0002 mD
Matrix permeability in regions, k_{im}	$k_{1m} = k_{2m} = k_{3m} = k_{4m} = 0.000005$ mD
Absorption factor, σ_m	4
Diffusion factor(Apparent permeability coefficient), β_t	1.5
Dimensionless stress-sensitive factor, γ_D^*	0.00008
Anomalous diffusion exponent, α	0.7
Tortuosity index, θ	0.86
Hausdorff index, d_f	1.85

6. Conclusions

In order to describe the flow retardation in complex fractures in a way that considers the SRV region with anomalous diffusion and fractal features, an improved five-region model was established in this work by introducing the time-fractional flux law. Based on the present model, type curves of pressure and pressure derivative without wellbore storage were plotted and five flow stages were identified: bilinear flow, first linear flow, inter-porosity and fractal-anomalous flow, second linear flow, and boundary control flow. The sensitivity analysis revealed that fractal-anomalous diffusion has a significant impact on pressure-transient behaviors. When the anomalous diffusion exponent decreased from 1 to 0.75, which indicates Darcy flow changing to anomalous diffusion, the pseudo-pressure had less depletion at the early linear flow stages, but this subsequently became greater. When the Hausdorff index changed from 2 to 1.8, greater pressure consumption was needed to achieve the same production. Additionally, stress sensitivity, absorption, and Knudsen diffusion showed non-negligible influences on the pressure-transient response. These effects cannot be ignored. Therefore, the typical five-region flow model which does not take the fractal-anomalous diffusion into account cannot be applied for heterogeneous multi-fractured systems. The present model can be used to provide a more accurate and appropriate interpretation of well-testing data to guide exploration and development.

Author Contributions: All authors have contributed to this work. Conceptualization, H.T., Q.L., and Y.Z.; software, Q.D.; validation, M.L.; formal analysis, H.T.; investigation, H.T.; resources, L.Z.; data curation, M.L.; writing—original draft preparation, H.T.; writing—review and editing, H.T., Q.L., and Q.D.; supervision, L.Z.

Funding: This research was funded by the National Natural Science Foundation of China (Key Program) (Grant No. 51534006), the National Natural Science Foundation of China (Grant No. 51704247), and the National Science and Technology Major Project (2017ZX05009-004).

Conflicts of Interest: The authors declare no conflict of interest.

Nomenclature

c	MPa^{-1}	gas compressibility
h	m	reservoir thickness
k	mD	permeability
$m(p)$	$MPa^2/(mPa{\cdot}s)$	pseudo-pressure
P	Mpa	gas pressure
P_L	Mpa	Langmuir pressure
q_{sc}	10^4 m^3/d	fracture production rate
R_m	m	spherical radius of matrix block
s	-	Laplace transform parameter
t	d	time
T	K	temperature

V_L	sm^3/m^3	Langmuir volume
x_f	m	fracture half length
z	-	gas factor
α	-	anomalous diffusion exponent
β_t	-	apparent permeability coefficient
γ_D^*	-	dimensionless stress-sensitive factor
η	cm^2/s	diffusivity
λ	-	inter-porosity flow coefficient
μ	mPa·s	viscosity
ρ	g/cm^3	gas density
σ_m	-	absorption factor
φ	-	porosity
ω	-	storage capacity coefficient

Appendix A Dimensionless Definitions

The parameters are as follows:

Dimensionless pseudo-pressure: $m_D = \frac{k_{nf}h}{0.01273q_{sc}T}\left(\Psi_i - \Psi_f\right)$

Dimensionless time: $t_D = \frac{3.6k_{nf}t}{\mu(\Phi_{1m}c_{1m}+\Phi_{1f}c_{1f})x_f^2}$

Dimensionless fracture conductivity: $\eta_{jD} = \frac{\eta_j}{\eta_{nf}}(j = 1,2,3,4,F)$

Dimensionless distance: $x_{eD} = \frac{x_e}{x_f}, y_{eD} = \frac{y_e}{x_f}, x_{nfD} = \frac{x_f}{x_f} = 1, y_{nfD} = \frac{y_{nf}}{x_f}$

Storage capacity ratio: $\omega_j = \frac{\Phi_{jf}c_{jf}}{\Phi_{jm}c_{jm}+\Phi_{jf}c_{jf}}\left\{\left(\frac{\eta_{nf}}{x_f^2}\right)^{\alpha-1}\right\}^{2-j}, j = 1$

Inter-porosity coefficient: $\lambda_1 = \frac{15k_{1m}x_f^2}{k_{nf}R_m^2}\left(\frac{\eta_{nf}}{x_f^2}\right)^{\alpha-1}$

Dimensionless stress sensitive factor: $\gamma_D^* = \frac{0.01273\gamma q_{sc}T}{k_{nf}h}$

Dimensionless width of the hydraulic fracture: $w_D = \frac{w_d}{x_f}$

Dimensionless fracture conductivity coefficient: $F_{CD} = \frac{k_F w_D}{k_{nf}}$

Appendix B Derivations for General Diffusivity Equation in the SRV

The general equation for the shale matrix in the SRV region is written as follows:

$$\nabla\cdot\left(k_{nf}\nabla m_{nf}\right) - \frac{3\beta_t k_m}{R_m}\frac{\partial m_m}{\partial r}|_{r=R_m} = \frac{\phi_f c_{gf}\mu}{3.6}\frac{\partial m_{nf}}{\partial t}. \tag{A1}$$

By employing the fractal permeability and porosity, the anomalous diffusion equation can be changed into

$$\frac{\partial^{1-\alpha}}{\partial t^{1-\alpha}}\nabla\cdot(k_{nf}(\frac{y}{x_f})^{d_f-\theta-2}\nabla m_{nf}) - \frac{3\beta_t k_m}{R_m}\frac{\partial m_m}{\partial r}|_{r=R_m} = \frac{\phi_f(\frac{y}{x_f})^{d_f-2}c_{gf}\mu}{3.6}\frac{\partial m_{nf}}{\partial t}. \tag{A2}$$

The stress sensitivity factor is substituted into Equation (A2):

$$\frac{\partial^{1-\alpha}}{\partial t^{1-\alpha}}\left\{e^{-\gamma(m_i-m_{nf})}\left[\nabla\cdot(k_{nf}(\frac{y}{x_f})^{d_f-\theta-2}\nabla m_{nf})\right]\right\} - \frac{3\beta_t k_m}{R_m}\frac{\partial m_m}{\partial r}|_{r=R_m} = \frac{\phi_f(\frac{y}{x_f})^{d_f-2}c_{gf}\mu}{3.6}\frac{\partial m_{nf}}{\partial t} \tag{A3}$$

Taking $\frac{\partial^{\alpha-1}}{\partial t^{\alpha-1}}$ of all terms and $\int_0^{x_f}$ both sides, multiplying by x_f^2, and applying the Pedrosa and zero order approximation in dimensionless form gives

$$\frac{\partial^2 \varphi_{nfD}}{\partial y_D^2} + \frac{d_f - \theta - 2}{y_D}\frac{\partial \varphi_{nfD}}{\partial y_D} + \frac{\partial \varphi_{nfD}}{\partial x_D}|_{x_D = x_{nfD}} - (y_D)^{\theta+2-d_f}\frac{\beta_t \lambda_1}{5}\frac{\partial^{\alpha-1}}{\partial t^{\alpha-1}}\frac{\partial \varphi_{1mD}}{\partial r_D}|_{r_D = r_{mD}} = (y_D)^{\theta}\frac{\omega_1}{\eta_{1D}}\frac{\partial^{\alpha}\varphi_{nfD}}{\partial t^{\alpha}} \tag{A4}$$

By utilizing the assumptions of the flow exchange in inter-porosity flow and interface flow directly related to the quality dimension, the general equation in the Laplace domain becomes

$$\frac{\partial^2 \varphi_{nfD}}{\partial y_D^2} + \frac{d_f - \theta - 2}{y_D}\frac{\partial \varphi_{nfD}}{\partial y_D} + (y_D)^{d_f-2}\frac{\partial \varphi_{nfD}}{\partial x_D}|_{x_D = x_{nfD}} - (y_D)^{\theta}\frac{\beta_t \lambda_1}{5}s^{\alpha-1}\frac{\partial \varphi_{1mD}}{\partial r_D}|_{r_D = r_{mD}} = (y_D)^{\theta}\frac{\omega_1}{\eta_{1D}}s^{\alpha}\frac{\partial \varphi_{nfD}}{\partial t}. \tag{A5}$$

The term $\frac{\partial \varphi_{1mD}}{\partial r_D}|_{r_D = r_{mD}}$ can be substituted from the spherical matrix solution as follows:

$$\frac{\partial \varphi_{1mD}}{\partial r_D}|_{r_D = r_{mD}} = \frac{1}{r_{mD}}[r_{mD}\sqrt{u_{1m}(s)}coth(r_{mD}\sqrt{u_{1m}(s)} - 1)]\overline{\varphi}_{nfD}|_{r_{mD}}. \tag{A6}$$

There is continuity of flux at $x_D = x_{nfD}$ in accordance with

$$\frac{\partial \varphi_{nfD}}{\partial x_D}|_{x_D = x_{nfD}} = \frac{k_{3a}}{k_{nf}(y_D)^{d_f-\theta-2}}\left(\frac{\eta_{nf}}{x_f}s\right)^{\alpha-1}\frac{\partial \varphi_{3D}}{\partial x_D}|_{x_D = x_{nfD}}. \tag{A7}$$

Finally, the general diffusion equation in the SRV region can be given as follows:

$$\frac{\partial^2 \overline{\varphi}_{nfD}}{\partial y_D^2} + \frac{d_f - \theta - 2}{y_D}\frac{\partial \varphi_{nfD}}{\partial y_D} = \left\{\frac{\omega_1}{\eta_{1D}}s^{\alpha} + \left\{\frac{\beta_t \lambda_1}{5}\left[\sqrt{u_{1m}(s)}coth\left(\sqrt{u_{1m}(s)} - 1\right)\right] - \left(\frac{\eta_{nf}}{x_f^2}\right)^{\alpha-1}\frac{k_{3a}}{k_{nf}}\sqrt{f_3(s)}tanh\left[\sqrt{f_3(s)}\left(x_{nfD} - x_{eD}\right)\right]\right\}s^{\alpha-1}\right\}(y_D)^{\theta}\overline{\varphi}_{nfD}. \tag{A8}$$

References

1. Taherdangkoo, R.; Tatomir, A.; Taylor, R.; Sauter, M. Numerical investigations of upward migration of fracking fluid along a fault zone during and after stimulation. *Energy Procedia* **2017**, *125*, 126–135. [CrossRef]
2. Tatomir, A.; McDermott, C.; Bensabat, J.; Class, H.; Edlmann, K.; Taherdangkoo, R.; Sauter, M. Conceptual model development using a generic Features, Events, and Processes (FEP) database for assessing the potential impact of hydraulic fracturing on groundwater aquifers. *Adv. Geosci.* **2018**, *45*, 185–192. [CrossRef]
3. Wang, J.; Jia, A.; Wei, Y.; Qi, Y. Approximate semi-analytical modeling of transient behavior of horizontal well intercepted by multiple pressure-dependent conductivity fractures in pressure-sensitive reservoir. *J. Pet. Sci. Eng.* **2017**, *153*, 157–177. [CrossRef]
4. Tang, C.; Chen, X.; Du, Z.; Yue, P.; Wei, J. Numerical Simulation Study on Seepage Theory of a Multi-Section Fractured Horizontal Well in Shale Gas Reservoirs Based on Multi-Scale Flow Mechanisms. *Energies* **2018**, *11*, 2329. [CrossRef]
5. Wang, H. Discrete fracture networks modeling of shale gas production and revisit rate transient analysis in heterogeneous fractured reservoirs. *J. Pet. Sci. Eng.* **2018**, *169*, 796–812. [CrossRef]
6. Zhang, Q.; Su, Y.; Wang, W.; Sheng, G. A new semi-analytical model for simulating the effectively stimulated volume of fractured wells in tight reservoirs. *J. Nat. Gas Sci. Eng.* **2015**, *27*, 1834–1845. [CrossRef]
7. Chen, P.; Jiang, S.; Chen, Y.; Zhang, K. Pressure response and production performance of volumetric fracturing horizontal well in shale gas reservoir based on boundary element method. *Eng. Anal. Boundary Elem.* **2018**, *87*, 66–77. [CrossRef]
8. Wang, M.; Fan, Z.; Xing, G.; Zhao, W.; Song, H.; Su, P. Rate Decline Analysis for Modeling Volume Fractured Well Production in Naturally Fractured Reservoirs. *Energies* **2018**, *11*, 43. [CrossRef]
9. Brown, M.; Ozkan, E.; Raghavan, R.; Kazemi, H. Practical solutions for pressure-transient responses of fractured horizontal wells in unconventional shale reservoirs. *SPE Reserv. Eval. Eng.* **2011**, *14*, 663–676. [CrossRef]
10. Stalgorova, K.; Mattar, L. Analytical model for unconventional multifractured composite systems. *SPE Reserv. Eval. Eng* **2013**, *16*, 246–256. [CrossRef]
11. Zhao, Y.-L.; Zhang, L.-H.; Luo, J.-X.; Zhang, B.-N. Performance of fractured horizontal well with stimulated reservoir volume in unconventional gas reservoir. *J. Hydrol.* **2014**, *512*, 447–456. [CrossRef]

12. Deng, Q.; Nie, R.-S.; Jia, Y.-L.; Huang, X.-Y.; Li, J.-M.; Li, H.-K. A new analytical model for non-uniformly distributed multi-fractured system in shale gas reservoirs. *J. Nat. Gas Sci. Eng.* **2015**, *27*, 719–737. [CrossRef]
13. Chen, D.; Pan, Z.; Ye, Z. Dependence of gas shale fracture permeability on effective stress and reservoir pressure: Model match and insights. *Fuel* **2015**, *139*, 383–392. [CrossRef]
14. Guo, J.; Zhang, L.; Zhu, Q. A quadruple-porosity model for transient production analysis of multiple-fractured horizontal wells in shale gas reservoirs. *Environ. Earth Sci.* **2015**, *73*, 5917–5931. [CrossRef]
15. Zhang, J.; Huang, S.; Cheng, L.; Xu, W.; Liu, H.; Yang, Y.; Xue, Y. Effect of flow mechanism with multi-nonlinearity on production of shale gas. *J. Nat. Gas Sci. Eng.* **2015**, *24*, 291–301. [CrossRef]
16. Zhang, L.; Gao, J.; Hu, S.; Guo, J.; Liu, Q. Five-region flow model for MFHWs in dual porous shale gas reservoirs. *J. Nat. Gas Sci. Eng.* **2016**, *33*, 1316–1323. [CrossRef]
17. Al-Rbeawi, S. Analysis of pressure behaviors and flow regimes of naturally and hydraulically fractured unconventional gas reservoirs using multi-linear flow regimes approach. *J. Nat. Gas Sci. Eng.* **2017**, *45*, 637–658. [CrossRef]
18. Yuan, B.; Su, Y.; Moghanloo, R.G.; Rui, Z.; Wang, W.; Shang, Y. A new analytical multi-linear solution for gas flow toward fractured horizontal wells with different fracture intensity. *J. Nat. Gas Sci. Eng.* **2015**, *23*, 227–238. [CrossRef]
19. Zeng, Y.; Wang, Q.; Ning, Z.; Sun, H. A Mathematical Pressure Transient Analysis Model for Multiple Fractured Horizontal Wells in Shale Gas Reservoirs. *Geofluids* **2018**, *2018*. [CrossRef]
20. Zeng, J. Analytical Modeling of Multi-Fractured Horizontal Wells in Heterogeneous Unconventional Reservoirs. Master's Thesis, University of Regina, Regina, Saskatchewan, 2017.
21. Zeng, J.; Wang, X.; Guo, J.; Zeng, F. Composite linear flow model for multi-fractured horizontal wells in heterogeneous shale reservoir. *J. Nat. Gas Sci. Eng.* **2017**, *38*, 527–548. [CrossRef]
22. Chen, C.; Raghavan, R. Transient flow in a linear reservoir for space–time fractional diffusion. *J. Pet. Sci. Eng.* **2015**, *128*, 194–202. [CrossRef]
23. Ren, J.; Guo, P. Anomalous diffusion performance of multiple fractured horizontal wells in shale gas reservoirs. *J. Nat. Gas Sci. Eng.* **2015**, *26*, 642–651. [CrossRef]
24. Albinali, A.; Ozkan, E. Analytical Modeling of Flow in Highly Disordered, Fractured Nano-Porous Reservoirs. In Proceedings of the SPE Western Regional Meeting, Anchorage, AK, USA, 23–26 May 2016.
25. Wang, W.; Su, Y.; Sheng, G.; Cossio, M.; Shang, Y. A mathematical model considering complex fractures and fractal flow for pressure transient analysis of fractured horizontal wells in unconventional reservoirs. *J. Nat. Gas Sci. Eng.* **2015**, *23*, 139–147. [CrossRef]
26. Fan, D.; Ettehadtavakkol, A. Semi-analytical modeling of shale gas flow through fractal induced fracture networks with microseismic data. *Fuel* **2017**, *193*, 444–459. [CrossRef]
27. Raghavan, R.S.; Chen, C.-C.; Agarwal, B. An analysis of horizontal wells intercepted by multiple fractures. *SPE J.* **1997**, *2*, 235–245. [CrossRef]
28. Al-Hussainy, R.; Ramey, H., Jr.; Crawford, P. The flow of real gases through porous media. *J. Pet. Technol.* **1966**, *18*, 624–636. [CrossRef]
29. King, G.R.; Ertekin, T. Comparative evaluation of vertical and horizontal drainage wells for the degasification of coal seams. *SPE Reserv. Eng.* **1988**, *3*, 720–734. [CrossRef]
30. Zhang, L.; Shan, B.; Zhao, Y.; Tang, H. Comprehensive Seepage Simulation of Fluid Flow in Multi-scaled Shale Gas Reservoirs. *Transp. Porous Media* **2017**, *121*, 263–288. [CrossRef]
31. Cai, J.; Yu, B. Prediction of maximum pore size of porous media based on fractal geometry. *Fractals* **2010**, *18*, 417–423. [CrossRef]
32. Cai, J.; Luo, L.; Ye, R.; Zeng, X.; Hu, X. Recent advances on fractal modeling of permeability for fibrous porous media. *Fractals* **2015**, *23*, 1540006. [CrossRef]
33. Sheng, M.; Li, G.; Tian, S.; Huang, Z.; Chen, L. A fractal permeability model for shale matrix with multi-scale porous structure. *Fractals* **2016**, *24*, 1650002. [CrossRef]
34. Cai, J.; Wei, W.; Hu, X.; Liu, R.; Wang, J. Fractal characterization of dynamic fracture network extension in porous media. *Fractals* **2017**, *25*, 1750023. [CrossRef]
35. Chang, J.; Yortsos, Y.C. Pressure transient analysis of fractal reservoirs. *SPE Form. Eval.* **1990**, *5*, 31–38. [CrossRef]
36. Acuna, J.; Ershaghi, I.; Yortsos, Y. Practical application of fractal pressure transient analysis of naturally fractured reservoirs. *SPE Form. Eval.* **1995**, *10*, 173–179. [CrossRef]

37. Acuna, J.A.; Yortsos, Y.C. Application of fractal geometry to the study of networks of fractures and their pressure transient. *Water Resour. Res.* **1995**, *31*, 527–540. [CrossRef]
38. Ozcan, O.; Sarak, H.; Ozkan, E.; Raghavan, R.S. A trilinear flow model for a fractured horizontal well in a fractal unconventional reservoir. In Proceedings of the SPE Annual Technical Conference and Exhibition, Amsterdam, The Netherlands, 27–29 October 2014.
39. Caputo, M. Linear models of dissipation whose Q is almost frequency independent—II. *Geophys. J. Int.* **1967**, *13*, 529–539. [CrossRef]
40. Molina, O.M.; Zeidouni, M. Analytical Model for Multifractured Systems in Liquid-Rich Shales with Pressure-Dependent Properties. *Transp. Porous Media* **2017**, *119*, 1–23. [CrossRef]
41. Pedrosa, O.A., Jr. Pressure transient response in stress-sensitive formations. In Proceedings of the SPE California Regional Meeting, Oakland, CA, USA, 2–4 April 1986.
42. Davies, B.; Martin, B. Numerical inversion of the Laplace transform: A survey and comparison of methods. *J. Comput. Phys.* **1979**, *33*, 1–32. [CrossRef]

MDPI

Article

Lattice Boltzmann Simulation of Fluid Flow Characteristics in a Rock Micro-Fracture Based on the Pseudo-Potential Model

Pengyu Wang, Zhiliang Wang *, Linfang Shen and Libin Xin

Faculty of Civil Engineering and Mechanics, Kunming University of Science and Technology, Kunming 650500, China; wangpengyu18@126.com (P.W.); shenlinfang@kmust.edu.cn (L.S.); 18904722465@163.com (L.X.)

* Correspondence: wangzhiliang@kmust.edu.cn

Received: 9 August 2018; Accepted: 26 September 2018; Published: 27 September 2018

Abstract: Slip boundary has an important influence on fluid flow, which is non-negligible in rock micro-fractures. In this paper, an improved pseudo-potential multi-relaxation-time (MRT) lattice Boltzmann method (LBM), which can achieve a large density ratio, is introduced to simulate the fluid flow in a micro-fracture. The model is tested to satisfy thermodynamic consistency and simulate Poiseuille flow in the case of large liquid-gas density ratio. The slip length is used as an index for evaluating the flow characteristics, and the effects of wall wettability, micro-fracture width, driving pressure and liquid-gas density ratio on the slip length are discussed. The results demonstrate that the slip length increases significantly with the increase of the wall contact angle in rock micro-fracture. And the liquid-gas density ratio has an important impact on the slip length, especially for the hydrophobic wall. Moreover, under the laminar flow regime the driving pressure and the micro-fracture width has little effect on the slip length.

Keywords: slip length; large density ratio; contact angle; pseudo-potential model; lattice Boltzmann method; micro-fracture

---

1. Introduction

The fluid flow in rock micro-fractures is a topic of great importance for a wide range of scientific problems, such as water conservancy, oil recovery in low-permeability oilfields, nuclear waste treatment and others [1–4]. Classical Navier-Stokes hydrodynamics with non-slip boundaries, in which the fluid velocity at a fluid-solid interface equals the solid velocity, is successfully applied to seepage research at the macro-scale. However, a series of experimental and numerical results indicate that the non-slip boundary notion is invalid at the micro-scale [5,6]. Besides, there are a large number of micro-fractures in the rock with the width < 1 mm [7]. The slip boundaries should be taken into account during the fluid flow in these micro-fractures [8,9]. Meanwhile, the surfaces of micro-fractures show different wettabilities due to its diverse mineral composition [10], and it has a strong influence on the wall slip. Therefore, great attention should be paid to studying the slip conditions in micro-fractures with different wettabilities.

Slip length is an important dynamic parameter for quantifying the slip condition of a liquid flowing over the solid surface, and it's difficult to measure directly through experiments due to the objective limits of the current techniques [11,12]. Thus, slip length studies in micro-fluidics have focused on numerical simulations. The lattice Boltzmann method (LBM) offers great potential for micro-scale flow simulations owing to its mesoscopic particle background [13]. Many scholars have studied micro-scale fluid flow with the LBM and proposed several typical multiphase models, such as the color model, the free-energy model, the pseudo-potential model, etc. Among these, the pseudo-potential model realizes the phase separation by implementing micro-molecular interactions. It has received considerable attention due to the fact the interface between different phases does not need to be tracked or captured. Based on this model, Zhang et al. [14] found that the non-slip boundary conditions could produce an apparent slip. Chen et al. [15] simulated the Couette flow and found that there was a direct relationship between the magnitude of slip length and the strength of solid-fluid interaction. Zhang et al. [16] investigated the droplet flow velocity in the micro-channels, and discussed the wall slip caused by the roughness. Kunert [17] studied the influence of fluid viscosity on the slip length. However, almost all the studies do not consider the effect of liquid-gas density ratio on the wall slip due to the difficulty of achieving a large liquid-gas density ratio with the original pseudo-potential model [18].

Based on the LBM, an improved pseudo-potential model is introduced to simulate the liquid flow in a rock micro-fracture, which could achieve a large liquid-gas density ratio. The validity of the proposed model is tested on two benchmark cases: the thermodynamic consistency and Poiseuille flow. The slip length is discussed considering the effects of wall wettability, micro-fracture width, driving pressure and liquid-gas density ratio.

2. Slip Condition

Bernoulli first proposed the non-slip boundary model (Figure 1a) where the fluid velocity was zero relative to the physical wall. Subsequently, some scholars have supposed that the fluid layer near the wall has a certain thickness (Figure 1b), in which no relative movement occurs between the fluid and the wall. In addition, some experimental results indicated that the fluid velocity was not zero at the wall (Figure 1c). The above three slip conditions have been confirmed, and they are deemed to be related to the wall wettability. When the fluid flow over a strong hydrophilic wall, it presents non-slip or negative slip, and a fluid usually shows a positive slip condition at a strong hydrophobic wall. As shown in Figure 1c, based on the assumption that the distance to the virtual wall, at which the extrapolated fluid velocity at a constant shear would be equal to the wall velocity, Navier defined the slip length l as:

$$l = v_s / \tau_{wall}, \tag{1}$$

where v_s is the slip velocity at the physical wall; τ_{wall} is the local shear near the physical wall.

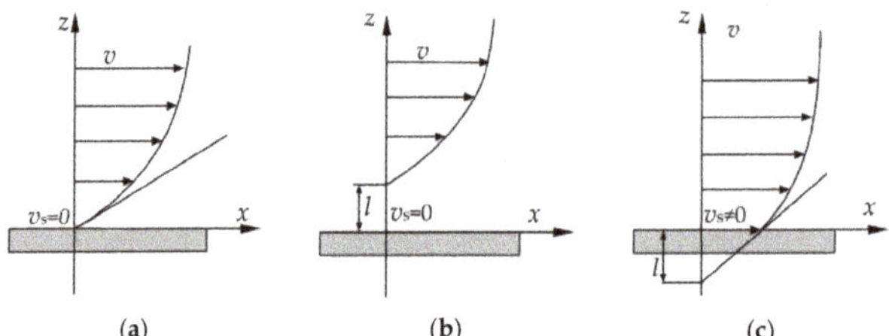

Figure 1. Slip boundary model (**a**) non-slip; (**b**) negative slip; (**c**) positive slip.

3. Coexistence Densities and Maxwell Construction Rule

In nature, the fluids in rock micro-fractures are mostly in a liquid-gas two phase situation. A two phase fluid could be described by its equation of state, which is an equation relating the fluid state variables in thermodynamics, such as pressure P, molar volume V, and temperature T. Figure 2 shows a series of $P - V$ curves of an equation of state with different temperatures, where the bulk (the region not close to any interface) pressure $P(V,T)$ is a function of the molar volume V and the temperature T. For different temperatures, the fluid may show subcritical, critical, and supercritical behaviors. At high temperature (above the critical temperature T_c) the fluid shows supercritical behavior and no distinct liquid and gas phases can be discerned. Below the critical temperature, phase separation into liquid and gas is possible, and different molar volumes V_l and V_g of the certain substance may coexist at a single pressure P_0 at equilibrium, where V_l and V_g are the liquid and gas phases molar volumes, respectively. At the critical temperature, the coexistence of different molar volumes is also impossible, nevertheless the first and second derivative of P at T_c should be zero [19]:

$$\begin{cases} \frac{\partial P(T_c)}{\partial V} = 0 \\ \frac{\partial^2 P(T_c)}{\partial V^2} = 0 \end{cases}, \tag{2}$$

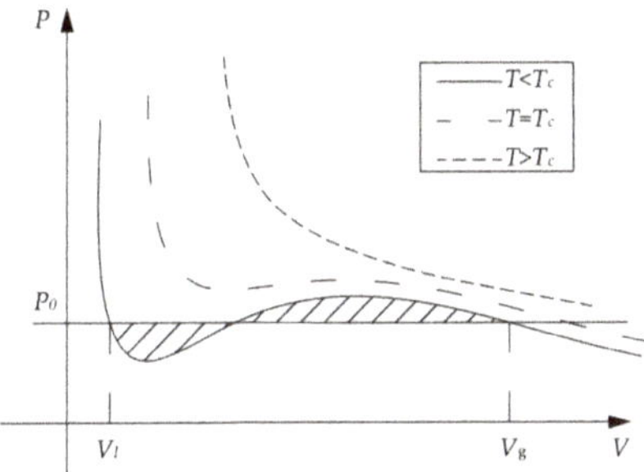

Figure 2. The equation of state of non-ideal fluid.

The Carnahan-Starling equation of state (C-S EOS) is one of classic equations of state for non-ideal fluid, which is given by [19]:

$$P(\rho, T) = \rho RT \frac{1 + b\rho/4 + (b\rho/4)^2 - (b\rho/4)^3}{(1 - b\rho/4)^3} - a\rho^2, \tag{3}$$

where a is the attraction parameter; b is the repulsion parameter;R is the gas constant and ρ is the fluid density. Rearranging and taking the derivatives of Equation (3), a and b can be presented as function of the critical pressure P_c and the T_c:

$$\begin{cases} a = 0.4963 R^2 T_c^2 / P_c \\ b = 0.18727 R T_c / P_c \end{cases}, \tag{4}$$

Then, T_c can be given as:

$$T_c = 0.3773 a/(bR), \tag{5}$$

In this paper, the parameters in Equation (3) are set as $a = 1$, $b = 4$, $R = 1$, respectively [20]. Then $T_c = 0.0943$. The reduced temperature coefficient t_r ($0 < t_r < 1$) is introduced to ensure the temperature $T = t_r T_c$ below the critical temperature, which makes the system at the subcritical behavior. The coexistence of different molar volumes at a single pressure could be found according to Maxwell construction rule [13]:

$$\begin{cases} \int_{V_g}^{V_l} [P_0 - P(V, T)] \mathrm{d}V = 0 \\ P_0 = P(V_g, T) = P(V_l, T) \end{cases}, \tag{6}$$

Replacing the molar volume V as the reciprocal of density $V = 1/\rho$, Maxwell construction rule Equation (6) can be rewritten as:

$$\begin{cases} \int_{\rho_g}^{\rho_l} [P_0 - P(\rho, T)] \frac{1}{\rho^2} \mathrm{d}\rho = 0 \\ P_0 = P(\rho_g, T) = P(\rho_l, T) \end{cases}, \tag{7}$$

where ρ_l and ρ_g are the liquid and gas phase coexistence densities, respectively. A series of coexistence densities with different temperatures could be directly obtained by numerical method according to Maxwell construction rule.

4. Numerical Model

4.1. The MRT-LBM

Using the Guo's force scheme [21], the evolution equation of LBM can be given as follows:

$$\text{Collision step}: \ f_\alpha(\mathbf{r}, t+\delta_t) - f_\alpha(\mathbf{r}, t) = \Omega_\alpha + \delta_t F_\alpha, \tag{8}$$

$$\text{Streaming step}: \ f_\alpha(\mathbf{r}+\boldsymbol{e}_\alpha\delta_t, t+\delta_t) = f_\alpha(\mathbf{r}, t+\delta_t), \tag{9}$$

where f_α is the density distribution function in the α direction; $\mathbf{r}$ represents the coordinate of the node;$\boldsymbol{e}_\alpha$ is the discrete velocity;δ_t is the time step; Ω_α is the collision operator; F_α is the forcing term.

For the D2Q9 model:

$$\boldsymbol{e}_\alpha = \begin{cases} (0,0) & \alpha = 0 \\ (\cos[(\alpha-1)\pi/2], sin[(\alpha-1)\pi/2]) & \alpha = 1,2,3,4 \\ \left(\sqrt{2}\cos[(\alpha-5)\pi/2+\pi/4], \sqrt{2}sin[(\alpha-5)\pi/2+\pi/4]\right) & \alpha = 5,6,7,8 \end{cases}, \tag{10}$$

The single-relaxation-time lattice Boltzmann method (SRT-LBM) can simplify the collision operator Ω_α of the Equation (8). However, it comes at a cost of reduced accuracy and stability. Compared with SRT-LBM, the multi-relaxation-time lattice Boltzmann method (MRT-LBM), which offers a series of relaxation-time can overcome these problems [22]. So the MRT-LBM is applied in the present work. The collision operator Ω_α and the Guo's forcing term F_α are respectively given by [23]:

$$\Omega_\alpha = -\left(\mathbf{M}^{-1}\boldsymbol{\Lambda}\boldsymbol{M}\right)_{\alpha\beta}\left(f_\beta(\mathbf{r},t) - f_\beta^{\text{eq}}(\mathbf{r},t)\right), \tag{11}$$

$$F_\alpha = \mathbf{S}_\alpha - 0.5\left(\mathbf{M}^{-1}\boldsymbol{\Lambda}\boldsymbol{M}\right)_{\alpha\beta}\mathbf{S}_\beta, \tag{12}$$

where $\mathbf{M}$ is an orthogonal transformation matrix; $\boldsymbol{\Lambda}$ is a diagonal matrix; f^{eq} is the equilibrium density distribution function;$\mathbf{S}$ is the forcing term in the moment space; α and β is the row and column number of a matrix, respectively. For the D2Q9 model, $\mathbf{M}$ and $\boldsymbol{\Lambda}$ can be written as:

$$\mathbf{M} = \begin{bmatrix} 1 & 1 & 1 & 1 & 1 & 1 & 1 & 1 & 1 \\ -4 & -1 & -1 & -1 & -1 & 2 & 2 & 2 & 2 \\ 4 & -2 & -2 & -2 & -2 & 1 & 1 & 1 & 1 \\ 0 & 1 & 0 & -1 & 0 & 1 & -1 & -1 & 1 \\ 0 & -2 & 0 & 2 & 0 & 1 & -1 & -1 & 1 \\ 0 & 0 & 1 & 0 & -1 & 1 & 1 & -1 & -1 \\ 0 & 0 & -2 & 0 & 2 & 1 & 1 & -1 & -1 \\ 0 & 1 & -1 & 1 & -1 & 0 & 0 & 0 & 0 \\ 0 & 0 & 0 & 0 & 0 & 1 & -1 & 1 & -1 \end{bmatrix}, \tag{13}$$

$$\boldsymbol{\Lambda} = diag\left(\tau_\rho^{-1}, \tau_e^{-1}, \tau_\zeta^{-1}, \tau_j^{-1}, \tau_q^{-1}, \tau_j^{-1}, \tau_q^{-1}, \tau_\upsilon^{-1}, \tau_\upsilon^{-1}\right), \tag{14}$$

where $\mathbf{M}$ contains all the relaxation rates; it is set as $\boldsymbol{\Lambda} = diag(1.0, 1.1, 1.1, 1.0, 1.1, 1.0, 1.1, 0.8, 0.8)$ [20] in this paper. The kinematic viscosity υ can be derived from one of the relaxation coefficients τ_υ: $\upsilon = c_s^2(\tau_\upsilon - 0.5)\delta_t$, where $c_s = c/\sqrt{3}$ is lattice sound speed, and c is the lattice constant, which is set as 1 for the present.

The moments $\mathbf{m}$ can be obtained by $\mathbf{m} = \mathbf{M}f$; the equilibrium moments $\mathbf{m}^{eq}$ can be straightforwardly computed by $\mathbf{m}^{eq} = \mathbf{M}f^{eq}$. Alternatively, it can be constructed more precisely and efficiently from the density ρ and the velocity $\mathbf{v}$ by [20]:

$$\mathbf{m}^{eq} = \rho\left[1, \left(-2+3|\mathbf{v}|^2\right), \left(1-3|\mathbf{v}|^2\right), v_x, -v_x, v_y, -v_y, \left(v_x^2 - v_y^2\right), v_x v_y\right]^{\mathrm{T}}, \quad (15)$$

where $\mathbf{v} = (v_x, v_y)$ is the macroscopic velocity given by $\mathbf{v} = \sum_{\alpha=1}^{N} e_\alpha f_\alpha/\rho + \mathbf{F}/2\rho$; and $\rho = \sum_{\alpha=0}^{N} f_\alpha$ is the macroscopic density; $\mathbf{F} = (F_x, F_y)$ is the total force.

Through Equations (11) and (12), the Equation (8) can be transformed as follows:

$$\mathbf{m}^* = \mathbf{m} - \mathbf{\Lambda}(\mathbf{m} - \mathbf{m}^{eq}) + \delta_t(\mathbf{I} - \mathbf{\Lambda}/2)\overline{\mathbf{S}}, \quad (16)$$

where $\mathbf{m}^*$ are the moments after the collision step; $\mathbf{I}$ is the unit tensor; $\overline{\mathbf{S}}$ is the forcing term in the moment space which is given by [20]:

$$\overline{\mathbf{S}} = \left[0, 6(v_x F_x + v_y F_y), -6(v_x F_x + v_y F_y), F_x, -F_x, F_y, -F_y, 2(v_x F_x - v_y F_y), (v_x F_y + v_y F_x)\right]^{\mathrm{T}}, \quad (17)$$

After the collision step, the moments $\mathbf{m}^*$ should be transformed to the density distribution function by $f^* = \mathbf{M}^{-1}\mathbf{m}^*$, then the f^* can be streamed to the next node through Equation (9).

4.2. The Original Pseudo-Potential Model

The total force in Equation (17) is given by $\mathbf{F} = \mathbf{F}_{\text{ex}} + \mathbf{F}_{\text{sc}}$, where $\mathbf{F}_{\text{ex}}$ is the external force such as gravitational force and buoyancy force. The inter-molecular force $\mathbf{F}_{\text{sc}} = \mathbf{F}_{\text{int}} + \mathbf{F}_{\text{w}}$ can be obtained by the pseudo-potential $\psi(\rho)$ presented in the following, which is a function to mimic the molecular interactions that cause phase separation. $\mathbf{F}_{\text{int}}$ is the inter-molecular force within fluids; $\mathbf{F}_{\text{w}}$ is the adhesive force between fluid particles and solid wall. Both $\mathbf{F}_{\text{int}}$ and $\mathbf{F}_{\text{w}}$ can be expressed as [24–26]:

$$\mathbf{F}_{\text{int}} = -G\psi(\rho(\mathbf{r}))\sum_{\alpha=1}^{N} w\left(|e_\alpha|^2\right)\psi(\rho(\mathbf{r}+e_\alpha))e_\alpha, \quad (18)$$

$$\mathbf{F}_w = -GS(\mathbf{r})\sum_{\alpha=1}^{N} w\left(|e_\alpha|^2\right)\psi(\rho(\mathbf{r}+e_\alpha))e_\alpha, \quad (19)$$

$$S(\mathbf{r}) = \begin{cases} 0 & (\mathbf{r} \neq \mathbf{r}_w) \\ \psi(\rho_w) & (\mathbf{r} = \mathbf{r}_w) \end{cases}, \quad (20)$$

where G is the interaction strength, which is set as 6.0 in the present paper; $w\left(|e_\alpha|^2\right)$ is the weight, $w(1) = 1/3$ and $w(2) = 1/12$; $\mathbf{r}_w$ represents the solid wall node; ρ_w is the wall density. It should be noticed that ρ_w is not a real physical density. It is merely an artificial parameter to acquire different wall affinities.

Using the Chapman-Enskog analysis, the following macroscopic equation can be derived from Equations (11), (15)–(17) [20]:

$$\nabla \cdot \mathbf{P} = \nabla \cdot \left(\rho c_s^2 \mathbf{I}\right) - \mathbf{F}, \quad (21)$$

Through the Taylor expansion, the Equation (18) can be rewritten as [26,27]:

$$\begin{aligned} \mathbf{F} &= -Gc^2\left[\psi\nabla\psi + \tfrac{c^2}{6}\psi\nabla(\nabla^2\psi)\right] + \cdots \\ &= -\tfrac{Gc^2}{2}\nabla\psi^2 - \tfrac{Gc^4}{6}\left[\nabla(\psi\nabla^2\psi) - \nabla^2\psi\nabla\psi\right] + \cdots \\ &= -\tfrac{Gc^2}{2}\nabla\psi^2 - \tfrac{Gc^4}{6}\nabla(\psi\nabla^2\psi) + \tfrac{Gc^4}{6}\left[\nabla\cdot(\nabla\psi\nabla\psi) - \tfrac{1}{2}\nabla|\nabla\,\psi|^2\right] + \cdots \end{aligned} \quad (22)$$

Combining Equations (21) and (22), the normal pressure tensor P_n of the interface can be given by [26]:

$$P_n = \rho c_s^2 + \frac{Gc^2}{2}\psi^2 + \frac{Gc^4}{12}\left[\alpha\left(\frac{d\psi}{dn}\right)^2 + \beta\psi\frac{d^2\psi}{dn^2}\right], \tag{23}$$

where the subscript n denotes the normal direction of the interface; α and β are constants for the D2Q9 model. According to Equation (23) and the physical requirement that the interface pressure should be equal to the bulk (the region not close to any interface) pressure at equilibrium, the following mechanical stability condition can be written as:

$$\int_{\rho_g}^{\rho_l}\left(P_0 - \rho c_s^2 - \frac{Gc^2}{2}\psi^2\right)\frac{\psi'}{\psi^{1+\varepsilon}}d\rho = 0, \tag{24}$$

where $\psi' = d\psi/d\rho$, $\varepsilon = -2\alpha/\beta$, and $P_0 = P(\rho_l) = P(\rho_g)$. In the pseudo-potential LB model, the coexistence densities (ρ_l and ρ_g) are determined by the mechanical stability condition Equation (24). Maxwell construction rule, which determines the thermodynamic coexistence, is built in the requirement of Equation (7). To satisfy the Equation (7), Sbragaglia and Shan [28] have proposed the following interaction potential $\psi(\rho)$:

$$\psi(\rho) = \begin{cases} \exp(-1/\rho) & ,\varepsilon = 0 \\ \left(\frac{\rho}{\varepsilon+\rho}\right)^{1/\varepsilon} & ,\varepsilon \neq 0 \end{cases}, \tag{25}$$

which gives $\psi'/\psi^{1+\varepsilon} = 1/\rho^2$. With the above potential function, Equations (7) and (22) will be same, thereby the thermodynamic consistency (Maxwell construction rule) will be obtained. But, to be consistent with the equation of state in the thermodynamic theory, the potential $\psi(\rho)$ must be chosen as [18,29]:

$$\psi(\rho) = \sqrt{\frac{2(P - \rho c_s^2)}{Gc^2}}, \tag{26}$$

Obviously, Equations (25) and (26) cannot be obtained at the same time, so this original pseudo-potential model cannot content the thermodynamic consistency.

4.3. The Improved Pseudo-Potential Model

To resolve the problem of thermodynamic inconsistency, the parameter ε in Equation (24) can be tuned to approximately achieve the thermodynamic consistency when $\psi(\rho)$ is defined by Equation (26). Through changing the original forcing term, Equation (17) is rewritten as follows [20]:

$$\overline{\mathbf{S}}^* = \begin{bmatrix} 0 \\ 6(v_x F_x + v_y F_y) + \frac{12\sigma|\mathbf{F}|^2}{\psi^2\delta_t(\tau_e - 0.5)} \\ -6(v_x F_x + v_y F_y) - \frac{12\sigma|\mathbf{F}|^2}{\psi^2\delta_t(\tau_\varsigma - 0.5)} \\ F_x \\ -F_x \\ F_y \\ -F_y \\ 2(v_x F_x - v_y F_y) \\ (v_x F_y - v_y F_x) \end{bmatrix}, \tag{27}$$

Meanwhile, Equation (23) could be given as:

$$P_n = \rho c_s^2 + \frac{Gc^2}{2}\psi^2 + \frac{Gc^4}{12}\left[(\alpha + 24G\sigma)\left(\frac{d\psi}{dn}\right)^2 + \beta\psi\frac{d^2\psi}{dn^2}\right], \tag{28}$$

which leads to $\varepsilon = -2(\alpha + 24G\sigma)/\beta$, where σ is a correction parameter which could tune the mechanical stability condition Equation (24) to approximately get the thermodynamic consistency.

4.4. Flowchart of the MRT-LBM

Based on the improved pseudo-potential model, the MRT-LBM has been proposed to simulate the fluid flow in the rock micro-fracture. The flowchart is shown in Figure 3.

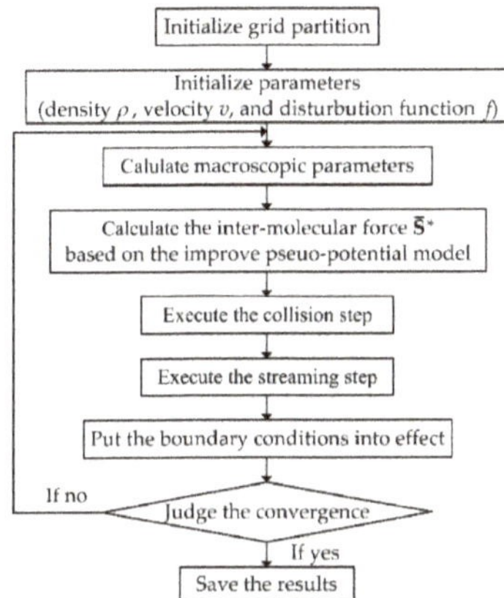

Figure 3. The flowchart of the improved pseudo-potential MRT-LBM.

4.5. Verification

Two reference problems are simulated to verify the improved pseudo-potential MRT-LBM in this section. One case is a bubble in gas, which is applied to test the thermodynamic consistency with large liquid-gas density ratio. The other is Poiseuille flow, in which the slip boundary with the slip length l close to zero is applied to simulate the non-slip behavior. It should be noticed that the units in this paper are all lattice units.

4.5.1. The Thermodynamic Consistency

A bubble in gas is simulated in a 100×100 lattice computational domain. The periodic boundary is set for its four sides, and the bubble is placed in the middle. The liquid-gas density coexistence curves are obtained by changing the temperature $T = t_r T_c$, which are compared with the theoretical curve given by Maxwell construction rule [30]. As shown in Figure 4, the results of the original forcing scheme ($\sigma = 0$), and the improved model with $\sigma = 0.09, 0.12$ are in good agreement with the theoretical results in the liquid phase, but there are large deviation between them in the gas phase. The numerical solutions of the improved forcing scheme with $\sigma = 0.11$ agrees well with the theoretical ones in both phases, which indicate that the proposed model could satisfy the thermodynamic consistency in the case of large liquid-gas density ratio, so the correction parameter σ is set as 0.11 in the following work.

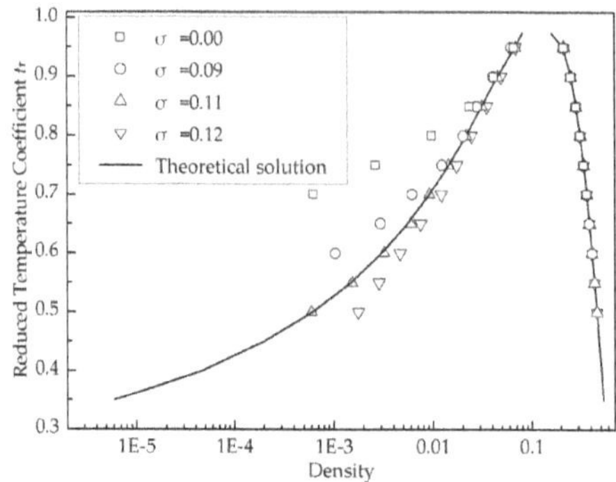

Figure 4. Comparison of the numerical coexistence curves with the theoretical solutions.

4.5.2. Poiseuille Flow

Poiseuille flow is simulated with the slip length l close to zero ($l = -0.02$). A 800 × 100 uniform lattice system is chosen in this case, the top and bottom boundaries are solid walls with the bounce-back boundaries, and the left and right sides are set as the periodic boundaries. To analyze the influence of different discretization features, we simulate the Poiseuille flow with the different mesh sizes (400 × 50, 800 × 100, and 1600 × 200). In this test, the liquid kinematic viscosity u is 0.1, the liquid is driven by the pressure of 3×10^{-4} along the horizontal direction. The wall density ρ_w is set as 0.12 to ensure the slip length l close to zero ($l = -0.02$). The reduced temperature coefficient t_r is 0.55. The correction parameter σ is 0.11. Other parameters are set as follows: lattice spaces are equal in horizontal and vertical direction, $\delta_x = \delta_y = 1$, and the time step is $\delta_t = 1$. Figure 5 shows the velocity profile obtained by the proposed model, which are compared with the theoretical solutions of Poiseuille flow (non-slip behavior) [13]. It is obvious that the fluid velocity given by numerical method shows a good match with the theoretical results and the grids number have little influence on the computed result. The maximum error is less than 3.24%. Hence, the improved pseudo-potential MRT-LBM can be used to simulate the fluid flow in the narrow channel successfully.

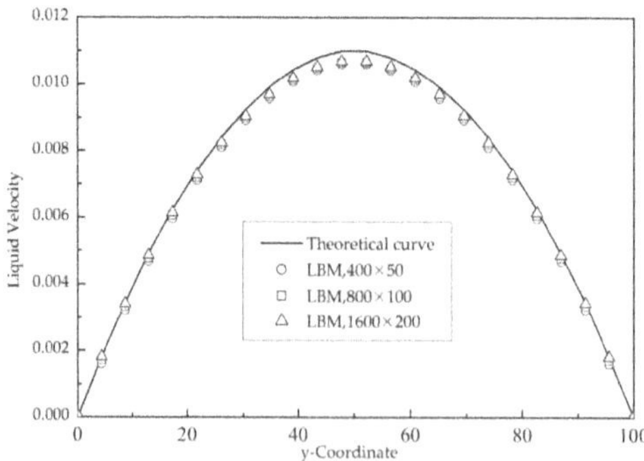

Figure 5. Comparison of the numerical velocity profile and the theoretical solutions.

5. Results and Discussion

In the present study, the slip condition is discussed in a rock micro-fracture with different wettabilities. The wall wettability is the macroscopic expression of the interaction between fluid particles and the solid wall, so the relationship between the contact angle and the wall density is tested based on the improved the improved pseudo-potential model and the slip length, as an important parameter for judging the slip condition, is investigated considering the effects of contact angle, driving pressure, and liquid-gas density ratio.

5.1. Contact Angle Test

To obtain the relationship between the contact angle and the wall density, a 200×200 lattice domain is used, the top and bottom boundaries are solid walls and the left and right sides are set as the periodic boundaries. A liquid droplet is placed at the solid surface. The liquid-gas density ratio is set as 293:1 in this test. Different wall densities are applied to describe the changing wall affinities, Figure 6 presents different contact angles, and the relationship between the contact angle and the wall density is listed in Table 1.

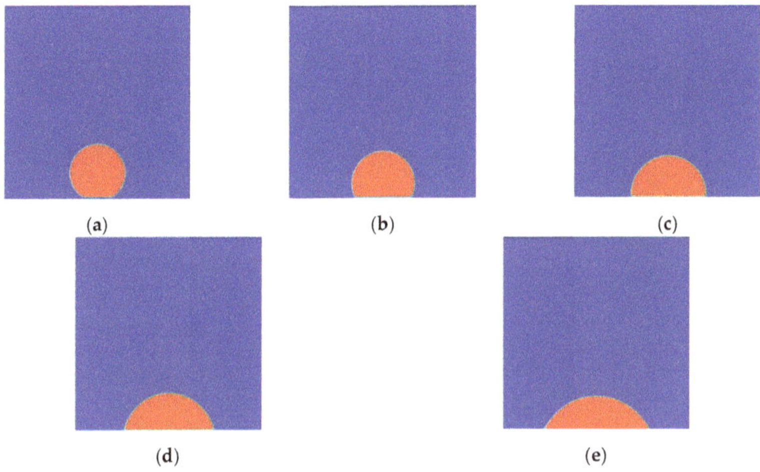

Figure 6. The wall wettabilities. (**a**) $\theta = 148.7°$; (**b**) $\theta = 115.9°$; (**c**) $\theta = 92.6°$; (**d**) $\theta = 76.0°$; (**e**) $\theta = 61.9°$.

Table 1. Relationship between the contact angle and the wall density.

Wall Density (ρ_w)	0.02	0.06	0.10	0.14	0.18
Contact Angle (θ)	148.7°	115.9°	92.6°	76.0°	61.9°

5.2. Discussion

Based on the improved pseudo-potential MRT-LBM, a size of 800×100 lattice domain is applied to simulate the rock micro-fracture with different wettabilities. As shown in Figure 7, the left and right sides of the micro-fracture are set as the periodic boundaries; the top and bottom of the model are the solid walls. The volume ratio of liquid to gas is 9:2 in the initial state, and the liquid phase is placed in the middle of the micro-fracture. To ensure a laminar flow regime, the driving pressure of 3×10^{-4} is applied in the horizontal direction after the liquid-gas phases separated, in which the calculated Reynolds number is 1.49. The slip length, as an important parameter for judging the slip degree, is used to discuss the effects of contact angle, driving pressure, and liquid-gas density ratio on the fluid flow state in the micro-fracture.

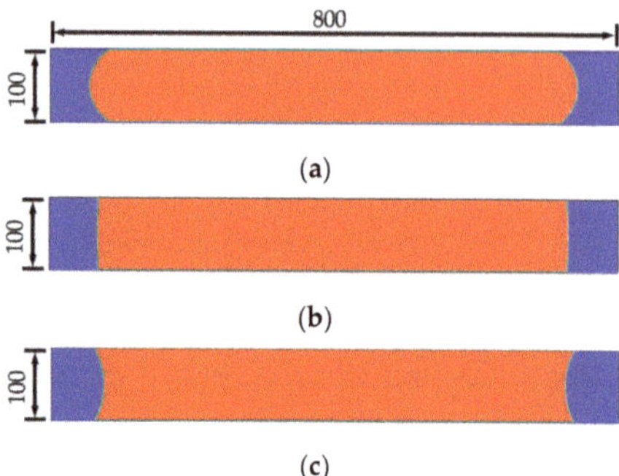

Figure 7. Fluids in a rock micro-fracture with different contact angles: (**a**) $\theta = 148.7°$; (**b**) $\theta = 92.6°$; (**c**) $\theta = 61.9°$.

5.2.1. Effect of the Contact Angle

In order to study the effect of the contact angle on the slip length, the liquid-gas density ratio is set as 293:1, and the micro-fracture widths are set as 25, 50, and 100, respectively. Figure 8 shows the slip length against the contact angle.

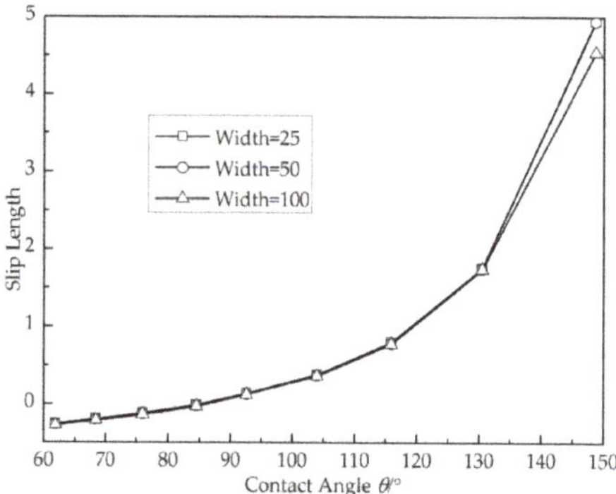

Figure 8. Slip length versus contact angle.

It can be observed that the slip length increases from −0.27 to 4.9 as the contact angle increasing from 61.9° to 148.7°. The larger the contact angle, the more significant the slip length changes. This finding is in agreement with that by Tsu-Hsu Yen [31] and Bladimir Ramos-Alvarado [32]. Moreover, in the case of different fracture widths, the variation trends of slip length with the contact angle are very close to each other, which indicate that the micro-fracture width has little effect on the slip length.

To analyze the physical mechanism of wall slip, the liquid density profiles (the fracture width is 100) for different contact angles are shown in Figure 9. As the contact angle increases, the liquid density near the wall decreases and the slip length increases, which confirms the existing slip theory [33,34]. The wall slip is induced by the changes of the liquid density near the solid surface, which is attributed to the varying interaction force between the liquid and the solid wall. With the increases of interaction force, the liquid density near the wall increases. Thereby it shows non-slip or even negative slip behaviors. As the interaction force decreases, the liquid density close to the wall surface decreases and the slip length increases. Even when the liquid density is less than a certain value, the liquid density near the wall surface is reduced to the gas density, so that a "gas layer" is formed at the solid surface.

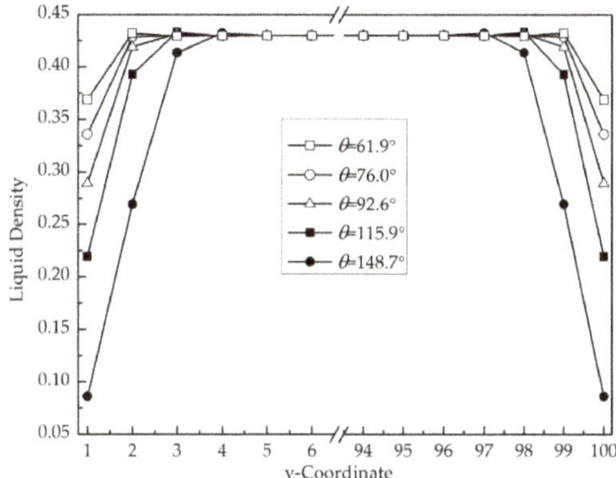

Figure 9. Liquid density profiles for different contact angles.

5.2.2. Effect of the Driving Pressure

To investigate the effect of the driving pressure on the slip length, the micro-fracture width W is set as 100, the liquid-gas density ratio is 293:1, and the contact angles are 61.9°, 76.0°, 92.6°, 115.9° and 148.7°, respectively. When the driving pressure ranges from 3×10^{-4} to 1.5×10^{-3}, the maximum Reynolds number is 7.48, which shows that the liquid flow is in a laminar flow regime. Figure 10 presents the relationship between the driving pressure and the average flow velocity, which indicates that the flow velocity linearly increases with the driving pressure. In Figure 11, the relationship between the driving pressure and the slip length shows that the driving pressure has almost no influence on the slip length. Xiang [35] and Huang [36] have also yielded the similar results by other numerical methods.

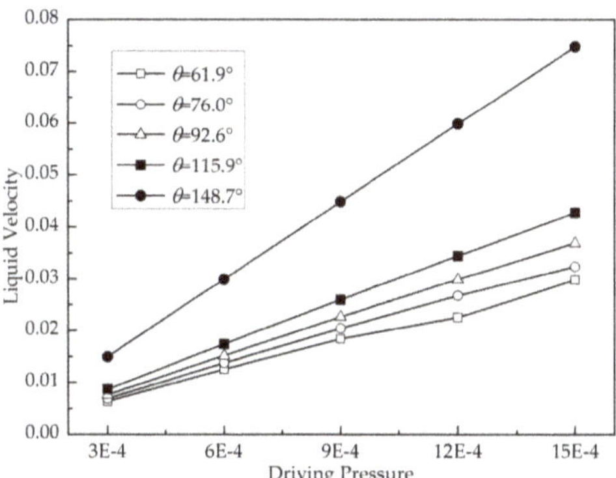

Figure 10. Liquid velocity against driving pressure.

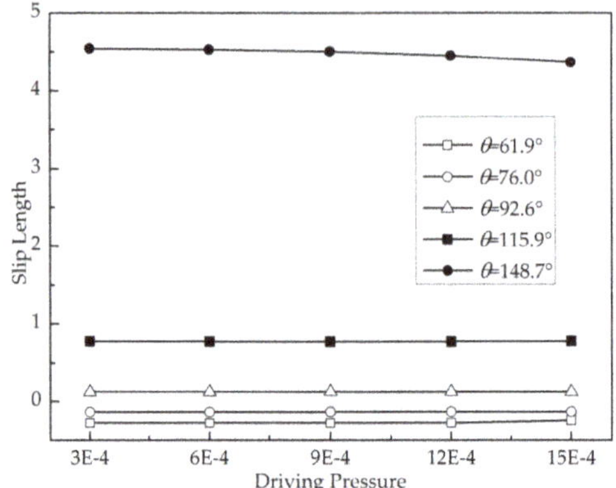

Figure 11. Slip length against driving pressure.

5.2.3. Effect of the Liquid-Gas Density Ratio

To study the influence of liquid-gas density ratio on the slip length, the liquid-gas density ratios are chosen in the range from 22:1 to 293:1. In this case, the micro-fracture width is 100, the driving pressure is 3×10^{-4}, and the contact angles are 61.9°, 76.0°, 92.6°, 115.9°, and 148.7°, respectively. It should be noticed that changing liquid-gas density ratio will affect the contact angle, so the wall density ρ_w should be lightly tuned to ensure the contact angle unchanged. Numerical results of the slip length against the density ratio are shown in Figure 12. In the case of $\theta = 148.7°$, the slip length increases from 0.18 to 4.54 with the liquid-gas density ratio from 22:1 to 293:1. The relationship between the slip length and the density ratio is close to linear. The above phenomena is induced by the strong changes of the liquid density near the wall and the formation of the "gas layer" at the solid surface. When the contact angles θ are 115.9°, 92.6°, 76.0°, and 61.9°, the slip lengths increase from 0.09 to 0.77, −0.09 to 0.12, −0.21 to −0.13, and −0.30 to −0.27, respectively, the growth slope is relatively smaller. Thus, the liquid-gas density ratio has an important influence on the fluid flow characteristics, especially for the hydrophobic wall.

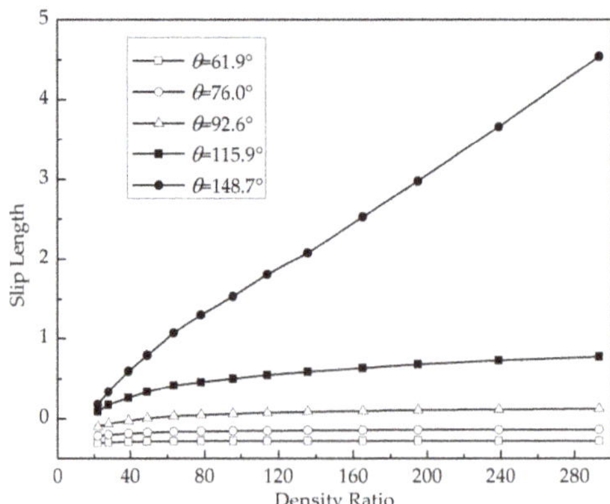

Figure 12. Slip length against liquid-gas density ratio.

6. Conclusions

Based on the improved pseudo-potential model, the MRT-LBM is proposed to investigate the fluid flow in the rock micro-fracture, and it is verified according to two benchmark cases. The slip length is used to evaluate the flow characteristics in the micro-fracture, and the effects of the contact angles, the driving pressure, and the liquid-gas density ratio on the slip length are discussed, the corresponding conclusions are summarized as follows:

(1) With increasing contact angle, the slip length increases at the wall, and the larger the contact angle, the more obvious the slip length changes.
(2) Under the laminar flow regime, the fluid flow velocity is proportional to the driving pressure, but there is almost no change in the slip length with the driving pressure increasing, so the driving pressure has almost no impact on the slip length.
(3) The slip length increases with the increasing of the liquid-gas density ratio, and the larger the wall contact angle, the more remarkable it shows. The liquid-gas density ratio has an important influence on the fluid flow characteristics, especially for the hydrophobic wall.
(4) The wall slip is induced by the changing of liquid density near the solid surface, which is attributed to the varying interaction force between fluid particles and the solid wall. With the decrease of interaction force, the liquid density near the wall decreases and the slip length increases.

And as the increase of interaction force, it shows no slip or even negative slip at the solid wall.

Author Contributions: Each author has made contributions to the present paper. W.Z. proposed this topic and designed the calculation verification. P.W. performed the simulation and the result analysis. General supervision was provided by L.S. L.X. provided simulation support. All authors have read and approved the final manuscript.

Funding: This research was funded by the National Natural Science Foundation of China (No. 51508253, 51668028) and the Yunnan Applied Basic Research Project (No. 2016FB077).

Conflicts of Interest: The authors declare no conflict of interest.

References

1. Wang, F.; Liu, Y.; Hu, C.; Shen, A.; Liang, S.; Cai, B.; Sciubba, E. A simplified physical model construction method and gas-water micro scale flow simulation in tight sandstone gas reservoirs. *Energies* **2018**, *11*, 1559. [CrossRef]
2. Zhu, W.; Li, B.; Liu, Y.; Song, H.; Wang, X. Solid-liquid interfacial effects on residual oil distribution utilizing three-dimensional micro network models. *Energies* **2017**, *10*, 2059. [CrossRef]
3. Di, Q.; Shen, C.; Wang, Z.H. Experimental research on drag reduction of flow in micro channels of rock using nano-particle adsorption method. *Acta Pet. Sin.* **2009**, *30*, 125–128.
4. Coli, N.; Pranzini, G.; Alfi, A.; Boerio, V. Evaluation of rock-mass permeability tensor and prediction of tunnel inflows by means of geostructural surveys and finite element seepage analysis. *Eng. Geol.* **2008**, *101*, 174–184. [CrossRef]
5. Voronov, R.S.; Papavassiliou, D.V.; Lee, L.L. Slip length and contact angle over hydrophobic surfaces. *Chem. Phys. Lett.* **2007**, *441*, 273–276. [CrossRef]
6. Burton, Z.; Bhushan, B. Hydrophobicity, adhesion, and friction properties of nanopatterned polymers and scale dependence for micro-and nanoelectromechanical systems. *Nano Lett.* **2005**, *5*, 1607. [CrossRef] [PubMed]
7. Hanna, R.B.; Rajaram, H. Influence of aperture variability on dissolutional growth of fissures in Karst Formations. *Water Resour. Res.* **1998**, *34*, 2843–2853. [CrossRef]
8. Cao, B.Y.; Chen, M.; Guo, Z.Y. Velocity slip of liquid flow in nanochannels. *Acta Phys. Sin. Chin. Ed.* **2006**, *55*, 5305–5310.
9. Xu, C.; He, Y.; Wang, Y. Molecular dynamics studies of velocity slip phenomena in a nanochannel. *J. Eng. Thermophys.* **2005**, *26*, 912–914.
10. Dekany, I. Liquid adsorption and immersional wetting on hydrophilic/hydrophobic solid surfaces. *Pure Appl. Chem.* **1992**, *64*, 1499–1509. [CrossRef]
11. Li, Y.; Pan, Y.; Zhao, X. Measurement and quantification of effective slip length at solid–liquid interface of roughness-induced surfaces with oleophobicity. *Appl. Sci.* **2018**, *8*, 931. [CrossRef]
12. Turner, S.E. Experimental investigation of gas flow in microchannels. *J. Heat Transf.* **2004**, *126*, 753–763. [CrossRef]

13. Sukop, M.C.; Thorne, D.T.J. *Lattice Boltzmann Modeling: An Introduction for Geoscientists and Engineers*; Springer: Heidelberg/Berlin, Germany, 2007; pp. 1490–1511.
14. Zhang, J.; Kwok, D.Y. Apparent slip over a solid-liquid interface with a no-slip boundary condition. *Phys. Rev. E* **2004**, *70*. [CrossRef] [PubMed]
15. Chen, Y.Y.; Yin, H.H.; Li, H.B. Boundary slip and surface interaction: A lattice boltzmann simulation. *Chin. Phys. Lett.* **2008**, *25*, 184–187. [CrossRef]
16. Zhang, R.L.; Di, Q.; Wang, X.L. Institute numerical study of wall wettabilities and topography on drag reduction effect in micro-channel flow by lattice boltzmann method. *J. Hydrodyn.* **2010**, *22*, 366–372. [CrossRef]
17. Kunert, C.; Harting, J. On the effect of surfactant adsorption and viscosity change on apparent slip in hydrophobic microchannels. *Prog. Comp. Fluid Dyn. Int. J.* **2006**, *8*, 197–205. [CrossRef]
18. Huang, H.; Krafczyk, M.; Lu, X. Forcing term in single-phase and shan-chen-type multiphase lattice boltzmann models. *Phys. Rev. E* **2011**, *84*. [CrossRef] [PubMed]
19. Yuan, P.; Schaefer, L. Equations of state in a lattice Boltzmann model. *Phys. Fluids* **2006**, *18*, 042101. [CrossRef]
20. Li, Q.; Luo, K.H.; Li, X.J. Lattice boltzmann modeling of multiphase flows at large density ratio with an improved pseudopotential model. *Phys. Rev. E* **2013**, *87*. [CrossRef] [PubMed]
21. Guo, Z.; Zheng, C.; Shi, B. Discrete lattice effects on the forcing term in the lattice boltzmann method. *Phys. Rev. E* **2002**, *65*. [CrossRef] [PubMed]
22. Pan, C.; Luo, L.S.; Miller, C.T. An evaluation of lattice boltzmann schemes for porous medium flow simulation. *Comput. Fluids* **2006**, *35*, 898–909. [CrossRef]
23. Mukherjee, S.; Abraham, J. A pressure-evolution-based multi-relaxation-time high-density-ratio two-phase lattice-Boltzmann model. *Comput. Fluids* **2007**, *36*, 1149–1158. [CrossRef]
24. Krüger, T.; Kusumaatmaja, H.; Kuzmin, A.; Shardt, O.; Silva, G.; Viggen, E.M. *The Lattice Boltzmann Method-Principles and Practice*; Springer International Publishing: New York, NY, USA, 2017.
25. Chen, L.; Kang, Q.; Mu, Y.; He, Y.; Tao, W. A critical review of the pseudopotential multiphase lattice boltzmann model: Methods and applications. *Int. J. Heat Mass Transf.* **2014**, *76*, 210–236. [CrossRef]
26. Shan, X. Pressure tensor calculation in a class of nonideal gas lattice Boltzmann models. *Phys. Rev. E* **2008**, *77*. [CrossRef] [PubMed]
27. Sbragaglia, M.; Benzi, R.; Biferale, L.; Succi, S.; Sugiyama, K.; Toschi, F. Generalized lattice boltzmann method with multirange pseudopotential. *Phys. Rev. E* **2007**, *75*. [CrossRef] [PubMed]
28. Sbragaglia, M.; Shan, X. Consistent pseudopotential interactions in lattice boltzmann models. *Phys. Rev. E* **2011**, *84*. [CrossRef] [PubMed]
29. Shan, X.; Doolen, G. Multicomponent lattice-boltzmann model with interparticle interaction. *J. Stat. Phys.* **1995**, *81*, 379–393. [CrossRef]
30. Li, Q.; Luo, K.H.; Li, X.J. Forcing scheme in pseudopotential lattice boltzmann model for multiphase flows. *Phys. Rev. E* **2012**, *86*. [CrossRef] [PubMed]
31. Yen, T.H.; Soong, C.Y. Effective boundary slip and wetting characteristics of water on substrates with effects of surface morphology. *Mol. Phys.* **2015**, *114*, 1–13. [CrossRef]
32. Ramos-Alvarado, B.; Kumar, S.; Peterson, G.P. Wettability transparency and the quasiuniversal relationship between hydrodynamic slip and contact angle. *Appl. Phys. Lett.* **2016**, *108*. [CrossRef]
33. Du, Y.P. The Study on Relationship of the Hydrophobic Surface Nanobubbles Characteristics and Solid-Liquid Boundary Slip Length. Master's Thesis, Harbin Institute of Technology, Harbin, China, June 2012.
34. Thompson, P.A.; Robbins, M.O. Shear flow near solids: Epitaxial order and flow boundary conditions. *Phys. Rev. A* **1990**, *41*, 6830–6837. [CrossRef] [PubMed]
35. Xiang, H. Heat Transport in Nanoparticle and Nanoporous Media and Nanoscale Liquid Flow. Ph.D. Thesis, Tsinghua University, Beijing, China, April 2008.
36. Huang, Y.D. Research on the Effect of Surface Micro Topography on Boundary Slip and Flow Resistance. Master's Thesis, Harbin Institute of Technology, Harbin, China, July 2015.

Article

Multiporosity and Multiscale Flow Characteristics of a Stimulated Reservoir Volume (SRV)-Fractured Horizontal Well in a Tight Oil Reservoir

Long Ren [1,2,*], Wendong Wang [3,*], Yuliang Su [3], Mingqiang Chen [1,2], Cheng Jing [1,2], Nan Zhang [1,2], Yanlong He [1,2] and Jian Sun [4]

[1] School of Petroleum Engineering, Xi'an Shiyou University, Xi'an 710065, China; mingqiangchen@163.com (M.C.); cjing@xsyu.edu.cn (C.J.); nanz@xsyu.edu.cn (N.Z.); ylhe@xsyu.edu.cn (Y.H.)
[2] Shaanxi Key Laboratory of Advanced Stimulation Technology for Oil & Gas Reservoirs, Xi'an 710065, China
[3] School of Petroleum Engineering, China University of Petroleum (East China), Qingdao 266580, China; suyuliang@upc.edu.cn
[4] School of Petroleum Engineering, China University of Petroleum (Beijing), Beijing 102249, China; xjkelsj@163.com
* Correspondence: renlong@xsyu.edu.cn (L.R.); wwdong@upc.edu.cn (W.W.)

Received: 28 August 2018; Accepted: 10 October 2018; Published: 11 October 2018

Abstract: There are multiporosity media in tight oil reservoirs after stimulated reservoir volume (SRV) fracturing. Moreover, multiscale flowing states exist throughout the development process. The fluid flowing characteristic is different from that of conventional reservoirs. In terms of those attributes of tight oil reservoirs, considering the flowing feature of the dual-porosity property and the fracture network system based on the discrete-fracture model (DFM), a mathematical flow model of an SRV-fractured horizontal well with multiporosity and multipermeability media was established. The numerical solution was solved by the finite element method and verified by a comparison with the analytical solution and field data. The differences of flow regimes between triple-porosity, dual-permeability (TPDP) and triple-porosity, triple-permeability (TPTP) models were identified. Moreover, the productivity contribution degree of multimedium was analyzed. The results showed that for the multiporosity flowing states, the well bottomhole pressure drop became slower, the linear flow no longer arose, and the pressure wave arrived quickly at the closed reservoir boundary. The contribution ratio of the matrix system, natural fracture system, and network fracture system during SRV-fractured horizontal well production were 7.85%, 43.67%, and 48.48%, respectively in the first year, 14.60%, 49.23%, and 36.17%, respectively in the fifth year, and 20.49%, 46.79%, and 32.72%, respectively in the 10th year. This study provides a theoretical contribution to a better understanding of multiscale flow mechanisms in unconventional reservoirs.

Keywords: tight oil reservoir; SRV-fractured horizontal well; multiporosity and multiscale; flow regimes; productivity contribution degree of multimedium

---

1. Introduction

It has been commonly recognized that tight oil reservoirs have threshold pressure gradient and medium deformation characteristics because of their great lithologic compaction, fine pore-throat, and high flow resistance [1–7]. In recent years, stimulated reservoir volume (SRV) fracturing has become the most efficient technology in tight reservoir formation treatment [8–15]. To enhance well production as much as possible, it is necessary to create complex fracture networks with a multiporosity medium by connecting hydraulic fractures with natural fractures away from the well bore, and then increasing the contact area with formations and reservoir stimulated volume [16–23]. Multiple porous media systems include network fractures, natural fractures, and matrix pore systems. Moreover, there exist different flowing states, i.e., multi-scale flow characteristics.

The research methods of the flow characteristics of SRV-fractured horizontal wells in a tight oil reservoir have been mainly focused on analytical, semi-analytical, and numerical methods. The analytical or semianalytical solution is mainly represented by the three linear flow model proposed by Brown [24], the five-zone model raised by Stalgorova [25], and the compound flow model presented by Su [26]. However, those models have relatively strict assumptions. Generally, the models need to idealize the complex fracture network to regular fracture network forms composed of orthogonal primary and secondary fractures and simplify the complex flow processes to specific flow regimes such as elliptic or linear flow regimes [27–29]. In terms of numerical models, Yao [30] and Fan [31] used the finite element method to carry out dynamic analysis of a horizontal well with a complex fractured continuous medium system, but those models did not consider the development degree of the natural fractures in tight oil reservoirs or the existence of the threshold pressure gradient in the matrix system. Therefore, it is a challenge to use these models to accurately describe the complex structures of actual network fractures and reveal the multiporosity and multiscale flow characteristics of an SRV-fractured horizontal well in tight oil reservoirs.

The objective of this work was to study the multiporosity and multiscale flow characteristics of SRV-fractured horizontal wells. Moreover, the innovation of this paper was to reveal the contribution of multiple porous media to horizontal well productivity by establishing a multiscale flow model. Enlightened by previous studies, a mathematical flow model was built to reflect the multiscale attributes of tight oil reservoirs based on the dual-porosity model (DPM) and discrete-fracture model (DFM), which were divided into three kinds of media systems. A reasonable solution of this numerical model was obtained and verified by the finite element method. Additionally, the flow mechanisms of an SRV-fractured horizontal well with the consideration of the multiporosity and multiscale effect were revealed, which were different to that of a conventional multifractured horizontal well without an SRV system. The findings of this research provide effective theoretical and methodological support for the prediction of the production performance prediction of unconventional hydrocarbon resources.

2. Physical Model and Assumed Conditions

SRV fracturing of a horizontal well in tight oil reservoirs with natural fractures has often induced complex fracture network growth, as revealed by microseismic monitoring [32–35]. Moreover, the complex fracture network divides the reservoir into multiple porous media systems. Furthermore, the physical properties and fluid flow rules of each system are different. Based on the network fracture propagation process and the final form in the tight oil reservoir, a physical model of an SRV fractured horizontal well was built that considered the structure characteristics of multiple porous media, as shown in Figure 1, where Δy_f is the interval between fracturing segments (m); and a and b are the band width and band length of single fracture network, respectively (m).

Complex fracture networks composed of primary and secondary fractures formed by SRV fracturing are integrated into both the natural fracture system and matrix system. A reservoir that has been subjected to SRV fracturing treatment can be represented by a combination of a complex fracture network system, a natural fracture system, and a matrix system. Assumptions of the physical model were made as follows: (1) the study area was a three-dimensional, box-shaped closed, and isotropic body with natural fractures; (2) the rock and fluid were slightly compressible bodies, and the nonlinear flow in the matrix system, Darcy flow in the fracture system, and pseudosteady crossflow between the matrix system and fracture system are also found in the multiple media; and (3) the simulated production process was a single-phase fluid flow in porous and isothermal media without considering the influence of gravity.

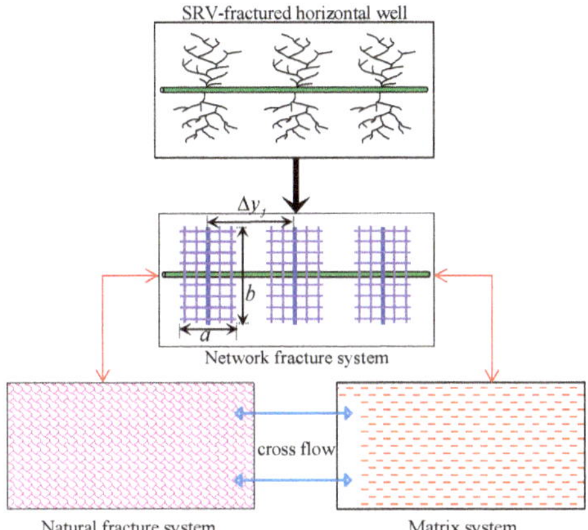

Figure 1. Physical model diagram of the SRV-fractured horizontal well with multi-porosity media.

3. Flow Mathematical Model Considering the Multiporosity

3.1. Nonlinear Flow in the Matrix System

The nonlinear flow equation in the matrix system can be given as [36,37]

$$v_m = -\frac{\boldsymbol{K}_m}{\mu}(\nabla p_m - \boldsymbol{\chi}) \tag{1}$$

where v_m is the flow velocity vector of fluid (10^{-3} m/s); $\boldsymbol{K}_m$ is the permeability tensor of the matrix (D); μ is the viscosity of fluids (mPa·s); ∇ is the Hamiltonian; p_m is the pore pressure in the matrix system (MPa); $\boldsymbol{\chi}$ is the threshold pressure gradient tensor (MPa/m) and can be defined as $\boldsymbol{\chi} = \chi\mathbf{E}$, where χ is the threshold pressure gradient of matrix (MPa/m), and $\mathbf{E}$ is the unit matrix.

Via a combination of the state equation and continuity equation, the surface source in the 3D space is equivalent to the superposition of line sources in the 2D space, and the mathematical flow model for the matrix system can be derived [38] as

$$\nabla^2 p_m - \chi C_L \nabla \cdot p_m - \frac{\phi_m \mu C_m}{K_m}\frac{\partial p_m}{\partial t} - \alpha(p_m - p_n) = 0 \tag{2}$$

where α is the shape factor of matrix; p_n is the pressure of natural fracture (MPa); C_L is the compression coefficient of fluid (MPa^{-1}); and C_m is the comprehensive compression coefficient of matrix system (MPa^{-1}).

Since $C_m = \phi_m C_L + (1 - \phi_m)C_{mf}$ [39], $\phi_m << 1$ and $C_{mf} = \phi_m C_p$, the comprehensive compressibility of the matrix system is defined as

$$C_m \approx \phi_m C_L + C_{mf} = \phi_m(C_L + C_p) \tag{3}$$

where ϕ_m is the porosity of the matrix; C_{mf} represents the compression coefficient of the matrix rock (MPa^{-1}); and C_p is the compression coefficient of the pore (MPa^{-1}).

The dimensionless pressure is defined as

$$p_{jD} = \frac{2\pi h_e K_n(p_i - p_j)}{\mu q_j} \tag{4}$$

where j represents m, n, or f; p_i is the initial formation pressure (MPa); p_j is the pressure of each system (MPa); K_n is the permeability of natural fracture (D); and q_j is the volume flow of each system (s^{-1}).

The dimensionless permeability of the matrix is defined as

$$K_{mD} = K_m / K_n \tag{5}$$

The dimensionless threshold pressure gradient is

$$\chi_D = \chi L C_L \tag{6}$$

The crossflow coefficient between the matrix system and natural fracture system is defined as

$$\lambda = \alpha L^2 K_{mD} \tag{7}$$

The elastic storativity ratio of the natural fracture system is

$$\omega_n = \frac{\phi_n C_n}{\phi_m C_m + \phi_n C_n} \tag{8}$$

where ϕ_n is the porosity of the natural fracture; and C_n is the comprehensive compression coefficient of the natural fracture system (MPa^{-1}).

The dimensionless production time is

$$t_D = \frac{K_n t}{\mu L^2 (\phi_m C_m + \phi_n C_n)} \tag{9}$$

Then, the dimensionless flow equation can be obtained [38] as

$$\nabla^2 p_{mD} - \chi_D \nabla \cdot p_{mD} - (1-\omega_n)\frac{\partial p_{mD}}{\partial t_D} - \lambda(p_{mD} - p_{nD}) = 0 \tag{10}$$

Accordingly, the initial and boundary condition for fluid flow in the matrix system are given by

$$\begin{cases} p_{mD}(x_D, y_D, z_D; t_D = 0) = 0 \\ \left.\frac{\partial p_{mD}}{\partial x_D}\right|_{x=x_{eD}} = \left.\frac{\partial p_{mD}}{\partial y_D}\right|_{y_D=y_{eD}} = \left.\frac{\partial p_{mD}}{\partial z_D}\right|_{z_D=z_{eD}} = 0 \end{cases} \tag{11}$$

3.2. Darcy Flow in the Natural Fracture System

Assuming that there exists fluid crossflow between the matrix system and natural fracture system in the formation as well only the natural fracture system instead of the matrix system for fluid exchange to the network fracture system [40], the dimensionless variables are defined as follows: the dimensionless distances are $M_D = M/L$, $M_{eD} = M_e/L$ ($M = x, y, z$), $a_D = a/L$, $b_D = b/L$, where the length, width, and height of the study area are x_e, y_e, and h_e, respectively (m); the horizontal well length is L (m); the dimensionless production rate is $q_{kD} = q_k/q_t$, where k represents n or f; and q_t is the total volume flow (s^{-1}).

Therefore, the dimensionless Darcy flow equation in the matrix system can be given [31] as

$$\nabla^2 p_{nD} - \omega_n \frac{\partial p_{nD}}{\partial t_D} + \lambda(p_{mD} - p_{nD}) + 2\pi h_{eD} q_{nD} \delta(M - M') = 0 \tag{12}$$

where $\delta(M - M')$ is the Dirac delta function.

The initial and boundary conditions for fluid flow are given by

$$\begin{cases} p_{nD}(x_D, y_D, z_D; t_D = 0) = 0 \\ \left.\frac{\partial p_{nD}}{\partial x_D}\right|_{x=x_{eD}} = \left.\frac{\partial p_{nD}}{\partial y_D}\right|_{y_D=y_{eD}} = \left.\frac{\partial p_{nD}}{\partial z_D}\right|_{z_D=z_{eD}} = 0 \end{cases} \tag{13}$$

3.3. Darcy Flow in the Network Fracture System

The discrete-fracture model (DFM) is used to characterize the fracture network stimulated system [41,42]. According to the fracture flow model of parallel plate openings (cubic law), the permeability of the network fracture is defined as $K_f = a^2{}_f/12$, where a_f is the fracture opening (mm). The dimensionless permeability of the network fracture is defined as $K_{fD} = K_f/K_n$; the elastic storativity ratio of the network fracture system is defined as $\omega_f = \phi_f C_f/(\phi_m C_m + \phi_n C_n)$, where ϕ_f is the porosity of the network fracture; and C_f is the comprehensive compression coefficient of the network fracture system (MPa^{-1}).

Similarly, the dimensionless Darcy flow equation in the network fracture system can be given [31] by

$$K_{fD}\nabla^2 p_{fD} - \omega_f \frac{\partial p_{fD}}{\partial t_D} + 2\pi h_{eD} q_{fD}\delta(M - M') = 0 \tag{14}$$

The initial and boundary conditions for fluid flow in the natural fracture system are given by

$$\begin{cases} p_{fD}(x_D, y_D, z_D; t_D = 0) = 0 \\ p_{mD}(x_D, y_D, z_D; t_D) = p_{nD}(x_D, y_D, z_D; t_D) = p_{fD}(x_D, y_D, z_D; t_D) \end{cases} \tag{15}$$

All of the above flow equations and the fixed solution conditions of the matrix, natural fracture, and network fracture systems together constitute the multiporosity and multiscale flow mathematical model for an SRV-fractured horizontal well in tight oil reservoir.

4. Numerical Solution with the Finite Element Method

4.1. Finite Element Method Meshing

The finite element integral equation is established by using Galerkin's weighted residual method and the continuous solving unit with an infinite degree of freedom is discretized into the finite element unit. The horizontal well, network fracture, and reservoir unit are described by a line, triangle, and tetrahedron, respectively. The dimensionless parameters of horizontal wells and hydraulic fractures in a box-shaped closed reservoir are the length of horizontal well L_D = 1; the reservoir domain x_{eD} = 6, y_{eD} = 6, and h_{eD} = 0.1; the coordinate of five fracturing sections in the x-direction (−0.4, 0.2, 0, 0.2, 0.4). It is assumed that all fractures are vertical, the mesh generation of the whole model is based on triangle forward algorithm, and local grid refinement (LGR) is performed at the horizontal well and network fracture. The three-dimensional gridding division of an SRV-fractured horizontal well can be obtained as shown in Figure 2.

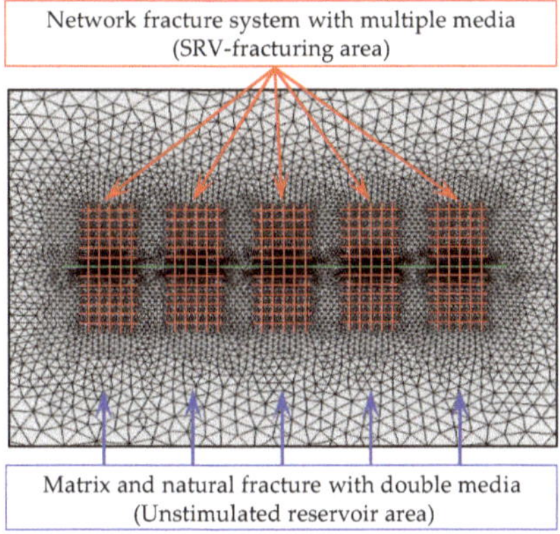

Figure 2. 3D gridding division near an SRV-fractured horizontal well area.

4.2. Finite Element Solution

Assuming that the study area node number is N_p, the node pressure of matrix system and natural fracture system can be written by $\mathbf{P}_m = [P_{m,1}, P_{m,2},\ldots, P_{m,Np}]^T$ and $\mathbf{P}_n = [P_{n,1}, P_{n,2},\ldots, P_{n,Np}]^T$. The equivalent integral transformation for control Equations (10), (12), and (14) is carried out by using the equilibrium condition and variation principle, and the characteristic matrix equation of the system element can be obtained. The element characteristic matrix of the network fracture system can be expressed [38] as

$$\begin{aligned} & a_{fD}K_{fD}\iint\limits_{\Omega_{e,f}} \nabla\mathbf{N}_{e,f}^T\nabla\mathbf{N}_{e,f}d\Omega_{e,f}\mathbf{P}_{e,f} + a_{fD}\omega_f\iint\limits_{\Omega_{e,f}} \mathbf{N}_{e,f}^T\mathbf{N}_{e,f}d\Omega_{e,f}\frac{\partial\mathbf{P}_{e,f}}{\partial t_D} \\ & = a_{fD}2\pi h_D\iint\limits_{\Omega_{e,f}} q_{fD}\mathbf{N}_{e,f}^T\delta(M_D - M'_D)d\Omega_{e,f} \end{aligned} \tag{16}$$

where a_{fD} is the dimensionless opening of the 2D fracture surface; $\mathbf{P}_{e,f}$ is the pressure matrix of the node in the network fracture system; $\Omega_{e,f}$ is the flow area of the network fracture located at the node; and $\mathbf{N}_{e,f} = [N_1, N_2, N_3]$ represents the shape function of two-dimensional triangular elements.

Finally, based on the element characteristic matrix of the matrix system and natural fracture system, the equilibrium equation of the reservoir system can be derived [38] as

$$\mathbf{A}_m\mathbf{P}_m + \mathbf{B}_m\frac{\partial\mathbf{P}_m}{\partial t_D} + \mathbf{C}(\mathbf{P}_m - \mathbf{P}_n) = 0 \tag{17}$$

$$\mathbf{A}_n\mathbf{P}_n + \mathbf{B}_n\frac{\partial\mathbf{P}_n}{\partial t_D} - \mathbf{C}(\mathbf{P}_m - \mathbf{P}_n) = \mathbf{Q}_n \tag{18}$$

where the expression of the coefficient matrix is

$$\mathbf{A}_m = \iiint\limits_{\Omega_{e,mn}} (\nabla\mathbf{N}_{e,mn}^T\nabla\mathbf{N}_{e,mn} + \chi_D\mathbf{N}_{e,mn}^T\nabla\mathbf{N}_{e,mn})d\Omega_{e,mn} + a_{fD}K_{fD}\iint\limits_{\Omega_{e,f}} \nabla\mathbf{N}_{e,f}^T\nabla\mathbf{N}_{e,f}d\Omega_{e,f}$$

$$\mathbf{A}_n = \iiint\limits_{\Omega_{e,mn}} \nabla\mathbf{N}_{e,mn}^T\nabla\mathbf{N}_{e,mn}d\Omega_{e,mn} + a_{fD}K_{fD}\iint\limits_{\Omega_{e,f}} \nabla\mathbf{N}_{e,f}^T\nabla\mathbf{N}_{e,f}d\Omega_{e,f}$$

$$\mathbf{B}_m = (1-\omega_n)\iiint\limits_{\Omega_{e,mn}} \mathbf{N}_{e,mn}^T\mathbf{N}_{e,mn}d\Omega_{e,mn} + a_{fD}\omega_f\iint\limits_{\Omega_{e,f}} \mathbf{N}_{e,f}^T\mathbf{N}_{e,f}d\Omega_{e,f}$$

$$\mathbf{B}_n = \omega_n\iiint\limits_{\Omega_{e,mn}} \mathbf{N}_{e,mn}^T\mathbf{N}_{e,mn}d\Omega_{e,mn} + a_{fD}\omega_f\iint\limits_{\Omega_{e,f}} \mathbf{N}_{e,f}^T\mathbf{N}_{e,f}d\Omega_{e,f}$$

$$\mathbf{C} = \lambda\iiint\limits_{\Omega_{e,mn}} \mathbf{N}_{e,mn}\mathbf{N}_{e,n}^T d\Omega_{e,mn}$$

$$\mathbf{Q}_n = 2\pi h_D\iiint\limits_{\Omega_{e,n}} q_{nD}\mathbf{N}_{e,mn}^T\delta(M_D - M'_D)d\Omega_{e,mn} + a_{fD}2\pi h_D\iint\limits_{\Omega_{e,f}} q_{fD}\mathbf{N}_{e,f}^T\delta(M_D - M'_D)d\Omega_{e,f}$$

Assuming that the fluid flows from the natural fracture system to the network fracture system in the initial time, by using the implicit backward difference method concerning time for the equilibrium Equation (18) of the natural fracture system, the governing equation of the finite element method corresponding to the $(k + 1)$th time of the fracture system can be obtained [38] by

$$\left\{\mathbf{A}_n + \frac{\mathbf{B}_n}{t_D^{k+1} - t_D^k} + \mathbf{C}\right\}\mathbf{P}_n^{k+1} = \mathbf{Q}_n^{k+1} + \frac{\mathbf{B}_n}{t_D^{k+1} - t_D^k}\mathbf{P}_n^k + \mathbf{C}\mathbf{P}_m^k \tag{19}$$

According to the Equation (17), the pressure of the matrix system at (k + 1)th time step can be calculated as

$$\left\{\mathbf{A}_m + \frac{\mathbf{B}_m}{t_D^{k+1} - t_D^k} + \mathbf{C}\right\}\mathbf{P}_m^{k+1} = \frac{\mathbf{B}_m}{t_D^{k+1} - t_D^k}\mathbf{P}_m^k + \mathbf{C}\mathbf{P}_n^k \tag{20}$$

When the coefficient matrix $\mathbf{A}_\mathrm{m}$ = 0, the model represents the triple-porosity, dual-permeability (TPDP) media. When $\mathbf{A}_\mathrm{m} \neq 0$, the abovementioned represents the triple-porosity, triple-permeability (TPTP) model. Using the abovementioned dominating Equations (12) and (13), the transient pressure and production performance of an SRV-fractured horizontal well under the conditions of constant productivity rate and stable bottomhole pressure can be calculated respectively.

5. Multiscale Flow Characteristics of SRV-Fractured Horizontal Well

In recent years, SRV fracturing technology has been widely used in the tight oil reservoirs of the Longdong oilfield, Ordos Basin, China. The Chang-7 oil reservoir in the mining area, which has an average depth of 1705 m, is a typical lithologically controlled oil reservoir characterized by tight pores, low pressure, and well-developed natural fractures. Therefore, complex fracture networks with multiple pores are easily developed in the formation after fracturing. According to the actual geological parameters and microseismic monitoring data of a ZP1 horizontal well with SRV fracturing of tight oil reservoirs in the Longdong oilfield, the basic parameters were determined (Table 1). The dimensionless variables used for the analysis and discussion of the results can be calculated, as shown in Table 2. The above parameters were substituted into the dominating Equations (12) and (13) to verify the finite element solution of the proposed model. Furthermore, the flow regimes and production performance of an SRV-fractured horizontal well with multiporosity media were analyzed.

Table 1. Geological and engineering parameters of the ZP1 well in the Longdong oilfield.

Geological and Engineering Parameters, Symbol (Unit)	Value
Reservoir size, $x_e \times y_e \times h_e$ (m)	2400 × 2400 × 40
Permeability, K_m, K_n, K_f (mD)	0.16, 160, 3.33 × 10^8
Porosity, ϕ_m, ϕ_n, ϕ_f	0.091, 0.27, 0.32
Compression coefficient of fluid and pore, C_L, C_p (MPa^{-1})	0.0014, 0.0042
Comprehensive compression coefficient of fracture system, C_n, C_f (MPa^{-1})	0.00061, 0.00061
Viscosity of fluids, μ (mPa·s)	1
Threshold pressure gradient, χ (MPa/m)	0.0025
Initial formation pressure, p_i (MPa)	20
Horizontal well length, L (m)	400
Number of fracturing segments, N	5
Segments spacing, Δy_f (m)	80
Network fracture size, $a_D \times b_D$ (m)	40 × 80
Fracture opening, a_f (mm)	2

Table 2. Dimensionless variables used for the analysis and discussion of the results.

Dimensionless Parameters, Symbol	Value
Reservoir size, $x_{eD} \times y_{eD} \times h_{eD}$	6 × 6 × 0.1
Network fracture size, $a_D \times b_D$	0.1 × 0.2
Fracture opening, a_{fD}	0.5 × 10^{-6}
Matrix permeability, K_{mD}	0.001
Network fracture permeability, K_{fD}	2.08 × 10^6
Threshold pressure gradient, χ_D	0.001
Elastic storativity ratio of the natural fracture system, ω_n	0.78
Elastic storativity ratio of network fracture system, ω_f	0.92
Crossflow coefficient, λ	60

5.1. Accuracy Verification of the Numerical Solution

To verify the accuracy of the numerical solution of our model, on the one hand, it was considered that the reservoir was a dual-porosity and single-permeability medium without threshold pressure. Moreover, only primary fractures exist in the reservoir after fracturing. The numerical solution of the finite element model was compared with the analytical solution of the Zerzar et al. 2004 model [43] for a conventional multistage fractured horizontal well, and the comparative curve of the pressure and pressure derivative behaviors were obtained, as shown in Figure 3. On the other hand, according to the actual geological parameters and fracturing parameters of a ZP1 well with 33 months of production history in the Longdong oilfield, the oil production rate and cumulative oil production of the ZP1 well with SRV fracturing could be calculated using the numerical model proposed in this paper, and the comparison curves are shown in Figure 4.

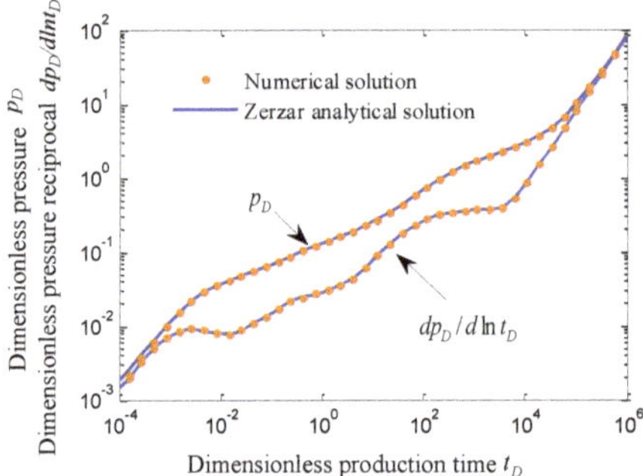

Figure 3. Pressure and pressure derivative behaviors in a multi-stage fractured horizontal well intercepted by the numerical solution and Zerzar [43] analytical solution.

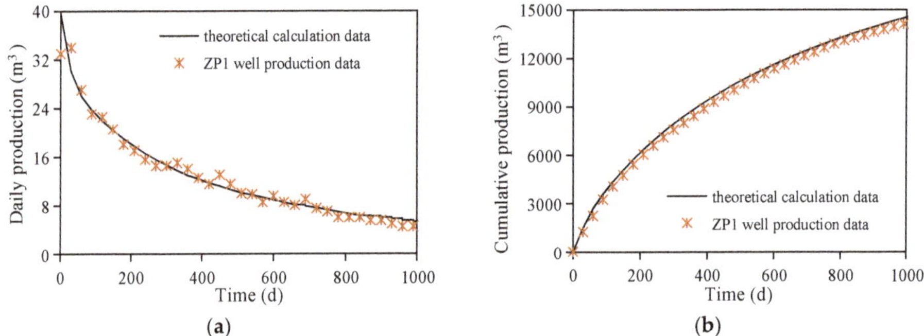

Figure 4. Comparison curve of the ZP1 well production data and theoretical calculation data. (**a**) Oil production rate. (**b**) Cumulative oil production.

Figure 3 shows that the pressure and pressure derivative behaviors of a multistage fractured horizontal well calculated by the two models were basically consistent. Figure 4 shows that the theoretical model had good degree of fit with the actual well production data. Therefore, the model established in this paper could not only be simplified as the Zerzar analytical solution model, but could also be used to accurately predict the production performance of an SRV-fractured well in tight oil reservoirs.

5.2. *Flow Regimes Division during Well Production*

Considering the effect of natural fractures inherent in tight formation and network fracture systems produced by SRV fracturing on the productivity of the horizontal well and using the TPDP and TPTP models to simulate the production performance of a horizontal well under the conditions of constant productivity rate, the pressure, and pressure derivative behaviors (type-curves of well testing) [44] for an SRV-fractured horizontal well in a tight oil reservoir could be obtained, as shown in Figure 5.

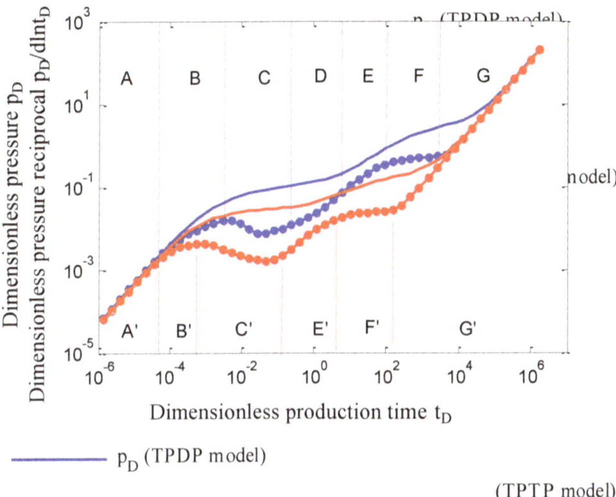

Figure 5. Type-curves of well testing for an SRV-fractured horizontal well with the TPDP and TPTP models.

For the TPDP model, the matrix system exhibited only the fluid crossflow phenomenon with the natural fracture system, but was not involved in the fluid flow process to the network fracture system. Under the assumption that the stimulated area was composed of triple-porosity media and the unstimulated area was composed of dual-porosity media, based on the pressure derivative curve, the TPDP model flow regimes during SRV-fractured horizontal well production in a tight oil reservoir could be divided into seven flow periods, as shown in Figure 6, where k is the slope of the pressure derivative curve; and both m and n are constants.

The TPDP model flow regimes can be divided into the following periods. Stage A: The initial pseudosteady flow around primary fractures; this stage mainly reflects the linear flow inside the primary fractures and the radial flow around the primary fractures, and the combination of the two causes the pressure derivative behavior to show a straight line with unit slope. Stage B: Linear flow inside the network fracture system; this stage reflects the linear flow from the secondary fractures to the primary fracture, and the pressure derivative behavior shows an oblique line with a near unit slope. Stage C: Pseudosteady crossflow between the matrix and natural fracture systems; as the pressure drop of the natural fracture system is greater than that of the matrix system, this stage mainly reflects the pseudosteady flow process from the matrix system into the natural fracture system, which leads to a concave part of the pressure derivative behavior. Stage D: Formation linear flow; this stage represents the linear flow around the network fracture, and the pressure derivative curve shows a straight line with a 1/2 slope. Stage E: Pseudosteady flow in the stimulated area; when the pressure wave propagates to the boundary of the stimulated area, the effective distance of fluid flow in the unstimulated area increases continuously, resulting in the formation of a moving sealed boundary with time changing around the stimulated areas. The pressure derivative behavior shows an oblique line with a near unit slope. Stage F: Pseudo radial flow near the SRV-fractured horizontal well; the flow characteristics at this stage are expressed as a pseudo radial flow centered on the horizontal well with the network fracture system, and the pressure derivative behavior is shown as a horizontal straight line. Stage G: Pseudosteady flow in the whole reservoir; the influence of the closed outer boundary is observed during the later stage of well production, i.e., when the pressure wave propagates to the reservoir boundary, the bottomhole pressure drops rapidly and pressure derivative behavior is shown as a straight line with unit slope.

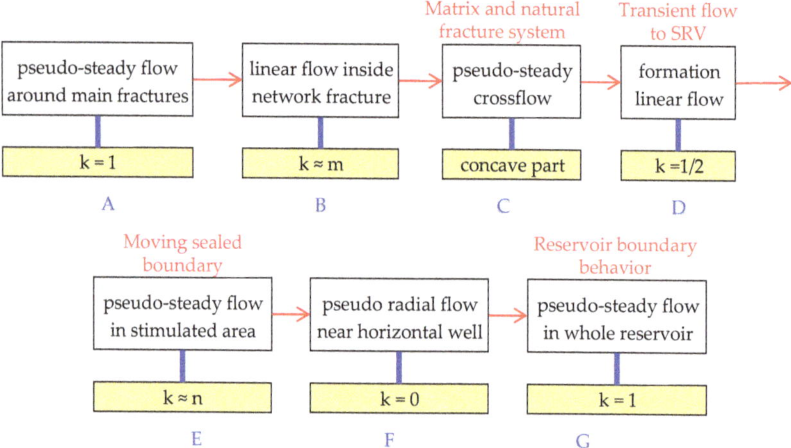

Figure 6. Flow regimes division during SRV-fractured horizontal well production in tight oil reservoir.

For the TPTP model, the fluid in the matrix system is involved in the flow to the network fracture system. Therefore, comparing with the TPDP model flow regimes, the bottomhole pressure drop of the horizontal well with the TPTP model becomes slower in the B′, C′, E′, and F′ stages. The linear flow in the formation (D) no longer arises and is covered by the pseudosteady crossflow (C′), which quickly changes the pseudosteady flow (E′). Then, the pressure wave propagates quickly to the closed reservoir boundary, and the bottomhole pressure drop increases rapidly during the pseudosteady flow in the whole reservoir (G′), which is consistent with the pressure and pressure derivative behaviors of the TPDP model gradually. According to the development experience of tight oil reservoirs, the TPTP model is more reasonable for tight oil reservoir simulation.

5.3. Productivity Contribution Degree of Multiporosity Systems

To further quantitatively analyze the contribution degree of multiporosity systems to well productivity, the TPTP model was used to simulate the production process of the SRV-fractured horizontal well (800 m in length and fracturing with 10 segments) under the following three cases: (1) there was only the matrix system in the reservoir; (2) there were only the matrix and natural fracture systems in reservoirs; (3) there were the matrix, natural fracture, and network fracture systems in the reservoirs. The productivity (including the daily production and cumulative production) contribution curves for the three systems (including the matrix, natural fracture, and network fracture systems) during SRV-fractured horizontal well production in tight oil reservoirs can be calculated respectively, as shown in Figure 7. Moreover, the daily production contribution ratio (DPCR) and cumulative production contribution ratio (CPCR) of the three systems to SRV-fractured horizontal well productivity can be obtained statistically, as shown in Table 3.

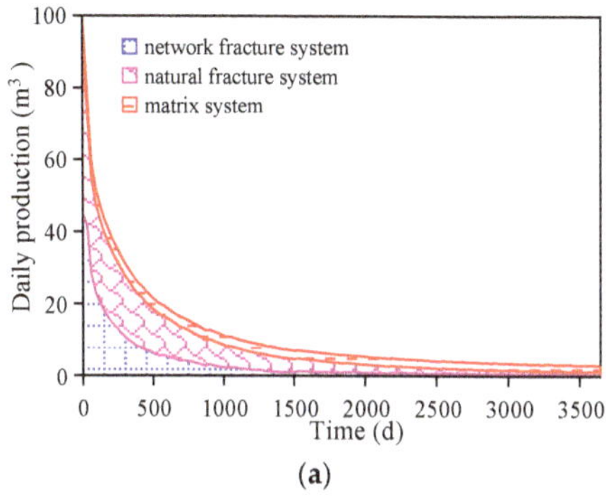

(a)

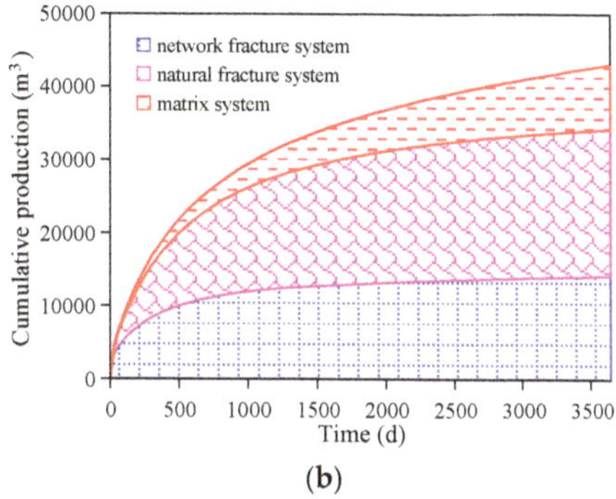

(b)

Figure 7. Productivity contribution curves of three systems during SRV-fractured horizontal well production in a tight oil reservoir. (**a**) Oil production rate. (**b**) Cumulative oil production.

Table 3. DPCR and CPCR of three systems to SRV-fractured horizontal well productivity in different development stages of tight oil reservoir

DPCR [1] (%) / CPCR [2] (%)	Matrix System	Natural Fracture System	Network Fracture System
1st year	11.73	56.36	31.91
	7.85	43.67	48.48
5th year	39.08	45.49	15.43
	14.60	49.23	36.17
10th year	59.12	26.80	14.08
	20.49	46.79	32.72

[1] DPCR is the daily production contribution ratio; [2] CPCR is the cumulative production contribution ratio.

The simulation results indicated that the proportion of productivity contribution for triple-porosity media systems during SRV-fractured horizontal well production varied at different stages of reservoir development. In the early stage of tight oil reservoir development, the productivity of the SRV-fractured horizontal well was mainly contributed to by natural fracture and network fracture systems with high conductivity. The daily production rate was large, but declined rapidly. After that stage, due to the fracture failure, the DPCR of the natural fracture and network fracture systems gradually decreased, and the latter was more serious; on the contrary, the DPCR of matrix system increased rapidly. In the late stage of reservoir development, the daily production of the horizontal well was maintained at a lower level, and the DPCR of the matrix system was more than half. The CPCR of the matrix system, natural fracture system, and network fracture system during SRV-fractured horizontal well production were 7.85%, 43.67%, and 48.48%, respectively in the 1st year; 14.60%, 49.23%, and 36.17%, respectively in the 5th year; and 20.49%, 46.79%, and 32.72%, respectively in the 10th year.

6. Conclusions

During the development of a tight oil reservoir after SRV fracturing, the flow characteristics are different from those of conventional reservoirs. This paper investigated the multiporosity and multiscale flow characteristics of an SRV-fractured horizontal well in a tight oil reservoir. Based on the dual-media theory and discrete-fracture network models, a mathematical flow model of an SRV-fractured horizontal well with multiporosity and multipermeability media was built, solved, and verified. It has been found that there exist different flow regimes and productivity characteristics in SRV-fractured horizontal wells. The TPDP model flow regimes during SRV-fractured horizontal well production in tight oil reservoirs could be divided into seven flow periods, which include the initial pseudosteady flow around the primary fractures, linear flow inside the network fracture system, pseudosteady crossflow, formation linear flow, pseudosteady flow in the stimulated area, pseudoradial flow near horizontal well, and pseudosteady flow in the whole reservoir. For the multiporosity and multiscale flowing states, the well bottomhole pressure drop became slower, the linear flow in the formation no longer arose, and the pressure wave arrived quickly at the closed reservoir boundary. The initial production rate of the SRV-fractured horizontal well was large but declined rapidly. The contribution ratio of the matrix system, natural fracture system, and network fracture system during SRV-fractured horizontal well production were 7.85%, 43.67%, and 48.48%, respectively in the 1st year; 14.60%, 49.23%, and 36.17%, respectively in the 5th year; and 20.49%, 46.79%, and 32.72%, respectively in the 10th year. The proposed research may provide valuable insight into understanding the multiporosity and multiscale flow mechanisms and unconventional hydrocarbon recovery maximization. For the actual oilfield, the change of the dynamic energy of the formation system can be predicted by the change of well productivity, which could guide managers in carrying out the development of regime adjustment and improvements in the management system in a timely manner.

Author Contributions: L.R. and W.W. conceived the strategy and designed the theoretical framework; Y.S., M.C., C.J. and N.Z. conducted the simulations and analyzed the data; L.R., Y.H., and J.S. wrote the manuscript; and all authors have read and approved the final manuscript.

Funding: This research was supported by the National Natural Science Foundation of China (no. 51704235, 51804328, 51874242), Young Talent fund of University Association for Science and Technology in Shaanxi, China (no. 20180417), Shandong Province Natural Science Foundation (ZR2018BEE008), Fundamental Research Funds for the Central Universities (18CX02168A) and Natural Science Basic Research Plan in Shaanxi Province of China (no. 2018JQ5208).

Acknowledgments: The authors appreciate the reviewers and editors for their critical and helpful comments.

Conflicts of Interest: The authors declare no conflict of interest.

References

1. Jia, C.Z.; Zheng, M.; Zhang, Y.F. Unconventional hydrocarbon resources in China and the prospect of exploration and development. *Pet. Explor. Dev.* **2012**, *39*, 129–136. [CrossRef]
2. Wang, W.D.; Zheng, D.; Sheng, G.L.; Zhang, Q.; Su, Y.L. A review of stimulated reservoir volume characterization for multiple fractured horizontal well in unconventional reservoirs. *Adv. Geo-Energy Res.* **2017**, *1*, 54–63. [CrossRef]
3. Singh, H.; Cai, J.C. A mechanistic model for multi-scale sorption dynamics in shale. *Fuel* **2018**, *234*, 996–1014. [CrossRef]
4. Ren, L.; Su, Y.L.; Zhao, G.Y. Non-Darcy flow pattern response and critical well spacing in tight oil reservoirs. *J. Cent. South Univ.* **2015**, *5*, 1732–1738. [CrossRef]
5. Wu, Q.; Xu, Y.; Zhang, S.L.; Wang, T.F.; Guan, B.S.; Wu, G.T.; Wang, X.Q. The core theories and key optimization designs of volume stimulation technology for unconventional reservoirs. *Acta Pet. Sin.* **2014**, *35*, 706–714. [CrossRef]
6. Liu, G.; Meng, Z.; Cui, Y.; Wang, L.; Liang, C.; Yang, S. A Semi-Analytical Methodology for Multiwell Productivity Index of Well-Industry-Production-Scheme in Tight Oil Reservoirs. *Energies* **2018**, *11*, 1054. [CrossRef]
7. Wang, W.D.; Su, Y.L.; Yuan, B.; Wang, K.; Cao, X.P. Numerical simulation of fluid flow through fractal-based discrete fractured network. *Energies* **2018**, *11*, 286. [CrossRef]
8. Tang, C.; Chen, X.; Du, Z.; Yue, P.; Wei, J. Numerical Simulation Study on Seepage Theory of a Multi-Section Fractured Horizontal Well in Shale Gas Reservoirs Based on Multi-Scale Flow Mechanisms. *Energies* **2018**, *11*, 2329. [CrossRef]
9. Zhang, Q.; Su, Y.; Wang, W.; Lu, M.; Ren, L. Performance analysis of fractured wells with elliptical SRV in shale reservoirs. *J. Nat. Gas Sci. Eng.* **2017**, *45*, 380–390. [CrossRef]
10. Yao, J.; Sun, H.; Li, A.F.; Yang, Y.F.; Huang, Z.Q.; Wang, Y.Y.; Zhang, L.; Kou, J.L.; Xie, H.J.; Zhao, J.L.; et al. Modern system of multiphase flow in porous media and its development trend. *Chin. Sci. Bull.* **2018**, *63*, 425–451. [CrossRef]

11. Mayerhofer, M.J.; Lolon, E.; Warpinski, N.R.; Cipolla, C.L.; Walser, D.W.; Rightmire, C.M. What is stimulated reservoir volume? *SPE Prod. Oper.* **2010**, *25*, 89–98. [CrossRef]
12. Chong, Z.; Li, X.; Chen, X.; Zhang, J.; Lu, J. Numerical investigation into the effect of natural fracture density on hydraulic fracture network propagation. *Energies* **2017**, *10*, 914. [CrossRef]
13. Wang, W.D.; Su, Y.L.; Zhang, X.; Sheng, G.L. Analysis of the complex fracture flow in multiple fractured horizontal wells with the fractal tree-like network models. Fractals Complex Geom. *Patterns Scaling Nat. Soc.* **2015**, *23*, 1550014. [CrossRef]
14. Wu, Q.; Xu, Y.; Wang, X.Q.; Wang, T.F.; Zhang, S.L. Volume fracturing technology of unconventional reservoirs: Connotation, optimization design and implementation. *Pet. Explor. Dev.* **2012**, *39*, 352–358. [CrossRef]
15. Yuan, B.; Zheng, D.; Moghanloo, R.G.; Wang, K. A novel integrated workflow for evaluation, optimization, and production predication in shale plays. *Int. J. Coal Geol.* **2017**, *180*, 18–28. [CrossRef]
16. Wang, Y.; Li, X.; He, J.; Zhao, Z.; Zheng, B. Investigation of fracturing network propagation in random naturally fractured and laminated block experiments. *Energies* **2016**, *9*, 588. [CrossRef]
17. Kaustubh, S.; Sharma, M.M. Mechanisms for the formation of complex fracture networks in naturally fractured rocks. In Proceedings of the SPE Hydraulic Fracturing Technology Conference and Exhibition, The Woodlands, TX, USA, 23–25 January 2018. [CrossRef]
18. Cai, J.C.; Wei, W.; Hu, X.Y.; Liu, R.C.; Wang, J.J. Fractal characterization of dynamic fracture network extension in porous media. *Fractals* **2017**, *25*, 1750023. [CrossRef]
19. Su, Y.; Sheng, G.; Wang, W.; Zhang, Q.; Lu, M.; Ren, L. A mixed-fractal flow model for stimulated fractured vertical wells in tight oil reservoirs. *Fractals* **2016**, *24*, 1650006. [CrossRef]
20. Zeng, F.; Cheng, X.; Guo, J.; Chen, Z.; Xiang, J. Investigation of the initiation pressure and fracture geometry of fractured deviated wells. *J. Pet. Sci. Eng.* **2018**, *165*, 412–427. [CrossRef]
21. Ren, L.; Su, Y.; Zhan, S.; Hao, Y.; Meng, F.; Sheng, G. Modeling and simulation of complex fracture network propagation with SRV fracturing in unconventional shale reservoirs. *J. Nat. Gas Sci. Eng.* **2016**, *28*, 132–141. [CrossRef]
22. Zhang, Q.; Su, Y.; Wang, W.; Sheng, G. A new semi-analytical model for simulating the effectively stimulated volume of fractured wells in tight reservoirs. *J. Nat. Gas Sci. Eng.* **2015**, *27*, 1834–1845. [CrossRef]
23. Weng, D.W.; Lei, Q.; Xu, Y.; Li, Y.; Li, D.Q.; Wang, W.X. Network fracturing techniques and its application in the field. *Acta Pet. Sin.* **2011**, *32*, 281–284. [CrossRef]
24. Brown, M.; Ozkan, E.; Raghavan, R.; Kazemi, H. Practical solutions for pressure transient responses of fractured horizontal wells in unconventional reservoirs. In Proceedings of the SPE Annual Technical Conference and Exhibition, New Orleans, LA, USA, 4–7 October 2009. [CrossRef]
25. Stalgorova, E.; Mattar, L. Analytical model for unconventional multifractured composite systems. *SPE Reserv. Eval. Eng.* **2013**, *3*, 246–256. [CrossRef]
26. Su, Y.L.; Wang, W.D.; Sheng, G.L. Compound flow model of volume fractured horizontal well. *Acta Pet. Sin.* **2014**, *35*, 504–510. [CrossRef]
27. Xu, W.Y.; Li, J.; Du, M. Quick estimate of initial production from stimulated reservoirs with complex hydraulic fracture network. In Proceedings of the SPE Annual Technical Conference and Exhibition, Denver, CO, USA, 30 October–2 November 2011. [CrossRef]
28. Jia, P.; Cheng, L.S.; Huang, S.J.; Xue, Y.C.; Ai, S. Production simulation of complex fracture networks for shale gas reservoirs using a semi-analytical model. In Proceedings of the SPE Asia Pacific Unconventional Resources Conference and Exhibition, Brisbane, Australia, 9–11 November 2015. [CrossRef]
29. Yang, R.Y.; Huang, Z.W.; Yu, W.; Lashgari, H.R.; Sepehrnoori, K.; Shen, Z.H. A semianalytical method for modeling two-phase flow in coalbed methane reservoirs with complex fracture networks. In Proceedings of the SPE/AAPG/SEG Unconventional Resources Technology Conference, San Antonio, TX, USA, 1–3 August 2016. [CrossRef]
30. Yao, J.; Yin, X.X.; Fan, D.Y.; Sun, Z.X. Trilinear flow well test model of fractured horizontal well in low permeability reservoir. *Well Test.* **2011**, *20*, 1–5. [CrossRef]
31. Fan, D.Y.; Yao, J.; Sun, H.; Zeng, H.; Wang, W. A composite model of hydraulic fractured horizontal well with stimulated reservoir volume in tight oil & gas reservoir. *J. Nat. Gas Sci. Eng.* **2015**, *24*, 115–123. [CrossRef]

32. Fisher, M.K.; Wright, C.A.; Davidson, B.M.; Steinsberger, N.P.; Buckler, W.S.; Goodwin, A.; Fielder, E.O. Integrating fracture mapping technologies to improve stimulations in the Barnett shale. *SPE Prod. Facil.* **2005**, *20*, 85–93. [CrossRef]
33. Daniels, J.L.; Waters, G.A.; Le Calvez, J.H.; Bentley, D.; Lassek, J.T. Contacting more of the Barnett shale through an integration of real-time microseismic monitoring petrophysics and hydraulic fracture design. In Proceedings of the SPE Annual Technical Conference and Exhibition, Anaheim, CA, USA, 11–14 November 2007. [CrossRef]
34. Maxwell, S.C.; Urbancic, T.I.; Steinsberger, N.; Zinno, R. Microseismic imaging of hydraulic fracture complexity in the Barnett shale. In Proceedings of the SPE Annual Technical Conference and Exhibition, San Antonio, TX, USA, 29 September–2 October 2002. [CrossRef]
35. Le Calvez, J.H.; Craven, M.E.; Klem, R.C.; Baihly, J.D.; Bennett, L.A.; Brook, K. Real-time microseismic monitoring of hydraulic fracture treatment: A tool to improve completion and reservoir management. In Proceedings of the SPE Hydraulic Fracturing Technology Conference, College Station, TX, USA, 29–31 January 2007. [CrossRef]
36. Prada, A.; Civan, F. Modification of Darcy's law for the threshold pressure gradient. *J. Petrol. Sci. Eng.* **1999**, *22*, 237–240. [CrossRef]
37. Cai, J.C. A fractal approach to low velocity non-Darcy flow in a low permeability porous medium. *Chin. Phys. B* **2014**, *23*, 044701. [CrossRef]
38. Ren, L.; Su, Y.L.; Hao, Y.M.; Zhang, Q.; Meng, F.K.; Sheng, G.L. Dynamic analysis of SRV-fractured horizontal wells in tight oil reservoirs based on stimulated patterns. *Acta Pet. Sin.* **2015**, *36*, 1272–1279. [CrossRef]
39. Hall, H.N. Compressibility of reservoir rocks. *J. Pet. Technol.* **1953**, *5*, 17–19. [CrossRef]
40. Sun, Z.X.; Yao, J.; Fan, D.Y.; Wang, Y.Y.; Zhang, K.S. Dynamic analysis of horizontal wells with complex fractures based on a discrete-fracture model. *J. China Univ. Pet.* **2014**, *38*, 109–115. [CrossRef]
41. Yan, B.; Mi, L.; Chai, Z.; Wang, Y.; Killough, J.E. An enhanced discrete fracture network model for multiphase flow in fractured reservoirs. *J. Pet. Sci. Eng.* **2018**, *161*, 667–682. [CrossRef]
42. Shakiba, M.; Sepehrnoori, K. Using embedded discrete fracture model (EDFM) in numerical simulation of complex hydraulic fracture networks calibrated by microseismic monitoring data. *J. Nat. Gas Sci. Eng.* **2018**, *55*, 495–507. [CrossRef]
43. Zerzar, A.; Tiab, D.; Bettam, Y. Interpretation of multiple hydraulically fractured horizontal wells. In Proceedings of the Society of Petroleum Engineers Abu Dhabi International Conference and Exhibition, Abu Dhabi, UAE, 10–13 October 2004. [CrossRef]
44. Zhao, Y.L.; Shan, B.C.; Zhang, L.H.; Liu, Q.G. Seepage flow behaviors of multi-stage fractured horizontal wells in arbitrary shaped shale gas reservoirs. *J. Geophys. Eng.* **2016**, *13*, 674–689. [CrossRef]

Article

Numerical Simulation Study on Seepage Theory of a Multi-Section Fractured Horizontal Well in Shale Gas Reservoirs Based on Multi-Scale Flow Mechanisms

Chao Tang [1], Xiaofan Chen [1,*], Zhimin Du [1], Ping Yue [1] and Jiabao Wei [2]

1 State Key Laboratory of Oil and Gas Reservoir Geology and Exploitation, Southwest Petroleum University, Chengdu 610500, Sichuan, China; 201511000111@stu.swpu.edu.cn (C.T.); duzhimin@swpu.edu.cn (Z.D.); 201331010025@swpu.edu.cn (P.Y.)

2 Hekou Production Plant of Shengli Oilfield, Dongying 257000, Shandong, China; jiabao.wei1988@gmail.com

* Correspondence: chenxf@swpu.edu.cn

Received: 3 August 2018; Accepted: 27 August 2018; Published: 4 September 2018

Abstract: Aimed at the multi-scale fractures for stimulated reservoir volume (SRV)-fractured horizontal wells in shale gas reservoirs, a mathematical model of unsteady seepage is established, which considers the characteristics of a dual media of matrix and natural fractures as well as flow in the large-scale hydraulic fractures, based on a discrete-fracture model. Multi-scale flow mechanisms, such as gas desorption, the Klinkenberg effect, and gas diffusion are taken into consideration. A three-dimensional numerical model based on the finite volume method is established, which includes the construction of spatial discretization, calculation of average pressure gradient, and variable at interface, etc. Some related processing techniques, such as boundedness processing upstream and downstream of grid flow, was used to limit non-physical oscillation at large-scale hydraulic fracture interfaces. The sequential solution is performed to solve the pressure equations of matrix, natural, and large-scale hydraulic fractures. The production dynamics and pressure distribution of a multi-section fractured horizontal well in a shale gas reservoir are calculated. Results indicate that, with the increase of the Langmuir volume, the average formation pressure decreases at a slow rate. Simultaneously, the initial gas production and the contribution ratio of the desorbed gas increase. With the decrease of the pore size of the matrix, gas diffusion and the Klinkenberg effect have a greater impact on shale gas production. By changing the fracture half-length and the number of fractured sections, we observe that the production process can not only pursue the long fractures or increase the number of fractured sections, but also should optimize the parameters such as the perforation position, cluster spacing, and fracturing sequence. The stimulated reservoir volume can effectively control the shale reservoir.

Keywords: shale gas; volume fracturing; finite volume method; production simulation; multi-scale flow; multi-scale fracture

---

1. Introduction

As an important part of unconventional oil and gas resources, shale gas resources have become a new hot spot in recent years. At present, numerical simulation models for shale gas mainly include dual media, multiple discrete media, and equivalent media, among which dual media models are widely used. Sawyer and Kucuk [1] first studied the pressure changes of shale gas reservoirs based on a dual-porosity continuous medium. Subsequently, Bumb and McKee [2] studied the effect of adsorption-desorption on transient behaviour by adding additional adsorption coefficients to the Langmuir isotherm equation. However, the above studies ignore the diffusion processes at the nano-microscales. Carlson and Mercer [3] investigated the pressure changes in vertical wells of a shale gas reservoir by introducing diffusion and desorption terms into a dual-porosity media. The model predicts the productivity of shale gas accurately in a short term. However, the long-term prediction of productivity of gas wells is inaccurate, due to the failure to consider the slippage effect. Swami et al. [4] established a dual media model that considers the Knudsen diffusion, slippage, and sorption-desorption processes, and is validated by laboratory data.

Some researchers [5–11] have pointed out that although dual media models are widely used in commercial software, due to their inherent shortcomings, full or partial encryption still suffers from poor adaptability to multi-scale fracture network systems. In addition, due to the micro-pores in shale gas reservoirs, it is usually necessary to conduct fracturing to obtain a commercial gas production rate. However, natural and artificial fractures have big differences in morphology and seepage ability. Kuuskraa et al. [12] propose to use multiple discrete media models to study the productivity of shale gas reservoirs. Based on the concept of multiple media, Schepers [13] and Dehghanpour et al. [14] established a Darcy flow model that couples diffusion and desorption processes with matrix flow, respectively. Wu et al. [15] established a multi-discrete medium model of dense fractured reservoirs, considering the stress sensitivity and slippage effects of fractures. The fractures are divided into natural micro-fractures and artificial fractures. The slippage effect in the matrix is considered and the differences between the multi-discrete and dual-media models are compared. Aboaba and Cheng [16] used a linear flow model to study the typical productivity curve, which describes changes of fractured horizontal wells in shale gas reservoirs without regard to adsorption and diffusion. In combination with the perturbation method and the point source function, a well test model for a horizontal well, considering diffusion and Darcy flow in a fracture, was proposed by Wang [17]. However, this model does not consider the influence of the reconstructed volume on the pressure change in a horizontal well. Fang et al. [18] considered the compressibility of tight reservoirs and the nonlinear seepage of matrix fluids, and established a multi-scale seepage discrete fracture model of two-dimensional volume fracturing.

In this paper, the authors summarize the law of flow in shale gas reservoirs and establish a three-dimensional (3D) composite model, which uses dual media to describe matrix-natural micro-fractures and utilizes discrete media to describe artificial fractures. The production of multi-section fractured horizontal wells in a rectangular shale gas reservoir is described, considering gas desorption, the Klinkenberg effect, and gas diffusion in the matrix. The stimulated volume is determined by parameter setting of the artificial fractures. The numerical solution is obtained by using the finite volume element method.

2. Mathematical Model

2.1. Assumptions

Figure 1a shows a multi-section fractured horizontal well in a shale gas reservoir. The x–y plane represents the horizontal plane, and the z-axis represents the vertical direction. Artificial fractures are represented by two-dimensional elemental bodies. Segments of the horizontal well are represented by one-dimensional, line-element entities. In order to simplify the model, we propose the following assumptions:

1. The gas reservoirs are rectangular, and the flow is an isothermal flow. The gas reservoirs are divided into artificial fractures, natural micro-fractures, and matrix;
2. Flows in artificial fractures and natural micro-fractures are described by Darcy's law. The gas desorption in a matrix pore is described by the Langmuir isotherm equation;
3. Horizontal wells produce at constant pressure. There is only a single-phase gas in gas reservoirs;
4. The fractures are perpendicular to the horizontal wellbore and symmetrical about the wellbore;
5. Permeability anisotropy and gravity effects are ignored, and natural gas can only flow into the horizontal wellbore through artificial fractures;

6. Shale gas consists of methane, and does not consider the effect of competitive adsorption on the adsorption-desorption process;
7. Gas diffusion process in shale gas matrix is a non-equilibrium, quasi-steady-state process, which obeys Fick's first law.

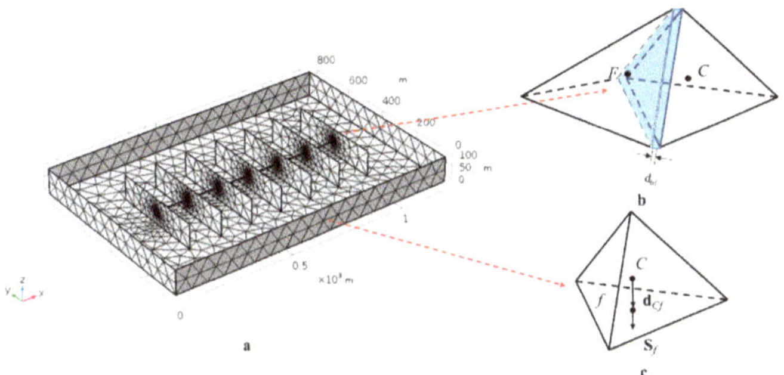

Figure 1. Diagram of a mathematical model. (**a**) Multi-section fractured horizontal well grid section diagram; (**b**) artificial fracture diagram; (**c**) grid of natural micro-fractures and matrix.

2.2. Governing Equation

2.2.1. Flow Equation

According to the real gas state equation, the shale gas density can be defined as

$$\rho_{\mathrm{gi}} = \frac{M_{\mathrm{g}} p_{\mathrm{i}}}{ZRT} \tag{1}$$

where i = m or f represents the matrix and the fractures, respectively; ρ_{g} is gas density (kg m^{-3}); M_{g} is the molecular mass (kg mol^{-1}); Z is the gas deviation factor (dimensionless); p is pressure (Pa); R is the universal gas constant (J mol^{-1} k^{-1}); and T is temperature (K).

Based on the above assumptions, the governing flow equation of shale gas in the matrix can be obtained from the law of conservation of mass, as follows:

$$\frac{\partial(\rho_{\mathrm{gm}}\phi_{\mathrm{m}})}{\partial t} + \nabla \cdot (\rho_{\mathrm{gm}}\mathbf{v}_{\mathrm{m}}) + (1-\phi_{\mathrm{m}})q_{\mathrm{m}} + q_{\mathrm{m-f}} = 0 \tag{2}$$

where ϕ_{m} is the shale matrix porosity (value), $\mathbf{v}_{\mathrm{m}}$ is the apparent gas velocity (m s^{-1}), q_{m} is the matrix desorption rate (m^3 s^{-1}), and $q_{\mathrm{m-f}}$ is the crow-flow rate from the matrix to the fracture (m^3 s^{-1}).

The first term in the formula represents the change of fluid mass in the unit volume element of the matrix. The second term is the flux flowing through the surface of the element, which must be modified by introducing the shale-gas-transport mechanisms in nanopores. In this paper, we consider gas molecular diffusion, slippage, and desorption. The third term is the desorption capacity of the matrix. The fourth item is the cross-flow from the matrix to the fracture. Generally speaking, water is not able to enter the micro-pores in the matrix of shale gas reservoirs. Therefore, it is reasonable to consider that there is only gas phase in the matrix. In other words, the gas in the micro-pores can be divided into adsorption gas adsorbing on the surface of the matrix and the free gas flowing in the micro-pores.

According to the study by Javadpour [19], the Knudsen number is in the transition zone of the viscous flow and Knudsen diffusion under shale gas formation conditions. At this time, the mass exchange of gas in the matrix is affected by viscous flow, Knudsen diffusion, and desorption. Therefore, corrections should be made to the mass flow in the matrix:

$$\rho_{\mathrm{gm}}\mathbf{v}_{\mathrm{m}} = \rho_{\mathrm{gm}}\left(\mathbf{v}_{\mathrm{m}}^{\mathrm{v}} + \mathbf{v}_{\mathrm{m}}^{\mathrm{k}}\right) \tag{3}$$

where $\mathbf{v}_{\mathrm{m}}^{\mathrm{v}}$ is the corrected gas velocity considering the Klinkenberg effect, and $\mathbf{v}_{\mathrm{m}}^{\mathrm{k}}$ is the corrected gas velocity considering the diffusive transport. The measure of pores in the matrix is usually tiny compared to other reservoir types. The additional contribution of the Klinkenberg effect to gas transport may be due to frequent collisions of gas molecules with the wall of the pores, causing the gas viscosity in the Knudsen layer to gradually deviate from the traditional gas viscosity. According to the study by Karniadakis et al., the gas effective viscosity can be expressed as

$$\mu_{eff} = \mu_{\mathrm{g}}\left(\frac{1}{1+8\alpha K_{\mathrm{n}}}\right) \tag{4}$$

where μ_{eff} is the gas effective viscosity (mPa·s), μ_{g} is gas viscosity (mPa·s), α is the rarefaction coefficient (dimensionless), and K_{n} is the Knudsen number (dimensionless).

Combing Equation (4) and Darcy's law, the corrected gas velocity with the Klinkenberg effect can be expressed as follows:

$$\mathbf{v}_{\mathrm{m}}^{\mathrm{v}} = -\frac{k_{\mathrm{m}}}{\mu_{\mathrm{g}}}(1+8\alpha K_{\mathrm{n}})\nabla p_{\mathrm{m}} \tag{5}$$

There are two fundamental modes: the advection and diffusion of fluid transport. The flow governing equation usually neglects the diffusive contribution, which is reasonable for most reservoirs—having a medium-high permeability, it may be unreasonable for shale gas. According to the mechanism of fluid dynamics [20], the gas diffusive velocity then can be expressed as

$$\mathbf{v}_{\mathrm{m}}^{\mathrm{k}} = -\frac{\delta^{\mathrm{m}}}{\rho_{\mathrm{gm}}\tau^{\mathrm{m}}}\phi_{\mathrm{m}}D_{\mathrm{g}}\nabla\rho_{\mathrm{gm}} = -\frac{\delta^{\mathrm{m}}}{\rho_{\mathrm{gm}}\tau^{\mathrm{m}}}\phi_{\mathrm{m}}D_{\mathrm{g}}\frac{\partial\rho_{\mathrm{gm}}}{\partial p_{\mathrm{m}}}\nabla p_{\mathrm{m}} = -\frac{\delta^{\mathrm{m}}}{\tau^{\mathrm{m}}}\phi_{\mathrm{m}}c_{\mathrm{g}}D_{\mathrm{g}}\nabla p_{\mathrm{m}} \tag{6}$$

where D_{g} is the Knudsen molecule diffusivity ($\mathrm{m^2\ s^{-1}}$), c_{g} is the gas compression factor ($\mathrm{Pa^{-1}}$; δ^{m}), and τ^{m} is the constrictivity and tortuosity of the shale matrix, respectively. The value of $\delta^{\mathrm{m}}/\tau^{\mathrm{m}}$ is always less than one. Therefore, Equation (3) can be rewritten as

$$\rho_{\mathrm{gm}}\mathbf{v}_{\mathrm{m}} = -\rho_{\mathrm{gm}}\left[\frac{k_{\mathrm{m}}}{\mu_{\mathrm{g}}}(1+8\alpha K_{\mathrm{n}}) + \frac{\delta^{\mathrm{m}}}{\tau^{\mathrm{m}}}\phi_{\mathrm{m}}c_{\mathrm{g}}D_{\mathrm{g}}\right]\nabla p_{\mathrm{m}} \tag{7}$$

Another contribution to gas production from shale reservoirs comes from the desorption of the gas (mostly to the kerogen) absorbed in shale, which is quantified via the change in the gas adsorption amount. The amount of gas adsorption per unit matrix volume at any pressure can be described by the Langmuir isotherm; then, the matrix desorption rate can be expressed as follows

$$q_{\mathrm{m}} = -\frac{\mathrm{d}V_{\mathrm{m}}}{\mathrm{d}t} = -\frac{\partial}{\partial t}\left(\frac{\rho_{\mathrm{m}}M_{\mathrm{g}}}{V_{\mathrm{std}}}\frac{V_{\mathrm{L}}p_{\mathrm{m}}}{p_{\mathrm{L}}+p_{\mathrm{m}}}\right) \tag{8}$$

where V_{m} is the adsorption capacity of per unit volume matrix ($\mathrm{m^3}$), V_{L} is the Langmuir volume ($\mathrm{m^3\ kg^{-1}}$), p_{L} is the Langmuir pressure (Pa), and V_{std} is the mole volume of gas at temperature (273.15 K) and pressure (101,325 Pa).

Substitute Equations (1), (5), (6) and (8) into Equation (2), the governing equation for gas transport in the shale matrix is given by

$$\begin{aligned}&\nabla\left\{\left[\frac{k_{\mathrm{m}}}{\mu_{\mathrm{g}}}(1+8\alpha K_{\mathrm{n}}) + \frac{\delta^{\mathrm{m}}}{\tau^{\mathrm{m}}}\phi_{\mathrm{m}}c_{\mathrm{g}}D_{\mathrm{g}}\right]\frac{M_{\mathrm{g}}p_{\mathrm{m}}}{ZRT}\nabla p_{\mathrm{m}}\right\} - \sigma_1\frac{k_{\mathrm{m}}}{\mu_{\mathrm{g}}}\frac{M_{\mathrm{g}}p_{\mathrm{m}}}{ZRT}(p_{\mathrm{m}}-p_{\mathrm{f}}) \\ &= \phi_{\mathrm{m}}\frac{M_{\mathrm{g}}}{ZRT}\frac{\partial p_{\mathrm{m}}}{\partial t} + (1-\phi_{\mathrm{m}})\frac{\partial}{\partial t}\left(\frac{\rho_{\mathrm{m}}M_{\mathrm{g}}}{V_{\mathrm{std}}}\frac{V_{\mathrm{L}}p_{\mathrm{m}}}{p_{\mathrm{L}}+p_{\mathrm{m}}}\right)\end{aligned} \tag{9}$$

Analogously, for natural micro-fractures systems:

$$\frac{\phi_f M_g}{ZRT}\frac{\partial p_f}{\partial t} - \nabla \cdot \left(\frac{M_g p_f}{ZRT}\frac{k_f}{\mu_g}\nabla p_f\right) - \sigma_1 \frac{k_m}{\mu_g}\frac{M_g p_m}{ZRT}(p_m - p_f) + q_{f-h} = 0 \tag{10}$$

where q_{f-h} is the crow-flow rate from the natural micro-fracture to the artificial fracture ($m^3 s^{-1}$). For artificial fractures system

$$d_{hf}\frac{\partial(\rho_g \phi_{hf})}{\partial t} - \frac{\partial}{\partial l} \cdot \left(d_{hf}\frac{\rho_g k_{hf}}{\mu_g}\frac{\partial p_{hf}}{\partial l}\right) - (q_{f-h} - q_{well}) = 0 \tag{11}$$

where q_{well} is the horizontal well productivity ($m^3\ s^{-1}$).

2.2.2. Initial Conditions

Under the initial conditions, the whole formation pressure is the original formation pressure, so

$$p_m(x,y,z,t)|_{t=0} = p_f(x,y,z,t)|_{t=0} = p_{hf}(x,y,z,t)|_{t=0} = p_i$$

2.2.3. Boundary Conditions

Γ_{out} represents the outside boundary, and Γ_{in} represents the inner boundary. It is assumed that the outer boundary of the model is the sealed boundary, and the production well produces at constant pressure. In that case, the well has the boundary conditions as follows:

$$\left.\frac{\partial p}{\partial n}\right|_{\Gamma_{out}} = 0$$
$$p|_{\Gamma_{in}} = p_{wf}$$

3. Discretization and Numerical Solution

3.1. Domain Discretization

Due to the geometric center coinciding with the centroid of the tetrahedral grid, in order to simplify the calculation, the whole reservoir region is discretized by the unstructured tetrahedron. Other types of grids are needed to recalculate the centroid of the control volume. As shown in Figure 1c, the matrix and micro-fracture system are expressed as a tetrahedron grid, considered as a dual continuous medium; the two-dimensional (2D) blue plane is decomposed in tetrahedral elements that are faces of the tetrahedron surrounding the artificial fracture interface, as shown in Figure 1b. Caumon G et al. [21] pointed out that fracture dimension reduction is a key method to improve the convergence of multi-scale simulation calculation. If the fracture is considered as three-dimensional, a large number of minimized grids will be generated, which will cause the subsequent calculations to fail to converge. The research of Juanes et al. [22] shows that the convergence of the two dimensions is significantly improved by considering the fracture in two dimensions. In addition, as shown in Figure 1c, unlike the traditional finite element method, we establish controlling volume on each grid to obtain the cell-centroid finite volume numerical calculation format. The computational domain Γ_d consists of two subdomains: Γ_{m-f} representing the matrix and micro-fractures system, and Γ_{hf} representing the artificial fractures system. In this paper, FGE is used to represent the flow governing equation. Therefore, the integration of the entire domain Γ_d can be written as

$$\iiint_{\Gamma_d} \mathrm{FGEd}\Gamma_d = \iiint_{\Gamma_{m-f}} \mathrm{FGEd}\Gamma_{m-f} + \mathrm{d}_{hf} \times \iiint_{\Gamma_{hf}} \mathrm{FGEd}\Gamma_{hf} \tag{12}$$

As shown in Figure 2a, different from the vertex-centered variable arrangement in conventional finite element methods, this paper uses a cell-centered variable arrangement to define a control volume.

As shown in Figure 2b, in a vertex-centered arrangement the flow variables are stored at the vertices, with elements constructed around the variables' locations.

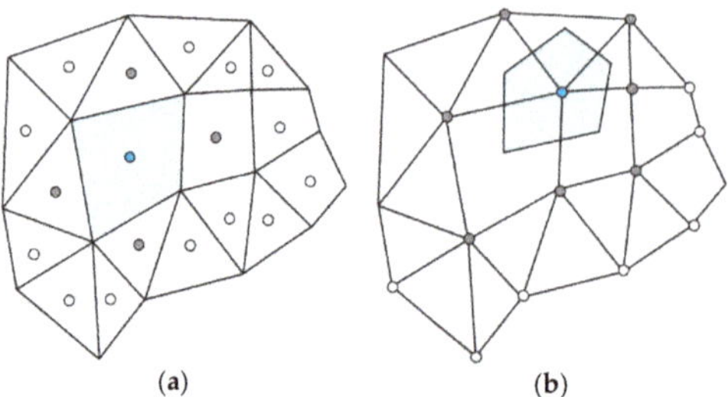

Figure 2. Control cell corresponding to the different variable arrangement. (**a**) Cell-centered; (**b**) vertex-centered.

Compared with vertex-centered variable arrangement, the cell-centered variable arrangement yields a high order accuracy of integrations. Moreover, it decreases the storage requirements. Furthermore, there is no additional treatment on the boundary to ensure a consistent solution. Another major advantage is that there is no need to pre-define a shape function based on element types.

3.2. Equation Discretization

This section uses the matrix system flow governing equation as an example to illustrate the equation discretization process in a tetrahedral mesh. Flow governing equations in micro-fractures and artificial fractures are similar. The process starts by integrating Equation (9) over element C, which enables the recovery of its integral balance, as

$$\begin{aligned}&\int\limits_{V_c} \phi_{\mathrm{m}} \frac{M_{\mathrm{g}}}{ZRT} \frac{\partial p_{\mathrm{m}}}{\partial t} \mathrm{d}V - \int\limits_{V_c} \nabla \left\{ \left[\frac{k_{\mathrm{m}}}{\mu_{\mathrm{g}}} (1 + 8\alpha K_{\mathrm{n}}) + \frac{\delta^{\mathrm{m}}}{\tau^{\mathrm{m}}} \phi_{\mathrm{m}} c_{\mathrm{g}} D_{\mathrm{g}} \right] \frac{M_{\mathrm{g}} p_{\mathrm{m}}}{ZRT} \nabla p_{\mathrm{m}} \right\} \mathrm{d}V \\ &+ \int\limits_{V_c} (1 - \phi_{\mathrm{m}}) \frac{\partial}{\partial t} \left(\frac{\rho_{\mathrm{m}} M_{\mathrm{g}}}{V_{\mathrm{std}}} \frac{V_{\mathrm{L}} p_{\mathrm{m}}}{p_{\mathrm{L}} + p_{\mathrm{m}}} \right) \mathrm{d}V + \int\limits_{V_c} \sigma_1 \frac{k_{\mathrm{m}}}{\mu_{\mathrm{g}}} \frac{M_{\mathrm{g}} p_{\mathrm{m}}}{ZRT} (p_{\mathrm{m}} - p_{\mathrm{f}}) \mathrm{d}V = 0\end{aligned} \tag{13}$$

The above formula shows that for any control volume, the change of gas mass flux in the matrix system within a certain time period is equal to the sum of the outflow through the control volume, as well as the desorption gas volume and the cross-flow rate, thus ensuring that the model still respects the conservation of mass in any local region. For the sake of mathematical simplicity, the following variable is chosen to be the apparent permeability of shale matrix:

$$\kappa = \frac{k_{\mathrm{m}}}{\mu_{\mathrm{g}}} (1 + 8\alpha K_{\mathrm{n}}) + \frac{\delta^{\mathrm{m}}}{\tau^{\mathrm{m}}} \phi_{\mathrm{m}} c_{\mathrm{g}} D_{\mathrm{g}} \tag{14}$$

Then, according to the divergence theorem, the volume integral of the advective-diffusive term in Equation (13) is transformed into a surface integral, yielding

$$\int\limits_{V_c} \nabla \left(\kappa \frac{M_{\mathrm{g}} p_{\mathrm{m}}}{ZRT} \nabla p_{\mathrm{m}} \right) \mathrm{d}V = \oint\limits_{\partial V_c} \left(\kappa \frac{M_{\mathrm{g}} p_{\mathrm{m}}}{ZRT} \nabla p_{\mathrm{m}} \cdot \mathbf{n} \right) \mathrm{dS} \tag{15}$$

In the presence of discrete faces, the surface integral in Equation (15) becomes

$$\oint\limits_{\partial V_c} \left(\kappa \frac{M_{\mathrm{g}} p_{\mathrm{m}}}{ZRT} \nabla p_{\mathrm{m}} \cdot \mathbf{n} \right) \mathrm{dS} = \sum_{f \sim faces(V_C)} \left[\int\limits_{f} \left(\kappa \frac{M_{\mathrm{g}} p_{\mathrm{m}}}{ZRT} \nabla p_{\mathrm{m}} \cdot \mathbf{n} \right) \mathrm{dS} \right] \tag{16}$$

where f is the integral point at the centroid of the boundary surface. Therefore, the integral in Equation (16) is numerically approximated to the flux at the centroids of the faces, which is a second-order approximation.

According to the trapezoidal integral formula, the surface integral in Equation (16) can be written as

$$\int_{f}\left(\kappa\frac{M_{\mathrm{g}}p_{\mathrm{m}}k_{\mathrm{m}}}{ZRT\mu_{\mathrm{g}}}\nabla p_{\mathrm{m}}\cdot\mathbf{n}\right)\mathrm{d}S\approx\kappa\frac{M_{\mathrm{g}}p_{\mathrm{m}}k_{\mathrm{m}}}{ZRT\mu_{\mathrm{g}}}\nabla p_{\mathrm{m}}\cdot\mathbf{S}_{f} \tag{17}$$

Therefore, the advective-diffusive term in Equation (13) can be written as

$$\int_{V_{\mathrm{c}}}\nabla\cdot\left(\kappa\frac{M_{\mathrm{g}}p_{\mathrm{m}}k_{\mathrm{m}}}{ZRT\mu_{\mathrm{g}}}\nabla p_{\mathrm{m}}\right)\mathrm{d}V=\sum_{f\sim faces(V_{\mathrm{C}})}\left[\left(\kappa\frac{M_{\mathrm{g}}p_{\mathrm{m}}}{ZRT}\lambda_{\mathrm{m}}\nabla p_{\mathrm{m}}\right)\cdot\mathbf{S}_{f}\right] \tag{18}$$

Similarly, the finite volume numerical calculation format for the unsteady term, gas desorption term, and the cross-flow term can be obtained as

$$\int_{V_{\mathrm{c}}}\phi_{\mathrm{m}}\frac{M_{\mathrm{g}}}{ZRT}\frac{\partial p_{\mathrm{m}}}{\partial t}\mathrm{d}V=V_{\mathrm{c}}\phi_{\mathrm{m}}\frac{M_{\mathrm{g}}}{ZRT}\frac{\partial p_{\mathrm{m}}}{\partial t} \tag{19}$$

$$\int_{V_{\mathrm{c}}}(1-\phi_{\mathrm{m}})\frac{\partial}{\partial t}\left(\frac{\rho_{\mathrm{m}}M_{\mathrm{g}}}{V_{\mathrm{std}}}\frac{V_{\mathrm{L}}p_{\mathrm{m}}}{p_{\mathrm{L}}+p_{\mathrm{m}}}\right)\mathrm{d}V=V_{\mathrm{c}}(1-\phi_{\mathrm{m}})\frac{\partial}{\partial t}\left(\frac{\rho_{\mathrm{m}}M_{\mathrm{g}}}{V_{\mathrm{std}}}\frac{V_{\mathrm{L}}p_{\mathrm{m}}}{p_{\mathrm{L}}+p_{\mathrm{m}}}\right) \tag{20}$$

$$\int_{V_{\mathrm{c}}}\sigma_{1}\frac{k_{\mathrm{m}}}{\mu_{\mathrm{g}}}\frac{M_{\mathrm{g}}p_{\mathrm{m}}}{ZRT}(p_{\mathrm{m}}-p_{\mathrm{f}})\mathrm{d}V=V_{\mathrm{c}}\sigma_{1}\frac{k_{\mathrm{m}}}{\mu_{\mathrm{g}}}\frac{M_{\mathrm{g}}p_{\mathrm{m}}}{ZRT}(p_{\mathrm{m}}-p_{\mathrm{f}}) \tag{21}$$

Since the capacity of the desorbed gas is a function of time, then according to Equation (2), the finite volume numerical calculation format of the flow governing equation in the matrix system can be rewritten as:

$$\begin{aligned}&\left[\frac{V_{\mathrm{c}}\phi_{\mathrm{m}}M_{\mathrm{g}}}{ZRT}-V_{\mathrm{c}}(1-\phi_{\mathrm{m}})\frac{\mathrm{d}V_{\mathrm{m}}}{\mathrm{d}p_{\mathrm{m}}}\right]\frac{\partial p_{\mathrm{m}}}{\partial t}-\sum_{f\sim faces(V_{\mathrm{C}})}\left(\kappa\frac{M_{\mathrm{g}}p_{\mathrm{m}}}{ZRT}\nabla p_{\mathrm{m}}\right)\cdot\mathbf{S}_{f}\\&+V_{\mathrm{c}}\sigma_{1}\frac{k_{\mathrm{m}}}{\mu_{\mathrm{g}}}\frac{M_{\mathrm{g}}p_{\mathrm{m}}}{ZRT}(p_{\mathrm{m}}-p_{\mathrm{f}})=0\end{aligned} \tag{22}$$

Similarly, the finite volume numerical calculation format for micro-fractures and artificial fractures can be obtained as

$$\frac{V_{\mathrm{c}}\phi_{\mathrm{f}}M_{\mathrm{g}}}{ZRT}\frac{\partial p_{\mathrm{f}}}{\partial t}-\sum_{f\sim faces(V_{\mathrm{C}})}\left(\frac{M_{\mathrm{g}}p_{\mathrm{f}}}{ZRT}\lambda_{\mathrm{f}}\nabla p_{\mathrm{f}}\right)\cdot\mathbf{S}_{f}-V_{\mathrm{c}}(q_{\mathrm{m-f}}-q_{\mathrm{f-h}})=0 \tag{23}$$

$$\begin{aligned}&d_{\mathrm{hf}}\left(\sum_{f\sim faces(V_{\mathrm{C}}-\mathrm{F})}\frac{S_{f}\phi_{\mathrm{hf}}M_{\mathrm{g}}}{ZRT}\right)\frac{\partial p_{\mathrm{hf}}}{\partial t}\\&-d_{\mathrm{hf}}\sum_{f\sim faces(V_{\mathrm{C}})}\sum_{b\sim bounds(f)}\left(\frac{M_{\mathrm{g}}p_{\mathrm{hf}}}{ZRT}\lambda_{\mathrm{hf}}\nabla p_{\mathrm{hf}}\right)\cdot\mathbf{S}_{\mathrm{b}}-V_{\mathrm{c}}(q_{\mathrm{f-h}}-q_{\mathrm{well}})=0\end{aligned} \tag{24}$$

3.3. Sequential Solution

The so-called sequential solution method means that each time step solves a variable firstly, and then it substitutes other variables expressions for an iterative solution. This method ensures that the amount of calculation is less than the overall solution method at each time step. Assuming that the current time step is k, then all variables related to the pressure of artificial fractures and matrix are implicitly solved using value at k + 1 time steps. Equations (22) and (24) can be written as:

$$\begin{aligned}&\left[\frac{V_c\phi_m M_g}{ZRT} - V_c(1-\phi_m)\frac{dV_m}{dp_m}\right]\frac{p_m^{k+1}-p_m^k}{\Delta t} = \sum_{f\sim faces(V_C)}\left(\kappa^{k+1}\frac{M_g p_m^{k+1}}{ZRT}\nabla p_m^{k+1}\right)\cdot\mathbf{S}_f\\&-V_c\sigma_1\lambda_m^{k+1}\frac{M_g p_m^{k+1}}{ZRT}\left(p_m^{k+1}-p_f^k\right)=0\end{aligned} \tag{25}$$

$$\begin{aligned}&d_{hf}\left(\sum_{f\sim faces(V_C-F)}\frac{S_f\phi_{hf}M_g}{ZRT}\right)\frac{p_{hf}^{k+1}-p_{hf}^k}{\Delta t}=\\&d_{hf}\sum_{f\sim faces(V_C)}\sum_{b\sim bounds(f)}\left(\frac{M_g p_{hf}^{k+1}}{ZRT}\lambda_{hf}^{k+1}\nabla p_{hf}^{k+1}\right)\cdot\mathbf{S}_b+V_c\lambda_f^{k+1}\left(p_f^k-p_{hf}^{k+1}\right)-V_c q_{well}^k\end{aligned} \tag{26}$$

Note that in the two expressions, p_f and q_{well} uses the value of the k time step—these are known values. Therefore, each of the two formulae contains an unknown variable p_{hf}^{k+1} and p_m^{k+1}. Therefore, we can iteratively solve the equation using the Newton–Raphson method. Then we substitute the results into the micro-fracture flow governing equation to solve p_f^{k+1} explicitly.

$$\begin{aligned}&\frac{V_c\phi_f M_g}{ZRT}\frac{p_f^{k+1}-p_f^k}{\Delta t}=\sum_{f\sim faces(V_C)}\left(\frac{M_g p_f^k}{ZRT}\lambda_f^k\nabla p_f^k\right)\cdot\mathbf{S}_f\\&+V_c\sigma_1\lambda_m^{k+1}\frac{M_g p_m^{k+1}}{ZRT}\left(p_m^{k+1}-p_f^k\right)-V_c\sigma_2\lambda_f^k\frac{M_g p_f^k}{ZRT}\left(p_f^k-p_{hf}^{k+1}\right)\end{aligned} \tag{27}$$

3.4. Gradient Computation

Obviously, to solve Equations (25)–(27), we need to calculate the gradient of an element field. The method adopted in this section is based on the Green-Gauss theorem, which is proven by Cengel [23] and Incropera [24] relatively straightforwardly and can be used for a variety of topologies and girds (structured/unstructured, orthogonal/non-orthogonal, etc.). The starting point is used to define the average pressure gradient over a finite volume element, as shown as Figure 3, of centroid C and volume V_C:

$$\overline{\nabla p_c} = \frac{1}{V_c}\int_{V_c}\nabla p dV \tag{28}$$

Then, using the divergence theorem, the volume integral is transformed into the surface integral

$$\overline{\nabla p_c} = \frac{1}{V_c}\int_{\partial V_c} p d\mathbf{S} \tag{29}$$

where $d\mathbf{S}$ is the surface vector pointing outward. In the case of discrete faces, Equation (29) can be rewritten as

$$\overline{\nabla p}_c V_c = \sum_{\partial V_c}\int_{face} p d\mathbf{S} \tag{30}$$

Next, the integral of a cell face is approximated by the mid-point integration rule, which is equal to the interpolated value of the field at the face centroid multiplied by the face area, resulting in

$$\overline{\nabla p}_c = \frac{1}{V_c}\sum_{f=nb(C)}\overline{p}_f\mathbf{S}_f \tag{31}$$

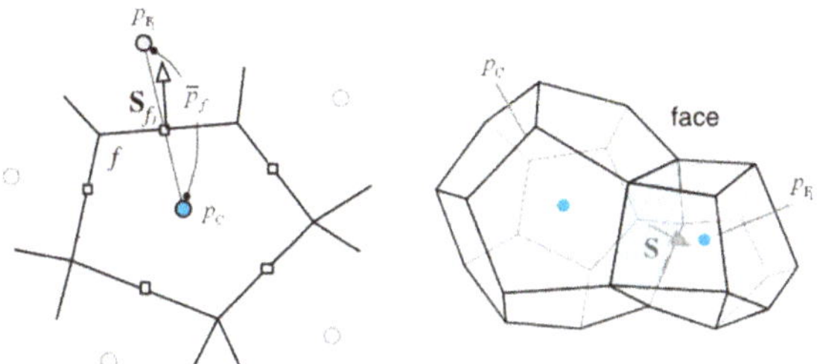

Figure 3. Gradient computation.

By reviewing Equations (30) and (31) in Figure 3, it is apparent that to calculate the average pressure gradient of the control element C, the information about the surface vector ($\mathbf{S}_f$) is needed, as well as information about the adjacent elements and the pressure values at the element centroids (p_C, p_{F_k}). This information is needed to calculate the pressure at the interface (p_f), which must be interpolated in some way.

Assuming that the pressure between the elements C and F straddling the interface f varies linearly, the approximate value for p_f, denoted by $\overline{p}_f$, can be calculated as

$$\overline{p}_f = g_F p_F + g_C p_C \tag{32}$$

Calculation of the weight factors g_F and g_C is given by

$$g_F = \frac{V_C}{V_C + V_F} g_C = \frac{V_F}{V_C + V_F} = 1 - g_F \tag{33}$$

Figure 4 considers the straddling elements; the surface vector cannot be outward at the same time, so the direction of the surface vector defined for this grid is determined by the grid index. The direction of the surface vector always points from the element which has a smaller index number to the element has a larger index number. In order to consider the vector direction, use a sign function to modify Equation (31) for the gradient as

$$\overline{\nabla p_k} = \frac{1}{V_k}\left(-\sum_{n \leftarrow \langle f = nb(C) \rangle < k} \overline{p}_n S_n + \sum_{n \leftarrow \langle f = nb(C) \rangle > k} \overline{p}_n S_n\right) \tag{34}$$

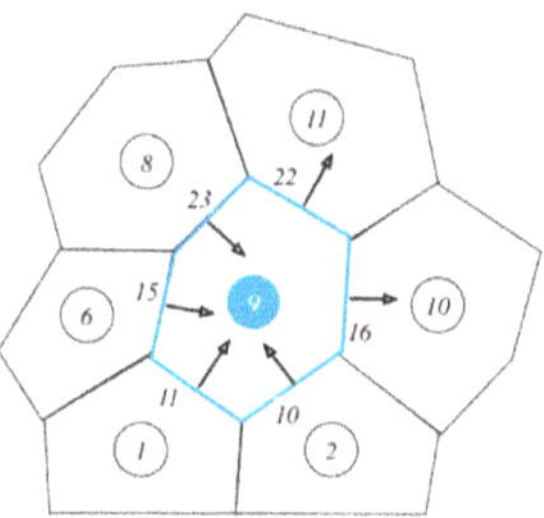

Figure 4. Element connectivity and face orientation using global indices.

3.5. Non-Orthogonality

Due to the non-structural grid used in this article, the grids are non-orthogonal. Therefore, the surface vector $\mathbf{S}_f$ and the vector **CF** connects the centroids of the elements, which straddle the interface and are not collinear. In this case, the pressure gradient perpendicular to the surface cannot be written as a function of p_C and p_F, because it has a component in the direction perpendicular to **CF**.

If **e** represents the unit vector along the direction defined by the vector **CF**, then the pressure gradient in the **e** direction can be written as

$$(\nabla p \cdot \mathbf{e})_f = \left(\frac{\partial p}{\partial e}\right)_f = \frac{p_F - p_C}{\|\mathbf{r}_F - \mathbf{r}_C\|} = \frac{p_F - p_C}{d_{CF}} \tag{35}$$

Thus, to achieve the linearization of the flux in non-orthogonal grids, the surface vector $\mathbf{S}_f$ should be written as the sum of two vectors $\mathbf{E}_f$ and $\mathbf{T}_f$, i.e.,

$$\mathbf{S}_f = \mathbf{E}_f + \mathbf{T}_f \tag{36}$$

where $\mathbf{E}_f$ is in the **CF** direction, such that part of the diffusion flux through face f can be written as a function of the nodal values p_C and p_F:

$$\begin{aligned}(\nabla p)_f \cdot \mathbf{S}_f &= (\nabla p)_f \cdot \mathbf{E}_f + (\nabla p)_f \cdot \mathbf{T}_f \\ &= E_f\left(\frac{\partial p}{\partial e}\right)_f + (\nabla p)_f \cdot \mathbf{T}_f \\ &= E_f \frac{p_F - p_C}{d_{CF}} + (\nabla p)_f \cdot \mathbf{T}_f\end{aligned} \tag{37}$$

Some researchers [25–27] give different options for the decomposition of $\mathbf{S}_f$, which are shown in Table 1.

Table 1. Different options for the decomposition of surface vector $\mathbf{S}_f$.

Option	Diagram	$\mathbf{E}_f$	$\mathbf{T}_f$
Minimum Correction Approach		$\mathbf{E}_f = (\mathbf{e}\cdot\mathbf{S}_f)\mathbf{e} = (S_f\cos\theta)\mathbf{e}$	$\mathbf{T}_f = (\mathbf{n} - \cos\theta\mathbf{e})S_f$
Orthogonal Correction Approach		$\mathbf{E}_f = S_f\mathbf{e}$	$\mathbf{T}_f = (\mathbf{n} - \mathbf{e})S_f$
Over-Relaxed Approach		$\mathbf{E}_f = \left(\frac{S_f}{\cos\theta}\right)\mathbf{e}$	$\mathbf{T}_f = \left(\mathbf{n} - \frac{1}{\cos\theta}\mathbf{e}\right)S_f$

The above methods are correct and satisfy Equation (9). These methods differ in their accuracy and stability on non-orthogonal grids. It has been found that the over-relaxed approach is the most stable, even when the grid is highly non-orthogonal.

Minimum Correction Approach

$$\mathbf{T}_f = (\mathbf{n} - \cos\theta\mathbf{e})S_f\mathbf{E}_f = S_f\mathbf{e}$$

Orthogonal Correction Approach

$$\mathbf{T}_f = (\mathbf{n} - \mathbf{e})S_f\mathbf{E}_f = \left(\frac{S_f}{\cos\theta}\right)\mathbf{e}$$

Over-Relaxed Approach

$$\mathbf{T}_f = \left(\mathbf{n} - \frac{1}{\cos\theta}\mathbf{e}\right)S_f$$

3.6. Model Verification

In this paper, a rectangular composite shale gas reservoir model considering a finite-conductivity fractured horizontal well is established. If the multi-scale flow mechanisms and the hydraulic fractures (SRV) region are ignored, the model can be applied to the multi-section fractured horizontal well of conventional dual-porosity gas reservoirs. To verify the accuracy of this method, comparisons are made with the solution using commercial software Eclipse [28]. Both simulations are applied for an 800 m long horizontal well with fifteen equally-spaced 200 m long transverse fractures in a bounded rectangular conventional reservoir. The data for the formation and well properties used in the simulations are shown in Table 2.

Table 2. Basic data of a multi-section fractured horizontal well in a single-porosity gas reservoir [27].

Parameter	Unit	Value
Reservoir length	m	2000
Reservoir width	m	2000
Reservoir height	m	50
Horizontal well length	m	800
Artificial fracture height	m	40
Wellbore radius, r_w	m	0.1
Matrix permeability, k_m	mD	0.5
Matrix porosity, ϕ_m	-	0.05
Artificial fracture number	-	15
Artificial fracture length, l	m	200
Artificial fracture spacing	m	50
Artificial fracture permeability	mD	100

As shown in Figure 5, the gas production curve for a constant wellbore pressure obtained from our method is in good agreement with those from commercial software.

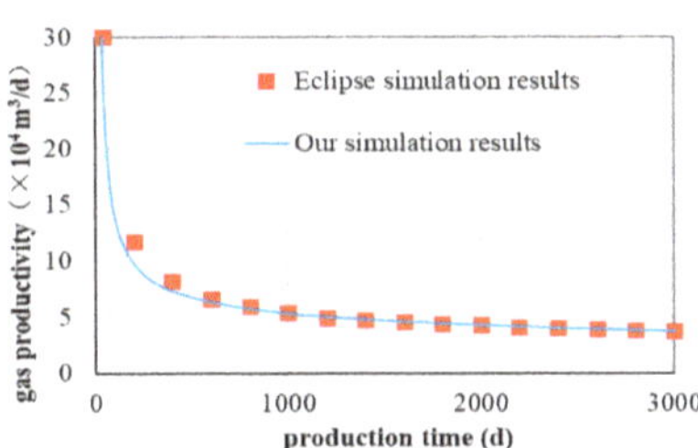

Figure 5. Comparison of the gas production calculated using finite volume method and commercial software Eclipse.

However, at the early non-steady flow period, the numerical method may produce a calculation error caused by the mesh precision. If dense grids around the artificial fracture system are used, the precision of this method can be further improved.

4. Example Simulation

4.1. Model Parameters

In this section, we simplify the shale gas reservoir with complex micro-scale fractures into a combination of as a dual porosity continuum media and a discrete fracture media. Based on the discrete fracture model, the artificial fracture can be simplified as a surface element by using a reduction dimensional method. The data for the formation and well properties used in the simulations are shown in Table 3.

Table 3. The parameter set for the shale gas reservoir model.

Reservoir Length	m	1200
Reservoir width	m	800
Reservoir height	m	100
Horizontal well length	m	800
Artificial fracture height	m	100
Bottom hole pressure, p_{wf}	MPa	5
Original formation pressure, p_{i}	MPa	30
Gas deviation factor, Z	-	0.93
Universal gas constant, R	J/(mol·K)	8.314
Formation temperature, T	K	343.15
Porosity of the matrix, ϕ_{m}	-	0.05
Porosity of the microfracture, ϕ_{f}	-	0.005
Porosity of the artificial fracture, ϕ_{hf}	-	0.1
Permeability of the matrix, k_{m}	mD	0.01
Permeability of the microfracture, k_{f}	mD	0.1
Permeability of the artificial fracture, k_{hf}	mD	150
Langmuir volume, V_{L}	m^3/kg	0.004
Langmuir pressure, p_{L}	MPa	5
Shale density, ρ_{m}	kg/m^3	2600
Gas molar mass, M_{g}	g/mol	16
Standard gas molar volume, V_{std}	$\mathrm{m}^3/\mathrm{mol}$	0.0024
Gas viscosity, μ_{g}	mPa·s	0.185
Constrictivity and of the shale matrix, δ^{m}	-	1.2
Tortuosity of the shale matrix, τ^{m}	-	1.5
Gas compression factor, c_{g}	1/Pa	4.39×10^{-8}

The spatial arrangement of multi-section fractured horizontal wells is shown in Figure 6. The half-length of the artificial fractures is 200 m.

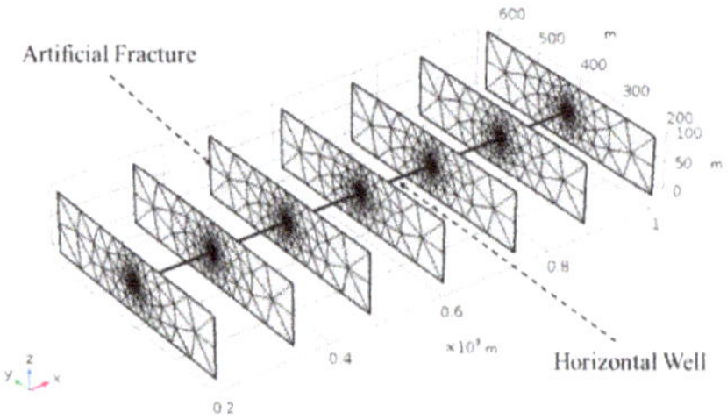

Figure 6. A numerical model of a multi-section fractured horizontal well.

4.2. Results Analysis

4.2.1. Pressure Distribution in Artificial Fractures

Figure 7 shows the pressure distribution in the artificial fractures at the beginning of production. The pressure distribution in the artificial fractures is related to parameters such as fracture aperture and permeability. As can be seen from the figure, due to the high conductivity of the artificial fractures, the pressure in the fracture rapidly decreases. A drawdown pressure is created between the artificial fractures and the matrix–micro-fracture system, so that gas flows from the matrix–micro-fracture system into artificial fractures and gas is produced by production well.

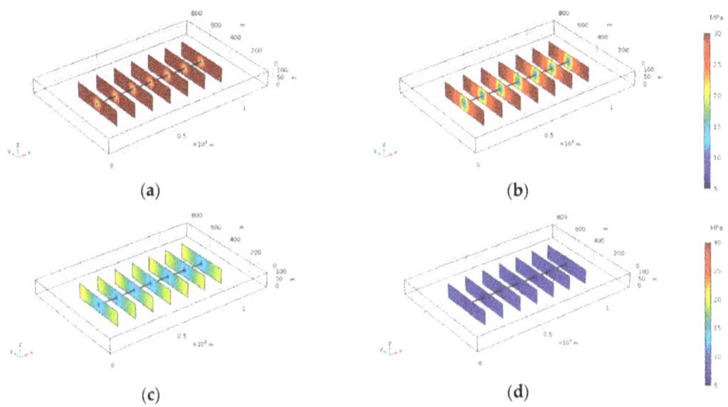

Figure 7. Pressure distribution in artificial fractures. (**a**) T = 0 d; (**b**) T = 1 d; (**c**) T = 2 d; (**d**) T = 3 d.

4.2.2. Gas Desorption Process

Based on the physical model parameters, the production of shale gas multi-section fractured horizontal wells is simulated. It has been shown in Figure 8 that during the first three years of production, the decline of pressure in the reservoir is mainly concentrated in the area that is near the wellbore and the hydraulic fracture faces, while the pressure drop in the outer area is very small. It shows that the produced gas mainly comes from free gas and desorption gas in the stimulated volume.

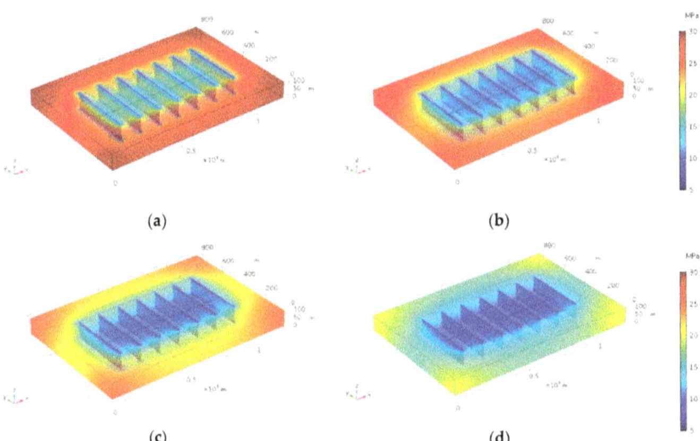

Figure 8. Reservoir pressure distribution at different production times. (**a**) T = 1a; (**b**) T = 3a; (**c**) T = 5a; (**d**) T = 10a.

Figure 9 shows the average reservoir pressure, gas production rate, and cumulative gas production at different Langmuir volumes. We found that the desorption process has the effect of supplementing the reservoir pressure, but the effect is not significant. Since the gas production rate is affected not only by the physical properties of the reservoir, but also by the pressure distribution, the gas desorption process has limited supplementary effects on pressure, and the impact on the gas production rate is not significant. At the same time, as the Langmuir volume increases, the cumulative gas production gradually increases.

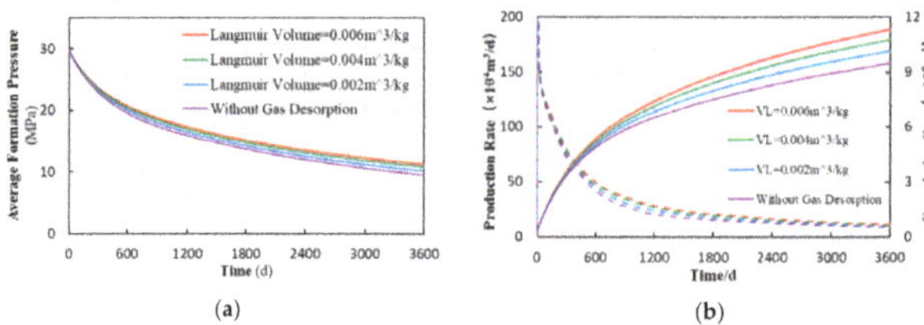

Figure 9. Influence of gas desorption on horizontal well productivity. (**a**) Average reservoir pressure at different Langmuir volumes; (**b**) Gas production rate and cumulative gas production at different Langmuir volumes.

4.2.3. The Klinkenberg effect and Diffusive Gas Transport

Figure 10a–c show the Knudsen number distribution of shale gas reservoirs in fractured horizontal wells at the same time of production, under different shale matrix permeabilities. As can be seen from the figure, the closer to the artificial fractures, the larger the Knudsen number. This is due to the negative correlation between Knudsen number and pressure, so the lower the pressure, the larger the Knudsen number. At the same time, when the shale permeability decreases, the pressure drop of the artificial fractures becomes larger and the pressure drop funnel becomes steeper. Therefore, the closer the pressure gets to artificial fractures, the greater the increase of the Knudsen number.

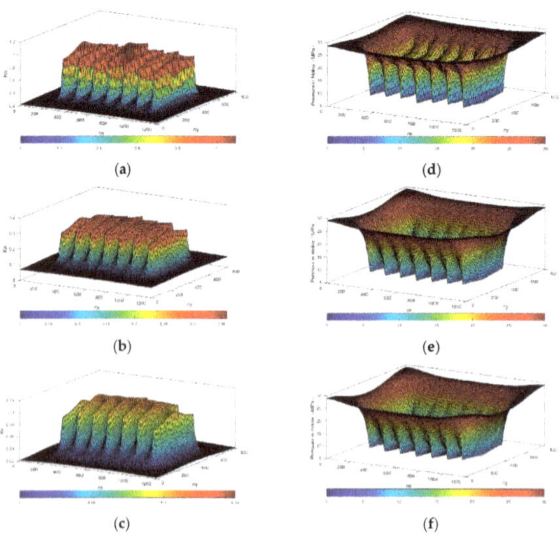

Figure 10. Field distribution of Knudsen number and matrix pressure in shale gas reservoir. (**a**) $k_m = 10^{-5}$ mD, Kn; (**b**) $k_m = 10^{-4}$ mD, Kn; (**c**) $k_m = 10^{-3}$ mD, Kn; (**d**) $k_m = 10^{-5}$ mD, p_m; (**e**) $k_m = 10^{-4}$ mD, p_m; (**f**) $k_m = 10^{-3}$ mD, p_m.

Figure 10d–f show the matrix pressure distribution of shale gas reservoirs in fractured horizontal wells at the same time of production, under different shale matrix permeabilities. It can be seen from the figure that the pressure of the artificial fractures falls fastest, and the closer to artificial fractures, the lower reservoir pressure. Comparing the reservoir pressures under different shale permeability conditions, the lower the shale permeability, the faster the pressure of the artificial fractures drops and the fewer reservoirs are used, resulting in steeper pressure drop funnels. This is because for shale reservoirs with low permeability, it is difficult for gas to flow in such dense porous media, so the gas stored in the shale cannot be added to the artificial fractures in time when the gas in the fracturing fractures. When the gas in the artificial fractures is recovered, the pressure in the fracture rapidly decreases. Compared to shales containing nano-micro pores, gases stored in the fractures and the region near fractures are more likely to be produced to make the pressure drop faster.

Figure 11 shows the curve of the gas production rate and cumulative gas production for multi-section fractured horizontal wells with different shale permeability. As can be seen from the figure, the gas production rate and cumulative gas production increase with the increase of shale permeability, and the growth rate also increases. However, compared with the production rate and cumulative production (without considering diffusion and slippage effects), the increment of gas production rate and cumulative production (considering the diffusion and slippage effects), decreases with increasing shale permeability. This shows that when shale permeability becomes smaller (pore size decreases), Knudsen diffusion and slippage effects have a greater impact on the daily gas production and the cumulative production of fracturing horizontal wells.

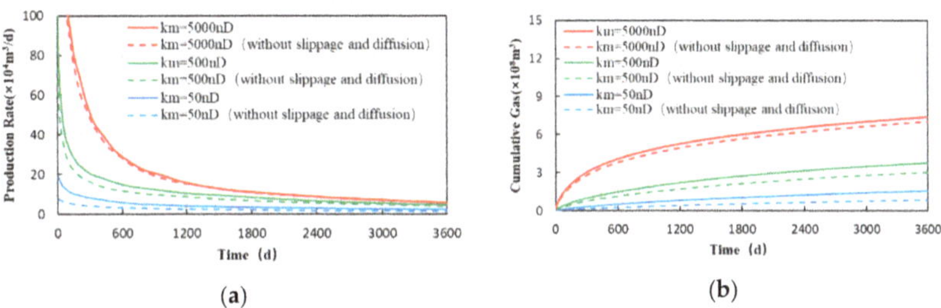

Figure 11. The effect of Knudsen diffusion and the Klinkenberg effect on productivity. (**a**) Production rate at different matrix permeability; (**b**) Cumulative gas at different matrix permeability.

4.2.4. Artificial Fracture Morphology

Based on the above numerical model, we change the number of fractured sections (Figure 12a–c) and the half-length of artificial fractures (Figure 12d–f) to simulate the production of shale gas. It can be seen from Figure 12g that as the half-length of artificial fractures increases, the gas production rate and cumulative gas production also increase. However, the increasing rate in the gas production rate and cumulative gas production has gradually decreased. The main reason for this is that as the half-length of the artificial fractures increases, the multi-fracture interference becomes severer. Therefore, as the half-length of artificial fractures increases, the increasing rate in the gas production rate and cumulative gas production decreases.

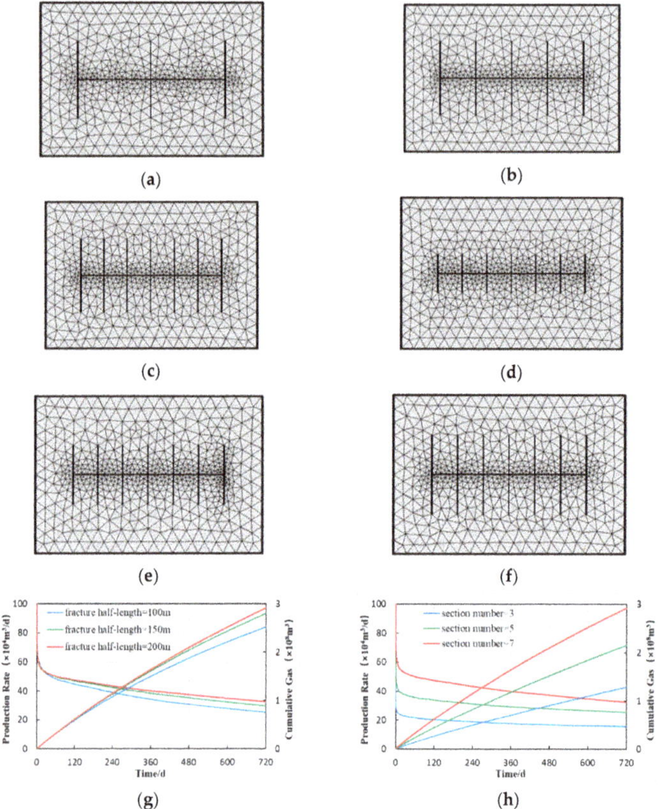

Figure 12. The effect of horizontal well parameters on productivity. (**a**) section number = 3; (**b**) section number = 5; (**c**) section number = 7; (**d**) fracture half-length = 100 m; (**e**) fracture half-length = 150 m; (**f**) fracture half-length = 200 m; (**g**) Three fractured sections, T = 5a; (**h**) Four fractured sections, T = 5a.

From Figure 12h, it can be seen that the number of sections has an important influence on the gas production rate and cumulative gas production. With the increase in the section number, the gas production rate and cumulative gas production also increase. It is worth noting that as the section number increases, the rate of decline in gas production rate also increases. Similar to the previous situation, the main reason is that with the increase of the section number, the multi-fracture interference becomes severer. As a result, the larger section number, the faster the gas production rate declines.

Through analysis, it is found that excessive half-length of fractures and section numbers will generate strong multi-fracture interference, which will have a negative impact on the productivity of horizontal wells. Therefore, for a horizontal well fracturing design, the half-length of fractures and section number should not be pursued blindly, but the parameters of horizontal wells should be optimized to reduce the multi-fracture interference.

5. Conclusions

In this paper, based on the matrix–micro-fracture continuous dual model and discrete fracture model, a mathematical model of the shale gas reservoir considering a multi-scale flow mechanisms is established.

The numerical calculation format using a cell-centered variable arrangement of shale gas three-dimensional flow based on the finite volume element method is deduced. In this case, the variables and their associated quantities are stored in the centroids of the control elements. Thus, the elements are the same as the discretization elements; in general, the method is second-order accurate, because all quantities are calculated at the element and face centroids. Talyor series expansion can be used to reconstruct the variations within the cell. Another advantage of the cell-centered formulation is that it allows the use of general polygonal elements without the need for pre-defined shape functions. This permits a straightforward implementation of a full multigrid strategy.

The artificial fracture is expressed by the two-dimensional surface, and the wellbore is expressed by a one-dimensional solid based on the dimension reduction method. The finite volume element method is used to solve the multi-section fractured horizontal well productivity and pressure distribution.

Through the analysis of the simulation results, it is found that the model can reflect the initial production of shale gas and its characteristics of rapid decline. The analysis shows that the gas desorption of shale gas has a great impact on reserves, which in turn have a supplementary effect on the reservoir pressure. On one hand, with the prolongation of production time, the proportion of desorption is increased. On the other hand, shale gas production is mainly affected by the scope of stimulated volume.

According to the development process of shale gas reservoirs, a numerical model of a stimulated reservoir volume fractured horizontal well is established. The analysis shows that the pressure will rapidly decrease in artificial fractures. The desorption process has a great influence on the geological reserves, but has a limited impact on the productivity of horizontal wells. With the decrease of the pore size of the matrix, the Klinkenberg effect and gas diffusion have a greater impact on shale gas productivity. When the matrix permeability is greater than 0.01 mD, those flow mechanisms has no significant effect on the productivity. Compared with the fracture half-length, the section number has a greater impact on the productivity of shale gas. However, the excessive half-length of the fracture and the section number all induce multi-fracture interference. Therefore, the horizontal well parameters need to be optimized.

The parameters of the artificial fracture network can be conveniently adjusted and the factors affecting the productivity can be analyzed. The research content of this paper has certain theoretical and practical significance for the volume fractured design of shale gas reservoirs and the reasonable evaluation of production capacity.

Author Contributions: Conceptualization, X.C.; Data curation, C.T.; Formal analysis, C.T.; Investigation, C.T.; Resources, X.C.; Software, P.Y.; Supervision, Z.D.; Validation, J.W.; Writing–original draft, C.T.; Writing–review & editing, J.W.

Funding: This work was supported by the visiting scholar program by the China Scholarship Council (No. 201608515035) and Texas Tech University, National Major Projects China (No. 2016ZX05048-002), National Major Projects China (No. 2016ZX05010-002-002), The Fund of SKL of Petroleum Resources and Prospecting, Beijing (No. PRP/open 1501), and the National Natural Science Foundation of China (Grant No. 51474179).

Conflicts of Interest: We declare that we do not have any commercial or associative interest that represents a conflict of interest in connection with the work submitted.

Nomenclature

p_{wf}	MPa	flowing bottom hole pressure
p_{i}	MPa	original reservoir pressure
ρ_{g}	$\mathrm{kg/m^3}$	gas density
ρ_{gm}	$\mathrm{kg/m^3}$	gas density in the matrix
ρ_{gf}	$\mathrm{kg/m^3}$	gas density in the fracture
Z	-	gas deviation factor
R	J/(mol·K)	universal gas constant
T	K	reservoir temperature
$\mathbf{v}_{\mathrm{m}}$	m/s	apparent gas velocity
$\mathbf{v}_{\mathrm{m}}^{\mathrm{v}}$	m/s	corrected gas velocity considering the Klinkenberg effect
$\mathbf{v}_{\mathrm{m}}^{\mathrm{k}}$	m/s	corrected gas velocity considering the diffusive transport
ϕ_{m}	-	porosity of matrix
ϕ_{f}	-	porosity of micro-fracture
ϕ_{hf}	-	porosity of artificial fracture
k_{m}	mD	permeability of matrix
k_{f}	mD	permeability of micro-fracture
k_{hf}	mD	permeability of artificial fracture
V_{L}	$\mathrm{m^3/kg}$	Langmuir volume
p_{L}	MPa	Langmuir pressure
ρ_{m}	$\mathrm{kg/m^3}$	density of shale
M_{g}	g/mol	molecular mass
V_{std}	$\mathrm{m^3/mol}$	standard molar volume
μ_{g}	mPa·s	gas viscosity
μ_{eff}	mPa·s	gas effective viscosity
α	-	rarefaction coefficient
K_{n}	-	Knudsen number
δ^{m}	-	constrictivity of the shale matrix
τ^{m}	-	tortuosity of the shale matrix
c_{g}	1/Pa	gas compression factor
D_{g}	$\mathrm{m^2/s}$	Knudsen molecule diffusivity
q_{m}	$\mathrm{m^3/s}$	desorption gas flow
$q_{\mathrm{m-f}}$	$\mathrm{m^3/s}$	cross-flow rate
V_{m}	$\mathrm{m^3}$	adsorption capacity of per unit volume matrix
p_{m}	MPa	pressure of matrix
p_{f}	MPa	pressure of micro-fracture
p_{hf}	MPa	pressure of artificial fracture
d_{hf}	-	dimensionless fracture aperture
q_{well}	$\mathrm{m^3/s}$	well production
f	-	faces of control element
b	-	bounds of face
$\mathbf{S}_f$	-	surface vector of faces
$\mathbf{S}_b$	-	surface vector of bounds
$\mathbf{n}$	-	unit normal vector

References

1. Kucuk, F.; Sawyer, W.K. Transient Flow in Naturally Fractured Reservoirs and Its Application to Devonian Gas Shales. In Proceedings of the SPE Annual Technical Conference and Exhibition, Dallas, TX, USA, 21–24 September 1980.
2. Bumb, A.C.; McKee, C.R. Gas-well testing in the presence of desorption for coalbed methane and devonian shale. *SPE Form. Eval.* **1988**, *3*, 179–185. [CrossRef]

3. Carlson, E.S.; Mercer, J.C. Devonian Shale Gas Production: Mechanisms and Simple Models. *J. Pet. Technol.* **1991**, *43*, 476–482. [CrossRef]
4. Swami, V. Shale Gas Reservoir Modeling: From Nanopores to Laboratory. In Proceedings of the SPE Annual Technical Conference and Exhibition, San Antonio, TX, USA, 8–10 October 2012.
5. Saputelli, L.; Lopez, C.; Chacon, A.; Soliman, M. Design Optimization of Horizontal Wells with Multiple Hydraulic Fractures in the Bakken Shale. In Proceedings of the SPE/EAGE European Unconventional Resources Conference and Exhibition, Vienna, Austria, 25–27 February 2014.
6. Rubin, B. Accurate Simulation of Non Darcy Flow in Stimulated Fractured Shale Reservoirs. In Proceedings of the SPE Western regional meeting, Anaheim, CA, USA, 27–29 May 2010.
7. Klimkowski, Ł.; Nagy, S. Key factors in shale gas modeling and simulation. *Arch. Min. Sci.* **2014**, *59*. [CrossRef]
8. Mirzaei, M.; Cipolla, C.L. A Workflow for Modeling and Simulation of Hydraulic Fractures in Unconventional Gas Reservoirs. *Geol. Acta* **2012**, *10*, 283–294.
9. Yao, J.; Wang, Z.; Zhang, Y.; Huang, Z.Q. Numerical simulation method of discrete fracture network for naturally fractured reservoirs. *Acta Pet. Sin.* **2010**, *31*, 284–288.
10. Hoteit, H.; Firoozabadi, A. Compositional Modeling of Fractured Reservoirs without Transfer Functions by the Discontinuous Galerkin and Mixed Methods. *SPE J.* **2004**, *11*, 341–352. [CrossRef]
11. Moinfar, A.; Narr, W.; Hui, M.H.; Mallison, B.T.; Lee, S.H. Comparison of Discrete-Fracture and Dual-Permeability Models for Multiphase Flow in Naturally Fractured Reservoirs. In Proceedings of the SPE Reservoir Simulation Symposium, The Woodlands, TX, USA, 21–23 February 2011.
12. Kuuskraa, V.A.; Wicks, D.E.; Thurber, J.L. Geologic and Reservoir Mechanisms Controlling Gas Recovery from the Antrim Shale. In Proceedings of the SPE Annual Technical Conference and Exhibition, Washington, DC, USA, 4–7 October 1992.
13. Schepers, K.C.; Gonzalez, R.J.; Koperna, G.J.; Oudinot, A.Y. Reservoir Modeling in Support of Shale Gas Exploration. In Proceedings of the Latin American and Caribbean Petroleum Engineering Conference, Cartagena de Indias, Colombia, 31 May–3 June 2009.
14. Dehghanpour, H.; Shirdel, M. A Triple Porosity Model for Shale Gas Reservoirs. In Proceedings of the Canadian Unconventional Resources Conference, Calgary, AB, Canada, 15–17 November 2011.
15. Wu, Y.S.; Moridis, G.; Bai, B.; Zhang, K. A Multi-Continuum Model for Gas Production in Tight Fractured Reservoirs. In Proceedings of the SPE Hydraulic Fracturing Technology Conference, The Woodlands, TX, USA, 19–21 January 2009.
16. Aboaba, A.L.; Cheng, Y. Estimation of Fracture Properties for a Horizontal Well With Multiple Hydraulic Fractures in Gas Shale. In Proceedings of the SPE Eastern Regional Meeting, Morgantown, WV, USA, 13–15 October 2010.
17. Wang, H.T. Performance of multiple fractured horizontal wells in shale gas reservoirs with consideration of multiple mechanisms. *J. Hydrol.* **2014**, *510*, 299–312. [CrossRef]
18. Fang, W.; Jiang, H.; Li, J.; Wang, Q.; Killough, J.; Li, L.; Peng, Y.; Yang, H. A numerical simulation model for multi-scale flow in tight oil reservoirs. *Pet. Explor. Dev.* **2017**, *44*, 415–422. [CrossRef]
19. Javadpour, F. Nanopores and Apparent Permeability of Gas Flow in Mudrocks (Shales and Siltstone). *J. Can. Pet. Technol.* **2009**, *48*, 16–21. [CrossRef]
20. Ferziger, J.H.; Perić, M. Computational Methods for Fluid Dynamics. *Phys. Today* **1997**, *50*, 80–84. [CrossRef]
21. Caumon, G.; Collon-Drouaillet, P.L.C.D.; De Veslud, C.L.C.; Viseur, S.; Sausse, J. Surface-Based 3D Modeling of Geological Structures. *Math. Geosci.* **2009**, *41*, 927–945. [CrossRef]
22. Juanes, R.; Samper, J.; Molinero, J. A general and efficient formulation of fractures and boundary conditions in the finite element method. *Int. J. Numer. Methods Eng.* **2002**, *54*, 1751–1774. [CrossRef]
23. Çengel, Y.A.Y. *Cengel, Heat and Mass Transfer: A Practical Approach*, 3rd ed.; McGraw-Hill: Singapore, 2006.
24. Incropera, F.P. *Fundamentals of Heat and Mass Transfer*; John Wiley & Sons: Jefferson City, MO, USA, 2007.
25. Patankar, S.V. *Numerical Heat Transfer and Fluid Flow*; CRC Press: New York, NY, USA, 1980.
26. Darwish, M.; Moukalled, F. A new approach for building bounded skew-upwind schemes. *Comput. Methods Appl. Mech. Eng.* **1996**, *129*, 221–233. [CrossRef]

27. Spekreijse, S. Multigrid solution of monotone second-order discretizations of hyperbolic conservation laws. *Math. Comput.* **1987**, *49*, 135–155. [CrossRef]
28. Fan, D.; Yao, J.; Sun, H.; Zeng, H. Numerical simulation of multi-fractured horizontal well in shale gas reservoir considering multiple gas transport mechanisms. *Chin. J. Theor. Appl. Mech.* **2015**, *47*, 906–915.

Article

A Numerical Study on the Diversion Mechanisms of Fracture Networks in Tight Reservoirs with Frictional Natural Fractures

Daobing Wang [1,2,†,‡], Fang Shi [3,‡], Bo Yu [1,*], Dongliang Sun [1], Xiuhui Li [4], Dongxu Han [1] and Yanxin Tan [4]

1 School of Mechanical Engineering, Beijing Key Laboratory of Pipeline Critical Technology and Equipment for Deepwater Oil & Gas Development, Beijing Institute of Petrochemical Technology, Beijing 102617, China; upcwdb@bipt.edu.cn (D.W.); sundongliang@bipt.edu.cn (D.S.); handongxubox@bipt.edu.cn (D.H.)

2 School of Aeronautic Science and Engineering, Beihang University, Beijing 100083, China

3 Jiangsu Key Laboratory of Advanced Manufacturing Technology, Huaiyin Institute of Technology, Huai'an 223003, China; shifang@hyit.edu.cn

4 The Conventional Natural Gas Research Institute, China Univeristy of Petroleum, Beijing 102249, China; lixiuhui031@foxmail.com (X.L.); tyx2285@foxmail.com (Y.T.)

* Correspondence: 0020150031@bipt.edu.cn; Tel.: +86-10-8129-2136

† Current address: 19 Qing-yuan North Road, Huang-cun, Da-xing District, Beijing 102617, China.

‡ These authors contributed equally to this work.

Received: 25 September 2018; Accepted: 31 October 2018; Published: 5 November 2018

Abstract: An opened natural fracture (NF) intercepted by a pressurized hydro-fracture (HF) will be diverted in a new direction at the tips of the original NF and subsequently form a complex fracture network. However, a clear understanding of the diversion behavior of fracture networks in tight reservoirs with frictional NFs is lacking. By means of the extended finite element method(XFEM), this study investigates the diversion mechanisms of an opened NF intersected by an HF in naturally fractured reservoirs. The factors affecting the diversion behavior are intensively analyzed, such as the location of the NF, the horizontal principal stress difference, the intersection angle between HF and NF, and the viscosity of the fracturing fluid. The results show that for a constant length of NF (7 m): (1) the upper length of the diverted fracture (DF) decreases by about 2 m with a 2 m increment of the upper length of NF (L_{upper}), while the length of DF increases 9.06 m with the fluid viscosity increased by 99 mPa·s; (2) the deflection angle in the upper parts increases by 30.8° with the stress difference increased by 5 MPa, while the deflection angle increases by 61.2° with the intersection angle decreased by 30°. It is easier for the opened NF in lower parts than that in upper parts to be diverted away from its original direction. It finally diverts back to the preferred fracture plane (PFP) direction. The diversion mechanisms of the fracture network are the results of the combined action of all factors. This will provide new insight into the mechanisms of fracture network generation in tight reservoirs with NFs.

Keywords: hydraulic fracturing; tight reservoirs; fracture diversion; extended finite element method; fracture network

1. Introduction

With the technological progress in petroleum industries, petroleum engineers are increasingly concerned with the exploration and development of tight reservoirs in recent years. Due to the ultra-low matrix permeability, hydraulic fracturing is a key technology for enhancing the recovery of tight hydrocarbon reservoirs [1–11]. Activation of preexisting natural fractures (NFs) during fracturing treatment is favorable for creating complex fracture networks. The interaction between a hydraulic fracture (HF) and an NF is a complex coupled process, which involves rock deformation, fluid flow, and fracture diversion [12–21].

When an HF intercepts an NF during a hydro-fracking treatment, three scenarios—arrest, offset, and cross—are observed. Renshaw and Pollard developed a criterion to describe the mechanical NF–HF interaction when they were perpendicular to each other [22]. Gu et al. afterwards proposed an extended Renshaw–Pollard criterion at nonorthogonal intersection angles on the basis of the experimental results of hydraulic fracturing for Colton sandstone [23]. This crossing criterion has been extensively applied in the mathematical models of stimulated reservoir volume (SRV) fracturing in shale gas wells. Various numerical techniques such as finite difference, discrete element, and finite element methods have been presented to investigate the mechanical interaction between HF and NF [16,24–26]. Based on the finite element software ABAQUS 6.14, Chen et al. developed a cohesive zone finite element-based model to investigate the NF–HF interaction complexity, which took into account the interface friction of weak planes [24]. Based on the discrete element method (DEM) model, Zou et al. numerically investigated HF network propagation in shale formations, and the plastic deformation in hydraulic fracturing was considered [25]. Wu and Wong used a numerical manifold method (NMM) to capture the strong discontinuity across the crack face, and this method could smoothly handle the problems of fracture network propagation [26]. However, the above-mentioned numerical methods have the drawback that crack paths should be predefined a priori. Therefore, crack cannot be freely extended on mesh grids if the direction of crack propagation is not known in advance. By means of a diffusive phase-field modeling approach, Heider et al. introduced a numerical framework of HF in tight rocks, but this simulation could be time-consuming because it requires a very fine mesh [27].

The extended finite element method (XFEM), which introduces additional enrichment functions to account for the jump across the crack surfaces and the singularity of stress in the vicinity of crack tips, provides a powerful tool to simulate the hydraulic fracturing problem. Its great advantage is that crack propagation is not mesh-dependent. Some scholars such as Dahi-Taleghani, Mao, and Gordeliy have done a great deal of innovative research on hydraulic fracturing simulation [28–32] in past decades, but some assumptions, such as a constant fluid pressure, were made to deal with the mechanical NF–HF interaction in order to simplify the complex precess. Recently, Shi and Wang successfully modeled the connection of two cracks by means of additional junction enrichment and by sharing pore pressure nodes at intersection points [33,34]. Using the combined method of XFEM and DEM, Ghaderi et al. concluded that the tensile and shear breakage of NFs were a function of angle and distance from an induced fracture [35]. Paul et al. developed a mixed linear cohesive law, which relies on a stable mortar formalism, and utilized the XFEM method to simulate the non-planar HF propagation [36]. Based on the XFEM technique, Remij et al. applied the enhanced local pressure (ELP) model to investigate crack interaction in hydraulic fracturing by assuming multiple discontinuities in the domain [37]. Vahab and Khalili used an XFEM penalty method, which was embedded in Kuhn–Tucker inequalities, to model multi-zone fracking treatments within saturated porous media [38].

It is well known that an HF usually has a non-planar crack growth (refereed to fracture diversion) by a stress shadow effect [39–44]. An opened NF intercepted by an HF will be diverted in a new direction at the tips of the original NF and subsequently form a complex fracture network [33,34,45,46]. However, a clear understanding of the diversion behavior of fracture network in tight reservoirs with frictional NFs is lacking [47]. In particular, the effect of factors such as the location of the NF, the horizontal stress difference, and the intersection angle between HF and NF on the mechanical diversion behavior of HFs is not clear at present. Therefore, with the XFEM technique [48–51], a numerical simulation on the diversion mechanisms of a fracture network in naturally fractured reservoirs was studied. This study focuses on the diversion propagation behavior in the vicinity of the two crack tips of the opened NF after an HF intersects with an NF. This will provide new insight on the mechanisms of fracture network formation in tight formations with pre-existed frictional NFs.

2. Problem Formulation

2.1. Governing Equations of Hydraulic Fracturing Problems

As shown in Figure 1, the domain Ω denotes a tight reservoir, which includes an HF and an NF. The injection point is located on the middle left of the domain Ω, and the corresponding pump rate is denoted by Q_0. As is known, the hydraulic fracturing problem is essentially a fluid–solid interaction process, so its governing equations consist of two parts: the stress equilibrium equation for rock skeleton and fluid pressure equation in the hydraulically driven fracture [52–54].

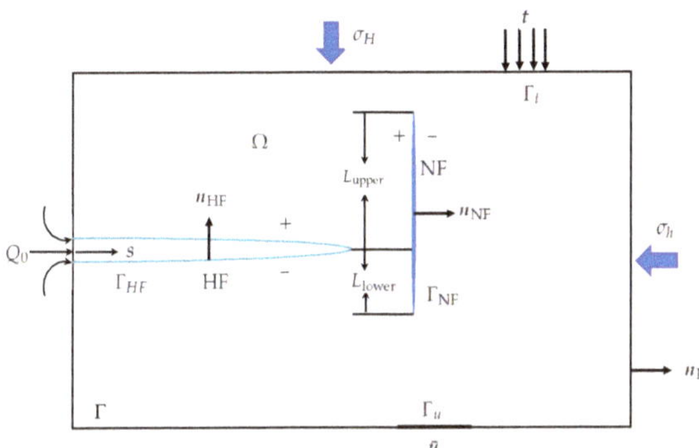

Figure 1. Schematic diagram of a hydro-fracture (HF) intersection with a natural fracture (NF).

(1) The Stress Equilibrium Equation: According to the theory of elasticity, the stress equilibrium equation is expressed:

$$\nabla \cdot \boldsymbol{\sigma} + \boldsymbol{b} = \boldsymbol{0} \quad in \Omega, \tag{1}$$

where $\boldsymbol{\sigma}$ denotes the stress tensor; $\boldsymbol{b}$ denotes the body force vector in the rock skeleton. As shown in Figure 1, the boundary conditions are composed of a displacement boundary condition (Γ_u) and a force boundary condition (Γ_t). They are expressed as

$$\begin{cases} \boldsymbol{u} = \bar{\boldsymbol{u}} & on \Gamma_u, \\ \boldsymbol{\sigma} \cdot \boldsymbol{n} = \boldsymbol{t}, & on \Gamma_t, \\ \boldsymbol{\sigma} \cdot \boldsymbol{n}_{HF} = p \boldsymbol{n}_{HF}, & on \Gamma_{HF}, \\ \boldsymbol{\sigma} \cdot \boldsymbol{n}_{NF} = \boldsymbol{t}_{NF}, & on \Gamma_{NF}, \end{cases} \tag{2}$$

where p denotes the fluid pressure on the artificial fracture; $\boldsymbol{t}_{NF}$ denotes the contact traction vector on the NF surface Γ_{NF}; $\bar{\boldsymbol{u}}$ denotes the displacement imposed on the boundary Γ_u; $\boldsymbol{t}$ denotes the traction vector imposed on the boundary Γ_t.

For brittle rocks, there is a linear relationship between stress tensor and strain tensor under a small deformation assumption, so the corresponding constitutive equation is expressed as

$$\boldsymbol{\sigma} = \boldsymbol{D} : \boldsymbol{\varepsilon}, \tag{3}$$

where $\boldsymbol{D}$ denotes the fourth elasticity tensor; $\boldsymbol{\varepsilon}$ denotes the strain tensor; the symbol ":" denotes the double dot product of the two tensors.

Under the assumption of small deformation, the relationship between displacement vector $\boldsymbol{u}$ and strain tensor $\boldsymbol{\varepsilon}$ are as follows:

$$\boldsymbol{\varepsilon} = \frac{1}{2}[\nabla \boldsymbol{u} + (\nabla \boldsymbol{u})^T]. \tag{4}$$

(2) Fluid Pressure in an HF: Under lubrication theory assumptions, the velocity profile of the fluid in the HF is that of a planar Poiseuille flow between two parallel plates. Therefore, the fluid pressure in the HF can be expressed as [15,33,34,55,56]

$$\frac{\partial w}{\partial t}-\frac{\partial}{\partial s}(k\frac{\partial p}{\partial s})=0, \tag{5}$$

where w denotes the fracture opening of HF; s denotes the crack propagation direction; t denotes the injection time; k denotes the fracture transmissivity.

According to the cubic law, the fracture transmissivity can be expressed as

$$k=\frac{w^3}{12\mu}, \tag{6}$$

where μ denotes the viscosity of fracturing fluid.

The corresponding initial and boundary conditions can be expressed as

$$\begin{cases} w(s,0)=0, \\ w(s_{tip},0)=0, \\ q(0,t)=Q_0, \\ q(s_{tip},0)=0, \end{cases} \tag{7}$$

where s_{tip} denotes the tips of HFs; q denotes the injection rate of fracturing fluid at the crack point s and time t.

It is can be seen that Equation (7) satisfies mathematically the Neumann boundary condition. In order to get a unique solution for the fluid pressure equation, the constraint condition should be additionally imposed. The necessary condition, i.e., the conservation of global mass in the HF can be written as

$$\int_0^{s_{tip}} w ds-\int_0^t Q_0 dt=0. \tag{8}$$

2.2. Crack Propagation Criterion

According to the theory of fracture mechanics, the maximum circumferential stress criterion is adopted to determine the propagation direction of a hydraulically driven fracture at every time t. The artificial fracture will propagate along a direction perpendicular to the maximum circumferential stress. If the stress intensify factor K is no less than the fracture toughness of rock skeleton K_{IC}, the crack will propagate along a certain direction. The interaction integral method in domain form is utilized to calculate the stress intensity factors K_{I} and K_{II}. The following equation of the interaction integral can be written as [57]

$$I^{(1,2)}=\int_A[\sigma_{ij}^{(1)}\frac{\partial u_i^2}{\partial x_1}+\sigma_{ij}^{(2)}\frac{\partial u_i^1}{\partial x_1}-W^{(1,2)}\delta_{1j}]\frac{\partial q_w}{\partial x_j}dA, \tag{9}$$

where $I^{(1,2)}$ denotes the interaction integral; $W^{(1,2)}$ denotes the interaction strain energy as follows; $q_w(\boldsymbol{x})$ denotes the smooth weighting function, which takes a value from 0 to 1; δ denotes the Kronecker symbol; the superscripts (1) and (2), respectively, denote the current state and the auxiliary state for the stress and strain field. The corresponding calculation procedures can be described in detail in [40,57,58]

$$W^{(1,2)}=\sigma_{ij}^{(1)}\varepsilon_{ij}^{(2)}=\sigma_{ij}^{(2)}\varepsilon_{ij}^{(1)}. \tag{10}$$

The direction of crack propagation θ can be computed in a local tip coordinate system:

$$\theta=2\arctan[\frac{1}{4}(\mathrm{K_I}/\mathrm{K_{II}}+\sqrt{\mathrm{K_I}/\mathrm{K_{II}})^2+8})], \tag{11}$$

where the symbol "arctan" denotes the arc-tangent function.

2.3. The Cross Criterion between HF and Frictional NF

As is known, when an HF encounters a frictional NF, there are three possible scenarios: arrested, direction-crossing, or a crossing with an offset [16,23,41]. Here, the extended Renshaw and Pollard rule is adopted to determine the interaction behavior between HF and NF. As shown in Figure 2, if a new fracture initiates on the opposite side of the NF, the maximum principle stress σ_1 will reach the rock tensile stress. Meanwhile, a no-slip condition should be satisfied along the NF surface. Otherwise, the HF will cross directly or branch into NF with an offset. The expressions of the combined shear stress and normal stress are shown in Equation (12), which is described in detail in other references [15,23].

$$\begin{cases} \sigma_1 = T_0, \\ \tau_\beta < S_0 - \mu_f \sigma_{\beta y}, \end{cases} \tag{12}$$

where β denotes the intersection angle between HF and NF; τ_β denotes the combined shear stress on the NF surface under the action of remote stress and the local crack tip stress; $\sigma_{\beta y}$ denotes the combined normal stress; T_0 denotes the rock tensile strength; S_0 denotes the cohesion force of the frictional NF; μ_f denotes the frictional coefficient of the NF surface.

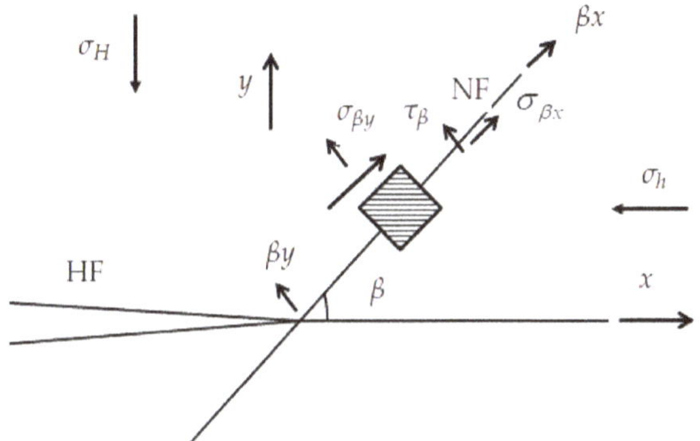

Figure 2. An HF approaching the NF.

2.4. XFEM and Discretization of the Governing Equations of the Hydraulic Fracturing Problem

The XFEM (extended finite element method) is utilized to approximate the displacement discontinuity on both sides of the HF. In order to represent the multiple cracks, a Junction enrichment function is introduced, as shown in Figure 3. The enriched displacement field can be written as [59].

$$\begin{aligned} \boldsymbol{u}(\boldsymbol{x}) &= \sum_{I\in N} N_I(\boldsymbol{x})\boldsymbol{u}_I + \sum_{j=1}^{M^{dis}} \sum_{J\in N^{dis}} N_J(\boldsymbol{x})[H(\boldsymbol{x}) - H(\boldsymbol{x}_J)]a_J \\ &+ \sum_{k=1}^{M^{tip}} \sum_{K\in N^{tip}} N_K(\boldsymbol{x}) \sum_{\alpha=1}^{4} [\Psi_{tip}^{\alpha}(\boldsymbol{x}) - \Psi_{tip}^{\alpha}(\boldsymbol{x}_K)]b_K^{\alpha} \\ &+ \sum_{l=1}^{M^{jun}} \sum_{L\in N^{jun}} N_L(\boldsymbol{x})[J^H(\boldsymbol{x}) - J^H(\boldsymbol{x}_L)]c_L, \end{aligned} \tag{13}$$

where N, N^{dis}, N^{tip} and N^{jun}, respectively, denote the set of standard nodes, Heaviside enrichment nodes, crack-tip nodes, and junction enrichment nodes; $\boldsymbol{u}$ denotes the standard nodal d.o.f. (degrees of freedom); a_J, $b_K^{\alpha}(\alpha = 1,4)$, and c_L, respectively, denote the corresponding enriched nodal d.o.f.; M^{dis} denotes the number of cracks including the main cracks and the secondary cracks; M^{tip} denotes the number of crack tips; M^{jun} denotes the number of junctions with $M^{jun} = M^{dis} - 1$; $H(\boldsymbol{x})$ denotes the Heaviside enrichment function; $J^{H}(\boldsymbol{x})$ denotes the junction enrichment function; $\Psi(\boldsymbol{x})$ denotes the crack-tip enrichment function; N_I, N_J,N_K, and N_L denote the standard shape function of node I, J, K, and L, respectively.

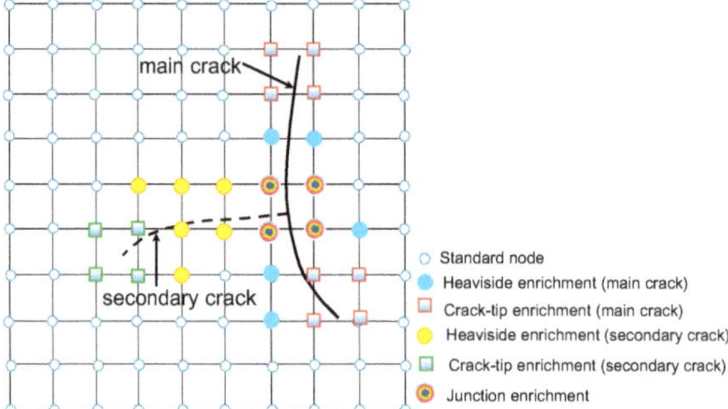

Figure 3. The discontinuous junction function for multiple cracks.

The Heaviside enrichment function is expressed as [59]

$$H(\boldsymbol{x}) = \begin{cases} 0, if & \boldsymbol{x} < 0, \\ 1, if & \boldsymbol{x} \geq 0. \end{cases} \tag{14}$$

The crack-tip enrichment function is defined as

$$\{\Psi^{\alpha}_{tip}(r,\theta)\}^{4}_{\alpha=1} = \{\sqrt{r}sin\frac{\theta}{2}, \sqrt{r}cos\frac{\theta}{2}, \sqrt{r}sin\theta sin\frac{\theta}{2}, \sqrt{r}sin\theta cos\frac{\theta}{2}\}, \tag{15}$$

where (r, θ) denotes the local crack-tip coordinate in the polar coordinate system.

The junction enrichment function $J^{H}(\boldsymbol{x})$ is defined as

$$J^{H}(\boldsymbol{x}) = \begin{cases} H(\varphi^{s}(\boldsymbol{x})), if & \varphi^{m}(\boldsymbol{x}) < 0, \\ 0, if & \varphi^{m}(\boldsymbol{x}) > 0, \end{cases} \tag{16}$$

where $\varphi^{m}(\boldsymbol{x})$ and $\varphi^{s}(\boldsymbol{x})$, respectively, denote the signed distance function of the main crack and the secondary crack. It can be seen that $J^{H}(\boldsymbol{x})$ is equal to 1, −1, or 0 on different sub-domains divided by the secondary cracks.

According to the finite element method, the pressure field p and the fracturing opening displacement vector $\boldsymbol{w}$ can be respectively approximated as

$$\boldsymbol{w} = \sum_{I \in S_w} N^{w}_{I} \boldsymbol{u}_I = \mathbf{N}^{w}(s)\boldsymbol{U}, \tag{17}$$

$$p(s) = \sum_{I \in S_{HF}} N^{p}_{I} \boldsymbol{p}_I = \mathbf{N}^{p}(s)\boldsymbol{P}, \tag{18}$$

where $\boldsymbol{U}$ and $\boldsymbol{P}$, respectively, denote the global nodal displacement vector and the nodal pressure vector; $\boldsymbol{N}^{p}(\boldsymbol{s})$ and $\boldsymbol{N}^{w}(\boldsymbol{s})$, respectively, denote the matrix of the shape function of the fracture opening and pressure.

By substituting the above XFEM formulation, displacement and pressure approximations into the weak form of stress equilibrium equation and lubrication equation, the corresponding discretization forms are written as

$$\boldsymbol{KU} - \boldsymbol{QP} - \boldsymbol{F}^{ext} = \boldsymbol{0}, \tag{19}$$

$$\boldsymbol{Q}^T\Delta\boldsymbol{U} + \Delta t\boldsymbol{HP} + \Delta t\boldsymbol{S} = \boldsymbol{0}, \tag{20}$$

where $\boldsymbol{K}$ denotes the global stiffness matrix; $\boldsymbol{Q}$ denotes the coupling matrix; $\boldsymbol{F}^{ext}$ denotes the external loading vector; $\boldsymbol{H}$ denotes the flow matrix; Δt denotes the time step; and $\boldsymbol{S}$ denotes the source term. They are, respectively, defined as follows [33,34]:

$$\begin{aligned}\boldsymbol{K} &= \begin{bmatrix} \int_\Omega (\boldsymbol{B}^{std})^T\boldsymbol{D}(\boldsymbol{B}^{std})d\Omega & \int_\Omega (\boldsymbol{B}^{std})^T\boldsymbol{D}(\boldsymbol{B}^{enr})d\Omega \\ \int_\Omega (\boldsymbol{B}^{enr})^T\boldsymbol{D}(\boldsymbol{B}^{std})d\Omega & \int_\Omega (\boldsymbol{B}^{enr})^T\boldsymbol{D}(\boldsymbol{B}^{enr})d\Omega + \int_{\Gamma_{NF}} (\boldsymbol{N}^w)^T\boldsymbol{D}^{cont}(\boldsymbol{N}^w)d\Gamma \end{bmatrix} \\ &= \begin{bmatrix} \boldsymbol{K}_{ss} & \boldsymbol{K}_{se} \\ \boldsymbol{K}_{es} & \boldsymbol{K}_{ee} + \boldsymbol{K}_{ee}^{cont} \end{bmatrix}.\end{aligned} \tag{21}$$

In the above Equation (21), $\boldsymbol{D}^{cont}$ denotes the contact stiffness matrix of fracture interfaces:

$$\boldsymbol{Q} = \int_\Omega (\boldsymbol{N}^w)^T\boldsymbol{n}_{\Gamma_{NF}}(\boldsymbol{N}^p)d\Omega, \tag{22}$$

$$\boldsymbol{F}^{ext} = \int_{\Gamma_t} (\boldsymbol{N}^u)^T\boldsymbol{t}d\Gamma, \tag{23}$$

$$\boldsymbol{H} = \int_{\Gamma_{HF}} \frac{\partial \boldsymbol{N}_p}{\partial s}^T k\frac{\partial \boldsymbol{N}_p}{\partial s}ds, \tag{24}$$

$$\boldsymbol{S} = \boldsymbol{N}_p(s)^T\ |_{s=0}\ Q_0. \tag{25}$$

As shown in Figure 4, the flow rate in the main crack and secondary crack satisfies the law of conservation, i.e., $Q_0 = Q_1 + Q_2$, where Q_1 and Q_2 denote the flow rate in Branches 1 and 2 of the secondary crack, respectively. The nonlinear fluid–solid coupling system of equations of hydraulic fracturing problems can be numerically solved by the Newton–Raphson method. More details are described in [33,34].

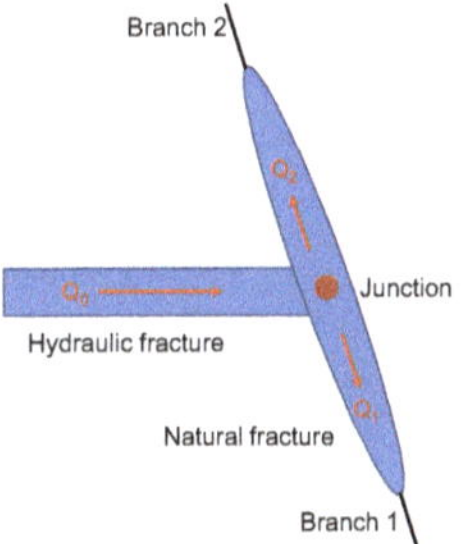

Figure 4. Schematic diagram of a T-shaped fluid-driven fracture.

3. Results and Discussion

3.1. Verification of the XFEM Model

For the model verification, the results from our models is summarized in Appendix A [33,34]. The verification model of this XFEM code is described in detail in [33,34], so the related process is not repeated in this article. It is shown that the numerical results have good agreement with the experimental results of true tri-axial hydraulic fracturing by TerraTek, Inc. (Salt Lake City, UT, USA).

For further details of numerical and experimental procedures and the corresponding results, we refer the reader to References [33,34,60,61].

3.2. Effect of the Location of Natural Fractures on the Diversion of Fracture Network Propagation

In this section, the effect of the location position of an NF on HF propagation paths is determined by numerical simulation using XFEM. The input parameter values of this model are as shown in Table 1. Under isotropic stress state conditions, the intersection angle between HF and NF is equal to 90°. As shown in Figure 1, the domain is a 25 m × 25 m square, where the injection point is located at the midpoint of the left edge. In this domain, the HF is 2.6 m in length, and the length of NF is equal to 7 m. Based on the above input parameters, the mechanical NF–HF interaction processes in hydraulic fracturing are numerically simulated at different lengths of NF in lower and upper parts, i.e., corresponding to L_{lower} and L_{upper} in Figure 1, respectively.

Table 1. Input parameter values of the hydro-fracking model.

Input Parameter	Value
Young's Modulus, E	20 GPa
Poisson's ratio, ν	0.2
Rock density, ρ	2460 kg/m^3
Friction coefficient of NF, μ_f	0.3
Cohesion of the NF, S_0	0 MPa
Fracture toughness, K_{IC}	1.0 MPa $\cdot$ m$^{\frac{1}{2}}$
Tensile strength, T_0	1.5 MPa
Unconfined compression strength, UCS	100 MPa
Apparent viscosity of fracturing fluid, μ	0.1 Pa $\cdot$ s
The consistency index of fracturing fluid, K	0.84 Pa $\cdot$ s^n
The flow behavior index of fracturing fluid, n	0.53
Dynamic viscosity index, m	2.0
Fluid pump rate, Q_0	0.001 m^2/s
Pore pressure, P_0	5 MPa
Maximum horizontal stress, σ_H	5 MPa
Minimum horizontal stress, σ_h	5 MPa

The corresponding crack propagation paths are as shown in Figure 5. It is obvious that fracture diversion occurs near the tips of the NF in all cases. In Figure 5a, when the HF intersects with the NF, the fracturing fluid flows into the opened NF. In the lower parts of the NF, the opened NF firstly propagates along a vertically downward path for a certain length, and it is then diverted along a new direction; in the upper parts of the NF, the opened NF is directly diverted at the upper tip of the original NF. In Figure 5b, both the lower and upper parts of the NF firstly extend vertically downward and upward for a short distance, respectively, and are then diverted to the right-hand side of the graph. However, the length of the lower parts of the diverted fracture (DF) is longer than that of the upper parts of the DF, corresponding to the red line in Figure 1. In Figure 5c, in the upper parts of the NF, the opened fracture can only propagate vertically upward, and cannot be diverted near the tip of the NF; in the lower parts of the NF, the opened fracture can be diverted away from the original NF. By making use of the data in Figure 5, the length of the DF propagation in the upper parts is calculated from the upper tip of the original NF. When the upper length of NF, i.e., L_{upper}, is equal to 4, 5, and 6 m, the corresponding length is 3.14, 2.15, and 1.14 m, respectively. If L_{upper} is increased by 2 m, the upper length of the DF will decrease by 2 m. Therefore, it is shown that the longer the upper parts of the original NF are, the more difficult it is for the opened NF to be diverted away from the upper tip of the NF under the conditions of an isotropic stress state, while the lower parts of the original NF is more easily diverted to the right-hand side than the upper parts of the original NF under this circumstance.

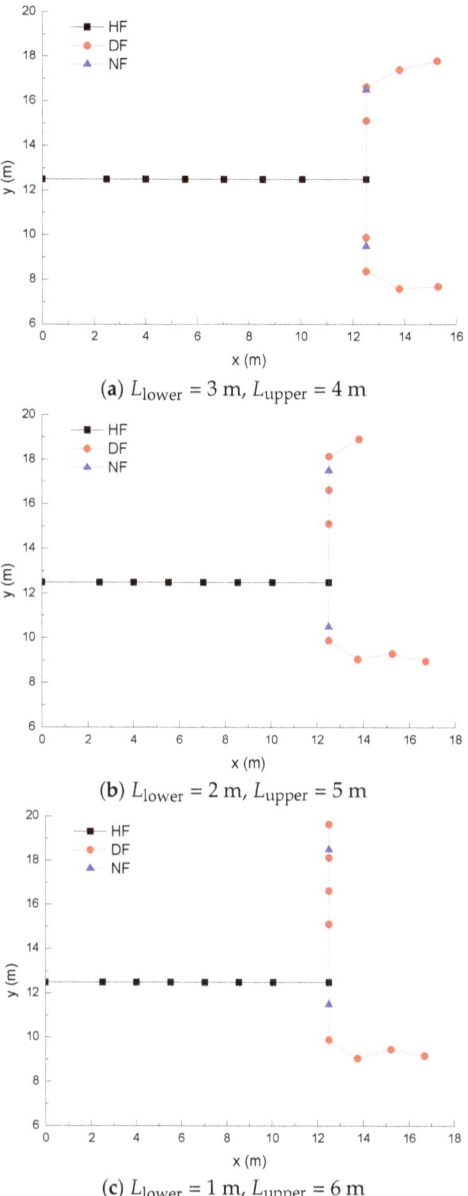

Figure 5. The crack propagation paths at different lengths of lower and upper parts of the NF. In this figure, the black, blue, and red dotted lines, respectively, denote the original HF, the initial NF, and the diverted fracture (DF).

The Von-Mises stress distributions are shown in Figure 6, where a blue color represent a relative stress value, while a yellow or red one represent a higher stress value. For all cases of the model, it is shown that there is a small region of stress concentration near the two tips of the original NF, which indicates that a higher pressure is required to divert the opened NF away from the original NF.

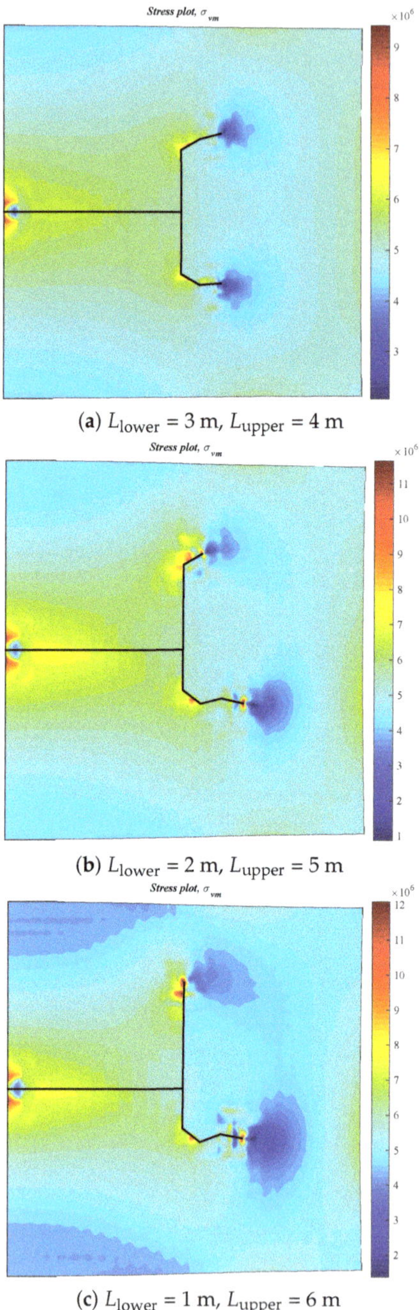

(a) L_{lower} = 3 m, L_{upper} = 4 m

(b) L_{lower} = 2 m, L_{upper} = 5 m

(c) L_{lower} = 1 m, L_{upper} = 6 m

Figure 6. Von-Mises stress distributions at different lengths of lower and upper parts of the NF.

The fracture aperture and net pressure curves of the diverted fracture are shown in Figures 7 and 8, respectively. With the decrease of L_{lower}, both curves have revealed an asymmetrical characteristic, which peak at the diverted point in the lower parts. In addition, at the intersection point between HF and NF, their corresponding values take second place. The fracture aperture and net pressure in the lower parts of the DF are much greater than those in the upper parts. By comparison, in the case of $L_{\text{lower}} = 3$ m and $L_{\text{lower}} = 4$ m, their curves are nearly symmetrical. This indicates that, under the combined action of remote stress and local crack-tip stress, the variation tendency of the fracture aperture and the net pressure is quite different from that in the case of only a single HF.

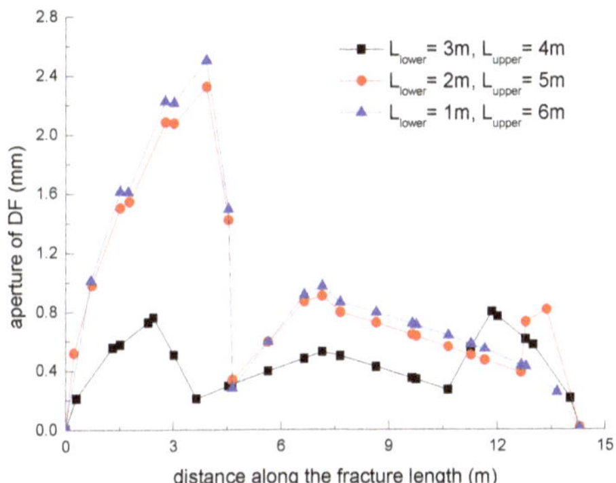

Figure 7. The fracture aperture curves of the DF along the fracture length at different lengths of the lower and upper parts of the NF. The distance in the *x*-axis is along the direction from the lower parts to the upper parts of the DF.

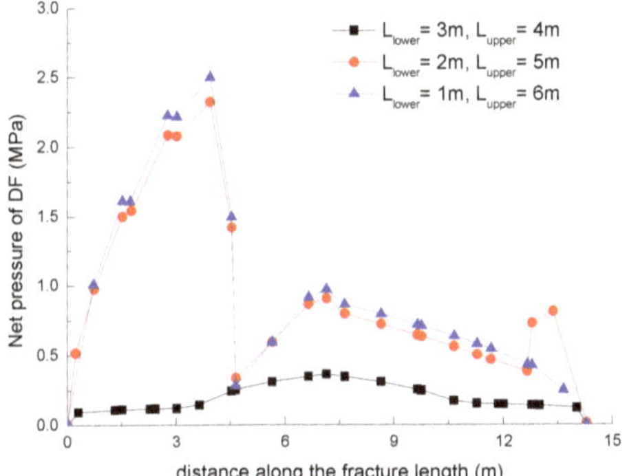

Figure 8. The net pressure curves of the DF along the fracture length at different lengths of the lower and upper parts of the NF. The distance in the *x*-axis is along the direction from the lower parts to the upper parts of the DF.

The flow rate in the diverted fracture is shown in Figure 9. When fluid flows into the intersection point between HF and NF, it will flow upward and downward, respectively. If the value of the flow rate is negative, fluid will flow upward; otherwise, it will flow downward. It is shown that the flow rate in the lower parts of the DF is much greater than that in the upper parts. Therefore, the fracture aperture and net pressure in the lower parts of the DF is greater than that in the upper parts under the condition of the same fluid viscosity. With the decrease of L_{lower}, fluid flows downward more easily. Thus, the lower parts of the original NF is more easily diverted than the upper parts of the original NF. This may explain the results in Figure 5.

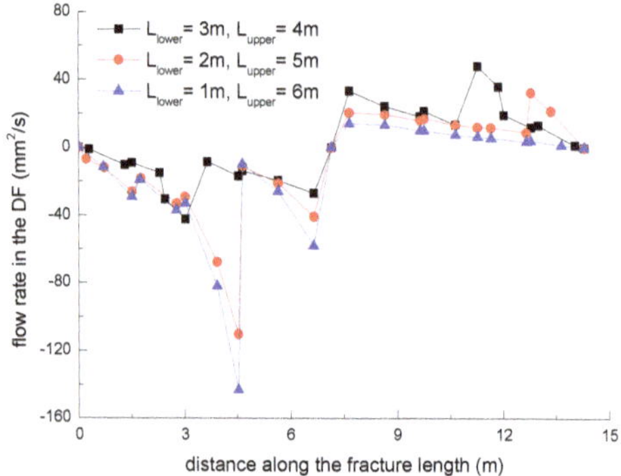

Figure 9. The flow rate curves in the DF along the fracture length at different lengths of the lower and upper parts of the NF. The distance in the x-axis is along the direction from the lower parts to the upper parts of the DF.

3.3. Effect of Horizontal Stress Differences on the Diversion of Fracture Network Propagation

Based on input parameters in Table 1, the effect of the remote stress difference on the diversion of fracture network propagation is numerically simulated at different levels of minimum horizontal stresses σ_h: 5, 4, and 0 MPa; the maximum horizontal stress σ_H is kept constant (5 MPa) in all cases of this model. At the same time, L_{lower} = 3 m, L_{upper} = 4 m, and the NF–HF intersection angle is equal to 90°.

The corresponding crack propagation paths are shown in Figure 10, in which the deflection angle is defined the angle between NF and DF at the tips of the original NF. According to the data in Figure 10, the deflection angle in the upper parts is calculated. When the horizontal stress difference is, respectively, equal to 0 and 5 MPa, the corresponding deflection angle is 59.2° and 90°, respectively. If the stress difference is increased by 5 MPa, the deflection angle in the upper parts is increased by 30.8°. The higher the horizontal stress difference is, the greater the deflection angle is. In Figure 10c, i.e., $\Delta\sigma$ = 5 MPa, the opened NF firstly diverts and propagates along the direction of minimum horizontal stress and finally tends to extend along the preferred fracture plane (PFP) direction in petroleum engineering. By contrast, when the stress state is approximately isotropic, crack propagation in both the lower and upper parts will extend along the minimum horizontal stress direction for some length. This indicates that crack propagation of the opened NF is a complex mechanical process under the combined action of the local crack-tip and the remote stress state.

As shown in Figure 11c, there is a lower Von-Mises stress region (corresponding to blue area on the contour) located on the right of DF for 5 MPa stress difference. This propagates the diverted fracture along the PFP. In Figure 11a,b, both cases correspond to a lower stress difference, and the lower stress region is mainly near the two crack tips. The local stress distribution is an explanation to interpret crack paths in Figure 10.

The fracture aperture and net pressure curves of DF are shown in Figures 12 and 13, respectively. The fracture aperture curve reveals an asymmetrical characteristic, while the net pressure curve reveals a nearly symmetrical characteristic. The fracture aperture peaks at a global maximum value at the inflection point, where the opened fracture in the upper parts diverts to the right side in Figure 10; at the NF–HF intersection point, the fracture aperture takes the second place; at the inflection point in the lower parts, it takes the third place. The net pressure reaches a maximum at the intersection point, and there are two inflection points on this curve. This means that the opened fracture diverts to the right side. The higher the stress difference is, the greater the net pressure it requires, which leads to a greater fracture aperture.

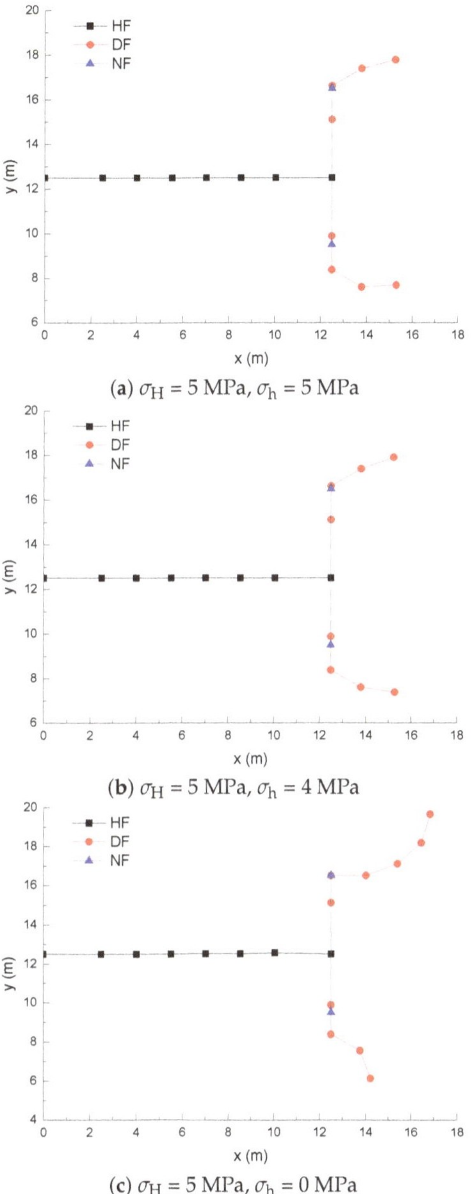

Figure 10. The crack propagation paths at different levels of remote horizontal principle stress difference.

The flow rate in the DF is shown in Figure 14. It is obvious that the flow rate in the upper parts is much greater than that in the lower parts. Therefore, the fracture aperture and net pressure in the upper parts are greater than those in the lower parts for the same fluid viscosity. This might explain the results of fracture aperture and net pressure in Figures 12 and 13.

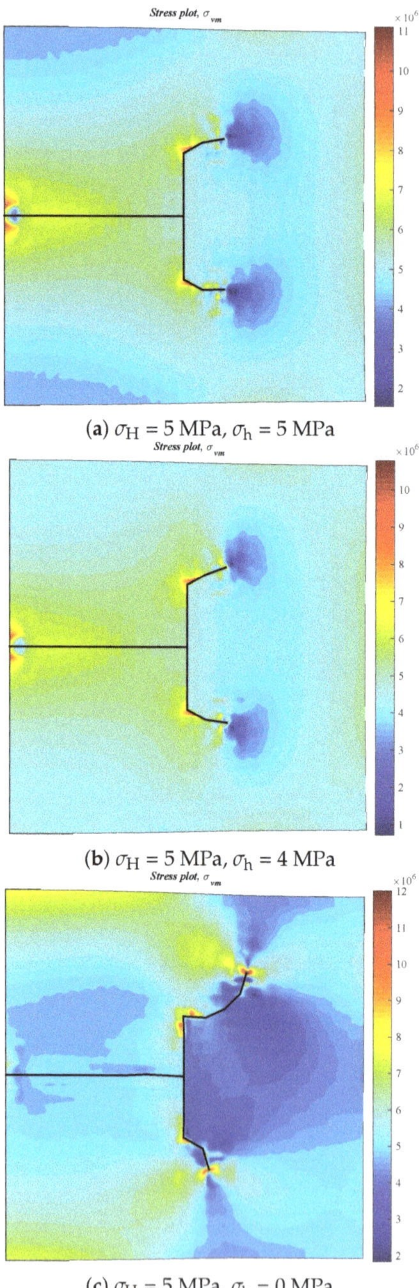

(**a**) $\sigma_H = 5$ MPa, $\sigma_h = 5$ MPa

(**b**) $\sigma_H = 5$ MPa, $\sigma_h = 4$ MPa

(**c**) $\sigma_H = 5$ MPa, $\sigma_h = 0$ MPa

Figure 11. Von-Mises stress distributions at different levels of remote horizontal principle stress difference.

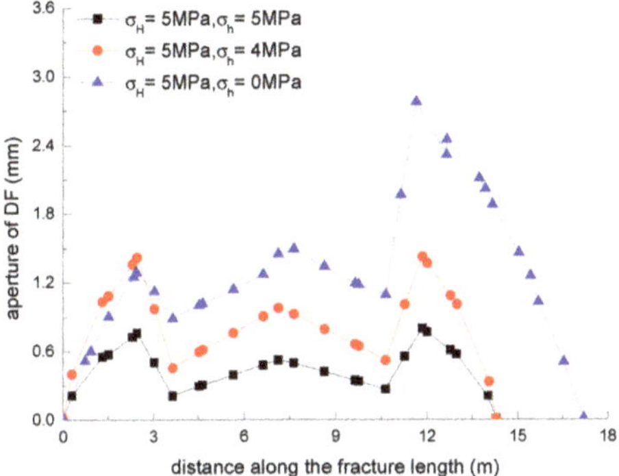

Figure 12. The fracture aperture curves of the DF along the fracture length at different levels of remote horizontal principle stress difference. The distance in the x-axis is along the direction from the lower part to the upper part of the DF.

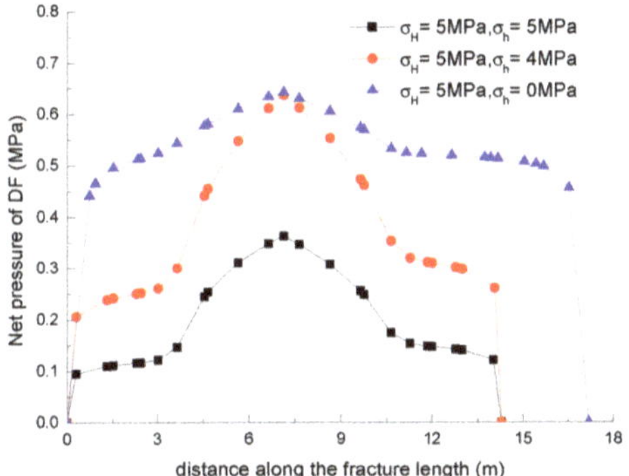

Figure 13. The net pressure curves of the DF along the fracture length at different levels of remote horizontal principle stress difference. The distance in the x-axis is along the direction from the lower part to the upper part of the DF.

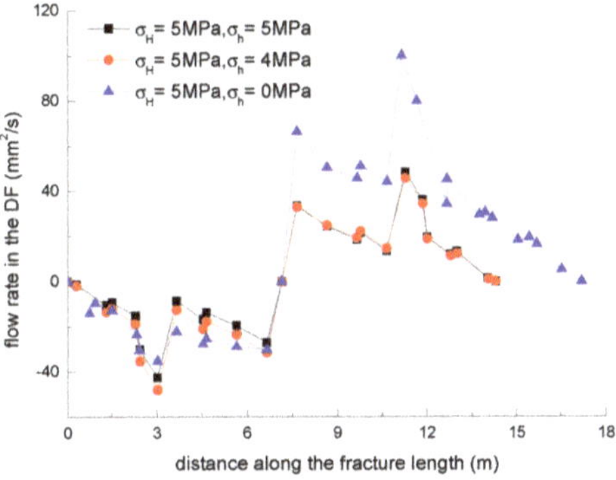

Figure 14. The flow rate curves of the DF along the fracture length at different levels of remote horizontal principle stress difference. The distance in the x-axis is along the direction from the lower part to the upper part of the DF.

3.4. Effect of the NF–HF Intersection Angle on the Diversion of Fracture Network Propagation

Based on the input parameters in Table 1, the effect of the NF–HF intersection angle on the diversion of fracture network propagation is numerically simulated at different levels of intersection angles β: 75°, 60°, and 45°; both the maximum and minimum horizontal principle stresses are equal to 5 MPa in all cases of this model. Meanwhile, L_{lower} and L_{upper} are, respectively, 3 m and 4 m.

The corresponding crack propagation paths are as shown in Figure 15. Under the condition of isotropic stress state, the NF–HF intersection angle will have a significant impact on the propagation direction of the primary HF, i.e., the black dotted line in Figure 15, when HF is approaching NF. With the decrease of the intersection angle, the primary HF deflects from the horizontal line. When the NF–HF intersection angle is greater than 60° (in Figure 15a,b), the opened NF in the upper parts is more easily diverted away from the original NF than that in the lower parts under the combined action of remote stress and the crack-tip stress field. However, the intersection angle is less than 60° (in Figure 15c). The opened NF in the lower parts is more easily diverted away from the original NF than that in the upper parts under the combined action of remote stress and the crack-tip stress field. By making use of the data in Figure 15, the deflection angle for the primary HF was calculated. When the intersection angle is decreased from 75° to 45°, the corresponding deflection angle is increased from 0° to 61.2°. This indicates that the NF–HF intersection angle will have a significant impact on the diversion propagation of the primary HF and the secondary opened NFs.

The Von-Mises stress distributions at different levels of NF–HF intersection angle are shown in Figure 16. In Figure 16a, there is a stress concentration region near the diversion point in the upper parts of the NF, which indicates that it will require a high net pressure to divert the opened fracture upward. In Figure 16b, the Von-Mises stress in the upper parts is greater than that in the lower parts, which causes the subsequent fracture to easily divert upward. In Figure 16c, the Von-Mises stress on the left of the NF is greater than that on the right, so it is more easily diverted downward.

The fracture aperture and net pressure curves of the DF are shown in Figures 17 and 18, respectively. In the case of $\beta = 75°$, the fracture aperture and net pressure in the upper parts is greater than that in the lower parts. In particular, the fracture aperture and net pressure near the diversion point in the lower parts is close to zero, and this is consistent with the results of Von-Mises stress at this point. In the other two cases, the maximum values of the fracture aperture and the net pressure are at the diversion point in the lower parts. The smaller the intersection angle is, the greater the net pressure it requires to divert the fracture.

The flow rate in DF is shown in Figure 19. It is obvious that, in the case of $\beta = 45°$, the flow rate in the upper parts is much greater than that in the lower parts. This indicates that it is easy for fracturing fluid to flow upward when the intersection angle is small. This is a possible reason for explaining the results in Figures 17 and 18.

3.5. Effect of Fluid Viscosity on the Diversion of Fracture Network Propagation

Based on the input parameters in Table 1, the effect of the viscosity of fracturing fluid on the diversion of fracture network propagation is numerically simulated at different levels of viscosity μ : 100 mPa·s, 10 mPa·s, and 1 mPa·s [62]. In this model, the maximum and minimum horizontal stresses are kept constant (5 MPa) for all cases. Meanwhile, the NF–HF intersection angle is equal to 90°.

The corresponding crack propagation paths are shown in Figure 20. By making use of the data in Figure 20, the length of DF is calculated. When the fluid viscosity is increased from 100 mPa·s to 1 mPa·s, the length of DF is increased from 14.29 to 5.23 m. It is clear that the smaller the viscosity is, the more difficult the opened fracture diverts into a new direction. When the viscosity is equal to 1 mPa·s, artificial fracture will propagate along the NF direction under given conditions. This indicates that more energy is required to divert the opened NF upward and downward.

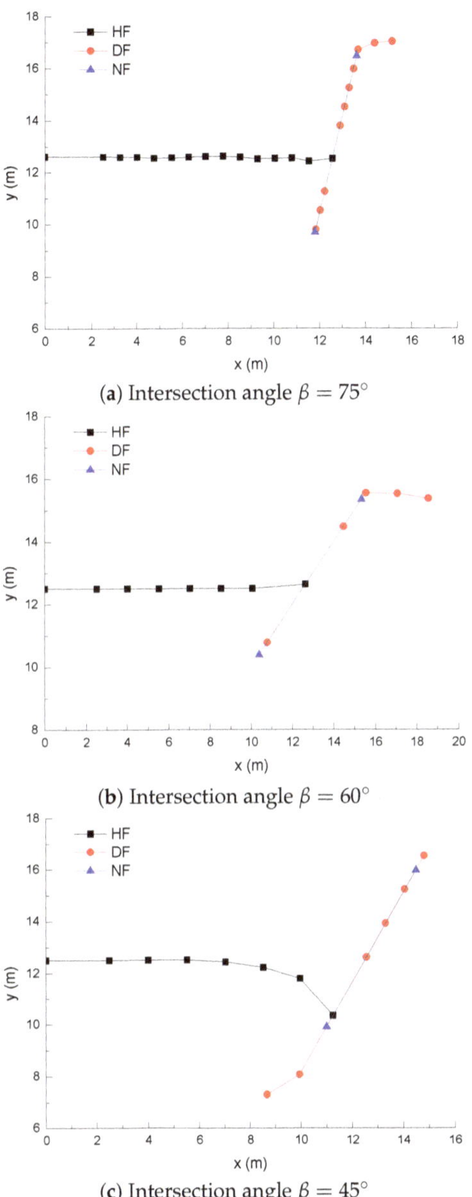

Figure 15. The crack propagation paths at different levels of intersection angle between HF and NF.

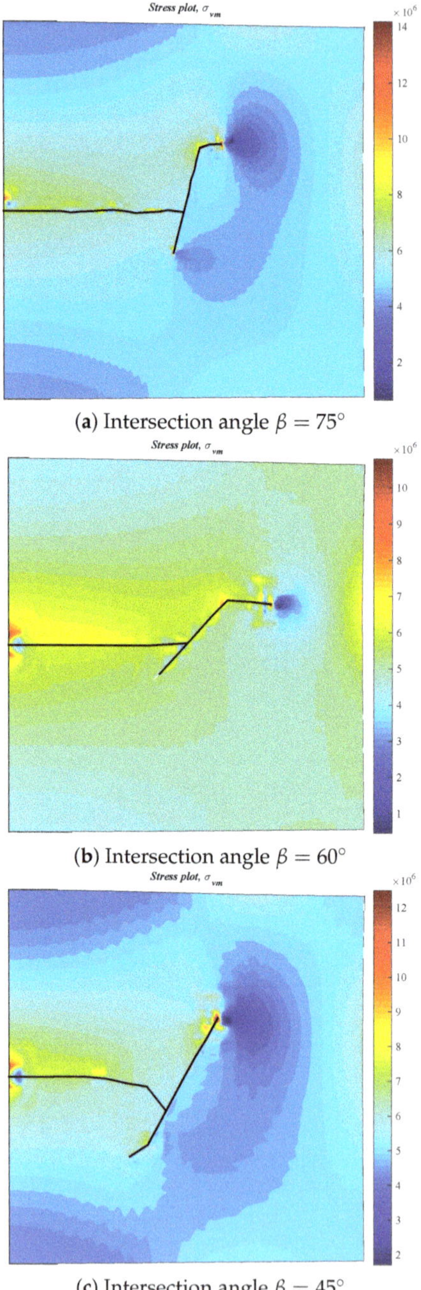

(a) Intersection angle $\beta = 75^\circ$

(b) Intersection angle $\beta = 60^\circ$

(c) Intersection angle $\beta = 45^\circ$

Figure 16. Von-Mises stress distributions at different levels of intersection angle between HF and NF.

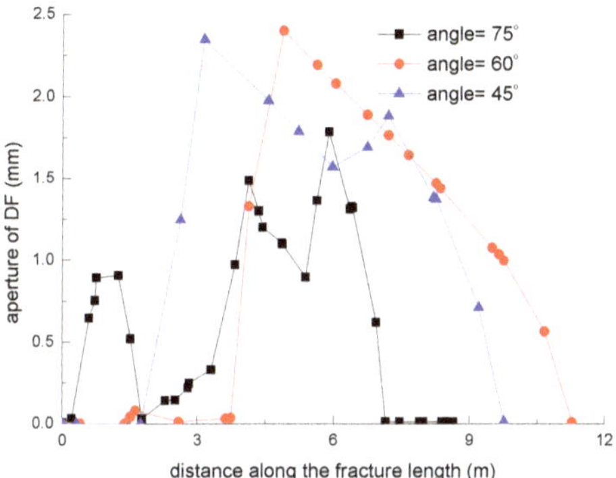

Figure 17. The fracture aperture curves of the DF along the fracture length direction at different levels of intersection angle between HF and NF. The distance in the x-axis is along the direction from the lower part to the upper part of the DF.

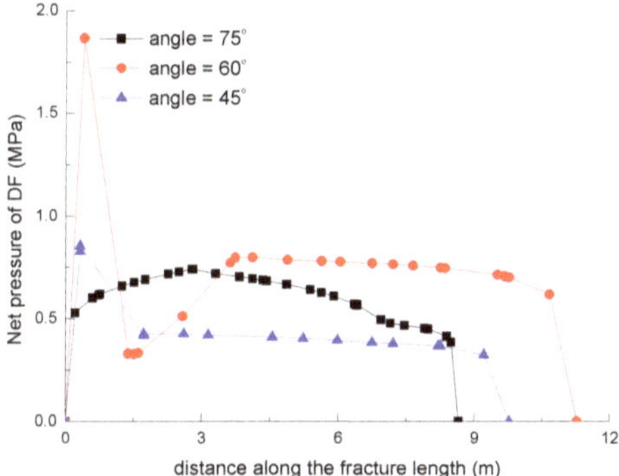

Figure 18. The net pressure curves of the DF along the fracture length direction at different levels of intersection angle between HF and NF. The distance in the x-axis is along the direction from the lower part to the upper part of the DF.

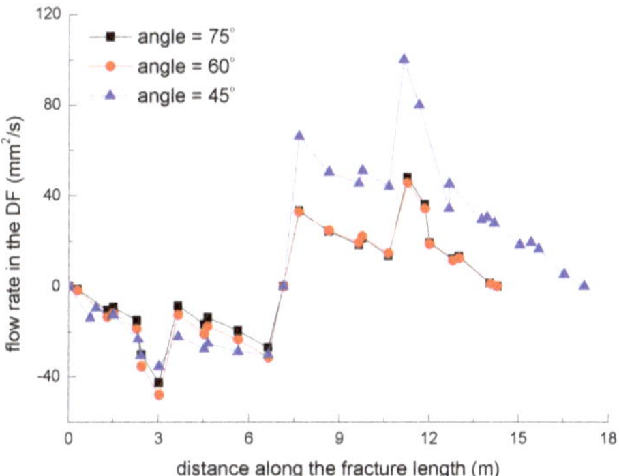

Figure 19. The flow rate curves of the DF along the fracture length direction at different levels of intersection angle between HF and NF. The distance in the x-axis is along the direction from the lower part to the upper part of the DF.

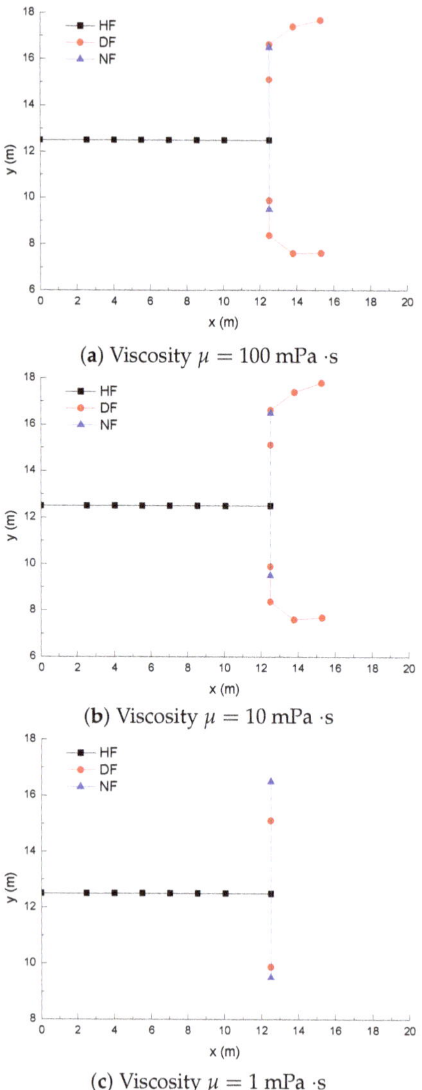

(a) Viscosity $\mu = 100$ mPa ·s

(b) Viscosity $\mu = 10$ mPa ·s

(c) Viscosity $\mu = 1$ mPa ·s

Figure 20. The crack propagation paths at different levels of viscosity of fracturing fluid.

The Von-Mises stress distributions at different levels of fluid viscosity are shown in Figure 21. In Figure 21c, there is a lower Von-Mises stress area on the right of the NF, which makes the opened NF propagate along the original NF direction. This is consistent with the results in Figure 20.

The fracture aperture and net pressure curves of the DF are shown in Figures 22 and 23, respectively. They are close to symmetrical about the axis of the original HF under the condition of an isotropic stress state. In the cases of $\mu = 100$ mPa ·s and $\mu = 10$ mPa ·s, there are two inflection points on the curves, which correspond to the diversion point in the lower and upper parts of the NF. The greater the fluid viscosity is, the greater the fracture aperture and net pressure are.

The flow rate in the DF is as shown in Figure 24. Obviously, the greater the viscosity is, the greater the flow rate is, and thus the easier it is for the secondary fracture to divert. This might explain the results in Figures 20 and 21 [63].

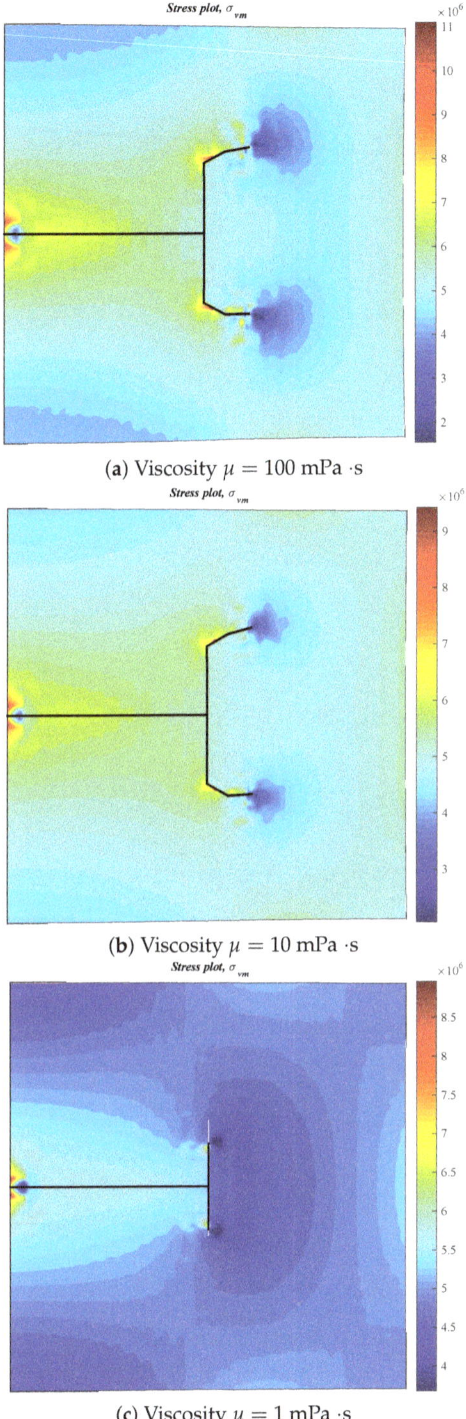

Figure 21. Von-Mises stress distributions at different levels of viscosity of fracturing fluid.

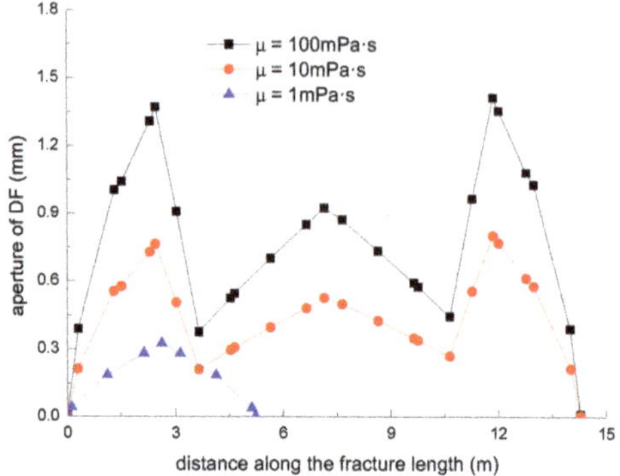

Figure 22. The fracture aperture curves of the DF along the fracture length at different levels of viscosity of fracturing fluid. The distance in the x-axis is along the direction from the lower part to the upper part of the DF.

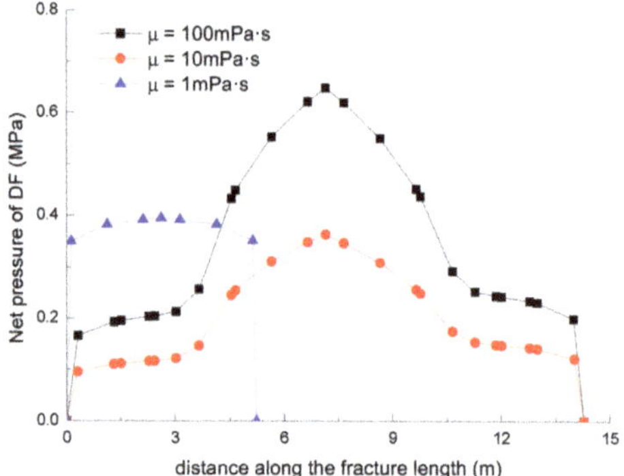

Figure 23. The net pressure curves of the DF along the fracture length at different levels of viscosity of fracturing fluid. The distance in the x-axis is along the direction from the lower part to the upper part of the DF.

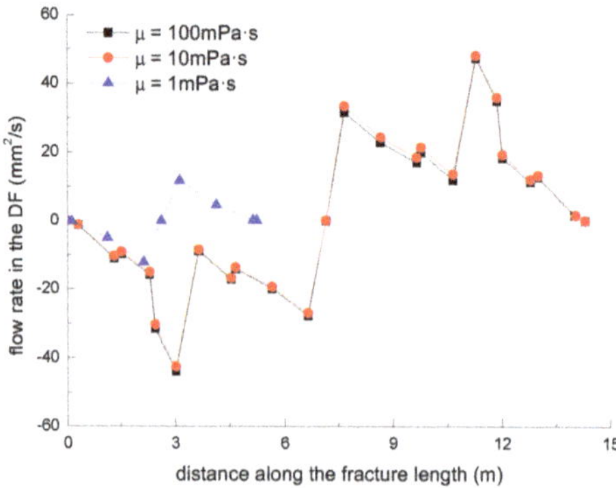

Figure 24. The flow rate curves of the DF along the fracture length direction at different levels of viscosity of fracturing fluid. The distance in the x-axis is along the direction from the lower part to the upper part of the DF.

4. Conclusions

This paper investigates the diversion mechanisms of a fracture network in tight formations with frictional NFs by means of the XFEM technique. The effects of some key factors such as the location of the NF, the intersection angle between the NF and HF, the horizontal stress difference, and the fluid viscosity on the mechanical diversion behavior of the HF were analyzed in detail. The following main conclusions can be drawn:

(1) Fracture diversion propagation will occur near the two tips of the opened NF after an HF is intersecting with an NF. The numerical results show that some key factors such as the NF position, the NF–HF intersection angle, the horizontal stress differences, and the fluid viscosity have a significant impact on the diversion propagation in the upper and lower parts of the opened NF.
(2) For a constant length of NF (7 m), the upper length of the DF decreases by about 2 m with a 2 m increment of the upper length of the NF (L_{upper}), while the length of the DF increases 9.06 m, with the fluid viscosity increased from 1 to 100 mPa.s; (2) the deflection angle in the upper parts increases by 30.8° with the stress difference increased by 5 MPa, while the deflection angle increases by 61.2° with the intersection angle decreased from 75° to 45°.
(3) The longer the upper parts of the original NF are, the more difficult it is for the opened NF to divert away from the upper tip of the NF under the conditions of an isotropic stress state, while the lower parts of the original NF is more easily diverted to the right-hand side than the upper parts of the original NF. The NF–HF intersection angle will have a significant impact on the diversion propagation of the primary HF and the secondary opened NFs.
(4) In general, the distributions of fracture aperture, net pressure, and flow rate reveal asymmetrical characteristics for the secondary hydraulically driven fractures. For the distribution of Von-Mises stress, there is usually a concentrated stress zone area near the turning point of the secondary cracks, which corresponds to the inflection points on the curves of the fracture aperture and net pressure.
(5) The diversion mechanisms of the fracture network are the results of the combined action of all factors. This will provide a new perspective on the mechanisms of fracture network generation. Future work should determine the primary and secondary relations of various factors by means of experiments and numerical calculation.

Author Contributions: D.W., F.S., and B.Y. conceived and designed the model of hydraulic fracturing problem using a XFEM technique. D.S., X.L., D.H. and Y.T. analyzed the data. D.W. wrote the paper.

Funding: The authors would like to give their sincere gratitude to the National Science Foundation of China (Nos.51804033 and 51706021), the Beijing Postdoctoral Research Foundation (2018-ZZ-045), the Project of Construction of Innovative Teams and Teacher Career Development for Universities and Colleges Under Beijing Municipality (No. IDHT20170507), the Program of Great Wall Scholar (No. CIT&TCD20180313), Jointly Projects of Beijing Natural Science Foundation and Beijing Municipal Education Commission (No. KZ201810017023), and the Natural Science Foundation of Jiangsu Province (No. BK20170457) for their financial support.

Conflicts of Interest: The authors declare no conflicts of interest.

Abbreviations

The following abbreviations are used in this manuscript:

XFEM	Extended Finite Element Method
DEM	Discrete Element Method
NMM	Numerical Manifold Method
SRV	Stimulated Reservoir Volume
HF	Hydraulic Fracture or Hydraulically Driven Fracture or Hydro-Fracture
NF	Natural Fracture
DF	Diverted Fracture
PFP	Preferred Fracture Plane
ELP	Enhanced Local Pressure

Appendix A

The result of XEFM is here compared with results of analytical solutions. As is known, depending on the dimensionless fracture toughness, the analytical solutions of the Kristianovich-Geertsma-de Klerk (KGD) model have different expressions. The dimensionless fracture toughness can be written as

$$K_m = 4\sqrt{\frac{2}{\pi}}\frac{K_{IC}(1-\nu^2)}{E}\left(\frac{E}{12\mu Q_0(1-\nu^2)}\right)^{1/4}, \tag{A1}$$

where K_m denotes the dimensional fracture toughness; K_{IC} denotes the rock fracture toughness; μ denotes the viscosity of fracturing fluid; Q_0 denotes the injection rate; E denotes the rock Young's modulus; ν denotes the Poisson's ratio of the rock matrix. If K_m is greater than 4, the fracture propagation regime is toughness dominated; if K_m is less than 1, the fracture propagation regime is viscosity dominated, which is much more common in most hydraulic fracturing treatments.

The input parameters of the verification model are listed in Table A1. According to Equation (A1), the dimensionless fracture toughness is equal to 0.313, which indicates that the fracture propagation regime is viscosity-dominated. In this model, the HF is located at the center of a symmetrical model with a length of 100 m and 180 m along the x- and y-direction directions, respectively. The domain is divided into 3080 bilinear quadrilateral elements.

Table A1. Input parameter values of hydro-fracturing.

Input Parameter	Value
Young's Modulus, E	20 GPa
Poisson's ratio, ν	0.2
Fracture toughness, K_{IC}	$0.1\ \text{MPa}\cdot\text{m}^{\frac{1}{2}}$
The consistency index of fracturing fluid, K	$0.84\ \text{Pa}\cdot\text{s}^n$
Injection rate, Q_0	$0.001\ \text{m}^2/\text{s}$
Viscosity, μ	$0.1\ \text{Pa}\cdot\text{s}$
Dimensionless fracture toughness, K_m	0.313
Injection time, t	30 s

The initial half-length of the HF is equal to 1.25 m, and it is assumed that a constant fluid pressure acting on the fracture wall is equal to 3.9 MPa. The curves of fluid pressure at the injection point and the fracture width at 30 s are shown in Figure A1a,b, respectively, which is compared with the corresponding analytical solutions. There is very good agreement between the numerical results and analytical solutions, which indicates that the XFEM model can obtain reliable results.

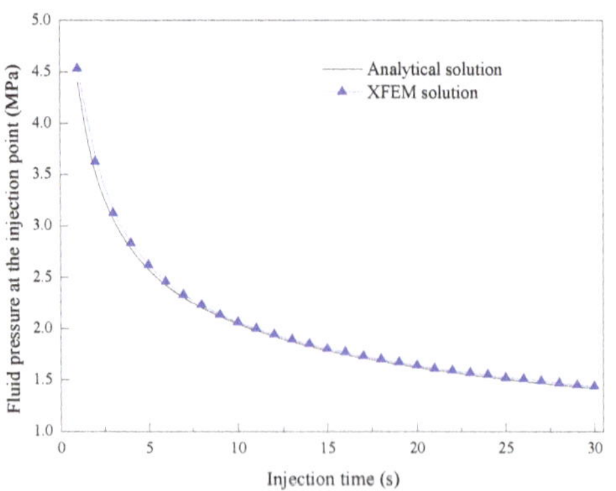

(**a**) Fluid pressure at the injection point.

Figure A1. *Cont.*

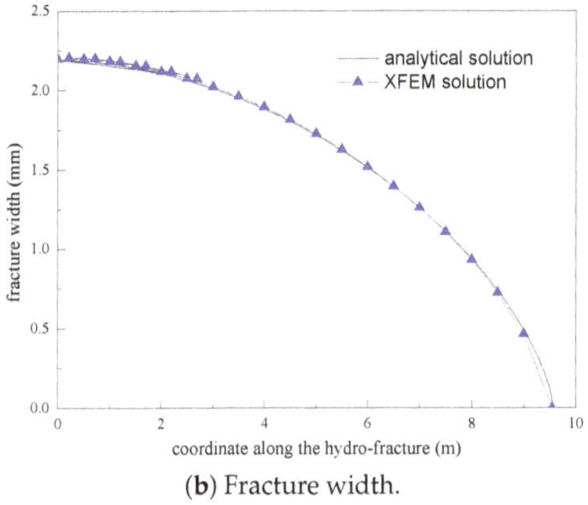

(**b**) Fracture width.

Figure A1. Results of the XFEM technique and of the analytical solutions.

References

1. Pang, Y.; Soliman, M.Y.; Deng, H.; Emadi, H. Analysis of effective porosity and effective permeability in shale-gas reservoirs with consideration of gas adsorption and stress effects. *SPE J.* **2017**, *22*, 1739–1759. [CrossRef]
2. Shen, Y.; Ge, H.; Li, C.; Yang, X.; Ren, K.; Yang, Z.; Su, S. Water imbibition of shale and its potential influence on shale gas recovery—A comparative study of marine and continental shale formations. *J. Nat. Gas Sci. Eng.* **2016**, *35*, 1121–1128. [CrossRef]
3. Pang, Y.; Soliman, M.; Deng, H.; Emadi, H. Effect of methane adsorption on stress-dependent porosity and permeability in shale gas reservoirs. In Proceedings of the SPE Low Perm Symposium, Denver, CO, USA, 5–6 May 2016.
4. Matsunaga, I.; Kobayashi, H.; Sasaki, S.; Ishida, T. Studying hydraulic fracturing mechanism by laboratory experiments with acoustic emission monitoring. In Proceedings of the 34th U.S. Symposium on Rock Mechanics (USRMS), Madison, WI, USA, 28–30 June 1993.
5. Wang, W.; Su, Y.; Sheng, G.; Cossio, M.; Shang, Y. A mathematical model considering complex fractures and fractal flow for pressure transient analysis of fractured horizontal wells in unconventional oil reservoirs. *J. Nat. Gas Sci. Eng.* **2015**, *23*, 139–147. [CrossRef]
6. Zhang, J.; Ouyang, L.; Zhu, D.; Hill, A. Experimental and numerical studies of reduced fracture conductivity due to proppant embedment in the shale reservoir. *J. Pet. Sci. Eng.* **2015**, *130*, 37–45. [CrossRef]
7. Wang, W.; Shahvali, M.; Su, Y. Analytical solutions for a quad-Linear flow model derived for multistage fractured horizontal wells in tight oil reservoirs. *J. Energy Resour. Technol.* **2017**, *139*, 012905. [CrossRef]
8. Wang, W.; Shahvali, M.; Su, Y. A semi-analytical model for production from tight oil reservoirs with hydraulically fractured horizontal wells. *Fuel* **2015**, *158*, 612–618. [CrossRef]
9. Cai, J.; Wei, W.; Liu, R.; Wang, J. Fractal characterization of dynamic fracture network extension in porous media. *Fractals* **2017**, *25*, 1750023. [CrossRef]
10. Wang, F.; Yang, K.; Cai, J. Fractal characterization of tight oil reservoir pore structure using nuclear magnetic resonance and mercury intrusion porosimetry. *Fractals* **2018**, *26*, 1840017. [CrossRef]
11. Singh, H.; Cai, J. Screening improved recovery methods in tight-oil formations by injecting and producing through fractures. *Int. J. Heat Mass Transf.* **2018**, *116*, 977–993. [CrossRef]
12. Detournay, E.; Cheng, A.D.; Roegiers, J.C.; McLennan, J. Poroelasticity considerations in In Situ stress determination by hydraulic fracturing. *Int. J. Rock Mech. Min. Sci. Geomech. Abstr.* **1989**, *26*, 507–513. [CrossRef]
13. Schmitt, D.; Zoback, M. Poroelastic effects in the determination of the maximum horizontal principal stress in hydraulic fracturing tests—A proposed breakdown equation employing a modified effective stress relation for tensile failure. *Int. J. Rock Mech. Min. Sci. Geomech. Abstr.* **1989**, *26*, 499–506. [CrossRef]
14. Mahrer, K.D. A review and perspective on far-field hydraulic fracture geometry studies. *J. Pet. Sci. Eng.* **1999**, *24*, 13–28. [CrossRef]

15. Yew, C.H.; Weng, X. *Mechanics of Hydraulic Fracturing*, 2nd ed.; Gulf Professional Publishing: Houston, TX, USA, 2014.
16. Hossain, M.M.; Rahman, M. Numerical simulation of complex fracture growth during tight reservoir stimulation by hydraulic fracturing. *J. Pet. Sci. Eng.* **2008**, *60*, 86–104. [CrossRef]
17. Yuan, B.; Zheng, D.; Moghanloo, R.G.; Wang, K. A novel integrated workflow for evaluation, optimization, and production predication in shale plays. *Int. J. Coal Geol.* **2017**, *180*, 18–28. [CrossRef]
18. Yuan, B.; Su, Y.; Moghanloo, R.G.; Rui, Z.; Wang, W.; Shang, Y. A new analytical multi-linear solution for gas flow toward fractured horizontal wells with different fracture intensity. *J. Nat. Gas Sci. Eng.* **2015**, *23*, 227–238. [CrossRef]
19. Wu, K.; Chen, Z.; Li, X.; Guo, C.; Wei, M. A model for multiple transport mechanisms through nanopores of shale gas reservoirs with real gas effect–adsorption-mechanic coupling. *Int. J. Heat Mass Transf.* **2016**, *93*, 408–426. [CrossRef]
20. Wei, B.; Li, Q.; Jin, F.; Wang, C. The potential of a novel nanofluid in enhancing oil recovery. *Energy Fuels* **2016**, *30*, 2882–2891. [CrossRef]
21. Zhang, Q.; Su, Y.; Wang, W.; Lu, M.; Sheng, G. Gas transport behaviors in shale nanopores based on multiple mechanisms and macroscale modeling. *Int. J. Heat Mass Transf.* **2018**, *125*, 845–857. [CrossRef]
22. Renshaw, C.; Pollard, D. An experimentally verified criterion for propagation across unbounded frictional interfaces in brittle, linear elastic materials. *Int. J. Rock Mech. Min. Sci. Geomech. Abstr.* **1995**, *32*, 237–249. [CrossRef]
23. Gu, H.; Weng, X. Criterion for fractures crossing frictional interfaces at non-orthogonal angles. In Proceedings of the 44th U.S. Rock Mechanics Symposium and 5th U.S.-Canada Rock Mechanics Symposium, Salt Lake City, UT, USA, 27–30 June 2010.
24. Chen, Z.; Jeffrey, R.G.; Zhang, X.; Kear, J. Finite-element simulation of a hydraulic fracture interacting with a natural fracture. *SPE J.* **2017**, *22*, 219–234. [CrossRef]
25. Zou, Y.; Zhang, S.; Ma, X.; Zhou, T.; Zeng, B. Numerical investigation of hydraulic fracture network propagation in naturally fractured shale formations. *J. Struct. Geol.* **2016**, *84*, 1–13. [CrossRef]
26. Wu, Z.; Wong, L.N.Y. Frictional crack initiation and propagation analysis using the numerical manifold method. *Comput. Geotech.* **2012**, *39*, 38–53. [CrossRef]
27. Heider, Y.; Reiche, S.; Siebert, P.; Markert, B. Modeling of hydraulic fracturing using a porous-media phase-field approach with reference to experimental data. *Eng. Fract. Mech.* **2018**, *202*, 116–134. [CrossRef]
28. Dahi-Taleghani, A.; Olson, J.E. Numerical modeling of multistranded-hydraulic-fracture propagation: Accounting for the interaction between induced and natural fractures. *SPE J.* **2011**, *16*, 575–581. [CrossRef]
29. Gordeliy, E.; Peirce, A. Enrichment strategies and convergence properties of the XFEM for hydraulic fracture problems. *Comput. Methods Appl. Mech. Eng.* **2015**, *283*, 474–502. [CrossRef]
30. Sheng, M.; Li, G.; Shah, S.; Lamb, A.R.; Bordas, S.P. Enriched finite elements for branching cracks in deformable porous media. *Eng. Anal. Bound. Elem.* **2015**, *50*, 435–446. [CrossRef]
31. Klimenko, D.; Taleghani, A.D. A modified extended finite element method for fluid-driven fractures incorporating variable primary energy loss mechanisms. *Int. J. Rock Mech. Min. Sci.* **2018**, *106*, 329–341. [CrossRef]
32. Sheng, M.; Li, G.; Sutula, D.; Tian, S.; Bordas, S.P. XFEM modeling of multistage hydraulic fracturing in anisotropic shale formations. *J. Pet. Sci. Eng.* **2018**, *162*, 801–812. [CrossRef]
33. Shi, F.; Wang, X.; Liu, C.; Liu, H.; Wu, H. An XFEM-based method with reduction technique for modeling hydraulic fracture propagation in formations containing frictional natural fractures. *Eng. Fract. Mech.* **2017**, *173*, 64–90. [CrossRef]
34. Wang, X.; Shi, F.; Liu, C.; Lu, D.; Liu, H.; Wu, H. Extended finite element simulation of fracture network propagation in formation containing frictional and cemented natural fractures. *J. Nat. Gas Sci. Eng.* **2018**, *50*, 309–324. [CrossRef]
35. Ghaderi, A.; Taheri-Shakib, J.; Mohammad Nik, A.S. The distinct element method (DEM) and the extended finite element method (XFEM) application for analysis of interaction between hydraulic and natural fractures. *J. Pet. Sci. Eng.* **2018**, *171*, 422–430. [CrossRef]
36. Paul, B.; Faivre, M.; Massin, P.; Giot, R.; Colombo, D.; Golfier, F.; Martin, A. 3D coupled HM–XFEM modeling with cohesive zone model and applications to non planar hydraulic fracture propagation and multiple hydraulic fractures interference. *Comput. Methods Appl. Mech. Eng.* **2018**, *342*, 321–353. [CrossRef]

37. Remij, E. Wand Remmers, J.J.C.; Huyghe, J.M.; Smeulders, D.M.J. On the numerical simulation of crack interaction in hydraulic fracturing. *Comput. Geosci.* **2018**, *22*, 423–437. [CrossRef]
38. Vahab, M.; Khalili, N. X-FEM modeling of multizone hydraulic fracturing treatments within saturated porous media. *Rock Mech. Rock Eng.* **2018**, *51*, 3219–3239. [CrossRef]
39. Wang, D.; Zhou, F.; Ding, W.; Ge, H.; Jia, X.; Shi, Y.; Wang, X.; Yan, X. A numerical simulation study of fracture reorientation with a degradable fiber-diverting agent. *J. Nat. Gas Sci. Eng.* **2015**, *25*, 215–225. [CrossRef]
40. Gravouil, A.; Moës, N.; Belytschko, T. Non-planar 3D crack growth by the extended finite element and level sets—Part II: Level set update. *Int. J. Numer. Methods Eng.* **2002**, *53*, 2569–2586. [CrossRef]
41. Taleghani, A.D.; Gonzalez, M.; Shojaei, A. Overview of numerical models for interactions between hydraulic fractures and natural fractures: Challenges and limitations. *Comput. Geotech.* **2016**, *71*, 361–368. [CrossRef]
42. Feng, Y.; Li, X.; Gray, K.E. Mudcake effects on wellbore stress and fracture initiation pressure and implications for wellbore strengthening. *Pet. Sci.* **2018**, *15*, 319–334. [CrossRef]
43. Olson, J.E.; Bahorich, B.; Holder, J. Examining hydraulic fracture: Natural fracture interaction in hydrostone block experiments. In Proceedings of the SPE Hydraulic Fracturing Technology Conference, Woodlands, TX, USA, 6–8 February 2012.
44. Zhang, Q.; Su, Y.; Wang, W.; Lu, M.; Sheng, G. Apparent permeability for liquid transport in nanopores of shale reservoirs: Coupling flow enhancement and near wall flow. *Int. J. Heat Mass Transf.* **2017**, *115*, 224–234. [CrossRef]
45. Wang, D.; Zhou, F.; Ge, H.; Shi, Y.; Yi, X.; Xiong, C.; Liu, X.; Wu, Y.; Li, Y. An experimental study on the mechanism of degradable fiber-assisted diverting fracturing and its influencing factors. *J. Nat. Gas Sci. Eng.* **2015**, *27*, 260–273. [CrossRef]
46. Cherny, S.; Lapin, V.; Esipov, D.; Kuranakov, D.; Avdyushenko, A.; Lyutov, A.; Karnakov, P. Simulating fully 3D non-planar evolution of hydraulic fractures. *Int. J. Fract.* **2016**, *201*, 181–211. [CrossRef]
47. Gupta, P.; Duarte, C.A. Coupled hydromechanical-fracture simulations of nonplanar three-dimensional hydraulic fracture propagation. *Int. J. Numer. Anal. Methods Geomech.* **2018**, *42*, 143–148. [CrossRef]
48. Zlotnik, S.; Díez, P.; Fernández, M.; Vergés, J. Numerical modelling of tectonic plates subduction using X-FEM. *Comput. Methods Appl. Mech. Eng.* **2007**, *196*, 4283–4293. [CrossRef]
49. Wang, T.; Liu, Z.; Zeng, Q.; Gao, Y.; Zhuang, Z. XFEM modeling of hydraulic fracture in porous rocks with natural fractures. *Sci. China-Phys. Mech. Astron.* **2017**, *60*, 084612. [CrossRef]
50. Haddad, M.; Sepehrnoori, K. XFEM-Based CZM for the simulation of 3D multiple-cluster hydraulic fracturing in quasi-brittle shale formations. *Rock Mech. Rock Eng.* **2016**, *49*, 4731–4748. [CrossRef]
51. Feng, Y.; Gray, K. Modeling of curving hydraulic fracture propagation from a wellbore in a poroelastic medium. *J. Nat. Gas Sci. Eng.* **2018**, *53*, 83–93. [CrossRef]
52. Chen, Z. A New Enriched Finite Element Method with Application to Static Fracture Problems with Internal Fluid Pressure. *Int. J. Appl. Mech.* **2015**, *7*, 1550037. [CrossRef]
53. Meschke, G.; Leonhart, D. A Generalized Finite Element Method for hydro-mechanically coupled analysis of hydraulic fracturing problems using space-time variant enrichment functions. *Comput. Methods Appl. Mech. Eng.* **2015**, *290*, 438–465. [CrossRef]
54. Mohammadnejad, T.; Khoei, A.R. Hydro-mechanical modeling of cohesive crack propagation in multiphase porous media using the extended finite element method. *Int. J. Numeri. Anal. Methods Geomech.* **2013**, *37*, 1247–1279. [CrossRef]
55. Taleghani, A.D.; Gonzalez-Chavez, M.; Yu, H.; Asala, H. Numerical simulation of hydraulic fracture propagation in naturally fractured formations using the cohesive zone model. *J. Pet. Sci. Eng.* **2018**, *165*, 42–57. [CrossRef]
56. Kumar, D.; Ghassemi, A. Three-Dimensional Poroelastic Modeling of Multiple Hydraulic Fracture Propagation from Horizontal Wells. *Int. J. Rock Mech. Min. Sci.* **2018**, *105*, 192–209. [CrossRef]
57. Moës, N.; Dolbow, J.; Belytschko, T. A finite element method for crack growth without remeshing. *Int. J. Numer. Methods Eng.* **1999**, *46*, 131–151. [CrossRef]
58. Feng, Y.; Gray, K.E. A fracture-mechanics-based model for wellbore strengthening applications. *J. Nat. Gas Sci. Eng.* **2016**, *29*, 392–400. [CrossRef]
59. Khoei, A.R. *Extended Finite Element Method: Theory and Applications*; John Wiley & Sons: Hoboken, NJ, USA, 2014.

60. Gu, H.; Weng, X.; Lund, J.B.; Mack, M.G.; Ganguly, U.; Suarez-Rivera, R. Hydraulic fracture crossing natural fracture at nonorthogonal angles: A criterion and its validation. *SPE Prod. Oper.* **2012**, *27*, 20–26. [CrossRef]
61. Jia, B.; Tsau, J.S.; Barati, R. A review of the current progress of CO2 injection EOR and carbon storage in shale oil reservoirs. *Fuel* **2019**, *236*, 404–427. [CrossRef]
62. McGinley, M.J. The Effects of Fracture Orientation and Anisotropy on Hydraulic Fracture Conductivity in the Marcellus Shale. Master's Thesis, Texas A&M University, College Station, TX, USA, 2015. Available online: https://doi.org/http://hdl.handle.net/1969.1/155300 (accessed on 1 September 2018).
63. Jia, B.; Tsau, J.S.; Barati, R. Experimental and numerical investigations of permeability in heterogeneous fractured tight porous media. *J. Nat. Gas Sci. Eng.* **2018**, *58*, 216–233. [CrossRef]

Article

Development of Chelating Agent-Based Polymeric Gel System for Hydraulic Fracturing

Muhammad Shahzad Kamal [1,*], Marwan Mohammed [2], Mohamed Mahmoud [2] and Salaheldin Elkatatny [2]

1 Center for Integrative Petroleum Research, King Fahd University of Petroleum & Minerals, Dhahran 31261, Saudi Arabia

2 Department of Petroleum Engineering, King Fahd University of Petroleum & Minerals, Dhahran 31261, Saudi Arabia; Marwan.morz@gmail.com (M.M.); mmahmoud@kfupm.edu.sa (M.M.); elkatatny@kfupm.edu.sa (S.E.)

* Correspondence: shahzadmalik@kfupm.edu.sa; Tel.: +966-13-860-8513

Received: 21 May 2018; Accepted: 4 June 2018; Published: 26 June 2018

Abstract: Hydraulic Fracturing is considered to be one of the most important stimulation methods. Hydraulic Fracturing is carried out by inducing fractures in the formation to create conductive pathways for the flow of hydrocarbon. The pathways are kept open either by using proppant or by etching the fracture surface using acids. A typical fracturing fluid usually consists of a gelling agent (polymers), cross-linkers, buffers, clay stabilizers, gel stabilizers, biocide, surfactants, and breakers mixed with fresh water. The numerous additives are used to prevent damage resulting from such operations, or better yet, enhancing it beyond just the aim of a fracturing operation. This study introduces a new smart fracturing fluid system that can be either used for proppant fracturing (high pH) or acid fracturing (low pH) operations in sandstone formations. The fluid system consists of glutamic acid diacetic acid (GLDA) that can replace several additives, such as cross-linker, breaker, biocide, and clay stabilizer. GLDA is also a surface-active fluid that will reduce the interfacial tension eliminating the water-blockage effect. GLDA is compatible and stable with sea water, which is advantageous over the typical fracturing fluid. It is also stable in high temperature reservoirs (up to 300 °F) and it is also environmentally friendly and readily biodegradable. The new fracturing fluid formulation can withstand up to 300 °F of formation temperature and is stable for about 6 h under high shearing rates (511 s^{-1}). The new fracturing fluid formulation breaks on its own and the delay time or the breaking time can be controlled with the concentrations of the constituents of the fluid (GLDA or polymer). Coreflooding experiments were conducted using Scioto and Berea sandstone cores to evaluate the effectiveness of the developed fluid. The flooding experiments were in reasonable conformance with the rheological properties of the developed fluid regarding the thickening and breaking time, as well as yielding high return permeability.

Keywords: fracturing fluid; rheology; chelating agent; viscosity; polymer

---

1. Introduction

Hydraulic fracturing and acid fracturing operations are currently considered as one of the most important stimulation methods in the oil and gas industry [1]. In acid fracturing, the acid is spent to create uneven etches (channels) in the rock (fracture face). In acid fracturing, the formation rock must contain minerals that are partially soluble in the acid used to create those etches. On the other hand, in hydraulic fracturing, single or multiple fractures are induced in the formation by injecting a high-pressure fluid to stimulate and enhance the producing wells. These fractures are then kept open using a proppant, thus preventing the closure of those fractures due to stresses that are acting on the formation. After the completion of the process, the injected fluids are broken into low viscosity liquids using breakers to enhance the flow back of the fluid to the surface [2–5].

Hydraulic Fracturing is prominent amongst permeability-impaired formations (low permeability reservoirs) i.e., shale-gas and tight-gas [6–9]. Hydraulic fracturing significantly improves the productivity of the wells and the overall recovery factor [10]. Hydraulic fracturing is also widely used in moderate permeability reservoirs (up to 50 mD for oil and 1 mD for gas) with the large skin around the vicinity of the wellbore by bypassing the damaged zone to further enhance the flow of hydrocarbon, allowing for accelerated production without negatively impacting the formation reserves. However, this case relies mostly on the economic feasibility of conducting such operations [11].

The fracturing fluid must be designed and tested carefully in order to avoid incompatibility with the formation. Especially, if the reservoir contains minerals that are water sensitive, such as clay minerals (smectite, illite) found in tight gas or shale gas reservoirs, which can cause fines migration or swelling that results in damaging the reservoir furthermore. Due to the large quantities of gas in those formations, any enhancement on their recovery is of great importance. Tight reservoirs are those reservoirs that are characterized by a low-permeability (i.e., less than 0.5 mD), they are either carbonate or sandstone reservoirs [12,13]. Problems that are associated with tight gas production in drilling or hydraulic fracturing operations include aqueous phase trapping, natural fractures (fluid leak-off), folding and faulting (making the prediction of fracture pressure difficult), and fluid incompatibility with the formation [14]. Water blockage or aqueous phase trapping (APT) is a serious problem in tight formations among others [15–17].

Several types of fracturing fluids have been used in oil & gas fields which include but not limited to linear polymer gel, viscoelastic surfactants, crosslinked polymer gels, and foam-based fracturing fluids [18–27]. Linear and crosslinked polymer fracturing fluids can achieve high viscosity, less fluid leak-off, and good proppant suspension capabilities for varying reservoir permeabilities. Polymer-based fracturing fluids are also thermally stable. At high pressure, filter cake formation further reduces the leak-off of fluids into the formation. However, high residue that is deposited within the fracture after the completion of fracturing process is a major disadvantage of the polymer-based fracturing fluid. Different types of breakers are used to break the viscosity after completion process. The viscoelastic surfactant-based fracturing fluid is thermodynamically stable and it causes less damage to the formation when compared to the polymer-based gel. However, the rheological properties of viscoelastic gels are severely affected by temperature, counterions, and surfactant concentration. The viscoelastic surfactant-based gels have more leak-off due to low molecular weight and absence of filter cake. Therefore, a fracturing fluid with better rheological properties, thermal stability, proppant suspension capability, and less leak-off is required.

In this work, we introduce a new smart fracturing fluid system that can be either used for proppant fracturing (high pH) or acid fracturing (low pH) operations in tight as well as conventional formations. The fluid system consists of glutamic acid diacetic acid (GLDA) that can replace cross-linker, breaker, biocide, and clay stabilizer from fracturing fluid formulation. GLDA could be manufactured in the form of sodium-GLDA or potassium-GLDA, and both sodium and potassium are considered as clay stabilizers. At the same time, GLDA at high pH is gentle to the clay minerals and does not break them like HCl [28,29]. Also, published literature showed that GLDA not only acts as a biocide, but also boosts the activity and efficiency of biocides as well [30,31]. GLDA is compatible and stable with both freshwater and seawater which is advantageous over other fracturing fluids. It is also stable in high-temperature reservoirs (up to 300 °F). GLDA (which is the main constituent of the newly proposed fracturing fluid) is a low-interfacial tension (IFT) fluid, which will reduce the IFT eliminating the APT. At low pH, GLDA reacts as an acid with the carbonate minerals in the formation producing CO_2 as a by-product, and at high pH, it will react with the rocks creating a lower IFT fluid than the initial value, which makes the fluid in both pH ranges effective in reducing the APT effect. The new fluid system was tested and evaluated in low and high permeability sandstones core samples (Scioto and Berea). The fracturing fluid was tested with several polymers at several concentrations and pH ranges.

2. Experimental

The fracturing fluid formulation was prepared by dissolving the polymer and chelating agent in fresh water. Five different polymers used are shown in Table 1. Partially hydrolyzed polyacrylamide (HPAM) and Copolymer of 2-acrylamido-2-methylpropane sulfonic acid and acrylamide (AMPS) were supplied by SNF FLOERGER, France [32–36]. Thermoviscofying polymer (TVP) was obtained from Hengju Polymer Co., Beijing, China. The structures of the chelating agents are given in Table 2. The GLDA was supplied by AkzoNobel, while other chelating agents were purchased from Sigma Aldrich (Saint Louis, MO, USA). Core sample characteristics and mineral compositions of core samples are given in Tables 3 and 4, respectively. Thermogravimetric analysis (TGA) was carried out using SDT-Q600 (TA Instruments, New Castle, DE, USA) at a heating rate of 9 °F/min under a nitrogen flow rate of 20 cm^3/min. Fourier Transform Infrared Spectroscopy (FTIR) of the solutions at a different pH was conducted using Bruker Tensor27 equipment (Bruker, Billerica, MA, USA). The rheological properties were determined using high temperature and high-pressure rheometer (Grace 5600, Grace Instrument Co., Houston, TX, USA).

Table 1. The structure of the polymers used in this study.

Polymer	Abbreviation	Structure
Partially hydrolyzed polyacrylamide	HPAM	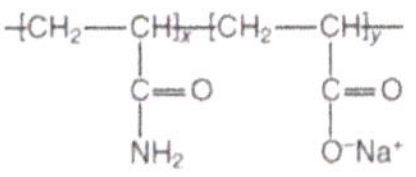
Xanthan Gum	XC	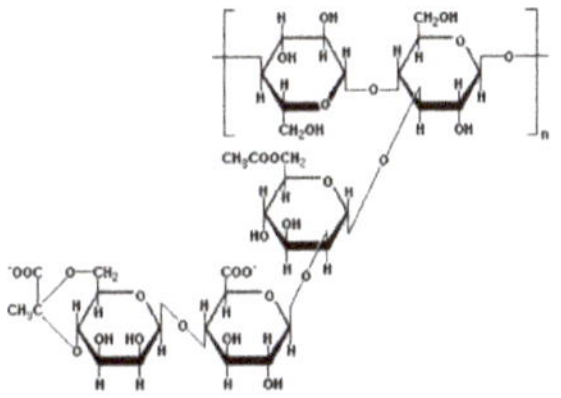
Guar Gum	HPG	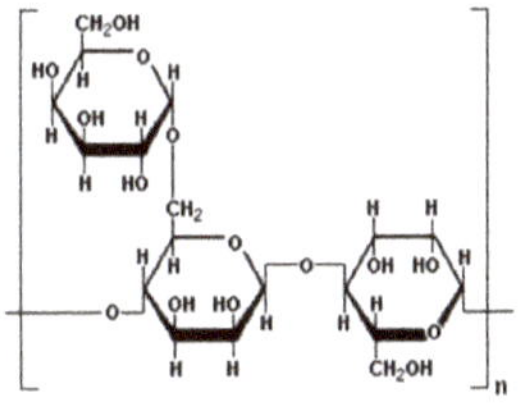
Thermoviscofying polymer	TVP	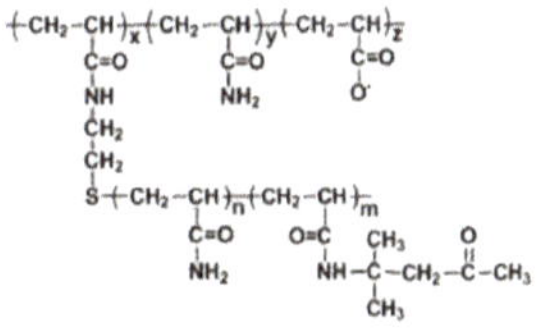
Copolymer of 2-acrylamido-2-methylpropane sulfonic acid and acrylamide	AMPS	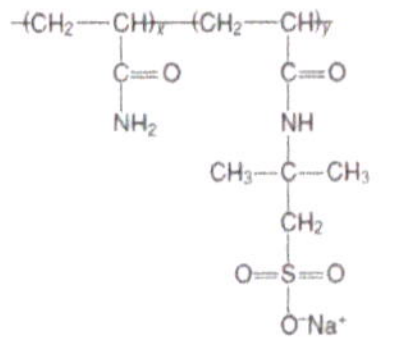

Table 2. The structure of the chelating agents used in this study.

Chelating Agent	Abbreviation	Structure
glutamic acid diacetic acid	GLDA	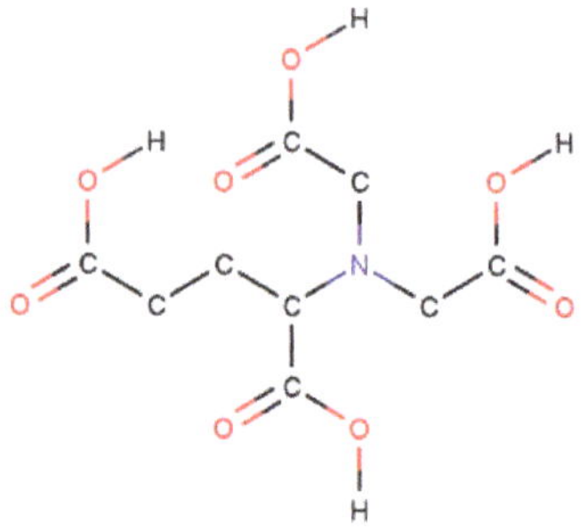
Ethylenediaminetetraacetic acid	EDTA	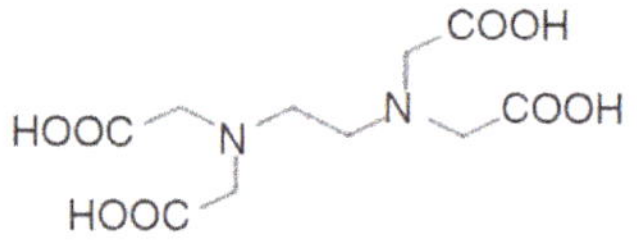
Diethylenetriaminepentaacetic acid	DTPA	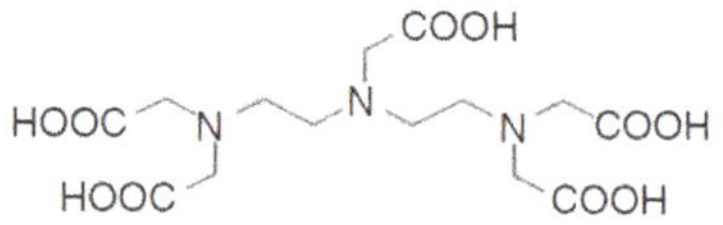

Table 3. Core Sample Characterization.

Sample	1	2
Type	Sandstone	Sandstone
Origin	Berea	Scioto
Diameter	6.35 cm	6.35 cm
Length	5.08 cm	5.08 cm
Pore Volume	35.4 cm^3	19.3 cm^3
Bulk Volume	160.8 cm^3	160.8 cm^3
Porosity	22%	12%
Pemreability	151.2 mD	3.837 mD

Table 4. Mineral composition of the core samples.

Minerals	Berea	Scioto
Quartz	86	70
Dolomite	1	-
Calcite	2	-
Feldspar	3	2
Kaolinite	5	Trace
Illite	1	18
Chlorite	2	4
Plagioclase	-	5

Two different sandstone cores with varying permeability (Table 3) were used in two coreflooding experiments. The cores were cut, polished, and the end faces were ground. The core samples were saturated with 3 wt % potassium chloride (brine water) to prevent damage occurring from clay minerals if contacted by fresh water. The preparations of core consisted of several steps. The cores were dried in an oven at 250 °F for 24 h. The dry cores were weighted and then saturated with brine under vacuum using a pump and a desiccator for 6 h. The saturated cores were weighted and the porosities of the cores were calculated. The permeabilities of the cores were calculated using Darcy's law. The schematic diagram of coreflooding setup is shown in Figure 1. For Scioto sandstone core samples, 20 wt % GLDA at pH 12 and 45 pounds per thousand gallons (pptg) of AMPS polymer diluted in deionized (DI) water were prepared for the continuous pumping experiment. For Berea sandstone core sample, 20 wt % GLDA at pH 12 and 70 pptg of XC-Polymer diluted in DI water were prepared for the continuous pumping experiment. The following procedure was adopted for coreflooding experiments:

1. Fill the cell from the top with the fracturing fluid, and tighten the cell top and connect the pressure lines coming from the transfer cells (Figure 2).
2. Insert the core sample into the cell and tighten the cell bottom of the cell against the core sample to prevent leaking and attach the pressure lines leading to the back-pressure system.
3. Set the temperature to the required value and allow enough time for the core sample to be heated (about 1 h).
4. Apply the required pressure on the transfer cells, and open the valves leading to the core cell, and apply the required back pressure to the system, and open the valves leading to the core cell.
5. Using the water pump, the injection rate was set to the required value and activated to start flooding the core sample, the pressure drop was monitored with time until the required pore volumes were injected. Effluents from some intervals were collected for analysis.

The inlet pressure, back pressure, and temperature was 500 psi, 200 psi, and 300 °F, respectively, for both cores. The injection rate for Scioto sandstone core was 1 cm^3/min and for the Berea sandstone core it was 20 cm^3/min.

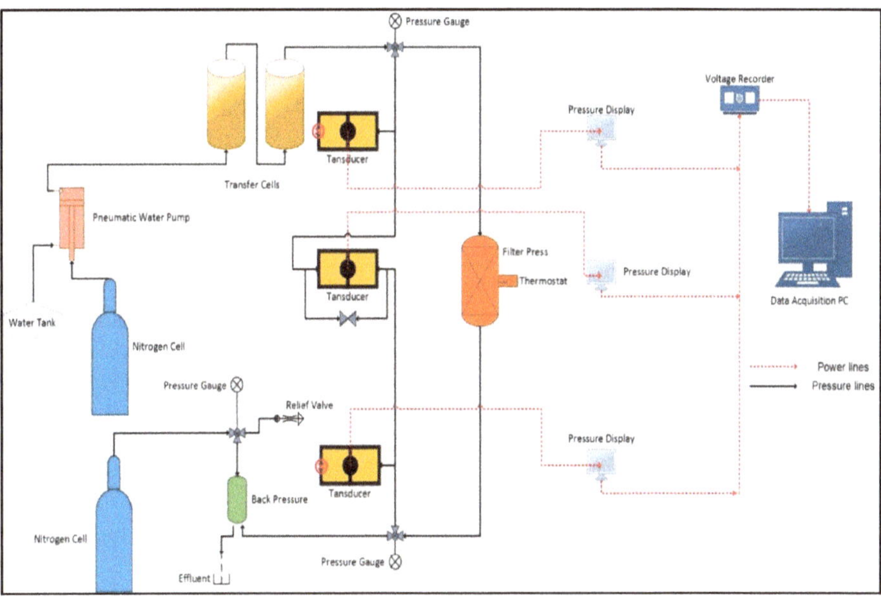

Figure 1. Filter-Press with continuous pumping set-up.

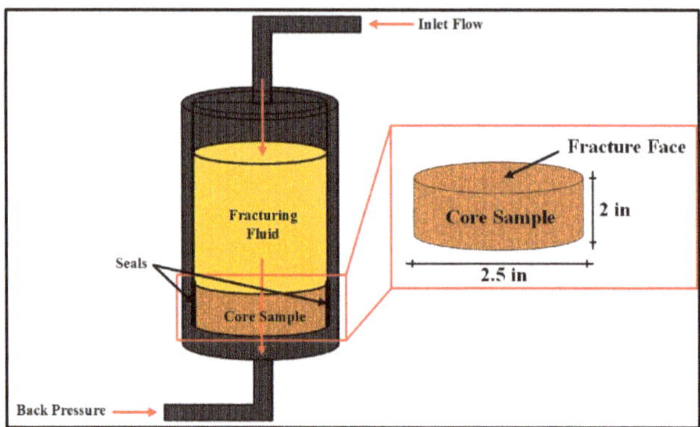

Figure 2. Cell and Core specifications of the continuous pumping set up.

3. Results & Discussion

The results and discussion section is divided into four different sections. The first section deals with the thermal stability of polymers that were used in this study. The second section describes the rheological properties of different fracturing fluids. The third section represents the FTIR analysis of the fracturing fluid formulations. Finally, the coreflooding results of the selected formulation are given in the fourth section.

3.1. Thermal Stability

Five different water-soluble polymers from different classes were selected to develop the optimum fracturing fluid formulation using polymer-chelating agent solution. The details of these polymers are given in Table 1. In the first step, the thermal stability of all the polymers was investigated using the thermogravimetric analyzer. Thermogravimetric analysis (Figure 3) showed that HPG polymer had the lowest mass loss of all the tested polymers (11.63%), followed by XC polymer (12.83%), AMPS (13.3%), HPAM (13.8%), and TVP (18.5%). However, the overall tolerance of the five polymers was good when subjected to high temperatures, a 10% average of mass loss of those polymers can be attributed to the residual humidity in the polymer powder and that is indicated by the sharp decline in the mass loss in temperatures up to 212 °F. No severe polymer degradation was noticed in the five polymer samples, which indicates that the polymers are resistive when subjected to temperatures similar to reservoir conditions (up to 350 °F).

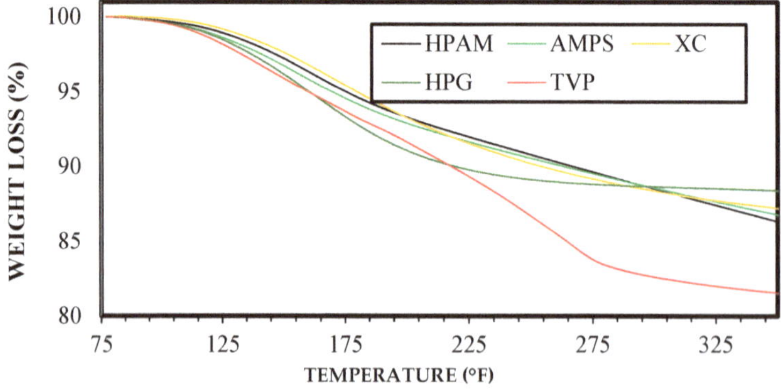

Figure 3. Thermogravimetric analysis of polymers used in this work.

3.2. Rheological Properties

The fracturing fluid formulation was developed by evaluating three different chelating agents and five polymers. The performance of three different chelating agents (DTPA, GLDA, and EDTA) with xanthan gum was determined. The apparent viscosity of the xanthan polymer solution in deionized water was measured by adding three different chelating agents at a fixed concentration (20 wt %). The concentration of the polymer was fixed to 0.43 wt % (typical field concentration). Figure 4 shows the apparent viscosity of xanthan gum with three different chelating agents. All of the investigated chelating agents (DTPA, GLDA, and EDTA) exhibited a thickening effect, however, only GLDA experienced breaking behavior without the addition of breakers. Owing to a constant viscosity with time (no breaking), DTPA and EDTA were excluded from further testing. It is, however, worth mentioning that the DTPA and EDTA can be used if a breaker is to be introduced to the system.

Five different polymers (TVP, HPG, XC, AMPS, and HPAM) at a fixed concentration (20 pptg) were mixed with GLDA (20 wt %) and the apparent viscosity was measured versus time for each sample. Figure 5 shows the viscosity of polymers-GLDA solution in deionized water at 300 °F and 300 psi. The maximum thickening effect was obtained using XC polymer followed by HPAM. The viscosity of the XC polymer was 2.9 cP after 370 min. The HPAM achieved the viscosity of water after 280 min, while the TVP and HPG approached the viscosity of water after 100 min. The minimum thickening effect was achieved using AMPS polymer.

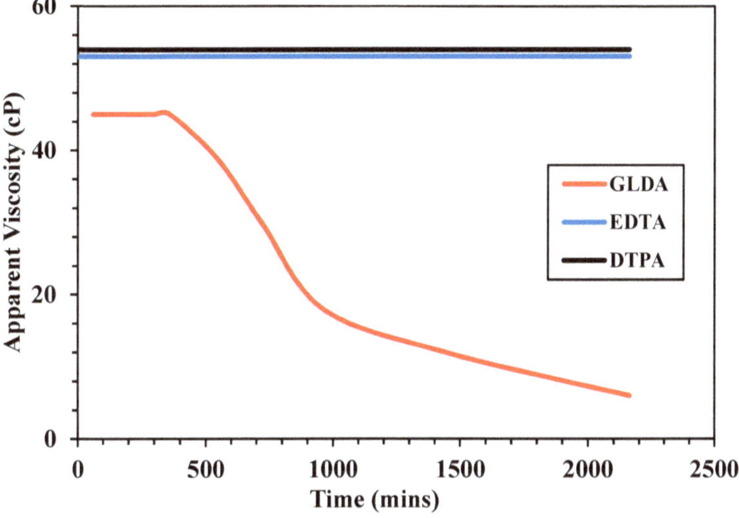

Figure 4. Apparent viscosity of xanthan gum (0.43 wt %) with three different chelating agents (20 wt %) in deionized water (Shear rate- 170.3 s^{-1}, T = 200 °F, P = 300 psi, pH = 12).

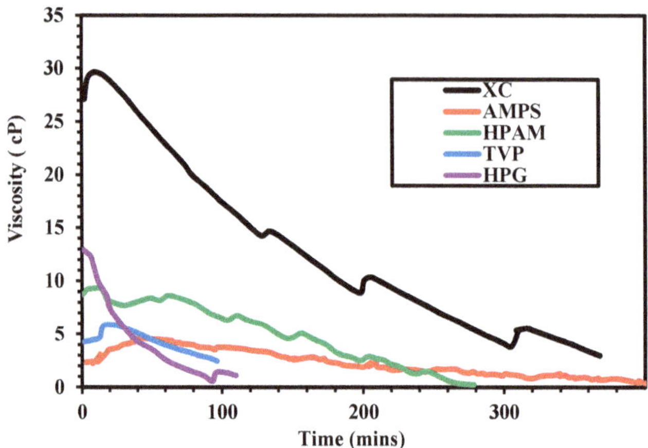

Figure 5. Apparent viscosity of different polymers (20 pptg) with glutamic acid diacetic acid (GLDA) (20 wt %) in deionized water (Shear rate = 511 s^{-1}, T = 300 °F, P = 300 psi, pH = 12).

Figure 6 shows the apparent viscosity of GLDA-XC polymer solution at different pH values. The mixing of GLDA with XC increased the apparent viscosity from 33 cP (the apparent viscosity of 0.43 wt % XC alone) to higher values at all investigated pH. At a pH of 4, the apparent viscosity of the GLDA-XC polymer solution increased to 55 cP, which was reduced to 50 cP after 10 h due to breakage of linked branches of the polymer. At a pH of 7, the apparent viscosity increased to 75 cP and reduced to 60 cP after 3.5 h. At a pH of 12, the apparent viscosity of the mixture increased to 45 cP. After 7 h, the viscosity of the mixture was reduced to below the initial value of XC polymer. Only at this pH, both thickening and breaking took place, which is the main requirement in fracturing fluids. This indicates breaking characteristics of GLDA at pH 12. The apparent viscosity of the GLDA-XC polymer solution at room temperature (pH = 12) increased 50 cP and remained intact throughout the entire time of mixing (approx. 40 h), which indicated the failure of breaking at room temperature.

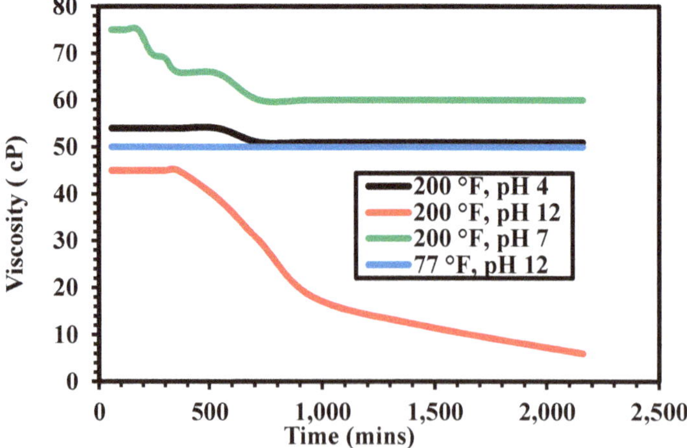

Figure 6. Apparent viscosity XC polymer (0.43 wt %) with GLDA (20 wt %) at different pH values (Shear rate- 170.3 s^{-1}, P = 300 psi).

Figure 7 shows the apparent viscosity of the AMPS polymer-GLDA solutions at different pH and temperatures. The viscosity of the AMPS-GLDA solutions is higher when compared to the viscosity of the AMPS solutions. The viscosity at pH 4 and pH 7 was almost constant throughout the experiment and no breakage of the solution viscosity was observed at both pH. However, at pH 12, the viscosity of the AMPS-GLDA solutions was increased initially and then decreased. At low temperature (77 °F), the viscosity of the AMPS-GLDA solution was fluctuating between 6 cP and 7 cP without any breaking. This indicates that, at room temperature, the GLDA thickens the polymer solution but it did not break it. This suggests that viscosity of chelating agent-polymer solution strongly depends on temperature and pH. When the polymer concentration was increased to 45 pptg, the initial viscosity was much higher when compared to the solution with 20 pptg solutions. However, the viscosity declined sharply after 30 min, which indicates the breaking of the polymer chains. As expected, increasing polymer concentration enhanced the viscosity of the thickened fluid. However, the stability of the fluid with time under constant shearing decreased.

The concentration of the GLDA was optimized using 45 pptg of AMPS polymer at 300 °F and pH of 12. The apparent viscosity of GLDA-polymer solutions at a different concentration of GLDA is shown in Figure 8. As observed from Figure, 5 wt % of GLDA yielded a very stable solution under high-temperature high-pressure conditions but the viscosity increase was minimal due to the small concentration of GLDA. The solution's viscosity is very close to the viscosity of the polymer alone, which indicates that the thickening effect was also minimal on this solution. The similar effect was observed at 10% of GLDA. At higher concentrations (20–40%), the thickening effect was increased significantly. At 40% GLDA, the thickening effect was less compared to the effect at 20% and 30%. The highest viscosity was obtained using a GLDA concentration between 20% to 30% and using 45 pptg of AMPS polymer in fresh water. The results clearly indicate that the viscosity thickening effect can be controlled with the concentration of GLDA, and for optimum conditions, it should not be more than 30 wt %. All of the stimulation operations in the field are performed with a concentration of 20 wt % because it was found to be the optimum in case of stimulation [37–39]. In this case, 20% of GLDA also showed the optimum results.

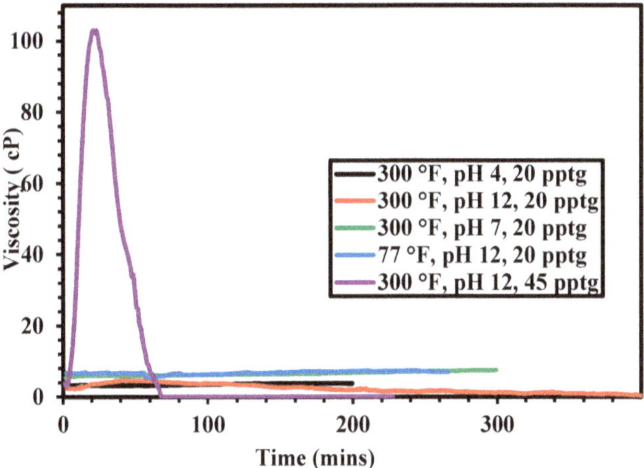

Figure 7. Apparent viscosity of acrylamide (AMPS) polymer with GLDA (20 wt %) at different conditions (Shear rate = 511 s^{-1}, P = 300 psi).

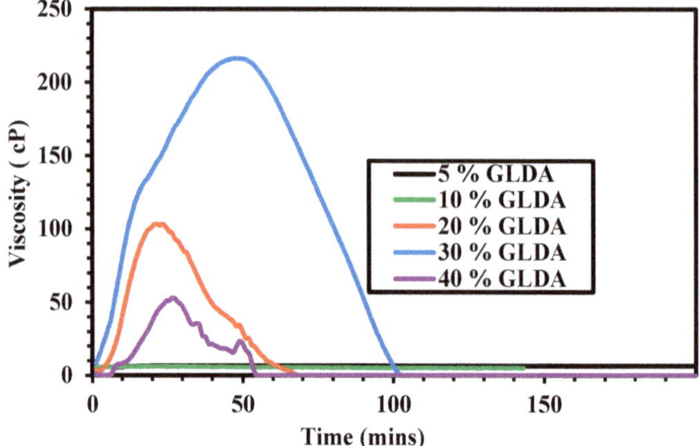

Figure 8. Apparent viscosity of AMPS polymer with a different concentration of GLDA at 300 °F (Shear rate = 511 s^{-1}, P = 300 psi, pH = 12).

3.3. FTIR Analysis

FTIR analysis was carried out to understand the thickening and breaking mechanism using GLDA. The FTIR analysis of GLDA was conducted at pH 4 and pH 12 (Figure 9). At pH 4, the carboxyl group was identified at the wavenumber 3477 cm^{-1}, which are the functional group of GLDA. It is characterized by a broad spectrum at 3477 cm^{-1} due to the OH group. The presence of C=O from the carboxyl group was identified at wavenumber 1641 cm^{-1}. At the wave number 1396 cm^{-1}, a peak was found and it was caused by the (C-N) group, however, this group is a non-functional group and it will not contribute to the thickening and breaking of the polymer. At pH 12, two peaks were identified at wavenumbers 1587 cm^{-1} and 1685 cm^{-1}; the first was associated with C=O of the carboxylate group (COO^-), while the later was associated with the C=O of a carboxyl group (C(=O) OH). The OH group was also identified at wavenumber 3610 cm^{-1}. Comparison between the two spectra at pH 4 and 12 shows that increasing the pH of GLDA resulted in a reaction between GLDA and base. The reaction between GLDA and base resulted in the partial loss of a proton from COOH group leaving behind the both COOH and COO^-, which is evident by the two peaks (1587 cm^{-1} & 1685 cm^{-1}).

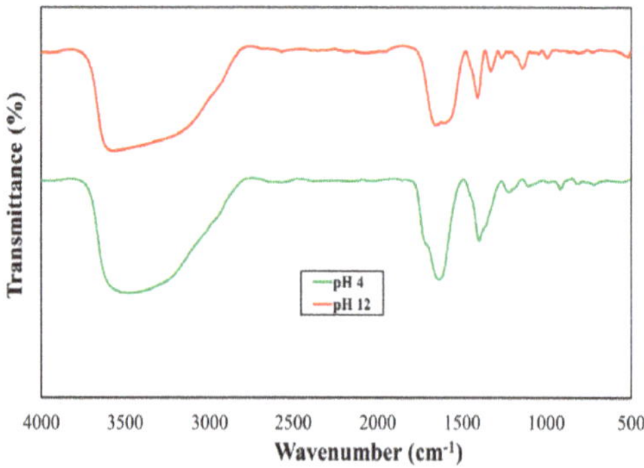

Figure 9. Fourier Transform Infrared Spectroscopy (FTIR) analysis of GLDA at different pH.

FTIR analysis of the polymer in fresh water and polymer/GLDA solution is given in Figure 10. From the spectrum, the amide group (O=C-NH2) was identified as the functional group, with the carbonyl group (C=O) at a wavenumber of 1631 cm^{-1} and (N-H) at wavenumber 3488 cm^{-1}. A mixture of GLDA (at pH 12) and the polymer in fresh water was prepared, and FTIR analysis was conducted on this fluid at the thickened and breaking stage in order to identify the functional groups responsible for the thickening-breaking effect. The (OH) from GLDA appeared at a wavenumber of 3621 cm^{-1}. The N-H from the AMPS also contributes to this broad peak. The peak around 1670 cm^{-1} is due to the contribution of carbonyl from amide group of the polymer and COOH from GLDA. The spectrum also shows two distinct (C-N) groups peaks forming at wavenumber 1403 cm^{-1} and 1322 cm^{-1}, one coming from the GLDA and the other from the polymer. The initial increase in the viscosity is associated with the partial loss of proton at high pH leaving behind the COO^{-}. This results in the formation of a complex of GLDA and the polymer that cause an increase in the viscosity. However, there is another competing reaction between the polymer and OH^{-}, which will result in the degradation of the polymer chain and viscosity reduction.

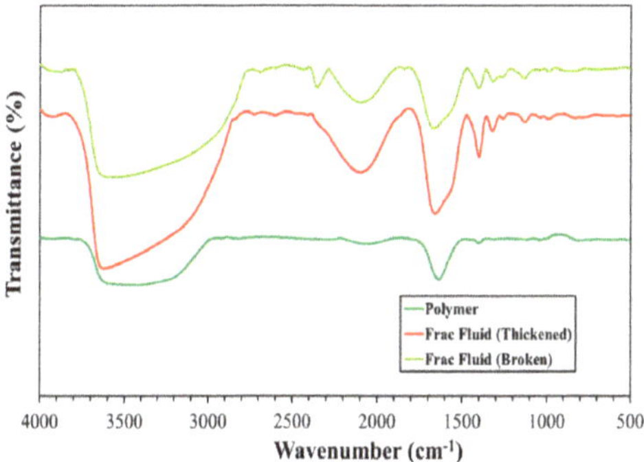

Figure 10. FTIR analysis of polymer in deionized water and developed fracturing fluid.

3.4. *Coreflooding*

Two sandstone core samples were cut and prepared for flooding using the continuous pumping setup. The porosity of the core samples was determined by measuring the dry and saturated weight of the core samples. The core samples were dried and weighted, followed by the saturation with 3 wt % KCl. After saturating the core with 3 wt % KCl, the core sample permeability has been measured using the set-up after the flow and pressure difference has been stabilized. The two cores selected were of different permeabilities and different fracturing fluids were evaluated. For high permeability core, 20% GLDA (at pH = 12) with 70 pptg XC polymer was used. For low permeability core, 20% GLDA (pH = 12) with 45 pptg AMPS polymer was injected.

3.4.1. High Permeability Coreflooding

The permeability was calculated using Darcy's law and the average permeability was found to be 151.2 mD. The fracturing fluid that was used in this core experiment consists of 20 wt % of GLDA (at pH 12) mixed with 70 pptg of XC polymer in fresh water. The reason behind using high polymer concentration is the high permeability of the core sample, which requires a thick fluid system. The viscosity of the developed fluid after thickening reached 200 cP. The experiment was conducted for approximately three hours and the pressure profile is shown in Figure 11. It can be seen from the pressure profile that the fracturing fluid did not flow at the beginning of the experiment due to the high viscosity of the fluid, which suggests that the thickening succeeded. The pressure difference that is required for this fluid to flow was 1048 psi using Darcy's law. Since the fluid was unable to flow through the core sample, the pressure started to build up until the fluid started to gradually break and hence allowing the fracturing fluid to flow through the core. The pressure started dropping after approximately two hours from the start of the flooding. The return permeability of the core sample was measured by reversing the core and flowing it back with 3 wt % KCl. The average return permeability was found to be 128 mD, and the regained permeability was found to be 85.2% of the original permeability.

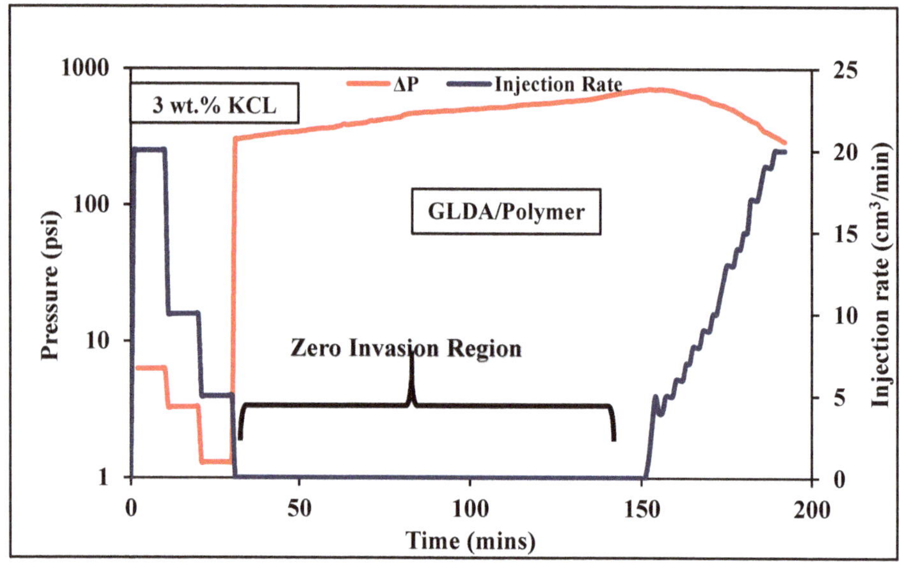

Figure 11. Coreflooding data for the GLDA-XC polymer solution.

3.4.2. Low Permeability Coreflooding

After saturating the core with 3 wt % KCl, the core sample permeability was measured and the average permeability of the core was found to be 3.837 mD. The fracturing fluid was prepared using 20 wt % of GLDA and 45 pptg of AMPS in fresh water. The experiment was conducted for approximately 4 h. It can be seen from the pressure profile (Figure 12) that the fracturing fluid did not flow at the beginning of the experiment due to the high viscosity of the fluid which suggests that the thickening succeeded. The pressure difference that is required for this fluid to flow was 1025 psi using Darcy's law. Since the fluid was unable to flow through the core sample, the pressure started to build up until the fluid started to gradually break, and hence allowing the fracturing fluid to flow through the core. The pressure started dropping after two hours from the start of the flooding. This result is not with great conformance with the rheology. This is because of the imposed shear rate in a rheological experiment that reduced the stability of the fluid. Whereas, in this case, the fluid was in the static state, which prolonged the breakage of the fluid. The return permeability of the core sample was then measured by reversing the core and flowing it back with 3 wt % KCl. The average return permeability was found to be 3.4067 mD, and the regained permeability was found to be 88.8% of the original permeability.

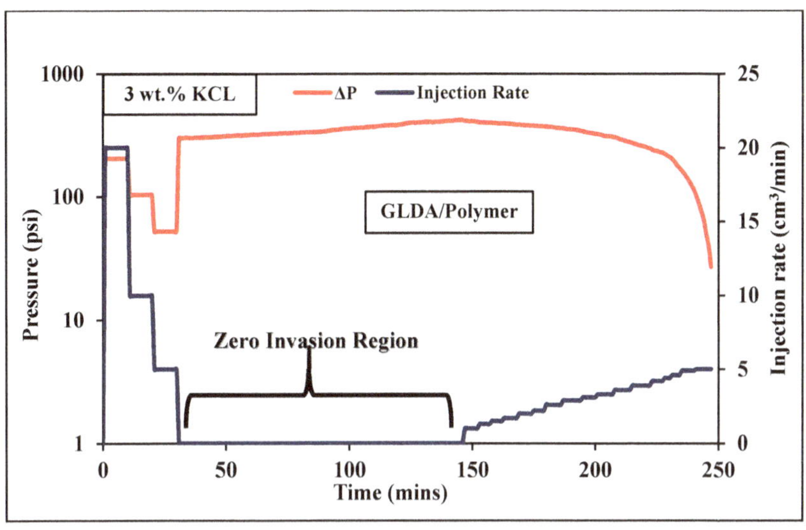

Figure 12. Coreflooding data for GLDA-AMPS polymer solution.

4. Conclusions

In this work, five different water-soluble polymers and three different chelating agents at various temperature, concentration, and pH were evaluated to develop a new, simple, smart, environmentally-friendly fracturing fluid for fracturing sandstone formations. The fracturing fluid mainly consists of a water-soluble polymer and chelating agent. Thermal stability, rheology, FTIR, and core flooding was performed to determine the optimum conditions and concentration of fracturing fluid. The thermogravimetric analysis reveals that all of the investigated polymers were thermally stable at reservoir temperature. The rheological properties were investigated by changing temperature, pH, shear rate, chelating agent type and concentration, and polymer type and concentration. Among investigated chelating agents, only GLDA shows both thickening and breaking profiles only at basic pH range. EDTA and DTPA showed the thickening behavior but could not break the viscosity. The optimum concentration of the GLDA was found to be between 20% and 30%, and the developed fluid will be more stable at high temperature. Fourier Transform Infrared Spectroscopy analysis was conducted to determine the functional groups that were responsible for the thickening and breaking of the developed fracturing fluid. The main groups that were responsible for the thickening and breaking effect are the amide group (present in the polymer) and the carboxyl group (present in the GLDA). Core flooding experiments were conducted on a low and a high permeability sandstones cores (Scioto & Berea) to prove the effectiveness of the developed fluid, by treating the core surface as the fracture face and studying the invasion of the fluid to the core. The coreflooding of Scioto (low permeability core) yielded a return permeability of 89% and the fluid used composed of 20 wt % of GLDA, 45 pptg of Co-polymer (AMPS) mixed in fresh water. The second coreflooding experiment on Berea sandstone yielded a return permeability of 85% and the fluid that was used to flood composed of 20 wt % GLDA, 70 pptg of XC polymer mixed with fresh water. The developed fluid could result in replacing several additives that are essential in the formulation of typical fracturing fluids, such as cross-linker, breaker, biocide, clay stabilizer, and friction reducer.

Author Contributions: Data curation, M.M. (Marwan Mohammed); Methodology, M.S.K. and S.E.; Project administration, M.M. (Mohamed Mahmoud); Supervision, M.M. (Mohamed Mahmoud); Writing—original draft, M.S.K., and M.M. (Marwan Mohammed); Writing—review & editing, S.E.

Acknowledgments: Mahmoud would like to acknowledge the College of Petroleum Engineering and Geosciences, King Fahd University of Petroleum & Minerals, Saudi Arabia for supporting this work under start-up grant.

Conflicts of Interest: The authors declare no conflict of interest.

References

1. Li, Q.; Xing, H.; Liu, J.; Liu, X. A review on hydraulic fracturing of unconventional reservoir. *Petroleum* **2015**, *1*, 8–15. [CrossRef]
2. Das, A.; Chauhan, G.; Verma, A.; Kalita, P.; Ojha, K. Rheological and breaking studies of a novel single-phase surfactant-polymeric gel system for hydraulic fracturing application. *J. Pet. Sci. Eng.* **2018**, *167*, 559–567. [CrossRef]
3. Zhao, J.; Fan, J.; Mao, J.; Yang, X.; Zhang, H.; Zhang, W. High performance clean fracturing fluid using a new tri-cationic surfactant. *Polymers* **2018**, *10*, 535. [CrossRef]
4. Fogang, L.T.; Sultan, A.S.; Kamal, M.S. Understanding viscosity reduction of a long-tail sulfobetaine viscoelastic surfactant by organic compounds. *RSC Adv.* **2018**, *8*, 4455–4463. [CrossRef]
5. Fogang, L.T.; Sultan, A.S.; Kamal, M.S. Comparing the Effects of Breakers on a Long-tail Sulfobetaine Viscoelastic Surfactant Solution for Well Stimulation. In Proceedings of the Abu Dhabi International Petroleum Exhibition & Conference, Abu Dhabi, United Arab Emirates, 7–10 November 2016. SPE-182915-MS.
6. Ahmed, S.; Elraies, K.A.; Hashmet, M.R.; Alnarabiji, M.S. Empirical modeling of the viscosity of supercritical carbon dioxide foam fracturing fluid under different downhole conditions. *Energies* **2018**, *11*, 782. [CrossRef]
7. Fan, M.; Li, J.; Liu, G. New method to analyse the cement sheath integrity during the volume fracturing of shale gas. *Energies* **2018**, *11*, 750. [CrossRef]
8. Zhang, C.; Ranjith, P.G. Experimental study of matrix permeability of gas shale: An application to CO_2-based shale fracturing. *Energies* **2018**, *11*, 702. [CrossRef]
9. Liu, R.; Jian, Y.; Haung, N.; Sugimoto, S. Hydraulic properties of 3d crossed rock fractures by considering anisotropic aperture distributions. *Adv. Geo Energy Res.* **2018**, *2*, 113–121. [CrossRef]
10. Parvizi, H.; Rezaei Gomari, S.; Nabhani, F.; Dehghan Monfared, A. Modeling the risk of commercial failure for hydraulic fracturing projects due to reservoir heterogeneity. *Energies* **2018**, *11*, 218. [CrossRef]
11. Holditch, S.A. Tight gas sands. *J. Pet. Technol.* **2006**, *58*, 86–93. [CrossRef]
12. Assiri, W.; Miskimins, J.L. The Water Blockage Effect on Desiccated Tight Gas Reservoir. In Proceedings of the SPE International Symposium and Exhibition on Formation Damage Control, Lafayette, LA, USA, 26–28 February 2014. SPE-168160-MS.
13. Bennion, D.; Thomas, F.; Schulmeister, B.; Sumani, M. Determination of true effective in situ gas permeability in subnormally water-saturated tight gas reservoirs. *J. Can. Pet. Technol.* **2004**, *43*. [CrossRef]
14. Wang, F.; Li, B.; Zhang, Y.; Zhang, S. Coupled thermo-hydro-mechanical-chemical modeling of water leak-off process during hydraulic fracturing in shale gas reservoirs. *Energies* **2017**, *10*, 1960. [CrossRef]
15. Civan, F. Analyses of Processes, Mechanisms, and Preventive Measures of Shale-gas Reservoir Fluid, Completion, and Formation Damage. In Proceedings of the SPE International Symposium and Exhibition on Formation Damage Control, Lafayette, LA, USA, 26–28 February 2014. SPE-168164-MS.
16. Middleton, R.S.; Carey, J.W.; Currier, R.P.; Hyman, J.D.; Kang, Q.; Karra, S.; Jiménez-Martínez, J.; Porter, M.L.; Viswanathan, H.S. Shale gas and non-aqueous fracturing fluids: Opportunities and challenges for supercritical CO_2. *Appl. Energy* **2015**, *147*, 500–509. [CrossRef]
17. Bertoncello, A.; Wallace, J.; Blyton, C.; Honarpour, M.; Kabir, C. Imbibition and Water Blockage in Unconventional Reservoirs: Well management implications during flowback and early production. In Proceedings of the SPE/EAGE European Unconventional Resources Conference and Exhibition, Vienna, Austria, 25–27 February 2014. SPE-167698-MS.
18. Kesavan, S.; Prud'Homme, R.K. Rheology of guar and (hydroxypropyl) guar crosslinked by borate. *Macromolecules* **1992**, *25*, 2026–2032. [CrossRef]
19. Jennings, A.R., Jr. Fracturing fluids-then and now. *J. Pet. Technol.* **1996**, *48*, 604–610. [CrossRef]
20. Shah, S.; Lord, D.; Rao, B. Borate-crosslinked Fluid Rheology under Various ph, Temperature, and Shear History Conditions. In Proceedings of the SPE Production Operations Symposium, Oklahoma City, Oklahoma, 9–11 March 1997. SPE-37487-MS.

21. Gaillard, N.; Thomas, A.; Favero, C. Novel Associative Acrylamide-based Polymers for Proppant Transport in Hydraulic Fracturing Fluids. In Proceedings of the SPE International Symposium on Oilfield Chemistry, The Woodlands, TX, USA, 8–10 April 2013. SPE-164072-MS.
22. Holtsclaw, J.; Funkhouser, G.P. A crosslinkable synthetic-polymer system for high-temperature hydraulic-fracturing applications. *SPE Drill. Complet.* **2010**, *25*, 555–563. [CrossRef]
23. Omeiza, A.; Samsuri, A. Viscoelastic surfactants application in hydraulic fracturing, it's set back and mitigation—an overview. *ARPN J. Eng. Appl. Sci.* **2006**, *9*, 25–29.
24. Samuel, M.; Polson, D.; Graham, D.; Kordziel, W.; Waite, T.; Waters, G.; Vinod, P.; Fu, D.; Downey, R. Viscoelastic Surfactant Fracturing Fluids: Applications in Low Permeability Reservoirs. In Proceedings of the SPE Rocky Mountain Regional/Low-Permeability Reservoirs Symposium and Exhibition, Denver, CO, USA, 12–15 March 2000. SPE-60322-MS.
25. Harris, P. Application of Foam Fluids to Minimize Damage during Fracturing. In Proceedings of the International Meeting on Petroleum Engineering, Beijing, China, 24–27 March 1992. SPE-22394-MS.
26. Ahmed, S.; Elraies, K.A.; Hanamertani, A.S.; Hashmet, M.R. Viscosity models for polymer free CO_2 foam fracturing fluid with the effect of surfactant concentration, salinity and shear rate. *Energies* **2017**, *10*, 1970. [CrossRef]
27. He, J.; Afolagboye, L.O.; Lin, C.; Wan, X. An experimental investigation of hydraulic fracturing in shale considering anisotropy and using freshwater and supercritical CO_2. *Energies* **2018**, *11*, 557. [CrossRef]
28. Mahmoud, M.A.; Nasr-El-Din, H.A.; De Wolf, C.A. High-temperature laboratory testing of illitic sandstone outcrop cores with hcl-alternative fluids. *SPE Prod. Oper.* **2015**, *30*, 43–51. [CrossRef]
29. Mahmoud, M.A.; Nasr-El-Din, H.A.; De Wolf, C.; Alex, A. Sandstone Acidizing Using a New Class of Chelating Agents. In Proceedings of the SPE International Symposium on Oilfield Chemistry, The Woodlands, TX, USA, 11–13 April 2011. SPE-139815-MS.
30. De Wolf, C.A.; Nasr-El-Din, H.; LePage, J.N.; Mahmoud, M.A.N.-E.-D.; Bemelaar, J.H. Environmentally Friendly Stimulation Fluids, Processes to Create Wormholes in Carbonate Reservoirs, and Processes to Remove Wellbore Damage in Carbonate Reservoirs. U.S. Patent 9150780B2, 6 October 2015.
31. Mahmoud, M.A.N.E.-D.; Gadallah, M.A. Method of Maintaining Oil Reservoir Pressure. U.S. Patent 20140345868A1, 27 November 2014.
32. Kamal, M.S.; Shakil Hussain, S.M.; Fogang, L.T. A zwitterionic surfactant bearing unsaturated tail for enhanced oil recovery in high-temperature high-salinity reservoirs. *J. Surfactants Deterg.* **2018**, *21*, 165–174. [CrossRef]
33. Ahmad, H.M.; Kamal, M.S.; Al-Harthi, M.A. High molecular weight copolymers as rheology modifier and fluid loss additive for water-based drilling fluids. *J. Mol. Liq.* **2018**, *252*, 133–143. [CrossRef]
34. Shahzad Kamal, M.; Sultan, A. Thermosensitive Water Soluble Polymers: A solution to High Temperature and High Salinity Reservoirs. In Proceedings of the SPE Kingdom of Saudi Arabia Annual Technical Symposium and Exhibition, Dammam, Saudi Arabia, 24–27 April 2017. SPE-188006-MS.
35. Malik, I.A.; Al-Mubaiyedh, U.A.; Sultan, A.S.; Kamal, M.S.; Hussein, I.A. Rheological and thermal properties of novel surfactant-polymer systems for eor applications. *Can. J. Chem. Eng.* **2016**, *94*, 1693–1699. [CrossRef]
36. Ahmad, H.M.; Kamal, M.S.; Al-Harthi, M.A. Rheological and filtration properties of clay-polymer systems: Impact of polymer structure. *Appl. Clay Sci.* **2018**, *160*, 226–237. [CrossRef]
37. Mahmoud, M.; Abdelgawad, K.Z.; Elkatatny, S.M.; Akram, A.; Stanitzek, T. Stimulation of seawater injectors by glda (glutamic-di acetic acid). *SPE Drill. Complet.* **2016**, *31*, 178–187. [CrossRef]
38. Ameur, Z.O.; Kudrashou, V.; Nasr-El-Din, H.; Forsyth, J.; Mahoney, J.; Daigle, B. Stimulation of High Temperature Sagd Producer Wells Using a Novel Chelating Agent (glda) and Subsequent Geochemical Modeling Using Phreeqc. In Proceedings of the SPE International Symposium on Oilfield Chemistry, The Woodlands, TX, USA, 13–15 April 2015. SPE-173774-MS.
39. Braun, W.; De Wolf, C.A.; Nasr-El-Din, H.A. Improved Health, Safety and Environmental Profile of a New Field Proven Stimulation Fluid. In Proceedings of the SPE Russian Oil and Gas Exploration and Production Technical Conference and Exhibition, Moscow, Russia, 16–18 October 2012. SPE-157467-MS.

Article

Experimental Study of Sulfonate Gemini Surfactants as Thickeners for Clean Fracturing Fluids

Shanfa Tang [1,2], Yahui Zheng [2,*], Weipeng Yang [3,*], Jiaxin Wang [2], Yingkai Fan [2] and Jun Lu [3]

1 Hubei Cooperative Innovation Center of Unconventional Oil and Gas in Yangtze University, Wuhan 430100, China; tangsf2005@126.com
2 School of Petroleum Engineering, Yangtze University, Wuhan 430100, China; 201672102@yangtzeu.edu.cn (J.W.); 201772084@yangtzeu.edu.cn (Y.F.)
3 McDougall School of Petroleum Engineering, The University of Tulsa, Tulsa, OK 74104, USA; jun-lu@utulsa.edu
* Correspondence: 201671182@yangtzeu.edu.cn (Y.Z.); weipeng-yang@utulsa.edu (W.Y.); Tel.: +86-130-2710-8092 (Y.Z.); +1-918-964-2735 (W.Y.)

Received: 24 October 2018; Accepted: 13 November 2018; Published: 16 November 2018

Abstract: Hydraulic fracturing is one of the important methods to improve oil and gas production. The performance of the fracturing fluid directly affects the success of hydraulic fracturing. The traditional cross-linked polymer fracturing fluid can cause secondary damage to oil and gas reservoirs due to the poor flow-back of the fracturing fluid, and existing conventional cleaning fracturing fluids have poor performance in high temperature. Therefore, this paper has carried out research on novel sulfonate Gemini surfactant cleaning fracturing fluids. The rheological properties of a series of sulfonate Gemini surfactant (DSm-s-m) solutions at different temperatures and constant shear rate (170 s^{-1}) were tested for optimizing the temperature-resistance and thickening properties of anionic Gemini surfactants in clean fracturing fluid. At the same time, the microstructures of solutions were investigated by scanning electron microscope (SEM). The experimental results showed that the viscosity of the sulfonate Gemini surfactant solution varied with the spacer group and the hydrophobic chain at 65 °C and 170 s^{-1}, wherein DS18-3-18 had excellent viscosity-increasing properties. Furthermore, the microstructure of 4 wt.% DS18-3-18 solution demonstrated that DS18-3-18 self-assembled into dense layered micelles, and the micelles intertwined with each other to form the network structure, promoting the increase in solution viscosity. Adding nano-MgO can increase the temperature-resistance of 4 wt.% DS18-3-18 solution, which indicated that the rod-like and close-packed layered micelles were beneficial to the improvement of the temperature-resistance and thickening performances of the DS18-3-18 solution. DS18-3-18 was not only easy to formulate, but also stable in all aspects. Due to its low molecular weight, the damage to the formation was close to zero and the insoluble residue was almost zero because of the absence of breaker, so it could be used as a thickener for clean fracturing fluids in tight reservoirs.

Keywords: sulfonate gemini surfactant; thickener; temperature-resistance; clean fracturing fluid

1. Introduction

The weakness of polymer crosslinking fracturing fluid in field applications are becoming more and more obvious with the development of unconventional oil reservoirs [1–5]. Compared with the traditional cross-linked polymer fracturing fluids, clean fracturing fluids consist of a viscoelastic surfactant and a corresponding salt solution. The have the advantages of high flowback, simple on-site preparation, simple injection process and less damage to the formation [6]. As is shown in Table 1, compared with the guanidine gum fracturing fluid, the cleaning fracturing fluid has many significant advantages. However, the increase of temperature is not conducive to the formation of closely packed worm-like micelles in solution for the conventional single-chain surfactants, which results in low viscosity and poor thickening performance [7]. Simultaneously, the amount of surfactant will obviously increase with the increase of temperature, and the cost of clean fracturing fluids will increase significantly [8,9]. Gemini surfactants have a special molecular structure including two hydrophilic groups and two hydrophobic chains. They have characteristics of lower oil-water interfacial tension, CMC, Krafft point, and higher viscosity compared with conventional single-chain surfactants [10–12]. Gemini surfactants can form stable micellar structures with obvious thickening effects at low concentration. Therefore, they have broad application potential in constructing a temperature-resistant thickening clean fracturing fluid system [13,14]. At present, the application of Gemini surfactants in cleaning fracturing fluids is mainly concentrated on cationic Gemini surfactants [15]. Zhu et al. [16] added 4 wt.% NaCl or 15 wt.% hydrochloric acid to 2–2.5 wt.% cationic Gemini surfactant solution to form a clean fracturing fluid. The temperature resistance of the mixed solution could be up to 95 °C and 80 °C, respectively. Yang et al. [17] found that the 1.0 wt.% cationic Gemini surfactant could maintain a high viscosity (50 mPa·s) at 120 °C and 100 s^{-1}. Yu [18] studied the viscosity characteristics and the micellar structures of cationic Gemini surfactants. It was found that by increasing the hydrophobic chain length at s = 2, the linear micelles in the solution tended to aggregate to form a network structure which led to an increase in viscosity.

Table 1. Comparison of clean fracturing fluid and guanidine gum fracturing fluid.

Parameter	Clean Fracturing Fluid	Guanidine Gum Fracturing Fluid
Molecular weight category	Molecular weight less than 500	Molecular weight greater than 50,000
With or without crosslinker	No	Yes
With or without a breaker	No	Yes
Sand carrying mechanism	Viscoelastic body carrying sand, viscosity greater than 30 mPa·s at 100 s^{-1}	Fracturing fluid carrying sand, viscosity greater than 100 mPa·s at 100 s^{-1}
Filtration loss	Low filtration	High filtration
Diversion capacity	More than 93%	More than 70%
Craftsmanship	Easy to make on site	Difficult to make on site

However, because of the positive charge-carrying properties of cationic Gemini surfactants, they are easily adsorbed on the surface of oil-bearing rocks, leading to alteration of the wettability of the oil-gas seepage channels and about 14 wt.% permeability damage to the reservoir [19]. Due to the fact that anionic Gemini surfactants have a lower adsorption than cationic Gemini surfactants, in recent years, scholars have studied anionic Gemini surfactant viscoelastic fracturing fluids. Several studies have shown that the viscosity of the sulfonate Gemini surfactant solutions is affected by the hydrophobic carbon chain and spacer length and additive concentration [20–22]. Moreover, the viscosity behavior of a sulfonate Gemini surfactant solution was investigated. It was concluded that the viscosity of the 0.6 wt.% surfactant solution could be 90 mPa·s at 30 °C and 6 s^{-1}. However, when the temperature exceeded 60 °C, the viscosity of the solution was close to that of water which indicated that the temperature resistance was poor. Tang et al. [23] formulated a carboxylate Gemini surfactant clean fracturing fluid containing 3 wt.% DC16-4-16 and 0.04 wt.% ZnO, and they found that it had high viscosity of 30 mPa·s at 100 °C. For application of anionic Gemini surfactants as thickeners for clean fracturing fluids, the temperature-thickening properties of the anionic Gemini surfactant solution are particularly important [24]. In addition, studies have shown that spacer group and hydrophobic carbon chain play the most important role in the viscosity of anionic Gemini surfactant solutions [25]. The bisulphonate structure in sulfonate Gemini surfactants makes them have good water solubility, coupling and high surface activity, and their raw material sources are relatively extensive, and the synthesis method is relatively simple, so sulfonate Gemini surfactants are most likely to achieve large-scale industrial production.

Clean fracturing fluid studies are now mainly focused on the study of cationic viscoelastic surfactants and have been used successfully in some oil fields. However, the cationic surfactant and the cationic Gemini surfactant cleaning fracturing fluid easily cause reservoir damage, and have the problems of poor degradability and pollution damage to the environment and increased consumption cost at high temperature. Simultaneously, the newly developed conventional anionic Gemini surfactant cleaning fracturing fluids have problems of high dosage of chemicals and poor high temperature stability. Therefore, a systematic investigation of the effects of hydrophobic carbon chain, spacer group, temperature, concentration, and additives on the viscosity of novel sulfonate Gemini surfactants is needed. Constructing a new non-invasive and temperature-resistant clean fracturing fluid to solve the problems of high dosage, poor temperature stability and inability to automatically break of the existing clean fracturing fluid has obvious scientific significance and practical value. The results have scientific, social and economic value for the efficient development of oil and gas reservoirs.

In this paper, the effects of hydrophobic chain, spacer group, concentration, temperature and addition of nano-MgO on the viscosity of sulfonate Gemini surfactant solution were investigated by measuring the solution viscosity. Then, their micellar microstructures were observed by Cryo-SEM. Finally, the thickening mechanism of sulfonate Gemini surfactant was investigated by correlating the relationship between solution viscosity and its microstructure. Therefore, this research studied the rheological properties of the sulfonate Gemini surfactant solution and provided experimental basis for the application of sulfonate Gemini surfactants in clean fracturing fluid.

2. Materials and Methods

The sulfonate Gemini surfactants shown in Table 2 were purified twice. By accurately weighing 4 g of each surfactant and dissolving in distilled water in a 50 mL volumetric flask, 8.0 wt.% solutions of different sulfonate Gemini surfactants were prepared. Other surfactant solution concentrations were prepared by appropriately diluting the 8.0 wt.% solution.

Table 2. Sulfonate Gemini surfactants and their classification.

Spacer	Sulfonate Gemini Surfactant (DSm-s-m)
s = 2	DS12-2-12, DS14-2-14, DS16-2-16, DS18-2-18
s = 3	DS18-3-18
s = 4	DS18-4-18

m: hydrophobic chain carbon number, s: spacer group carbon number.

By accurately weighing 0.4 g of nano-MgO and dissolving in distilled water in a 100 mL volumetric flask, a 0.4 wt.% solution of nano-MgO was prepared. Other solution concentrations were prepared by diluting the 0.4 wt.% solution.

Viscosity tests of sulfonate Gemini surfactant solutions were performed using a MCR-301 rheometer (Anton-Paar, Graz, Austria). The temperature for analysis ranged from 50 to 90 °C (the experimental temperature was 65 °C if no special instructions are given), and the shear rate was fixed with 170 s^{-1} according to the Oil and Gas Industry Standard SY/T5107-2005. The viscosity data in the figures below were all averaged after three times unless otherwise stated, and the viscosity test error was 1 mPa·s ± 0.005 mPa·s.

The micellar structure of sulfonate Gemini surfactant solution was observed using a S4800 FESEM (Hitachi, Toyko, Japan). The test solution was placed in a liquid nitrogen tank for quick freezing. After freezing, the sample was quickly placed in a freeze dryer for 24 h. The sample was coated with a conductive plastic on a scanning electron microscope, and the sample was sprayed using ion sputtering apparatus. Finally, the sample was observed by the S4800 field emission scanning electron microscope.

This experimental method can directly link the macroscopic solution viscosity to the microstructure, and then the effect of the micelle on the viscosity of the solution can be investigated, which is beneficial to the study of the thickening mechanism.

The disadvantage of this experimental method is that the scanning electron microscope can only qualitatively observe the microstructure of the solution micelles but cannot quantitatively describe the size and diameter of the micelles.

3. Results and Discussion

3.1. Effect of Hydrophobic Chain on the Viscosity of Sulfonate Gemini Surfactant Solution

The effect of hydrophobic chain on the viscosity of sulfonate Gemini surfactant (DS-m-2-m) solution (8.0 wt.%) was studied at 65 °C and 170 s^{-1}. As can be seen from Figure 1, the viscosity of sulfonate Gemini surfactant solution increases with increasing carbon number of the hydrophobic chain. The reason for this phenomenon is the hydrophobicity of the molecule increases as the length of the hydrophobic chain increases, which promotes the formation of micelles in the solution, and the micelles become tightly entangled with each other, leading to the increase in solution viscosity [26]. The DS-16-2-16 solution's viscosity is slightly lower, probably because the long hydrophobic chain is easily deformed, making for less effective hydrophobic groups in the solution and a decrease of solution viscosity [27]. To better explain this phenomenon, the micellar structures for selected sulfonate Gemini surfactant solution are shown in Figure 2. DS-12-2-12 in solution self-assembles to form distinct spherical micelles with a diameter of about 200 nm. The spherical micelles are closely arranged forming a plate shape. DS-14-2-14 in solution self-assembles into rod micelles. Even though the diameters of rod micelles are about 100 nm, these micelles are tightly packed together. In addition, DS-16-2-16 self-assembles into spherical and rod-shaped micelles, and their diameters are about 100 to 200 nm. More importantly, these micelles are packed closer than in DS12-2-12. DS-18-2-18 also self-assembles in solution to form spherical and ellipsoid (like rods) micelles, but their diameters are approximately 300 nm. Unlike DS12-2-12 and DS16-2-16, these micelles are loosely arranged, and only a few micelles form the network structure. Therefore, DS-18-2-18 has a better viscosity-increasing effect, and the maximum viscosity of the micelle solution is 14.7 mPa·s.

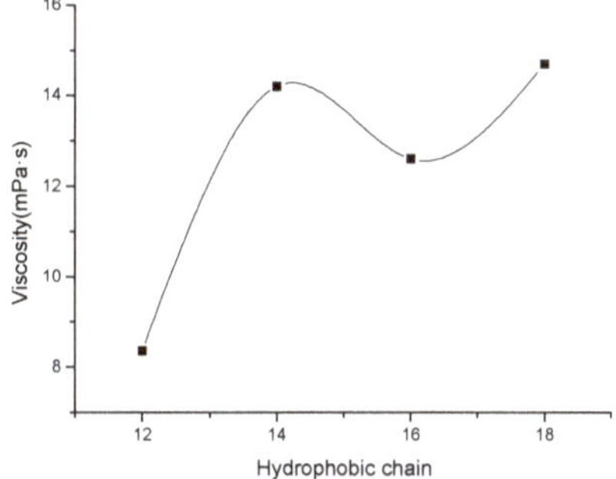

Figure 1. Effect of hydrophobic chain on the viscosity of DSm-2-m solution.

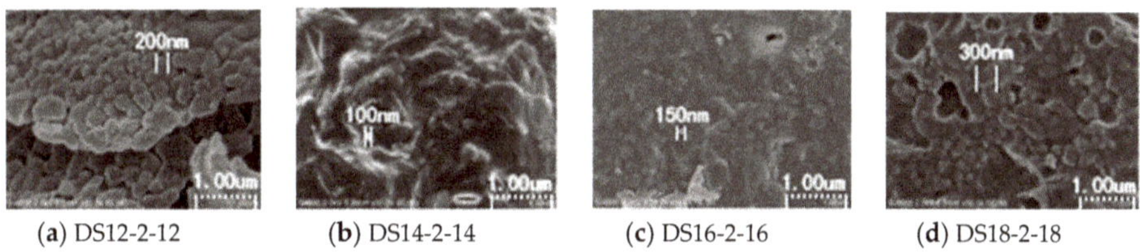

(**a**) DS12-2-12 (**b**) DS14-2-14 (**c**) DS16-2-16 (**d**) DS18-2-18

Figure 2. Effect of hydrophobic chain on micelle structure of DSm-2-m solution.

3.2. Effect of Spacer Group on the Viscosity of Sulfonate Gemini Surfactant Solution

The effect of spacer group on the viscosity of sulfonate Gemini surfactant (DS-18-s-18) solution (8.0 wt.%) was studied at 65 °C and 170 s^{-1}. It can be seen from Figure 3 that the viscosity of the sulfonate Gemini surfactant solution increases dramatically when the spacer group carbon number increases from 2 to 3, and the solution viscosity reaches a maximum of 56.53 mPa·s. The viscosity decreases after the spacer group carbon number increases to 4. Figure 4 shows the micelle structures of the selected sulfonate Gemini surfactants. DS-18-2-18 self-assembles in solution to form spherical and ellipsoid (like rods) micelles, and their diameter are approximately 300 nm. These micelles are loosely arranged, and only a few micelles form the network structure. DS-18-3-18 self-assembles in solution to form layered micelles, which are intertwined to form a similar network structure. In addition, DS-18-4-18 self-assembles into a flat-shaped structure with spherical micelles, but these micelle structures are loosely arranged and cannot fill the whole space. Thus, as the spacer group changes, the microstructures of selected DS18-s-18 solution undergoes a transition from spherical micelles to a network of layered micelles and finally to spherical/layered mixed micelles. DS-18-3-18 has a better thickening behavior under the experimental conditions.

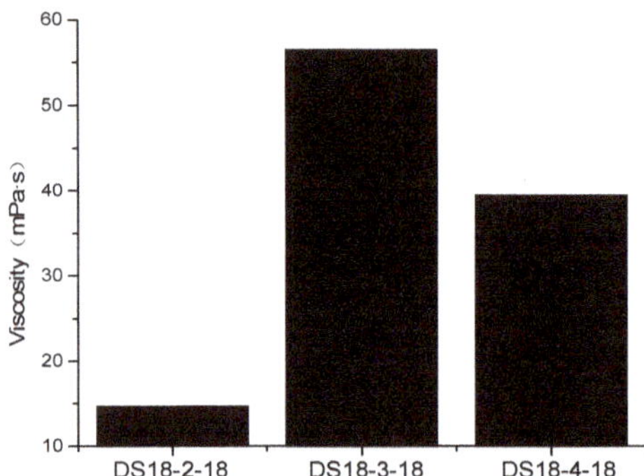

Figure 3. Effect of spacer group on the viscosity of DS18-s-18 solution(s = 2, 3, 4).

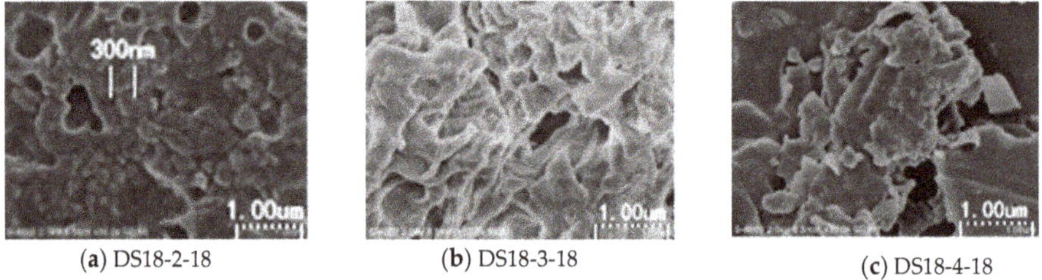

(**a**) DS18-2-18 (**b**) DS18-3-18 (**c**) DS18-4-18

Figure 4. Effect of spacer group on micelle structure of DS18-s-18 solution(s = 2, 3, 4).

3.3. Effect of Concentration on the Viscosity of DS18-3-18 Solution

According to the results in Sections 3.1 and 3.2, DS18-3-18 was selected as thickener for our clean fracturing fluid. The effect of concentration on DS-18-3-18 solution viscosity was studied at 65 °C and 170 s^{-1}. It can be observed from Figure 5 that the solution viscosity increases with the concentration, and the viscosity increases sharply when concentration increases from 3.0 wt.% to 4.0 wt.%. When the solution concentration is higher than 4.0 wt.%, the increase tendency of the viscosity becomes slower, and the thickening ability becomes weak. The reason for this phenomenon is that the sulfonate Gemini surfactant will form the dense network structure in solution with increasing concentration, which is beneficial to the increase of solution viscosity. However, with the increase of concentration, the micelle structure ultimately reaches a stable state, resulting in little change in viscosity [28]. The effect of concentration on microstructure of DS18-3-18 was investigated, and the results are exhibited in Figure 6. DS18-3-18 self-assembles into incomplete sheet micelles, and these micelles can't cover the whole space when the concentration is 1.0 wt.%. Then the complete sheet structures appeared in the DS18-3-18 solution and occupy the entire space when the concentration was 2.0 wt.% (Figure 6b). The number of sheet micelles increased, and some of the sheet micelles aggregated to form closely-coupled rod micelles, forming a network structure, and the density of the micelles in the solution became greater when the concentration was 4.0 wt.% (Figure 6c). It can be seen that the micelle structure gradually changed from an incomplete sheet micelle to tightly wound rod-like and sheet structures as the concentration increased. Thus, 4.0 wt.% DS-18-3-18 was selected as thickener's concentration for the next research of thickening effect.

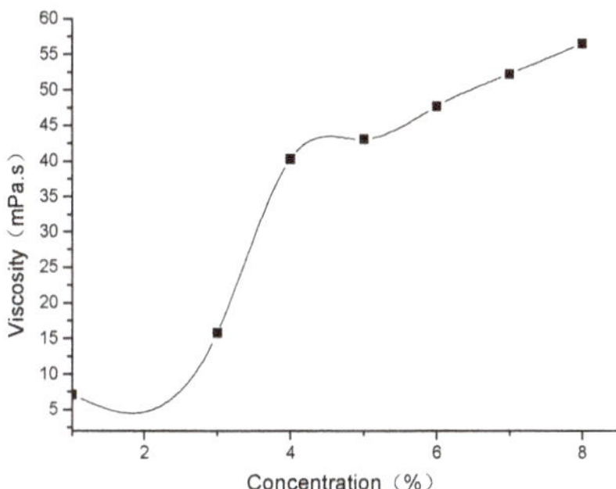

Figure 5. Effect of concentration on the viscosity of DS18-3-18.

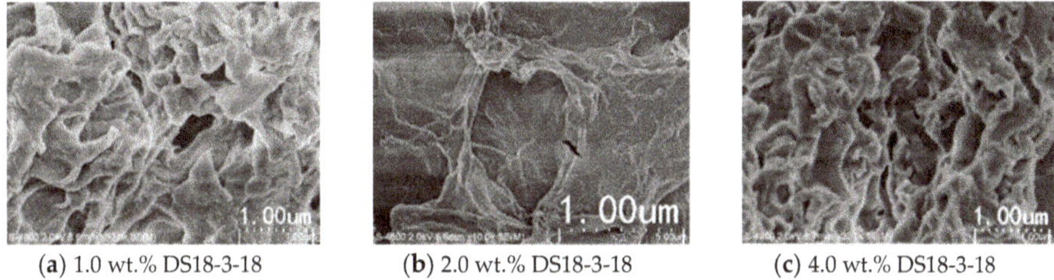

(**a**) 1.0 wt.% DS18-3-18 (**b**) 2.0 wt.% DS18-3-18 (**c**) 4.0 wt.% DS18-3-18

Figure 6. Effect of concentration on micelle structure of DS18-3-18.

3.4. *Effect of Temperature on Viscosity of DS18-3-18 Solution*

The relationship between DS-18-3-18 solution viscosity and temperature was studied, and the results are shown in Figure 7. As the temperature increases, the solution viscosity of 4.0 wt.% DS18-3-18 shows a downward trend. In particular, there is a significant decrease in DS18-3-18 solution viscosity when temperature increases from 60 °C to 70 °C. Then the tendency slows down when the temperature exceeds 70 °C. This is because the network micelles in DS18-3-18 solution will be destroyed as temperature increases [29,30]. Moreover, lamellar micelles, reticular micelles, and worm-like micelles can help to keep the viscosity stability of DS18-3-18 solution at higher temperatures. To further investigate the mechanism of temperature-resistance thickening of DS18-3-18 solution, the self-assembly morphology of DS18-3-18 micelles were characterized by scanning electron microscopy, and the electronic micrographs are shown in Figure 8, where it can be seen that DS18-3-18 self-assembles into a complete layered micelle, and the layered micelles have large pieces of sheet micelles attached at 50 °C. When the temperature raised to 70 °C, the lamellar micelles gradually disintegrate to form sheet micelles, and these sheet micelles became smaller (Figure 8b). Then the sheet micelle structures in the solution became sparse, and part of the sheet structure disintegrated to form irregular spherical micelles and rod-like micelles at 90 °C. But these micelles still formed the network structure in solution (Figure 8c).

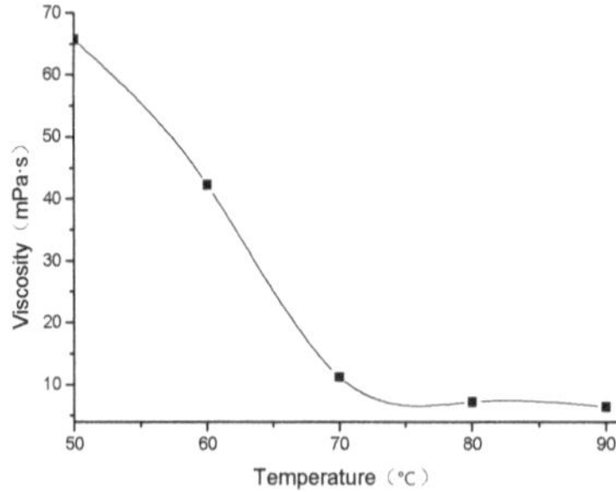

Figure 7. Effect of temperature on the viscosity of 4 wt.% DS18-3-18.

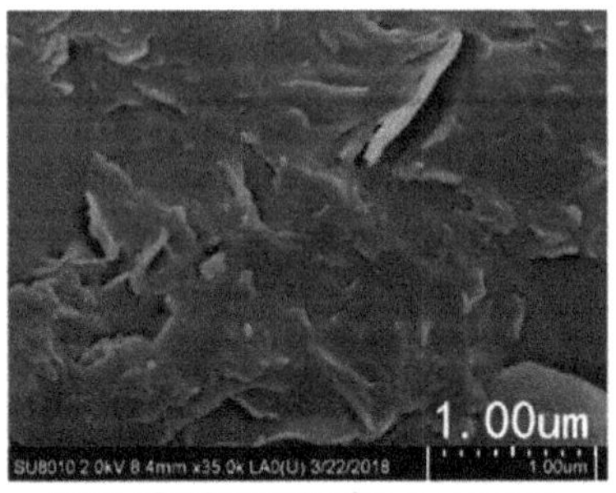

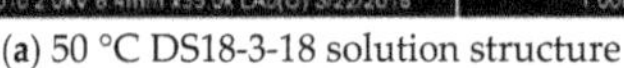

(**a**) 50 °C DS18-3-18 solution structure

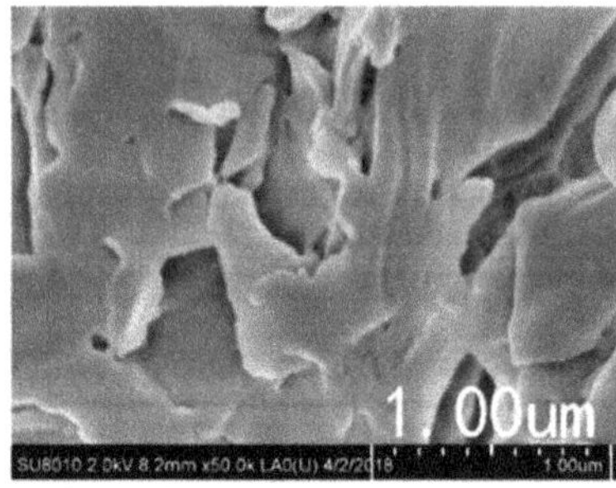

(**b**) 70 °C DS18-3-18 solution structure

(**c**) 90 °C DS18-3-18 solution structure

Figure 8. Effect of temperature on the micelle structure of 4.0 wt.% DS18-3-18.

This indicates that the density of network lamellar micelles in DS18-3-18 solution decreases as the temperature increase, and the morphology of the micelles gradually becomes incomplete, resulting in a decrease in solution viscosity. But the spatial network structure is conducive to the DS18-3-18 solution viscosity when temperature is higher than 70 °C. Thus, DS18-3-18 solution viscosity was still 6.42 mPa·s at 90 °C, which exhibited prominent thickening ability and tolerance of DS18-3-18 at high temperature (>80 °C).

3.5. *Effect of Nano-MgO on the Viscosity of 4.0 wt.% DS18-3-18*

The influences of nano-MgO concentration on the 4.0 wt.% solution viscosity were studied at 90 °C. Figure 9 shows that the DS18-3-18 solution viscosity increases firstly and then decreases as the nano-MgO concentration increases at high temperature. When 0.02 wt.% nano-MgO is added, the solution viscosity reaches its maximum of 21.87 mPa·s, which meet the required viscosity of a clean fracturing fluid. This is because the nano-MgO with extremely high specific surface area and can be easily adsorbed on the micelle end plane or surface in the DS18-3-18 solution when the nano-MgO concentration is low. Therefore, nano-MgO particles shield the micelles from electrostatic repulsion, which is beneficial to form the network micelles and increase solution viscosity [31]. Due to the high content (4.0 wt.%) of DS18-3-18, double-layered structures were formed with the nano-MgO through non-covalent bonds. The hydrophilic heads of the surfactants faced outward, and the nanoparticles were in the center [32]. As the mass fraction of nano-MgO continued to increase, it was more conducive to the formation of this double-layer structure, which promoted the repulsion between DS18-3-18 solution micelles and weakened the stability of the network structure. Therefore, the viscosity and temperature-resistance of the solution were reduced when the nano-MgO reached a certain mass fraction. To better explain the observed phenomena, the micellar structures of the DS18-3-18 solution with different concentration of nano-MgO were investigated. As can be seen from Figure 10, 4.0 wt.% DS18-3-18 self-assembles into sparse sheet micelles. Then the aggregated rod-like micelle structure appears on the sheet-like micelle structure in solution when adding 0.01 wt.% nano-MgO. The rod-like micelles accumulated through the structure of the sheet micelles in the DS18-3-18 solution gradually became larger and aggregated into a compact layer when the concentration of nano-MgO was 0.02 wt.% (Figure 10c). Finally, the compact layer structure gradually became sparse, and the rod-like micelles gradually transformed into worm-like micelle structure when the concentration was 0.03 wt.% (Figure 10d). Also, the viscosity of DS18-3-18 solution firstly increases and then decreases with the increase of nano-MgO concentration. Therefore, nano-MgO affect viscosity of 4.0 wt.% DS18-3-18 solution in high temperature though changing micellar structure morphology of DS18-3-18 solution, and it has the best temperature-resistant thickening effect on solution viscosity at the concentration of 0.02 wt.%.

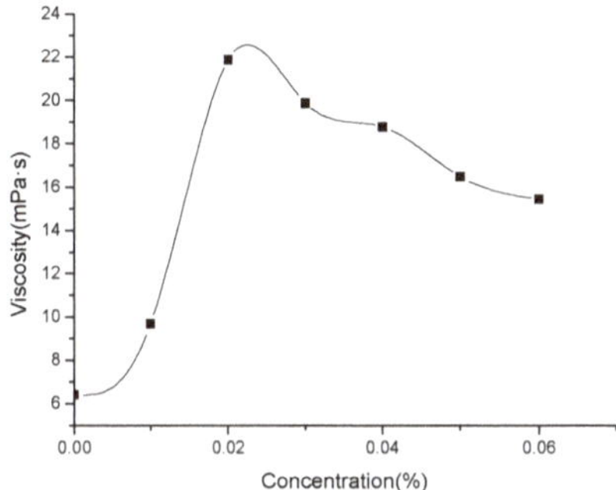

Figure 9. Effect of nano-MgO concentration on the viscosity of 4.0 wt.% DS18-3-18.

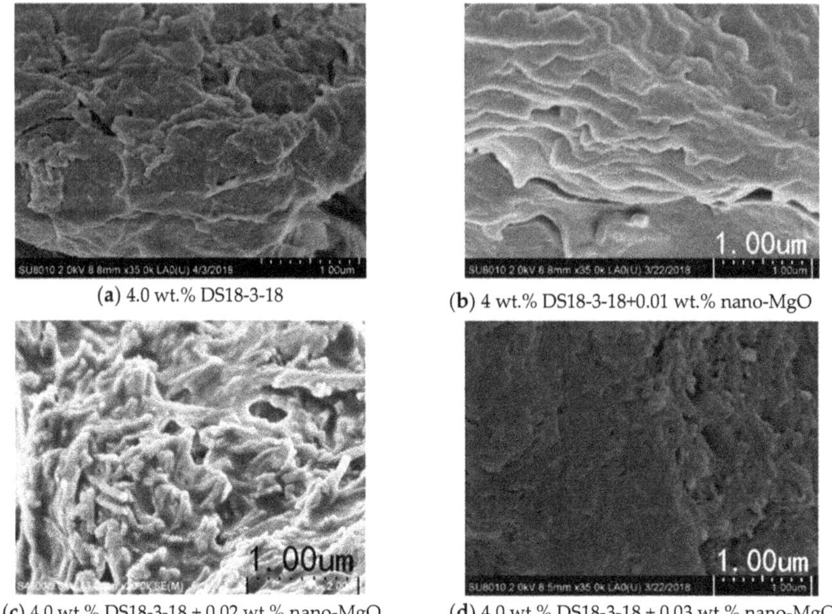

(a) 4.0 wt.% DS18-3-18
(b) 4 wt.% DS18-3-18+0.01 wt.% nano-MgO
(c) 4.0 wt.% DS18-3-18 + 0.02 wt.% nano-MgO
(d) 4.0 wt.% DS18-3-18 + 0.03 wt.% nano-MgO

Figure 10. Effect of different concentration of nano-MgO on the micelle structure of DS18-3-18 solution.

4. Conclusions

The effects of molecular structure, temperature and addition of nano-MgO on the viscosity of sulfonate Gemini surfactants were investigated by testing the solution viscosity. The micelles of the solution were observed by Cryo-SEM, and the mechanism of increasing the viscosity in surfactant solution was explored. The most important conclusions for this work are summarized as below:

(1) The viscosity of the DSm-s-m solution showed a fluctuating upward trend with increasing the length of hydrophobic chain at s = 2. Meanwhile, the viscosity of the DS18-s-18 solution increased firstly and decreased later with increasing spacer group length. Moreover, DS18-3-18 showed prominent viscosity behavior.

(2) The viscosity of the DS18-3-18 solution increased as the increase of surfactant concentration, and the tendency of the increase became slow when the solution concentration exceeded 4.0 wt.%.

(3) 4.0 wt.% DS18-3-18 solution had better temperature-resistance, and the viscosity of the solution kept above 6 mPa·s at 90 °C and 170 s^{-1}, which satisfied the experimental requirements for a clean fracturing fluid thickener.

(4) The addition of nano-MgO improved the temperature-resistance of 4.0 wt.% DS18-3-18 solution. The viscosity of the solution increased firstly and decreased later with increasing nano-MgO concentration. The viscosity of the solution reached its maximum of 21.87 mPa·s at 90 °C when 0.02 wt.% nano-MgO was added, which achieved the required viscosity of the clean fracturing fluid.

(5) DS18-3-18 self-assembled intodense layered micelles, and the micelles entangled with each other to form a network structure, contributing to the better temperature-resistance and higher viscosity. Nano-MgO further enhanced the temperature-resistance of 4.0 wt.% DS18-3-18 solution by changing the micellar morphology.

(6) The limitation of this experimental study was that the temperature resistance of the sulfonate Gemini surfactant can only reach 90 °C, and the cost was slightly higher.

(7) Compared with other studies, DS18-3-18 had a temperature resistance of 90 °C, and the viscosity after adding nano-MgO could reach up to 21.87 mPa·s, which meet the viscosity requirements of a clean fracturing fluid and has good application prospects.

Author Contributions: S.T. and Y.Z., designed the experimental framework and wrote the paper. J.W. and Y.F. conducted experiments and collected datas. W.Y. and J.L. checked and revised the paper.

Funding: This work is funded by the National Natural Science Foundation of China (51474035, 51774049) and Innovation Fund Project of Hubei Cooperative Innovation Center of Unconventional Oil and Gas (HBUOG-2014-2).

Conflicts of Interest: The authors declare no conflicts of interest.

References

1. Zhang, L.; Kou, Z.; Wang, H.; Zhao, Y.; Dejam, M.; Guo, J.; Du, J. Performance analysis for a model of a multi-wing hydraulically fractured vertical well in a coalbed methane gas reservoir. *J. Pet. Sci. Eng.* **2018**, *166*, 104–120. [CrossRef]
2. Dejam, M.; Hassanzadeh, H.; Chen, Z. Semi-analytical solution for pressure transient analysis of a hydraulically fractured vertical well in a bounded dual-porosity reservoir. *J. Hydrol.* **2018**, *565*, 289–301. [CrossRef]
3. Seright, R.S. How Much Polymer Should Be Injected During a Polymer Flood? In Proceedings of the SPE Improved Oil Recovery Conference, Tulsa, OK, USA, 11–13 April 2016.
4. Saboorian-Jooybari, H.; Dejam, M.; Chen, Z. Heavy oil polymer flooding from laboratory core floods to pilot tests and field applications: Half-century studies. *J. Pet. Sci. Eng.* **2016**, *142*, 85–100. [CrossRef]
5. Rui, Z.; Wang, X.; Zhang, Z.; Lu, J.; Chen, G.; Zhou, X.; Patil, S. A realistic and integrated model for evaluating oil sands development with Steam Assisted Gravity Drainage technology in Canada. *Appl. Energy* **2018**, *213*, 76–91. [CrossRef]
6. Yan, Z.; Dai, C.; Zhao, M.; Feng, H.; Gao, B.; Li, M. Research and Application Progress of Cleaning Fracturing Fluid. *Oilfield Chem.* **2015**, *32*, 141–145.
7. Yan, Z.; Dai, C.; Zhao, M.; Sun, Y. Rheological characterizations and molecular dynamics simulations of self-assembly in an anionic/cationic surfactant mixture. *Soft Matter* **2016**, *12*, 6058–6066. [CrossRef] [PubMed]
8. Yan, J.; Wang, F. Wettability alteration of silica induced by surfactant adsorption. *Oilfield Chem.* **1993**, *10*, 195–200.
9. Zhang, J.; Zhang, M.; Zhang, S.; Bai, B.; Gao, B. Development and Field Pilot Test of a Novel Viscoelastic Anionic-Surfactant Fracturing Fluid. In Proceedings of the Society of Petroleum Engineers Western North American Regional Meeting—In Collaboration with the Joint Meetings of the Pacific Section AAPG and Cordlleran Section GSA, Anaheim, CA, USA, 27–29 May 2010.
10. Lai, L.; Mei, P.; Wu, X.; Hou, C.; Zheng, Y.; Liu, Y. Micellization of anionic gemini surfactants and their interaction with polyacrylamide. *Colloid Polym. Sci.* **2014**, *292*, 2821–2830. [CrossRef]
11. Pei, X.; Zhao, J.; Ye, Y.; You, Y.; Wei, X. Wormlike micelles and gels reinforced by hydrogen bonding in aqueous cationic gemini surfactant systems. *Soft Matter* **2011**, *7*, 2953–2960. [CrossRef]
12. Zana, R. Dimeric and oligomeric surfactants. Behavior at interfaces and in aqueous solution: A review. *Adv. Colloid Interface Sci.* **2002**, *97*, 205–253. [CrossRef]
13. Li, G.; Zhao, L.; Wang, P.; Ning, X.; Qin, P.; Xue, M. Synthesis and Surface Activity of Cationic Gemini Surfactant, N,N′-bis(Octadecyl/tetradecyl dimethyl)-1,2-dichloro-1,4-benzene Dimethyl Ammonium Salt. *Fine Chem.* **2016**, *33*, 519–523.
14. Tang, S.; Cui, Y.; Xiong, X.; Lei, X. Viscosity and Influencing Factors of Anionic Gemini Surfactant (GA-16) Solution. *J. Oil Gas Technol.* **2014**, *36*, 146–149. (In Chinese)
15. Akbas, H.; Elemenli, A.; Boz, M. Aggregation and Thermodynamic Properties of Some Cationic Gemini Surfactants. *J. Surfactants Deterg.* **2012**, *15*, 33–40. [CrossRef]
16. Zhu, H.; Niu, H.; Lou, P.; Ding, H.; Zhang, H. Synthesis and Performance of Gemini Surfactant—Containing Acidic Clean Fracturing Fluid. *Adv. Fine Petrochem.* **2011**, *12*, 5–8.
17. Yang, J.; Guan, B.; Lu, Y. Viscoelastic Evaluation of Gemini Surfactant Gel for Hydraulic Fracturing. In Proceedings of the SPE European Formation Damage Conference and Exhibition, Noordwijk, The Netherlands, 5–7 June 2013.

18. Yu, H.; Shen, Y.; Yang, X.; Liu, G.; Zhang, L. Interfacial activity and rheological behavior of N,N′-bis (hexadecyldimethyl)-1,2-dibro-mide-ethanediyl ammonium salt. *Acta Pet. Sin.* **2014**, *30*, 542–547.
19. Liu, Y.; Guo, L.; Bi, K.; Ren, W.; Liu, L.; Li, Y.; Han, L. Synthesis of Cationic Gemini Surfactant and Its Application in Water Blocking. *Fine Spec. Chem.* **2011**, *19*, 8–10. (In Chinese)
20. Tang, S.; Liu, Z.; Liu, S.; Hu, X.; Wu, R. Research on Rheological Behavior of Anion Gemini Surfactant Solution. *Adv. Mater. Res.* **2011**, *146–147*, 536–541. [CrossRef]
21. Pi, Y.; Zhang, L.; Liu, Z.; Gao, D.; Tang, S. Research on Rheological Properties of Sulfuric Acid Ester Salt Gemini Surfactant Solution. *J. Oil Gas Technol.* **2011**, *33*, 135–138.
22. Du, X.; Li, L.; Lu, Y.; Yang, Z. Unusual viscosity behavior of a kind of anionic gemini surfactant. *Colloid Surf. A* **2007**, *308*, 147–149. [CrossRef]
23. Tang, S.; Zhao, C.; Tian, L.; Zou, T. Temperature-resistance clean fracturing fluid with carboxylate gemini surfactant: A case study of tight sandstone gas reservoirs in the Tarim Basin. *Nat. Gas Ind.* **2016**, *36*, 45–51.
24. Mao, J.; Yang, X.; Chen, Y.; Zhang, Z.; Zhang, C.; Yang, B.; Zhao, J. Viscosity reduction mechanism in high temperature of a Gemini viscoelastic surfactant (VES) fracturing fluid and effect of counter-ion salt (KCl) on its heat resistance. *J. Pet. Sci Eng.* **2018**, *164*, 189–195. [CrossRef]
25. Zana, R.; Talmon, Y. Dependence of aggregate morphology on structure of dimeric surfactants. *Nature* **1993**, *362*, 228–230. [CrossRef]
26. Gao, D.; Yu, M. The Comparative Study of Influence Factors of Ionic Gemini Surfactant Solution's Rheology. *J. Chongqing Univ. Sci. Technol. (Nat. Sci. Ed.)* **2012**, *14*, 101–104.
27. Zhu, Q. Study on Synthesis, Properties and Applications in Enhanced Oil Recovery (EOR) of Novel Gemini Surfactant. Master's Thesis, Chengdu University of Technology, Chengdu, China, 2009.
28. Han, L.J.; Chen, H.; Luo, P.Y. Viscosity behavior of cationic gemini surfactants with long alkyl chains. *Surf. Sci.* **2004**, *564*, 141–148. [CrossRef]
29. Chen, H.; Ye, Z.; Han, L.; Luo, P.; Zhang, L. Temperature-induced micelle transition of gemini surfactant in aqueous solution. *Surf. Sci.* **2007**, *601*, 2147–2151. [CrossRef]
30. Yu, H. Preparation and Performance of Quaternary Ammonium Gemini Surfactant. Master's Thesis, Shaanxi University of Science & Technology, Xi'an, China, 2010.
31. Yang, J. Study on the Turbulent Flow and Heat Transfer Characteristics of Viscoelastic Fluid Based Nanofluid. Master's Thesis, Harbin Institute of Technology, Harbin, China, 2013.
32. Helgeson, M.E.; Hodgdon, T.K.; Kaler, E.W.; Wagner, N.J.; Vethamuthu, M.; Ananthapadmanabhan, K.P. Formation and Rheology of Viscoelastic "Double Networks" in Wormlike Micelle-Nanoparticle Mixtures. *Langmuir* **2010**, *26*, 8049–8060. [CrossRef] [PubMed]

Article

Experimental Investigation of Oil Recovery from Tight Sandstone Oil Reservoirs by Pressure Depletion

Wenxiang Chen [1], Zubo Zhang [1], Qingjie Liu [1], Xu Chen [1], Prince Opoku Appau [2] and Fuyong Wang [2,*]

1 State Key Laboratory of Enhanced Oil Recovery, Research Institute of Petroleum Exploration and Development, Beijing 100083, China; 2002160062@cugb.edu.cn (W.C.); zzbo@petrochina.com.cn (Z.Z.); lqj@petrochina.com.cn (Q.L.); chx9500@petrochina.com.cn (X.C.)

2 Research Institute of Enhanced Oil Recovery, China University of Petroleum, Beijing 102249, China; 2017290109@student.cup.edu.cn

* Corresponding author: wangfuyong@cup.edu.cn

Received: 26 July 2018; Accepted: 2 October 2018; Published: 7 October 2018

Abstract: Oil production by natural energy of the reservoir is usually the first choice for oil reservoir development. Conversely, to effectively develop tight oil reservoir is challenging due to its ultra-low formation permeability. A novel platform for experimental investigation of oil recovery from tight sandstone oil reservoirs by pressure depletion has been proposed in this paper. A series of experiments were conducted to evaluate the effects of pressure depletion degree, pressure depletion rate, reservoir temperature, overburden pressure, formation pressure coefficient and crude oil properties on oil recovery by reservoir pressure depletion. In addition, the characteristics of pressure propagation during the reservoir depletion process were monitored and studied. The experimental results showed that oil recovery factor positively correlated with pressure depletion degree when reservoir pressure was above the bubble point pressure. Moreover, equal pressure depletion degree led to the same oil recovery factor regardless of different pressure depletion rate. However, it was noticed that faster pressure drop resulted in a higher oil recovery rate. For oil reservoir without dissolved gas (dead oil), oil recovery was 2–3% due to the limited reservoir natural energy. In contrast, depletion from live oil reservoir resulted in an increased recovery rate ranging from 11% to 18% due to the presence of dissolved gas. This is attributed to the fact that when reservoir pressure drops below the bubble point pressure, the dissolved gas expands and pushes the oil out of the rock pore spaces which significantly improves the oil recovery. From the pressure propagation curve, the reason for improved oil recovery is that when the reservoir pressure is lower than the bubble point pressure, the dissolved gas constantly separates and provides additional pressure gradient to displace oil. The present study will help engineers to have a better understanding of the drive mechanisms and influencing factors that affect development of tight oil reservoirs, especially for predicting oil recovery by reservoir pressure depletion.

Keywords: dissolved gas; experimental evaluation; reservoir depletion; recovery factor; tight oil

1. Introduction

Over the years, the oil and gas Exploration and Production (E&P) industry has shifted their focus from conventional oil and gas resources to unconventional resources due to decline of conventional resources and increasing need for energy. Until now fossil fuels still remain the world's leading source of energy, therefore unconventional resources like tight oil and shale oil provides a means to supplement our energy demand for the years to come [1–6]. Data from bulletin of United States Geological Survey (USGS), International Energy Agency (IEA) and British Petroleum (BP) indicate that the recoverable resources of tight oil in the world is about 472.8×10^8 t [7]. Therefore, with fast depletion of conventional oil resources, it is imperative to find ways to exploit these oil trapped in tight formations [8,9]. However, developing tight oil reservoirs are very challenging due to each unique formation characteristics, making field development based solely on its individual petrophysical attributes. Pore structures of tight formations are inherent factors affecting the storage and development oil tight oil reservoirs. This makes a comprehensive characterization of tight oil pore structures a great importance for their overall development [10–12]. Generally, there is no formal definition of tight oil, nonetheless several researchers define tight oil as those found in reservoir with ultra-low permeability and porosity (less than 0.1 md and 10% matrix porosity) [13–17]. Large reserve distribution across the world and better output potential of tight oil has led to the increasing exploitation of these resources in countries like United States, Canada, and Australia [18–22]. Tight oil reservoirs are also widely distributed in China, such as in the Ordos, Sichuan, Songliao, Junggar and Tuha basins, albeit their exploration and development remain in the pilot stage [23–25]. The Chang 7 tight reservoir in the Ordos Basin of the Changqing Oilfield has become the largest experimental area for developing tight oil in China [26]. Horizontal well technology and multi-stage hydraulic fracturing technology provide a basis for commercial exploitation of tight oil [27–30]. Even so, the recovery factor of tight oil reservoirs obtained by relying on formation energy is 3–10% due to its tight lithology, large seepage resistance and poor pressure conduction ability [31–34].

Reserve estimation is crucial for every oil and gas E&P venture and recovery factor is a key parameter that aid in calculating the reserve of a new oil and gas asset [35,36]. Recovery factor is usually defined as the ratio of geological reserves to economically extracted quantities. The recovery factor of tight oil is uncertain because it takes into account the original oil in place, natural and hydraulic fractures, crude oil properties and the formation's low permeability and porosity. The resource potential of reservoir is typically assessed by Decline Curve Analysis (DCA) [37–40]. Though, published estimates of recovery factor in tight oil is mainly by the following three methods (1) Production data analysis; (2) Numerical simulation; (3) Laboratory experimental evaluation. When reservoir properties are known and production data is available, the first method would be the most accurate. Yet, this method has many uncertainties, many indexes and poor universality, and most importantly cannot reveal the factors affecting oil recovery in essence. In Bakken reservoir, the recovery factor according to Reisz et al., [41], Brohrer et al., [42] and Clark [43] with production data analysis method are 15–20%, 0.7–3.7 % and 6.1–8.7 % respectively. The results showed that the production analysis method cannot accurately evaluate the recovery factor of the Bakken reservoir. With the application of numerical simulation, Ghaderi et al., built a black oil simulator by using ECLIPSE 100^{TM} simulator to evaluate the factors affecting primary recovery in multi-fractured horizontal wells [44]. Xu et al., used nonlinear seepage numerical simulation software of the low permeability reservoir to give the optimum fracture parameters of the specific block and analyze the effect of the fracture parameters on the depletion [45]. Xu et al., established a numerical model and experimental method to study the seepage law of tight oil from the microscopic study and analyze the relationship between the pressure, permeability, core size and depletion recovery factor in one dimensional space [46]. Kabir et al., generated a synthetic example with a finite-difference simulator to demonstrate the use of various analytical and numerical tools to learn about both short and long-term reservoir behaviors [47]. Dechongkit and Prasad used deterministic and probabilistic methods to calculate the Antelope, Pronghorn and Parshall oilfield recovery to be 9.2–16% [48]. Clark estimated the recovery factor to be 4–6 % with the material balance equation at the dissolved gas saturation pressure in Bakken shale reservoir [43]. The accuracy of the aforementioned three adopted methods to estimate the recovery rate is compared with the Perm reservoirs in Russia, which were evaluated by the three dimensional recovery factor rate model, the chart method and the (American Petroleum Institute) API recovery formula [49]. At the same time, according to the influencing factors of these methods, the recovery level of depletion were analyzed [50]. Recovery factor estimation methods of newly added measured reserves in present petroleum reserves standardization of China are inapt for sandstone oil reservoirs of extra-low permeability. For this reason, there are inadequate mature standard reference in the experimental study of tight oil depletion in China, hence few reports on experimental evaluation have been published [51]. Therefore, further research needs to be done experimentally in order to fully grasp the factors affecting recovery of tight oil and how recovery factor can be improved significantly before a decision whether or not to exploit specific tight oil reservoir can be made.

In this paper, a novel depletion laboratory experimental platform and its evaluation method about tight oil reservoir is developed. To simulate the actual conditions of horizontal flow in a well in the Chang 7 tight oil reservoir of the Yanchang Formation, three horizontal core samples were used. The depletion experiment at different temperature, formation pressure coefficient and oil properties were conducted to measure the recovery factor, as well as a real-time monitoring of the pressure propagation in the process of reservoir depletion. At the end of the experiment, the drive mechanism and recovery factor of tight oil reservoirs depletion were revealed.

2. Materials and Experiments

2.1. Materials

The target tight oil reservoir for this research is the Yanchang Formation, located in the Erdos Basin of China, and the reservoir temperature is 60 °C. The reservoir formation depth is 2000 m and hydrostatic pressure is 20 MPa. However, the reservoir formation pressure is only 16 MPa, with a formation pressure coefficient of 0.8. The reservoir GOR is 54.1 m^3/m^3 with saturation pressure of 8.85 MPa. Kerosene was used as the experimental oil whereas methane as the dissolved gas. Because of low permeability and porosity of tight oil reservoir cores, using conventional core sample which has 1 in. diameter and pore volume usually smaller than 0.008 L leads to large measurement error. To reduce these systematic errors, three horizontal core samples with a total length of 88.3 cm and a total pore volume of 0.776 L were used during the experiment, as shown in Figure 1. The core samples were collected from the outcrop of the tight sandstone formation. Tables 1 and 2 present the detailed oil and core sample properties used respectively.

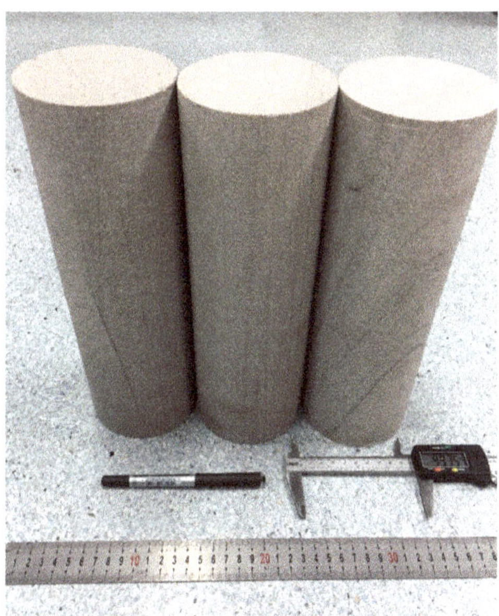

Figure 1. Core samples used in the experiment.

Table 1. The density and viscosity of kerosene used in the experiments at room and reservoir temperature.

Temperature (°C)	Density (g/cm^3)	Viscosity (mPa·s)
20.1	0.754	1.44
60	0.725	0.86

Table 2. The parameters of core samples used in the experiments.

Sample	Length (cm)	Diameter (cm)	Porosity (%)	Porosity Volume (L)	Air Permeability (mD)
A2	28.959	9.979	10.420	0.236	0.350
A3	29.264	9.975	10.670	0.235	0.320
AB1	30.06	9.906	13.180	0.305	0.300
Total	88.3	-	-	0.776	-

The conventional method of saturating cores is unsuitable to saturate low permeability cores samples in the laboratory because they are normally restricted to small core samples which are easy to be destroyed in the saturation process. In this study, two centrifugal pumps were used to vacuum the 3 core samples in the core holder for 24 hours and at a final pressure of 0.01 MPa. The samples were later saturated with kerosene to measure the volume of the saturated kerosene. The saturation degree of the kerosene (ratio of the volume of kerosene to the cores pore volume) was more than 96%.

2.2. Experimental Platform and Methods

A novel experimental platform for studying tight oil reservoir depletion was developed in this paper. A schematic diagram of the experimental apparatus is shown in Figure 2. The core holder's length was 1 m and seven piezometric points were placed along the holder from inlet to the end. The pressure propagation during the coring process was monitored in real time by sensors. A constant confining pressure of 46 MPa was applied to simulate the overburden pressure. The oil production by pressure depletion was measured with pump. As shown in Figure 3, the core pressure decreases with the oil expansion out of the core holder. Assuming the pore volume is V_P, and the oil production at time t_i is V_i, the oil recovery R_i at the same time t_i can be calculated as:

$$R_{\mathrm{i}} = \frac{V_i}{V_p} \times 100\% \quad (1)$$

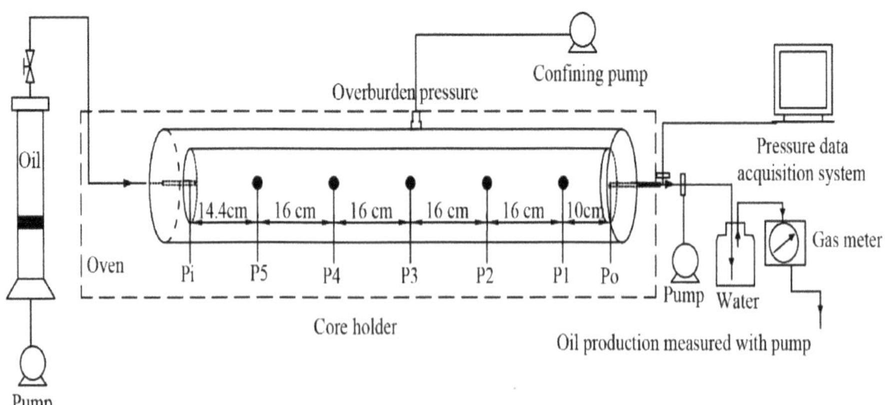

Figure 2. Schematic diagram of the experimental setup.

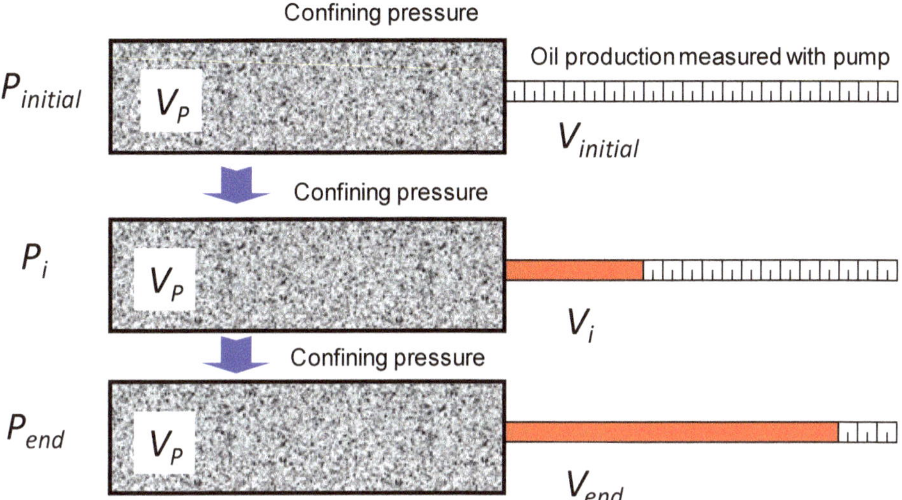

Figure 3. A schematic diagram for the oil production measurement during pressure depletion.

2.3. Experimental Scheme

The pressure depletion in tight sandstones saturated with dead oil and live oil were studied respectively. Besides, the effects of different experiment temperature, formation pressure coefficient, pressure depletion type (linear or step-like) and pressure depletion range on oil recovery and pressure propagation were investigated.

3. Experimental Results and Analysis

3.1. Depletion Characteristics of Tight Oil without Dissolved Gas

3.1.1. Depletion Experiments with Formation Pressure Coefficient of 1

At room temperature of 20.1 °C and formation pressure coefficient of 1, six set of linear pressure depletion experiments with different depletion rate were conducted, as shown in Table 3. The output of the experiment is shown in Figure 4. Although the depletion rate were different, the ultimate oil recovery was the same, around 2%. However, higher depletion rate resulted in higher oil production rate. Therefore, it can be concluded that depletion rate does not affect the ultimate oil recovery but directly affect oil production rate.

Table 3. Linear pressure depletion experiment with six different depletion rates.

No.	Temperature (°C)	Pressure Depletion Range (MPa)		Depletion Time (min)	Pressure Depletion Rate (MPa/min)
		Initial Pressure	Final Pressure		
1	20.1	20	5	10	1.50
2	20.1	20	5	20	0.75
3	20.1	20	5	30	0.50
4	20.1	20	5	40	0.38
5	20.1	20	5	50	0.30
6	20.1	20	5	60	0.25

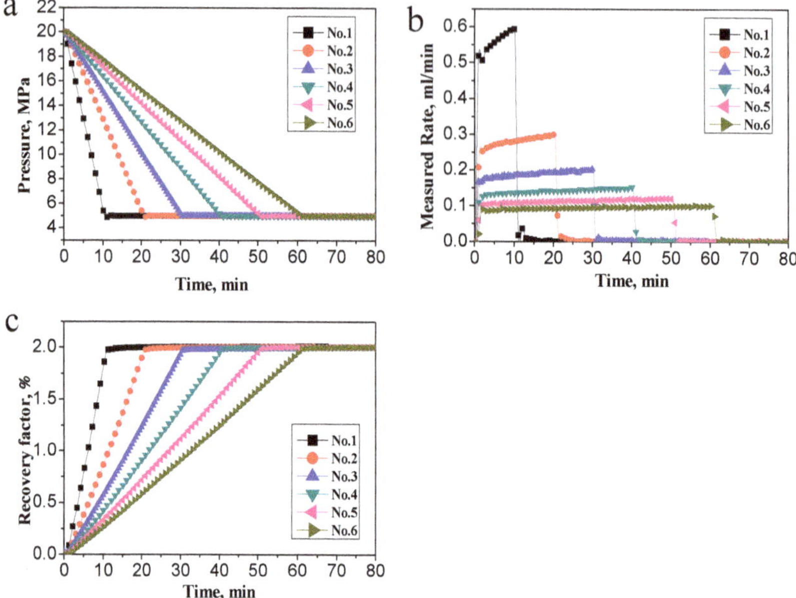

Figure 4. Pressure, oil production rate and recovery factor of six different kinds of linear pressure depletion. (**a**) Pressure depletion with different depletion rate; (**b**) Oil production rate with different depletion rate; (**c**) Oil recovery factor with different depletion rate.

Other than the continuous linear pressure depletion, oil recovery by step-like pressure depletion was also investigated. As shown in Table 4, the depletion process was divided into four stages, and during each stage the pressure was kept constant then followed by linear pressure depletion of 5 MPa with a constant pressure depletion rate. The results of the step-like pressure depletion experiment are depicted in Figure 5. Comparatively, the results of the linear pressure depletion and step-like pressure depletion revealed a similar oil recovery trend irrespective of the change in pressure depletion type. This is because under the same initial formation pressure and final pressure conditions, the elastic recovery is basically the same. The depletion by means of formation pressure is mainly used to characterize the elastic energy of rocks and fluids.

When the formation pressure decreases, the fluid expands and the pore size shrinks causing elastic energy of the fluid in the rock pore to be released from the pore spaces into the wellbore. For these pressure depletion experiment with formation pressure coefficient of 1 and without dissolved gas, depletion mainly by the elastic deformation of rock and the release of fluid elastic energy resulted in a recovery factor of about 2%.

Table 4. Experimental conditions of step-like pressure depletion.

No.	Temperature (°C)	Pressure Depletion Range (MPa)		Depletion Time (min)	Pressure Depletion Rate (MPa/min)
		Initial Pressure	Final Pressure		
1	20.1	20	15	60	8.33×10^{-2}
2	20.1	15	10	60	8.33×10^{-2}
3	20.1	10	5	60	8.33×10^{-2}
4	20.1	5	0	60	8.33×10^{-2}

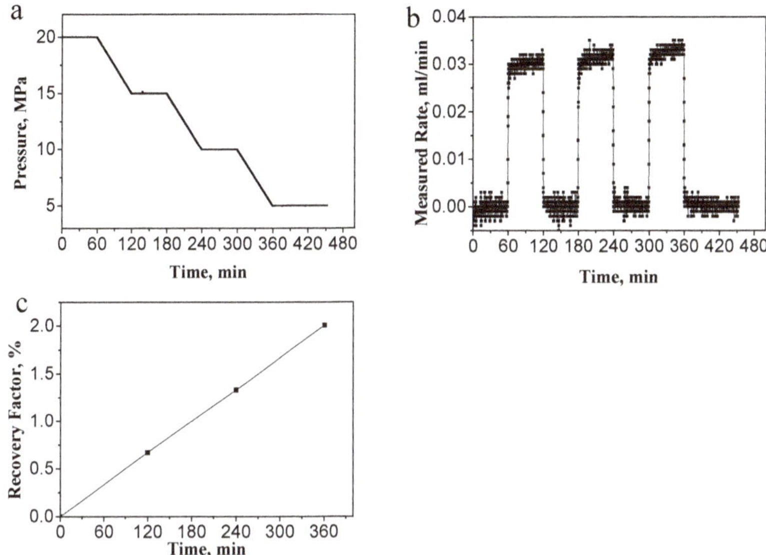

Figure 5. Pressure, production rate and oil recovery of the step-like pressure depletion experiment. (**a**) Pressure change with time; (**b**) Oil production rate change with time; (**c**) Oil recovery factor change with time.

3.1.2. Depletion Experiments with Formation Pressure Coefficient of 1.5

The formation pressure coefficient was adjusted to a value of 1.5 resulting in an increase in the formation pressure from 20 MPa to 30 MPa. The significance of the increment in formation pressure coefficient was to ascertain how much the recovery factor can be improved if the reservoir's pore pressure is increased. Eight linear pressure depletion experiments with different depletion rate were carried out on the core samples. The detailed experimental conditions are shown in Table 5.

Table 5. Experimental conditions of linear pressure depletion with different pressure depletion rate for the cases of formation pressure coefficient equal to 1.5.

No.	Temperature (°C)	Pressure Depletion Range (MPa)		Depletion Time (min)	Pressure Depletion Rate (MPa/min)
		Initial Pressure	Final Pressure		
1	20.1	30	5	0	∞
2	20.1	30	5	1	25.0
3	20.1	30	5	10	2.5
4	20.1	30	5	60	4.2×10^{-1}
5	20.1	30	5	120	2.1×10^{-1}
6	20.1	30	5	180	1.4×10^{-1}
7	20.1	30	5	240	1.0×10^{-1}
8	20.1	30	5	480	5.2×10^{-2}

Figure 6 presents the pressure, oil production rate and oil recovery versus time for the linear pressure depletion with eight different pressure depletion rates for the cases of formation pressure coefficient equal to 1.5. It is noticeable that the ultimate oil recovery is the same, however the increase in depletion speed, resulted in a faster oil recovery. Nonetheless, with the different depletion types, the ultimate recovery factor of tight oil reservoir was almost 3%. Compared with the cases of formation pressure coefficient equal to 1, the enhanced ultimate oil recovery by increasing initial formation pressure is proportional to the increased formation pressure coefficient. According to the theory of reservoir engineering, the elastic energy of the formation determines the final oil recovery by pressure depletion before dissolved gas comes out of crude oil, as given by [52]:

$$E_R = \frac{B_{oi}}{B_{ob}} \frac{\left[C_f + \phi[C_o(1 - S_{wc}) + C_w S_{wc}]\right]}{\phi(1 - S_{wc})} (P_{\text{initial}} - P_{\text{bubble}}) \quad (2)$$

where E_R is the recovery factor by pressure depletion; C_f is the rock compressibility coefficient; C_o is the compression coefficient of crude oil; C_w is the formation water compression coefficient; ϕ is porosity; S_{wc} is the connate water saturation; P_{initial} is the initial formation pressure; P_{bubblr} is the Crude oil bubble point pressure; B_{oi} is the initial oil formation volume factor; B_{ob} is the oil formation volume factor at bubble point. Equation (2) is generally adopted for estimating the recovery factor of primary oil recovery by depletion, and the oil recovery by pressure depletion is proportional to the pressure depletion range. When the formation pressure coefficient was 1.5, the recovery factor was higher compared to that with formation pressure coefficient of 1 due to the increase in reservoir's energy. This indicates that recovery factor of the depletion process positively correlates with reservoir pressure. Under different pressure coefficients, the reservoir rock and fluid have different elastic energies. The larger the pressure coefficient, the higher elastic energy and the recovery factor will increase when the elastic energy is released.

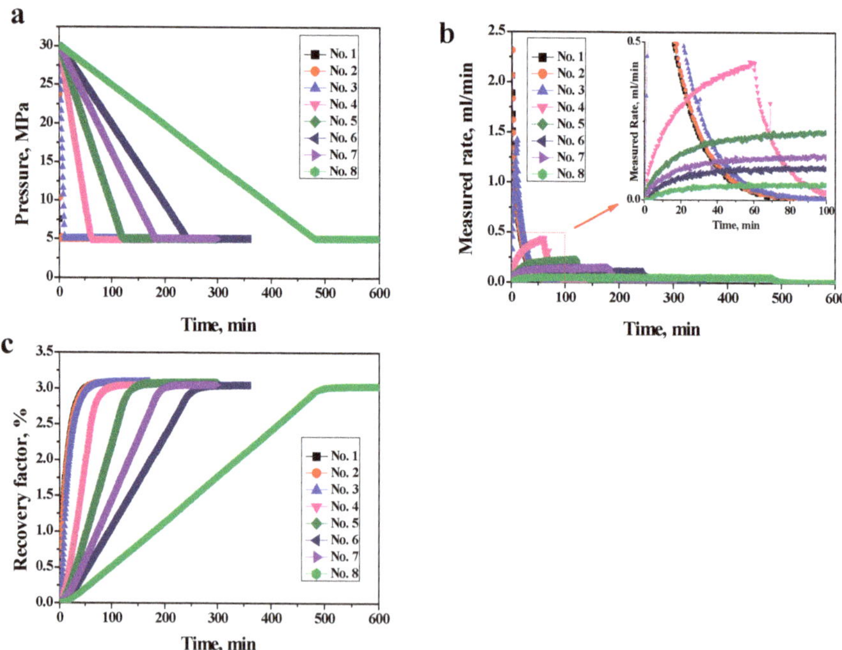

Figure 6. Results of the linear pressure depletion experiments. (**a**) Pressure depletion with different depletion rate; (**b**) Oil production rate with different depletion rate; (**c**) Oil recovery factor with different depletion rate.

3.1.3. Characteristics of Pressure Propagation of Dead Oil Depletion

It should be noted that although pressure can instantly deplete from the initial pressure to the ending pressure, it still takes some time to reach the ultimate oil recovery. For example, the pressure depletion of the No. 1 case in Figure 6 finishes instantly, oil recovery reaches the ultimate value but takes more than 50 min. It may be that pressure takes some time to propagate in the tight formations, hence the pressure cannot be balanced instantaneously. During the depletion process, the pressure propagation was analyzed from the real-time data collected from the pressure points distributed along the core holder.

Figure 7 depicts the pressure distribution along the core holder during the depletion experiments with different pressure depletion rate. Taking Figure 7c as an example, when the pressure depleted from 30 MPa to 5 MPa in 10 min, the pressure near the oil outlet dropped immediately. Otherwise, if the pressure points are far away from the oil outlet, the slower the rate of pressure drop, signifying that pressure propagation rate become much slower from the inlet to the outlet. With the pressure depletion rate increasing, the asynchrony of pressure depletion at different location becomes more significant. The pressure propagation can be explained by the radius of investigation of the reservoirs as given by [53]:

$$t = 0.0872\frac{\phi\mu C_t r^2}{k} \tag{3}$$

where C_t is the total compressibility in 1/MPa; ϕ is the formation porosity in fraction; μ is the fluid viscosity in mPa·s; k is the formation permeability in mD; r is the reservoir radius in m; t is the time for the transient pressure propagating to the reservoir radius r. Although Equation (3) is mainly used to estimate the time pseudo-steady flow begins in a homogenous reservoir with radial flow, it can also reflect how pressure propagation speed negatively correlates to formation permeability. For the tight formation with ultra-low permeability, pressure needs more time to propagate in the formation compared with highly permeable formation.

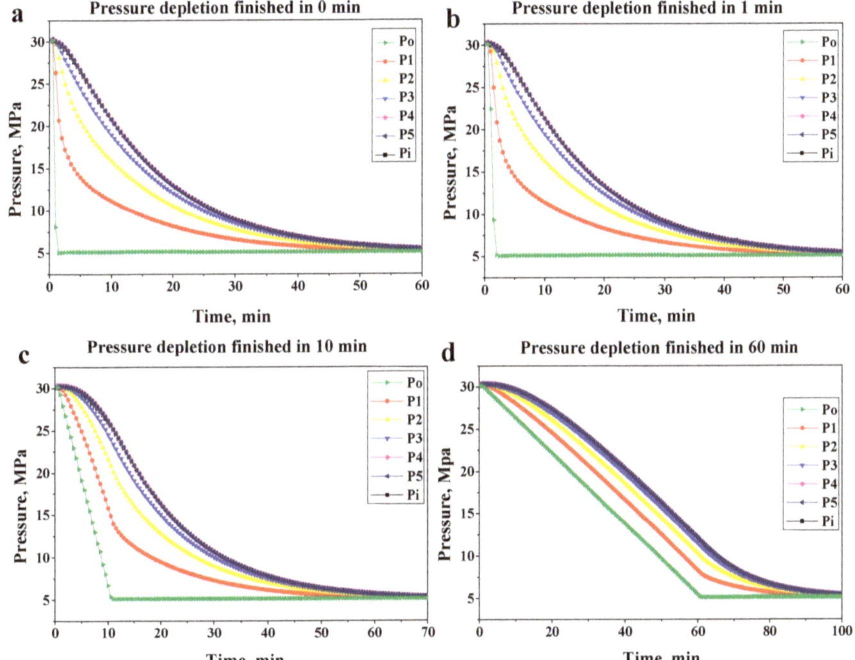

Figure 7. *Cont.*

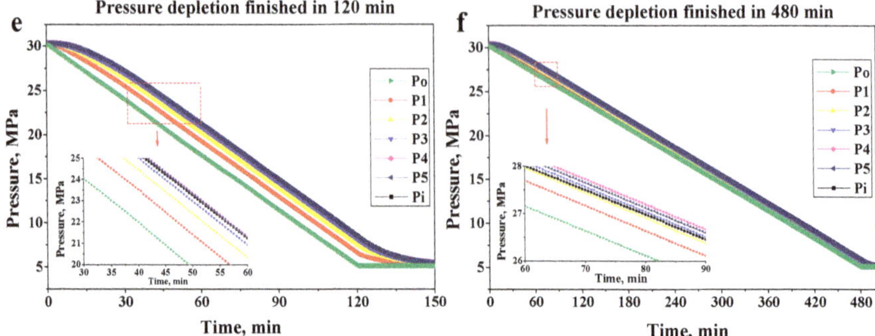

Figure 7. Characteristics diagram of pressure propagation during depletion experiments. (**a**) Pressure depletion finished in 0 min; (**b**) Pressure depletion finished in 1 min; (**c**) Pressure depletion finished in 10 min, (**d**) Pressure depletion finished in 60 min; (**e**) Pressure depletion finished in 120 min; (**f**) Pressure depletion finished in 480 min.

3.2. Characteristics of Tight Oil Depletion with Dissolved Gas

3.2.1. Depletion Experiments at Room Temperature (20.1 °C)

Dissolved gas usually exists in-situ in reservoir, therefore there is the need to carry out the depletion experiment with the oil containing dissolved gas. Four sets of different depletion experiments were carried out at 20.1 °C temperature with and without dissolved gas. The experimental conditions are shown in Table 6. The GOR of the live oil used in the experiment was 90 m^3/m^3 with saturation pressure of 9.1 MPa. The initial formation pressure was set to 20 MPa, and the linear pressure depletion was used to reduce the outlet pressure from 20 MPa to 5 MPa. The experimental conditions of the four groups were the same except the first group which did not contain dissolved gas (dead oil).

Table 6. The experimental parameters with dissolved gas.

No.	Oil Type	GOR (m^3/m^3)	Saturation Pressure (MPa)	Pressure Depletion Range (MPa)		Depletion Time (min)	Pressure Depletion Rate (MPa/min)
				Initial Pressure	Final Pressure		
1	dead oil			20	5	240	0.0625
2	live oil	90	9.1	20	5	240	0.0625
3	live oil	90	9.1	20	5	240	0.0625
4	live oil	90	9.1	20	5	240	0.0625

Figure 8 presents the oil recovery versus pressure depletion of experimental results. During the depletion process, at formation pressure higher than bubble point pressure, the recovery curve of the four groups basically coincided. However, it can still be identified that for the same pressure depletion range, the oil recovery of depletion experiments with dissolved gas (live oil) was a bit higher than that of the depletion experiment without dissolved gas (dead oil). This is ascribed to the fact that, the formation with dissolved gas has larger elastic energy to expand. However, when the formation pressure dropped below the bubble point pressure, there was an abrupt rise in the recovery degree of the live oil groups, though the rising trend of the dead oil recovery continued. As the formation pressure decreased from 20 MPa to 5 MPa, the ultimate recovery factor of dead oil group was only 2%. On the other hand, the recovery factor of the three groups of live oil (1–3) were 14.1%, 11.9% and 11.6% respectively. Therefore, from the output of the depletion process, dissolved gas has a significant effect on the recovery of tight oil depletion since recovery rate of live oil reached 14.1% compared to that of dead oil which was only 2%.

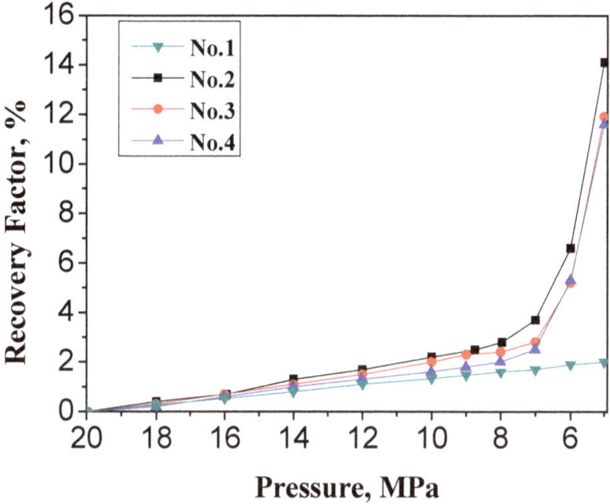

Figure 8. Recovery factor of live and dead oil depletion.

3.2.2. Depletion Experiments at Reservoir Temperature (60 °C)

Five groups of pressure depletion experiments with live oil at reservoir temperature with different initial formation pressure were conducted, and experimental conditions are shown in Table 7. There were three different formation pressure coefficients, i.e. 0.8, 1.25 and 1.5, and the depletion range and time vary from each other.

Table 7. Experimental conditions of live oil depletion under different initial formation pressure.

No.	Formation Pressure Coefficient	Pressure Depletion Range (MPa)		Depletion Time (min)	Pressure Depletion Rate (MPa/min)	Oil Type	GOR (m^3/m^3)	Saturation Pressure (MPa)
		Initial Pressure	Final Pressure					
1	1.5	30	5	1440	0.0173	live oil	50	8.85
2	1.25	25	6	1440	0.0132	live oil	50	8.85
3	0.8	16	6	750	0.0133	live oil	50	8.85
4	0.8	16	6	750	0.0133	live oil	50	8.85
5	0.8	16	6	750	0.0133	live oil	50	8.85

Figure 9 shows the oil recovery versus pressure depletion degree for the cases with the different initial formation pressure, and the detailed produced oil and gas volume are shown in Table 8. The experimental results demonstrate that ultimate oil recovery increases with increasing initial formation pressure. For the experiments with formation pressure coefficient equal to 0.8, the average ultimate oil recovery by pressure depletion from 16 MPa to 6 MPa was 11.41%. When the formation pressure coefficient was increased to 1.25, the oil recovery increased to 12.35%, but less than 1% oil recovery increment was seen. This is due to the limited elastic energy of fluids and rock. However, when the initial formation pressure was increased to 30 MPa and the final pressure decreased to 5 MPa, the ultimate oil recovery increased to 18.18%. It can be inferred that most of the oil recovery increment is due to the expansion of dissolved gas. When the formation pressure was above the bubble point pressure, the oil recovery was proportional to the pressure depletion degree, and the oil recovery lines were parallel which indicates the same total compressibility of formation. When the formation pressure dropped below the bubble point pressure, the oil recovery increased sharply, also the lower the final pressure, the high oil recovery factor will be.

Table 8. Experimental results of live oil depletion under different initial formation pressure.

No.	Produced Gas Volume (mL)	Produced Oil Volume (mL)	Ultimate Oil Recovery Factor (%)
1	11,827.8	133.72	18.18
2	5660.28	88.62	12.35
3	5496.11	77.6	11.27
4	6364.32	79.6	11.56
5	5234.4	78.42	11.39

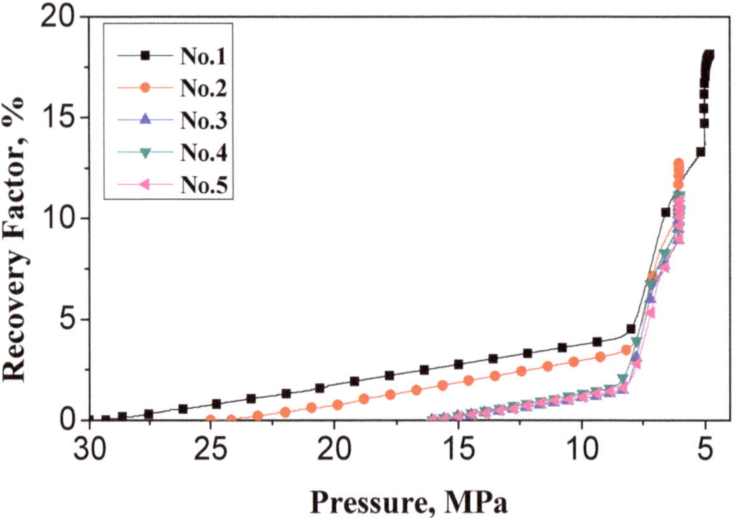

Figure 9. Recovery factor of live oil depletion under different pressure coefficient.

3.2.3. Characteristics of Pressure Propagation of Live Oil Depletion

The pressure distribution along the core holder was monitored during the experiments of pressure depletion using live oil. Figure 10 shows the pressure propagation during the live oil depletion experiments at different formation pressure coefficients. When formation pressure is higher than the bubble point pressure, only single-phase flow exists in the reservoir and the pressure drop of the pressure measurement points are synchronized as the pressure depletion rate is low. The initial stage is similar to the pressure propagation of dead oil depletion with low pressure depletion rate as depicted in Figure 7f. However, when the pressure is lower than the bubble point pressure, there exist pressure differences along the core holder, and the pressure of the measurement point far from the outlet is high than the outlet pressure due to the dissolved gas that comes out of solution. The dissolved gas provides additional pressure gradient to drive oil to the outlet, which in turn increases ultimate oil recovery. The pressure difference along the core holder after gas comes out of solution in Figure 10a is more significant than that in Figure 10b, due to the lower final pressure (5 MPa) which can account for the higher ultimate oil recovery (18.18%). The large pressure differences between measurement points due to more gas coming out of the oil provide larger pressure gradient to enhance oil recovery. The pressure propagation curves in Figure 10 can explain the mechanism of dissolved gas enhancing oil recovery.

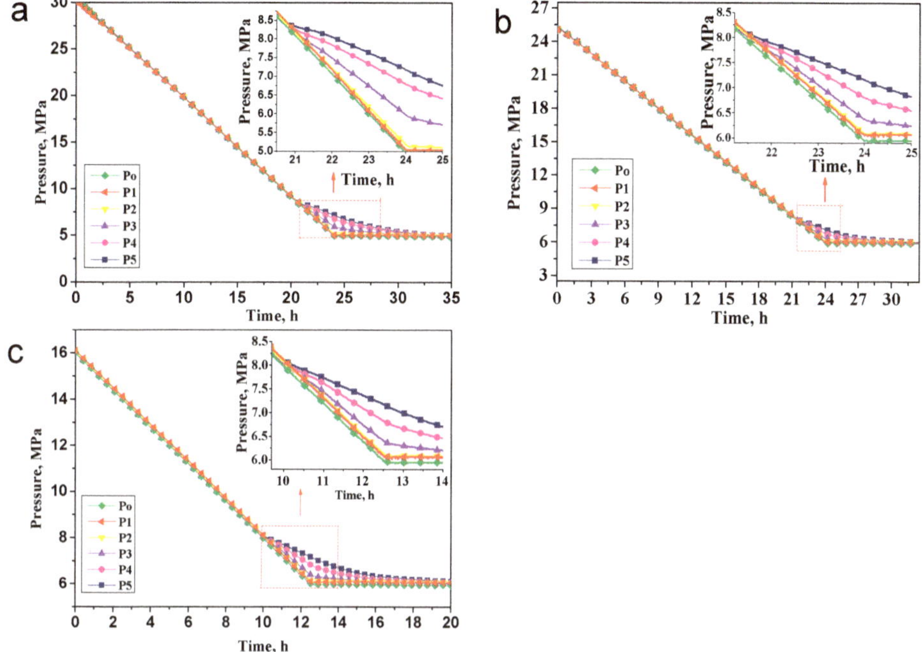

Figure 10. Pressure propagation characteristic of depletion experiments with different pressure depletion range (**a**) pressure depletion from 30 MPa to 5 MPa; (**b**) pressure depletion from 25 MPa to 6 MPa; (**c**) pressure depletion from 16 MPa to 6 MPa.

4. Discussion

The recovery factor of tight oil depletion is mainly affected by the following factors.

4.1. Effect of Formation Pressure Coefficient on Recovery Factor

Formation pressure coefficient will affect the physical properties of reservoir rock and fluid, mainly reflected on the elastic energy. From the elastic energy equation, the higher the formation pressure coefficient, the higher the elastic energy of the rock and fluid. The recovery factor is higher with depletion that has higher elastic energy. The effect was seen in the experimental results as the recovery factor of dead oil estimated with formation pressure coefficient of 1 was only 2% but 3% for formation pressure coefficient of 1.5, regardless of the different depletion time.

4.2. Effect of the Type of Depletion on Recovery Factor

Many depressurization ways can be used in tight reservoir depletion. Under the same formation pressure, depletion range can be reduced to different pressure values under different depletion times. From our experimental results, both the linear pressure depletion and step-like depletion method had a recovery factor of 2% despite the different depletion method. This means that, different ways of lowering pressure will only affect the rate of oil production, but not the final recovery factor since the elastic energy of the reservoir that releases the fluids is much dependent on the pressure rather than the depressurization method used. The way of lowering the pressure can be determined according to production requirements and equipment conditions.

4.3. Effect of Dissolved Gas on Recovery Factor

The presence of dissolved gas in tight oil reservoirs has a major effect on the recovery of tight oil. This is because when the pressure is higher than the bubble point pressure, gas is completely dissolved in the oil and the fluid in the reservoir is a single phase. In this case, the flow characteristics and the recovery factor of the dead oil and the live oil are basically the same. However, when the reservoir pressure is lower than the bubble point pressure, the dissolved gas is separated from the oil and expanded to form gas-liquid two-phase flow. The continuous expansion of the gas due to pressure decline tends to fills greater portion of the ultra-low rocks pores continuously thereby forcing the oil out of the pore spaces. This will accordingly improve the recovery factor greatly as it was seen that the recovery factor of live oil improved significantly to 18% compared to that of the deal oil which was only about 2–3%.

5. Conclusions

A novel experimental platform for modelling the pressure depletion process in tight oil reservoirs was developed. The developed experimental platform can effectively measure the oil recovery over pressure depletion, and the pressure propagation during depletion can also be recorded. The experimental results showed that pressure depletion without dissolved gas has limited elastic energy and the oil recovery was about 2–3%. In addition, the ultimate oil recovery was dependent on pressure depletion range but not pressure depletion types. The transient pressure propagates slowly in tight formations, and obvious pressure lags exist especially for the reservoir depletion with high pressure depletion rate. Dissolved gas can greatly enhance tight oil recovery when pressure depletes below bubble point pressure, since the ultimate oil recovery reached 11–18%, and it will continue to increase with decreasing final pressure. Pressure propagation curves of live oil depletion experiments demonstrated that the additional pressure gradients due to the evolution of gas out of the oil can account for the significant improvement in the oil recovery.

Author Contributions: Writing-Original Draft Preparation, W.C.; Experiment and Methodology, Z.Z. and X.C.; Project Administration, Q.L.; Writing-Review & Editing, P.O.A.; Data Analysis & Editing, F.W.

Acknowledgments: This work was supported by the National Science and Technology Special Project (Nos. 2016ZX05046 and 2017ZX05013-003), the National Natural Science Foundation of China (Nos. 51604285 and 41330319), and Scientific Research Foundation of China University of Petroleum, Beijing (No. 2462017BJB11).

Conflicts of Interest: The authors declare no conflict of interest.

References

1. Chew, K.J. The future of oil: Unconventional fossil fuels. *Philos. Trans. R. Soc. A* **2014**, *372*, 34. [CrossRef] [PubMed]
2. Wang, J.; Feng, L.; Steve, M.; Tang, X.; Gail, T.E.; Mikael, H. China's Unconventional Oil: A review of its resources and outlook for long-term production. *Energy* **2015**, *82*, 31–42. [CrossRef]
3. Leimkhler, J.; Leveille, G. Unconventional Resources. *JSPE* **2012**, *8*, 27–28. [CrossRef]
4. Qiu, Z.; Zou, C.; Li, J. Unconventional Petroleum Resources Assessment: Progress and Future Prospects. *NGG* **2013**, *24*, 238–246.
5. Rich, J.; Ammerman, M. Unconventional Geophysics for Unconventional Plays. In Proceedings of the SPE Unconventional Gas Conference, Pittsburg, PA, USA, 23–25 February 2010.
6. Zou, C.; Yang, Z.; Zhang, G. Conventional and Unconventional Petroleum "Orderly Accumulation": Concept and Practical Significance. *PED* **2014**, *41*, 14–27. [CrossRef]
7. Zou, C.N.; Zhai, G.M.; Zhang, G.Y.; Wang, H.J.; Zhang, G.S.; Li, J.Z. Formation, distribution, potential and prediction of global conventional and unconventional hydrocarbon resource. *J. Pet. Explor. Dev.* **2015**, *42*, 13–25. [CrossRef]

8. Miller, R.G.; Sorrell, S.R. The future of oil supply. *Philos. Trans. R. Soc. A* **2014**, *372*, 20130179. [CrossRef] [PubMed]
9. Zhang, K. Potential technical solutions to recover tight oil. Unpublished Master's Thesis, Norwegian University of Science and Technology (NTNU), Trondheim, Norway, 2014; p. 147.
10. Zhou, Q.F.; Yang, G.F. Definition and application of tight oil and shale oil terms. *Oil Gas Geol.* **2012**, *3*, 541–544.
11. Wang, F.; Yang, K.; Cai, J. Fractal Characterization of Tight Oil Reservoir Pore Structure Using Nuclear Magnetic Resonance and Mercury Intrusion Porosimetry. *Fractals* **2018**, *26*, 1840017. [CrossRef]
12. Wang, F.; Liu, Z.; Jiao, L.; Wang, C.; Guo, H. A Fractal Permeability Model Coupling Boundary-Layer Effect For Tight Oil Reservoirs. *Fractals* **2017**, *25*, 1750042. [CrossRef]
13. Law, B.E.; Curtis, J.B. Introduction to Unconventional Petroleum Systems. *AAPG Bull.* **2002**, *86*, 1851–1852.
14. Lin, S.H.; Zou, C.N.; Yuan, X.J. Status quo of tight oil exploitation in the United States and its implication. *Lithol. Reserv.* **2011**, *23*, 25–30.
15. Zou, C.N.; Zhu, R.K.; Wu, S.T. Types, characteristics, genesis and prospects of conventional and unconventional hydrocarbon accumulations: Taking tight oil and tight gas in China as an instance. *Acta Pet. Sin.* **2012**, *33*, 173–187.
16. Jia, C.Z.; Zheng, M.; Zhang, Y.F. Unconventional hydrocarbon resources in China and the prospect of exploration and development. *Pet. Explor. Dev.* **2012**, *39*, 129–136. [CrossRef]
17. Jia, C.Z.; Zou, C.N.; Tao, S.Z. Assessment criteria, main types, basic features and resource prospects of the tight oil in China. *Acta Pet. Sin.* **2012**, *33*, 343–350.
18. Jing, D.; Ding, F. The Exploration and Development of Tight Oil in USA. *Land Resour. Inf.* **2012**, *67*, 18–19.
19. Yan, C.Z.; Li, L.G.; Wang, B.F. *New Progress of Shale Gas Exploration and Development in North America*; Petroleum Industry Press: Beijing, China, 2009.
20. Yu, C.; Guan, P.; Zou, C.; Wei, H.; Deng, K.; Wang, P.; Wu, Y. Formation conditions and distribution patterns of N1 tight oil in Zhahaquan Area, Qaidam Basin, China. *Energy Explor. Exploit.* **2016**, *34*, 339–359. [CrossRef]
21. Azari, M.; Hamza, F.; Hadibeik, H.; Ramakrishna, S. Well testing Challenges in Unconventional and Tight Gas Reservoirs. In Proceedings of the SPE Western Regional Meeting, Garden Grove, CA, USA, 22–27 April 2018.
22. Zhao, X.; Liao, X.; Chen, Z.; Mu, L.; Zhang, F.; Zhou, Y.; Zhou, Z.; Zhao, N. A Well Testing Analysis Methodology and Application for Tight Reservoirs. In Proceedings of the Offshore Technology Conference Asia, Kuala Lumpur, Malaysia, 22–25 March 2016.
23. Zou, C.N.; Zhang, G.Y.; Tao, S.Z. Geological features, major discoveries and unconventional petroleum geology in the global petroleum exploration. *Pet. Explor. Dev.* **2010**, *37*, 129–145.
24. Sun, Z.D.; Jia, C.Z.; Li, X.F. *Unconventional Oil & Gas Exploration and Development (Upper Volume)*; Petroleum Industry Press: Beijing, China, 2011.
25. Yang, H.; Li, S.; Liu, X. Characteristics and resource prospects of tight oil in Ordos Basin, China. *Pet. Res.* **2016**, *1*, 27–38. [CrossRef]
26. Du, J.H.; He, H.Q.; Yang, T.; Li, J.Z.; Huang, F.X.; Guo, B.J.; Yan, W.P. Progress in China's tight oil exploration and challenges. *J. China Pet. Explor.* **2014**, *19*, 1–9.
27. Wei, Y.; Lashgari, H.R.; Wu, K.; Sepehrnoori, K. CO_2 injection for enhance oil recovery in Bakken tight reservoirs. *J. Fuel* **2015**, *159*, 354–363.
28. Clark, R.; Husain, A.; Rainey, S. Successful Post-fracture Stimulation Well Cleanup and Testing of Tight Gas Reservoir in the Sultanate of Oman. In Proceedings of the SPE Middle East Unconventional Resources Conference and Exhibition, Muscat, Oman, 26–28 January 2015.
29. Jia, H.; Sheng, J.J. Discussion of the feasibility of air injection for enhanced oil recovery in shale oil reservoirs. *Petroleum* **2017**, *3*, 249–257. [CrossRef]
30. Hui, P.; Qiquan, R.; Yong, L.; Youngjun, W. Development Strategy Optimization for Different Kinds of Tight oil Reservoirs. In Proceedings of the SPE Kingdom of Saudi Arabia Annual Technical Symposium and Exhibition, Dammam, Saudi Arabia, 24–27 April 2017.
31. Manrique, E.J.; Thomas, C.P.; Ravikiran, R.; Kamouei, M.I.; Lantz, M.; Romero, J.L.; Alvarado, V. EOR: Current Status and Opportunities. In Proceedings of the SPE Improved Oil Recovery Symposium, Tulsa, OK, USA, 24–28 April 2010.

32. Christensen, J.R.; Stenby, E.H.; Skauge, A. Review of WAG Field Experience. *SPE Reserv. Eval. Eng.* **2001**, *4*. [CrossRef]
33. Huffman, B.T. Comparison of various gases for enhance recovery from shale oil reservoirs. In Proceedings of the SPE Improved Oil Recovery Symposium, Tulsa, OK, USA, 14–18 April 2012.
34. Hawthorne, S.B.; Gorecki, C.D.; Sorensen, J.A.; Steadman, E.N.; Harju, J.A.; Melzer, S. Hydrocarbon Mobilization Mechanism from Upper, Middle, and Lower Bakken Reservoirs Rocks Exposed to CO_2. In Proceedings of the SPE Unconventional Resources Conference Canada, Calgary, AB, Canada, 5–7 November 2013.
35. Noureldien, D.M.; El-Banbi, A.H. Using Artificial Intelligence in Estimating Oil Recovery Factor. In Proceedings of the SPE North Africa Technical Conference and Exhibition, Cairo, Egypt, 14–16 September 2015.
36. Demirmen, F. Reserves Estimation: The Challenge for the Industry. *J. Pet. Technol.* **2007**, *59*, 10. [CrossRef]
37. Male, F.; Marder, M.; Browning, J.; Gherabati, A.; Ikonnikova, S. Production Decline Analysis in the Eagle Ford. In Proceedings of the SPE/AAPG/SEG Unconventional Resources Technology Conference, San Antonio, TX, USA, 1–3 August 2016.
38. Swindell, G.S. Eagle Ford Shale—An Early Look at Ultimate Recovery. In Proceedings of the SPE Annual Technical Conference and Exhibition, San Antonio, TX, USA, 8–10 October 2012.
39. Kanfar, M.; Wattenbarger, R. Comparison of Empirical Decline Curve Methods for Shale Wells. In Proceedings of the SPE Canadian Unconventional Resources Conference, Calgary, AB, Canada, 30 October–1 November 2012.
40. Moridis, N.; Soltanpour, Y.; Medina-Cetina, Z.; Lee, W.J.; Blasingame, T.A. A Production Characterization of the Eagle Ford Shale, Texas—A Bayesian Analysis Approach. In Proceedings of the SPE Canadian Unconventional Resources Conference, Calgary, AB, Canada, 30 October–1 November 2017.
41. Reisz, M.R. Reservoir Evaluation of Horizontal Bakken Well Performance on the Southwestern Flank of the Williston Basin. In Proceedings of the International Meeting on Petroleum Engineering, Beijing, China, 24–27 March 1992.
42. Bohrer, M.; Fried, S.; Helms, L.; Hicks, B.; Juenker, B.; McCusker, D.; Anderson, F.; LeFever, J.; Murphy, E.; Nordeng, S. State of North Dakota Bakken Formation Resource Study Project. *Appendix C*. Available online: https://www.legis.nd.gov/assembly/60-2007/docs/pdf/ts063008appendixc.pdf (accessed on 21 June 2018).
43. Clark, A.J. Determination of recovery factor in the Bakken formation, Mountrail County, ND. In Proceedings of the SPE Annual Technical Conference and Exhibition, New Orleans, LA, USA, 4–7 October 2009.
44. Ghaderi, S.M.; Clarkson, C.R.; Kaviani, D. Investigation of Primary Recover in Tight Oil Formations: A Look at the Cardium Formation, Alberta. In Proceedings of the Canadian Unconventional Resources Conference, Calgary, AB, Canada, 15–17 November 2011.
45. Xu, Q.; Zhu, D.; Ling, H.; Gao, T.; Wang, X. Fractures Parameters Optimization of the Depletion in Fractured Horizontal Wells for Ultra-low Permeability Reservoir. *J. Unconv. Oil Gas* **2014**, *1*, 37–42.
46. Xu, Y.; Yang, S.; Zhang, Z.; Han, W. Study on Mining Failure Law of Tight reservoir. *J. Liaoning Shihua Univ.* **2017**, *37*, 37–41.
47. Kabir, C.S.; Rasdi, M.F.; Igboalisi, B.O. Analyzing Production Data from Tight-oil Wells. In Proceedings of the Canadian Unconventional Resources and International Petroleum Conference, Calgary, AB, Canada, 19–21 October 2010.
48. Dechongkit, P.; Prasad, M. Recovery Factor and Reserves Estimation in the Bakken Petroleum System (Analysis of the Antelope, Pronghorn and Parshall fields). In Proceedings of the Canadian Unconventional Resources and International Petroleum Conference, Calgary, AB, Canada, 15–17 November 2011.
49. Turbakov, M.; Shcherbakov, A. Determination of Enhanced Oil Recovery Candidate Fields in the Volga-Ural Oil and Gas Region Territory. *Energies* **2015**, *8*, 11153–11166. [CrossRef]
50. Zhao, W.Q.; Yao, C.J.; Wang, X.D.; Zhou, R.P. Determination of oil recovery factor under the condition of exhaustion mining. *J. Foreign Oil Field Eng.* **2010**, *26*, 1–2.
51. Yuan, Z.; Wang, J.; Li, S.; Ren, J.; Zhou, M. A new approach to estimating recovery factor for extra-low permeability water-flooding sandstone reservoirs. *Pet. Explor. Dev.* **2014**, *41*, 1–10. [CrossRef]

52. Li, A.; Zhang, Z.; Cui, C.; Sun, R.; Yao, T. *Petrophysics*; China University of Petroleum Press: Qingdao, China, 2011; p. 331.
53. Jiang, H.; Yao, J.; Jiang, R. *Principles and Methods of Reservoir Engineering*; China University of Petroleum Press: Qingdao, China, 2006; p. 141.

Article

Investigation on the Application of NMR to Spontaneous Imbibition Recovery of Tight Sandstones: An Experimental Study

Chaohui Lyu [1,2], Qing Wang [1,2,*], Zhengfu Ning [1,2,*], Mingqiang Chen [1,2], Mingqi Li [1,2], Zhili Chen [1,2] and Yuxuan Xia [1,3]

1 State Key Laboratory of Petroleum Resources and Prospecting in China University of Petroleum, Beijing 102249, China; 2016312050@student.cup.edu.cn (C.L.); 2016312031@student.cup.edu.cn (M.C.); 2016212572@student.cup.edu.cn (M.L.); 2016212140@student.cup.edu.cn (Z.C.)

2 Ministry of Education Key Laboratory of Petroleum Engineering in China University of Petroleum, Beijing 102249, China

3 Hubei Subsurface Multi-Scale Imaging Key Laboratory, Institute of Geophysics and Geomatics, China University of Geosciences, Wuhan 430074, China; xiayx@cug.edu.cn

* Correspondence: wq2012@cup.edu.cn (Q.W.); nzf@cup.edu.cn (Z.N.); Tel.: +86-010-897-323-18 (Q.W.); +86-010-897-321-98 (Z.N.)

Received: 1 August 2018; Accepted: 5 September 2018; Published: 6 September 2018

Abstract: In this paper, the nuclear magnetic resonance (NMR) technique is applied to exploring the spontaneous imbibition mechanism in tight sandstones under all face open (AFO) boundary conditions, which will benefit a better understanding of spontaneous imbibition during the development of oil & gas in tight formations. The advantages of nuclear magnetic resonance imaging (NMRI) and NMR T_2 are used to define the distribution of remaining oil, evaluate the effect of micro structures on imbibition and predict imbibition recovery. NMR T_2 results show that pore size distributions around two peaks are not only the main oil distributions under saturated condition but also fall within the main imbibition distributions range. Spontaneous imbibition mainly occurs in the first 6 h and then slows down and even ceases. The oil signals in tiny pores stabilize during the early stage of imbibition while the oil signal in large pores keeps fluctuating during the late stage of imbibition. NMRI results demonstrate that spontaneous imbibition is a replacement process starting slowly from the boundaries to the center under AFO and ending with oil-water mixing. Furthermore, the wetting phase can invade the whole core in the first 6 h, which is identical with the main period of imbibition occurring according to NMR T_2 results. Factors influencing the history of oil distribution and saturation differ at different periods, while it is dominated by capillary imbibition at the early stage and allocated by diffusion at later time. Two imbibition recovery curves calculated by NMRI and NMR T_2 are basically consistent, while there still exists some deviations between them as a result of the resolutions of NMRI and NMR T_2. In addition, the heterogeneity of pore size distributions in the two samples aggravates this discrepancy. The work in this paper should prove of great help to better understand the process of the spontaneous imbibition, not only at the macroscopic level but also at the microscopic level, which is significant for oil/gas recovery in tight formations.

Keywords: tight sandstones; spontaneous imbibition; remaining oil distributions; imbibition front; imbibition recovery; NMR

1. Introduction

Tight oil reservoirs show typical low porosity and ultralow permeability characteristics, caused by a wide pore size distribution and complex pore throat structures [1–3]. Large volumes of slick water during multistage hydraulic fracturing are pumped into the tight formation to improve petrophysical properties by creating complex fracture networks [4,5]. Therefore, the spontaneous imbibition mechanism in tight sandstones is a key issue that needs to be focused upon to prevent channeling in the development of tight oil reservoirs. Hence, oil recovery by spontaneous imbibition is of special importance in tight formations, particularly when the formation is characterized by developed fractures [6,7]. For decades, researchers have carried out a lot of research on imbibition and obtained important conclusions [8–10]. The characterization of the imbibition process, especially in tight formations, has become a research hot topic. Many authors have focused on evaluating the imbibition recovery based on conventional experiments, while few of them have examined the imbibition recovery and residual oil distributions at microscopic level [11–15]. Indeed, acquiring a better understanding of spontaneous imbibition in tight oil reservoirs can be tricky because of a wide pore size distribution with a significant portion of it being nanoscale porosity [3]. Furthermore, an accurate measurement of oil imbibed from tight samples is crucial for calculating the oil recovery of spontaneous imbibition. Traditional measurement methods such as the volume method and the mass method, are not suitable for imbibition of tight oil [16–18]. Therefore, a new method for characterizing spontaneous imbibition is desperately needed. NMR has become a common experimental method in light of its fast, visual and non-destructive properties [19–22]. On the one hand, NMRI can obtain images of the residual oil distribution and imbibition front at relatively large scales [23,24]. On the other hand, NMR T_2 can reflect quantitatively the residual oil distributions at microscopic level [25,26]. Therefore, not only the imbibition front can be observed, but also the fluids change at pore scale during spontaneous imbibition can be obtained [27,28].

In this paper, spontaneous imbibition experiments and NMR tests are performed in tight sandstone samples. This study aims to explore the potential of the NMR technique in characterizing the microscopic imbibition mechanism in tight oil reservoirs. First of all, The changing characteristics of the wetting and non-wetting phase at the microscopic level during imbibition are described using NMR T_2; Secondly, imbibition front advancing characteristics are observed by NMRI; lastly, the applicability of two methods for calculating the imbibition recovery is evaluated and the reasons for the deviations of imbibition recovery based on the two methods are respectively discussed.

2. Methodology

Imbibition recovery is an important parameter when evaluating imbibition effect. During imbibition, oil is always adhered to the core surface and not easy to separate from core plugs. Moreover, the oil content in tight sandstone core is low and the volume of oil that can be imbibed from the core is less, which is difficult to measure. As a result, traditional methods of predicting imbibition recovery present defects. Therefore, simple and efficient methods of predicting imbibition recovery in tight sandstones are desperately needed. Herein, two methods for predicting imbibition recovery are recommended based on NMR T_2 distributions and NMR 2D-images, which is respectively called the NMR T_2 method and the NMRI method.

2.1. NMR T_2

Under the action of radio frequency (RF) pulses hydrogen nuclei not only change in phase, but also absorb energy to transition to higher energy states. After the RF excitation stops, the phase and energy of hydrogen nuclei are restored to the original state. This process is called relaxation. There are two relaxation times: T_1 (longitudinal relaxation time) and T_2 (transverse relaxation time). Although, T_2 contains the same information as T_1 (T_1 = kT_2, k is constant), but it has more collected points than T_1. Moreover, the time for obtaining T_2 is much shorter [29,30]. Therefore, T_2 is usually choosen for core analysis. The expression for T_2 can be expressed as follows [19]:

$$\frac{1}{T_2} = (\frac{1}{T_2})_S + (\frac{1}{T_2})_D + (\frac{1}{T_2})_B \quad (1)$$

where $(\frac{1}{T_2})_S$ is the relaxation contribution from the surface of the rock particles (1/ms), $(\frac{1}{T_2})_D$ is the relaxation contribution from diffusion in magnetic gradients (1/ms), $(\frac{1}{T_2})_B$ is the relaxation contribution from bulk (1/ms). Equation (1) can also be described as follows:

$$\frac{1}{T_2} = \rho_2(\frac{S}{V}) + \frac{D(\gamma G T_E)^2}{12} + \frac{1}{T_{2B}} \tag{2}$$

Since the value of T_{2B} is always 2~3 s, which is much larger than T_2. $\frac{1}{T_{2B}}$ in Equation (2), it can be ignored. G is the magnetic gradient (gauss/cm), whose value is 0 in uniform magnetic field. Therefore, $\frac{D(\gamma G T_E)^2}{12}$ in Equation (2) can also be ignored. Hence, Equation (1) can be converted to Equation (3):

$$\frac{1}{T_2} = \rho_2(\frac{F_S}{\mathrm{r}}) \tag{3}$$

where F_S is the shape factor for irregular balls, for example, F_S equals 3 in a spherical model and equals 2 in a column model. r is the pore and throat radius (um), ρ_2 is the transverse relaxation strength (constant). Therefore, a relationship between T_2 and the pore radius can be further built in Equation (4):

$$\mathrm{r} = C T_2 \tag{4}$$

where $C = \frac{1}{F_S \rho_2}$, μm/ms C is a constant since F_S and ρ_2 are constant. Thus, there is a one to one relationship existing between T_2 and r. The coefficient factor C in Equation (4) can be obtained by contrasting the T_2 distribution and the pore throat size by PCP [31]. Consequently, the larger pores have a longer relaxation time. And the larger the pore volume of the corresponding pore, the larger the area of pore radius.

With the NMR T_2 method, the remaining oil percentage can be obtained by dividing the sum of the signal amplitudes at a measurement time by that at the initial time. The imbibition recovery can be calculated using the remaining oil percentage and the specific expression is shown in Equation (5):

$$R_{NMR} = (1 - \frac{A_\mathrm{i}}{A_0})100\% \tag{5}$$

where R_{NMR} is the imbibition recovery calculated by Equation (5), %, A_i is the acreage enclosed by T_2 curves of core after imbibition and on the X-axis, A_0 is the acreage enclosed by T_2 curves of core under saturation conditions and on the X-axis.

2.2. NMRI

There are three factors contributing to the signal in the process of general NMRI, including the longitudinal relaxation time, the transversal relaxation time and the hydrogen proton density [19]. The signal amplitude in NMRI can be expressed as follows:

$$A = A_O \rho (1 - e^{-T_R/T_1}) e^{-T_E/T_2} \tag{6}$$

where A is the signal amplitude, A_O is the original signal amplitude, ρ is the hydrogen proton density, T_R is the repeat time, T_1 is the longitudinal relaxation time, and T_E is the echo time. As can be seen from Equation (6), we can highlight the effect of one factor and restrain the effects of two other factors by adjusting the imaging parameters. In this paper, a larger T_R and a smaller

. $\mathrm{T_E}$ were set, weighing the impact of proton density and reducing the impact of T_1 and T_2 on the image, which is also called the proton density-weighted imaging method. Consequently, Equation (6) can be converted into Equation (7):

$$A = A_O \rho \tag{7}$$

As can be seen from Equation (7), the signal amplitude is determined only by ρ. The stronger the signal intensity, the larger the proton density and the more the oil content. Therefore, images of oil distributions can be obtained by Equation (7), which is called the NMR proton density-weighted imaging method. The specific process of obtaining NMRI images is described below: select a slice of the sample parallel to the gradient magnetic field in the Z-axis direction. At the same time, the Y-axis of a pixel can be determined by phase encoding and the X-axis determined by rate encoding. 2D-images can be reconstructed based on signal intensity and the information above, which reflects oil distributions in a core section.

With the NMRI method, the remaining oil percentage can be obtained by dividing the mean of pixels at a measurement time by the mean of pixels at the initial time. The imbibition recovery can also be calculated using the remaining oil percentage and the specific expression is shown in Equations (8) and (9):

$$R_{NMRI} = (1 - \frac{\overline{P_1}}{\overline{P_0}}) \times 100\% \tag{8}$$

$$\overline{P} = \frac{\sum_1^{5500} P_n}{5500} \tag{9}$$

where R_{NMRI} is the imbibition recovery calculated by Equation (8),%, $\overline{P_i}$ is the pixel mean of 5500 voxels at a certain imbibition moment, $\overline{P_0}$ is the pixel mean at the initial time, P_n is the pixel value of one pixel, $\overline{P_1}$ and $\overline{P_0}$ can be calculated by Equation (9). 5500 pixels are randomly selected from the NMR 2D-images.

3. Experimental Materials

3.1. Samples

Two cylindrical outcrop tight samples were collected to perform spontaneous imbibition and NMR tests in this study. Petrophysical properties of two samples are shown in Table 1. The length and radius of core plugs were measured, and then helium porosity and nitrogen permeability tests were performed. Contact angle was measured by the contact angle measuring instrument JC2000D3. It should be noted that oil saturation in Table 1 represents the oil percentage of a blank core after being saturated by white oil-5 and the detailed calculation can be referred in Section 3.4.

Table 1. Mineral contents, porosity, permeability and contact angle of two samples.

Sample	Petrophysical Property						X-ray Results/%				
	L/cm	R/cm	K/mD	ϕ/%	S_o/%	θ/°	Qz	Pl	Cal	Dol	TCCM
SX-5	3.99	2.52	0.21	9.7	96.5	49.3	69.4	8.5	1.1	5.0	16
YL-1	3.83	2.52	0.89	10.2	98.3	40.5	82.7	-	-	-	17.3

Note: L—core length, R—core radius, K—permeability, ϕ—porosity, S_o—oil saturation, θ—contact angle; Qz—quartz, Pl—plagioclase, Cal—calcite, Dol—dolomite; TCCM represents the total content of clay minerals.

3.2. Fluids

In order to eliminate the hydrogen signal of water, $MnCl_2$ solution with the weight percentage of 40% is selected as the wetting phase by comparing the shield effect of various $MnCl_2$ solutions with different concentrations. So as to avoid volatilization in the magnetic chamber, white oil-5 is used as the non-wetting phase instead of kerosene, which is characterized by a density of 0.82 g/cm^3 and a viscosity of 3.5 mPa·s.

3.3. Experimental Instrument

A PANalytical diffractometer was used to acquire the relative mineral percentages, estimated by a semi-quantitative method. It was performed on powdered tight sandstone at room temperature under a relative humidity (RH) of 66 %.

NMR tests were performed on a MacroMR23-60H-I instrument (Suzhou Niumag Analytical Instrument Corporation, Suzhou, China) with a constant magnetic field strength of 0.55 T and a resonance frequency of 23.408 MHz. The measurement parameters are as follow: echo spacing, 0.12 ms; waiting time, 2500 ms; echo numbers, 6000; numbers of scans, 64. All experiments were conducted at room temperature under a relative humidity of 60%. The temperature of the magnetic chamber was 305.15 K.

3.4. Experimental Procedure

Firstly, core samples were dried at 378.15 K to a constant weight. Then the core samples were weighed and measured (both the length and the diameter). Secondly, two core samples were saturated by displacement until 8 pore volume white oil-5 are obtained at the end of the core holder. The weights of the saturated cores were recorded. The oil saturation (Table 1) of each core can be calculated using the dry weight, the wet weight and the volume of the sample. Thirdly, raw NMR data of oil-saturated cores were obtained. At last, the oil-saturated cores were immersed in $MnCl_2$ solution to simulate spontaneous imbibition under AFO boundary conditions (Figure 1). Boundary conditions are complicated due to the existence of both co-current and counter-current imbibition, among which AFO condition is the most common one. The boundary condition AFO was chosen to further investigate in this study [32–34]. Six sets of NMR experiments were conducted at 3 h, 6 h, 19 h, 26 h, 60 h, and 180 h after being immersed into $MnCl_2$ solution (The rate of spontaneous imbibition slow down with time in previous research, measurement intervals were small at the early stage of spontaneous imbibition and then measurement intervals were large at the later stage of spontaneous imbibition, so our NMR measurements were set at 3 h, 6 h, 19 h, 26 h, 60 h, and 180 h). Then, NMR T_2 and NMR images for each set of the NMR experiment were obtained.

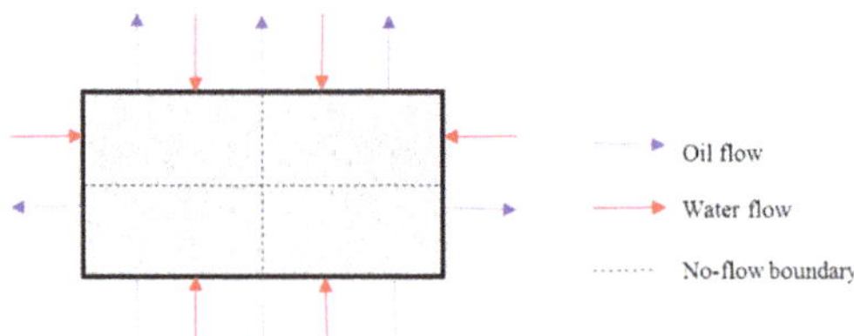

Figure 1. Counter-current imbibition in oil-saturated cores under AFO boundary condition.

4. Results and Discussion

4.1. NMR T_2 under Saturated Condition

The top curves in Figures 2 and 3 show the T_2 distributions of samples under saturation condition. Each T_2 distribution curve shows two peaks with a higher right peak and a lower left peak. The larger the value of T_2 is, the bigger the pore is. Pores ranging 10 ms–1000 ms and 0.1 ms–10 ms can be defined as large pores and tiny pores respectively, which is around right peak and left peak. The two peaks of SX-5 are located at about 5 ms and 100 ms whereas the two peaks of YL-1 are located at about 1 ms and 150 ms. The height of the left peak is obviously lower than that of the right one in SX-5, while the divergence is smaller inYL-1. The area enclosed by left peak is apparently smaller than that of the right peak for SX-5, while the relationship is opposite for YL-1. In conclusion, oil content in tiny pores is less than oil content in large pores for SX-5 and the relationship in YL-1 is just opposite. One possible reason for the difference is probably that YL-1 has more tiny pores than SX-5. The other probable reason is that YL-1 has a better connectivity. However, considering the high oil saturation of the two samples (Table 1). We attribute the difference to the first one.

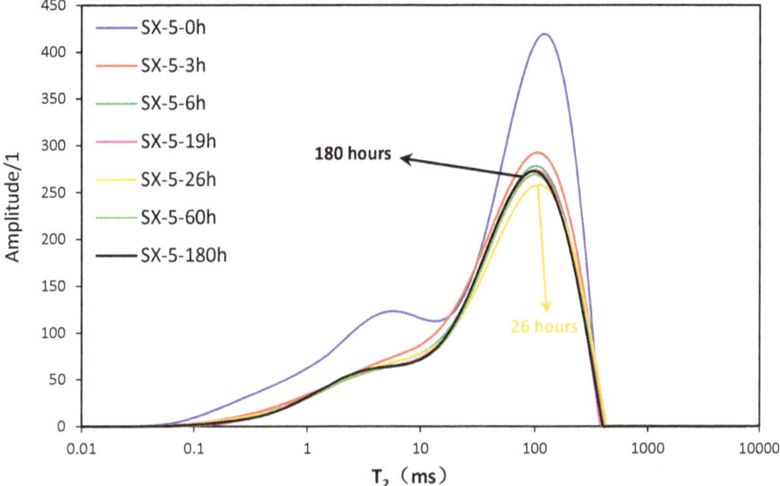

Figure 2. T_2 distributions of core SX-5 during imbibition.

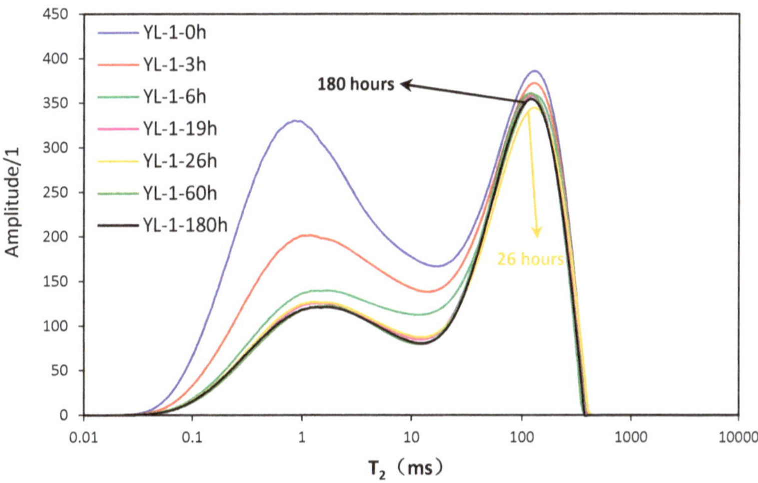

Figure 3. T_2 distributions of coreYL-1 during imbibition.

4.2. Remaining Oil Distributions after Imbibition

The other six curves from top to bottom in Figure 2 or Figure 3 represent the T_2 distribution curves of the remaining oil in the core after imbibition for 3 h, 6 h, 19 h, 26 h, 60 h and 180 h, for simplicity, they are recorded as C_3, C_6, C_{19}, C_{26}, C_{60} and C_{180}, respectively. When taking the overall process of imbibition of SX-5 into account, the left peak of C_3 descends significantly and part of the left peak of C_6 descends slightly. Left peaks of C_6 and C_{19} almost coincide with each other. Then the left peak of C_{26} rises up and then the left peaks of C_{60} and C_{180} are almost identical. When it comes to the right peaks at different imbibition time, it can be seen that the right peaks of C_{19} and C_6 almost coincide. The right peak of C_{26} falls down slightly and then the right peaks of C_{60} and C_{180} both rise up in an apparent manner. The change of T_2 distributions of SX-5 indicates that the main imbibition time of tiny pores and large ones is relatively 0~3 h and 0~6 H. In contrast with SX-5, YL-1 shows a more obvious main imbibition pore size interval around the left peak. The left peak falls down continuously at early stage and then stays unchanged during the 19–26 h period, then slightly falls down at 60 h and again stays unchanged at 180 h. The right peak falls down slightly during the first 26 h, then rises up a little at 60 h and again falls down slightly at 180 h. However, the right peak of C_{180} is still lower than that of C_{60}, but still higher than that of C_{26}. As for the reasons why the time of experiment sets 26 h is fallen as the lowest curve and 180 h located in the middle in both Figures 2 and 3, and the oil signal amplitude in pores, especially in large pores, fluctuates during the later period of imbibition process, the detailed explanation can be found in Section 4.3.

The changes of the two peaks of the two samples during the imbibition process are not synchronized, which indicates that the difference of imbibition between tiny pores and large pores is obvious. The left peaks of the two samples move right during imbibition which does not happen in the right peak, indicating that tiny pores exert a strong advantage in imbibition. At the early stage of imbibition process of two samples, oil in tiny pores is displaced by the wetting phase and oil in large pores is also discharged by counter-current imbibition under the AFO boundary condition. Therefore, the oil signal of tiny pores and large pores is weakened just as in the T_2 results above. Tiny pores play a stronger role than large pores in imbibition at the early stage. Reasons for this are as follows: on the one hand, as the main driving force for imbibition, the capillary force of large pores is smaller than that of tiny pores. As a result, imbibition in large pores is slower than that in tiny pores. On the other hand, both oil originally stored in large pores and oil imbibed from tiny pores should be discharged from the large pores at the same time. Therefore, the oil signal in tiny pores decreases more quickly than in large pores at the beginning. Since tiny pores in SX-5 are less developed, the difference of imbibition intensity between tiny pores and large pores in SX-5 is little. Also, the above reasons as well as the existence of abundant tiny pores inYL-1 leads to a significantly main imbibition interval. As imbibition continues, the oil signal in tiny pores and maintains a slightly downward trend and even tends to remain unchanged while the oil signal in large pores fluctuates during the later stage.

At the same time, the common point of two samples is that tiny pores show a strong ability of imbibition. However, by comparing the two samples, we can find that there are some differences between them. The first one is that imbibition in tiny pores of YL-1 lasts longer than that of SX-5. The other is that large pores of SX-5 show a stronger ability of imbibition than those of YL-1. Since right peaks of SX-5 and YL-1 are located at 5 ms and 1ms when left ones are at 100 ms and 150 ms, tiny pores distributed in the left peak of YL-1 have a larger capillary force than those of SX-5 while the capillary force in large pores in the right peak shows the opposite result. Hence, tiny pores ofYL-1 show a stronger imbibition than those of SX-5 when large pores display the opposite result. Therefore, more oil is imbibed from tiny pores to large ones. It is thus not imbibed out of large pores. Consequently, oil is trapped in large pores, which leads to a slow fall of the right peak of T_2 distribution. As can be seen from all the above statements, the pore size distribution has a great effect on imbibition. Pore size distribution can account for the two differences.

4.3. Imbibition Front

2D-images in Figures 4 and 5 are reconstructed by employing the NMR proton density-weighted imaging method, which represent fluid distributions in the center section of cylindrical cores at different stages (0 h, 3 h, 6 h, 19 h) of imbibition. It is noted that NMR-2D images at 26 h, 60 h, and 180 h were left out because few changes happen after 26 h. The red area means that core is saturated 100% by oil and the blue region on the edge of the core is affected by a noise-signal ratio. The green color corresponds to the oil-water two phase region and the lighter green is related to areas with a larger water saturation. As can be seen from Figure 5, oil is evenly distributed throughout the overall section before imbibition. The oil signal around the border then weakens. Furthermore, the oil-water mixing zone approaches the center of 2D-image with imbibition carrying on taking place.

Finally, the whole core is filled with oil-water two phase. The center of the 2D-image corresponds to a no-flow boundary, which is invaded by water at last. The core surface is directly contacted with the wetting phase under AFO boundary condition, which can be considered as the situation where the matrix is coming into contact with the fractures, in this paper. It can also be seen from Figures 4 and 5 that although the resolution of NMRI 2D-image is low, the imbibition front advancing is relatively clear. Spontaneous imbibition is a replacement process starting slowly from the boundaries towards the center, in which a radial displacement is formed during counter-current imbibition under AFO boundary conditions.

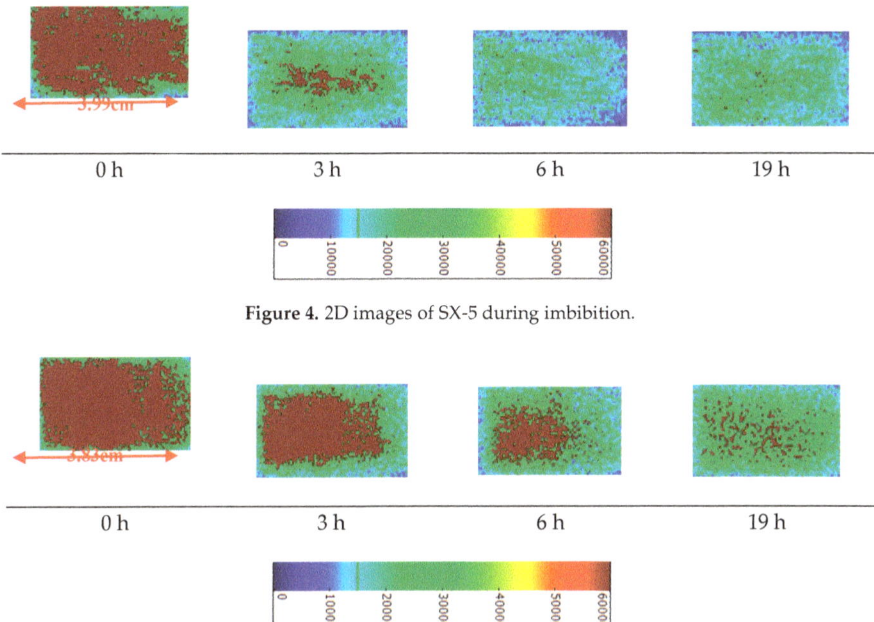

Figure 4. 2D images of SX-5 during imbibition.

Figure 5. 2D images of YL-1 during imbibition.

The results demonstrate that the wetting phase can invade the whole core in 6 h. The change of fluid distribution is no longer obvious when time passes from 6 h to 19 h. As concluded in Section 4.2, NMR T_2 curves of the remaining oil distributions after 6 h also change a little, which makes good agreement. The oil signal inYL-1 changes obviously from 0 h to 19 h and pure oil areas are sporadically distributed near the no-flow boundary until 19 h, which coincides with the changes of the left peak of YL-1 T_2 curves. Therefore, the imbibition front advancing trend concluded from NMRI makes good accordance with NMR T_2. According to the results of NMRI and NMR T_2, continuous oil is cut off once water invades the whole core. The reason why the oil signal amplitude in pores, especially in large pores, fluctuates during the later period of the experiment is that diffusion plays a dominant role in the distribution change of discontinuous oil. Based on observations above, we concluded that the early imbibition time represents a period dominated by the capillary force, whereas the later period towards the end of the experiment is controlled by diffusion. During the later period of imbibition, the oil signal in large pores increases in an obvious manner and the oil signal in tiny pores fluctuates slightly or even stays unchanged. The oil in tiny pores invaded into larger pores under the process of diffusion and the oil is not discharged out timely, which causes an increase of the oil signal in large pores. It can be seen that the remaining oil is mainly distributed in the larger pores and cannot be fully imbibed by water.

4.4. Recommended Methods to Calculate Imbibition Recovery

Using the two methods recommended in Methodology (Section 2), imbibition recovery curves of the two samples above are obtained in Figure 6 or Figure 7. The imbibition recoveries of two samples exceed 30% at 6 h after imbibition and keep stable after 19 h. The ultimate recoveries of the NMR T_2 method are 35% and 42% respectively for SX-5 andYL-1, which is consistent with the recovery results using the imbibition bottle in our previous work. However, the ultimate recovery results of the NMRI method are 51% (SX-5) and 38% (YL-1). The recovery results show no obvious relationship with the permeability. Even the recovery results show differences. The imbibition recovery results still indicate that imbibition is critical for the development of tight oil reservoirs.

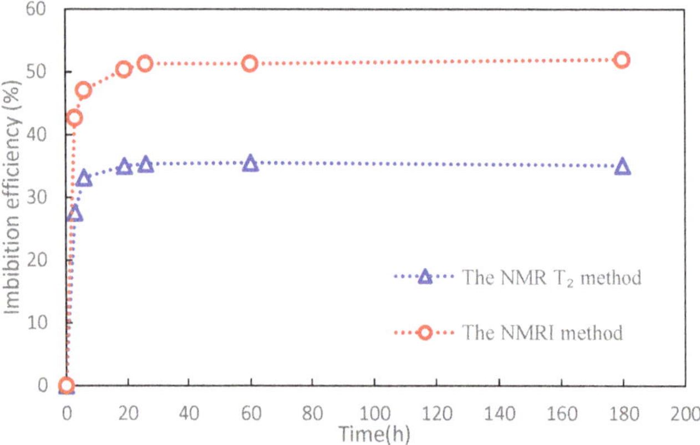

Figure 6. The imbibition recovery of SX-5 predicted by the NMR T_2 method and the NMRI method.

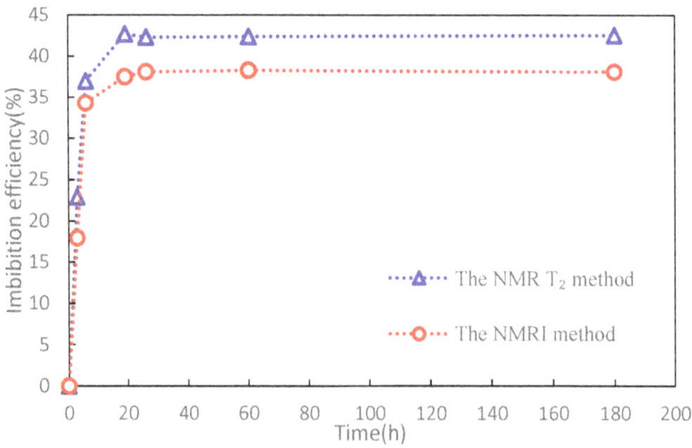

Figure 7. The imbibition recovery of YL-1 predicted by the NMR T_2 method and the NMRI method.

Comparisons between two methods in Figure 6 or Figure 7 were analyzed. On the one hand, as can be seen from Figure 6 or Figure 7, the two curves show the same trend with time. On the other hand, the imbibition recovery obtained by the NMRI method for SX-5 is always larger than that acquired by the NMR method, while the relationship is just opposite for YL-1. Reasons for the discrepancies are mainly attributed to the resolution of two methods and the heterogeneity of pore size distributions. The resolution of NMR T_2 is much higher than that of 2D-images. The NMRI method cannot effectively identify oil changes in tiny pores during spontaneous imbibition. The NMR T_2 and NMRI results can lead us to make the following remarks. Firstly, both tiny pores and large pores are abundant inYL-1 whereas the large pore content is high and tiny pore content is low in SX-5. Furthermore, the imbibition ability in large pores of YL-1 is weak while tiny pores show a strong imbition ability inYL-1. Not only tiny pores, but also large pores show a strong imbibition ability in SX-5. Here, the dominant imbibition pores are large pores whereas the dominant imbibition pores in YL-1 are tiny pores. Therefore, the imbibition recovery of SX-5 can be mostly contributed by the dominant imbibition pores when the imbibition recovery of YL-1 can be mostly contributed by a poor pore behavior. What's more, the heterogeneity of pore size distributions in two samples aggravates this dicrepancy in the two methods.

Based on the results above, the NMR T_2 method can effectively identify tiny and large pores, we recomend it as an effective tool for predicting imbibitoin recovery in tight sandstones. However, with the improvement of resolutions in NMRI, the imbibiton recovery predicted by the NMRI method will get more and more close to the recovery calculated by the NMR T_2 method. In our future work, the relationship between the resolution of NMRI and predicted imbibition recovery will be studied.

5. Conclusions

In this study, two outcrop samples were collected to perform spontaneous imbibition experiments under AFO boundary condition. NMRI and NMR T_2 tests were conducted during spontaneous imbibition, and employed to evaluate the spontaneous imbibition mechanism and the oil displacement recovery by imbibition. The main conclusions are as follows: (1) Spontaneous imbibition slows down and even ceases after 6 h. The oil signal in tiny pores stabilizes during the early stage of imbibition while the oil signal in large pores keeps fluctuating during the late stage of imbibition. (2) According to the results of NMRI and NMR T_2, we conclude that continuous oil is cut off once water invades the whole core and then imbibition slows down or even ceases. The reason why the oil signal amplitude in pores, especially large pores, fluctuates at a late period of the experiment is that diffusion plays a dominant role in determining the distributions of discontinuous oil at the late stage. (3) The imbibition recovery of two samples using two methods exceeded 35%. Even differences occurred, the results still indicate that imbibition is critical for the development of tight oil reservoirs. (4) The two imbibition recovery curves that were predicted by the two methods are basically consistent. Discrepancies, however, still exist. Due to the differences of the resolutions of the two methods, there are some discrepancies in the imbibition recovery values. Furthermore, the heterogeneity of pore size distributions in the two samples aggravates this discrepancy between the two methods.

Author Contributions: Each author has made contributions to the present paper. Conceptualization, C.L., M.C., and Q.W.; Data curation, C.L. and Q.W.; Investigation, C.L., M.C., Q.W., and Z.N.; Writing—review & editing, C.L., Z.C., M.L. All authors have read and approved the final manuscript.

Acknowledgments: This research was funded by National Natural Science Foundation of China (51474222 and 51504265), PetroChina Innovation Foundation (2017D50070205) and the Foundation of State Key Laboratory of Petroleum Resources and Prospecting, China University of Petroleum, Beijing (No. PRP/open-1601).The authors will also gratefully appreciate the anonymous reviewers' suggestions.

Conflicts of Interest: The authors declare no conflict of interest.

References

1. Jia, C.Z.; Zou, C.N.; Li, J.Z.; Li, D.H.; Zheng, M. Assessment criteria, main types, basic features and resource prospects of the tight oil in Chin. *Acta Petrol. Sin.* **2012**, *33*, 343–350.
2. Chen, M.Q.; Cheng, L.S.; Cao, R.Y.; Lyu, C.H. A Study to Investigate Fluid-Solid Interaction Effects on Fluid Flow in Micro Scales. *Energies* **2018**, *11*, 2197. [CrossRef]
3. Lyu, C.; Ning, Z.; Wang, Q.; Chen, M. Application of NMR T_2 to Pore Size Distribution and Movable Fluid Distribution in Tight Sandstones. *Energy Fuels* **2018**, *32*, 1395–1405. [CrossRef]
4. Novlesky, A.; Kumar, A.; Merkle, S. Shale Gas modeling workflow: From microseismic to simulation-A horn river case study. In Proceedings of the Canadian Unconventional Resources Conference, Calgary, AB, Canada, 15–17 November 2011.

5. Yang, L.; Ge, H.; Shi, X. The effect of microstructure and rock mineralogy on water imbibition characteristics in tight reservoirs. *J. Nat. Gas Sci. Eng.* **2016**, *34*, 1461–1471. [CrossRef]
6. Meng, M.; Ge, H.; Ji, W. Monitor the process of shale spontaneous imbibition in co-current and counter-current displacing gas by using low field nuclear magnetic resonance method. *J. Nat. Gas Sci. Eng.* **2015**, *27*, 336–345. [CrossRef]
7. Olafuyi, O.A.; Cinar, Y.; Knackstedt, M.A. Spontaneous imbibition in small cores. In Proceedings of the 2007 SPE Asia Pacific Oil & Gas Conference and Exhibition, Jakarta, Indonesia, 30 October–1 November 2007.
8. Wei, W.; Cai, J.; Xiao, J.; Meng, Q.; Xiao, B.; Han, Q. Kozeny-Carman constant of porous media: Insights from fractal-capillary imbibition theory. *Fuel* **2018**, *234*, 1373–1379. [CrossRef]
9. Li, C.X.; Shen, Y.H.; Ge, H.K.; Su, S.; Yang, Z.H. Analysis of spontaneous imbibition in fractal tree-like network system. *Fractals* **2016**, *24*. [CrossRef]
10. Cai, J.; Perfect, E.; Cheng, C. Generalized modeling of spontaneous imbibition based on Hagen-Poiseuille flow in tortuous capillaries with variably shaped apertures. *Langmuir* **2014**, *30*, 5142–5151. [CrossRef] [PubMed]
11. Alomair, O.A. New Experimental Approach for Measuring Drainage and Spontaneous Imbibition Capillary Pressure. *Energy Fuels* **2009**, *23*, 260–271. [CrossRef]
12. Lyu, C.; Ning, Z.; Chen, M.; Wang, Q. Experimental study of boundary condition effects on spontaneous imbibition in tight sandstones. *Fuel* **2019**, *235*, 374–383. [CrossRef]
13. Li, C.X.; Shen, Y.H.; Ge, H.K.; Yang, Z.H.; Su, S.; Ren, K.; Huang, H.Y. Analysis of capillary rise in asymmetric branch-like capillary. *Fractals* **2016**, *24*, 15–22. [CrossRef]
14. Mason, G.; Fernø, M.A.; Haugen, Å.; Morrow, N.R.; Ruth, D.W. Spontaneous Counter-Current Imbibition outwards from a HemiSpherical Depression. *J. Pet. Sci. Eng.* **2012**, *90–91*, 131–138. [CrossRef]
15. Cai, J.C.; Yu, B.M.A. Discussion of the Effect of Tortuosity on the Capillary Imbibition in Porous Media. *Transp. Porous. Med.* **2011**, *89*, 251–263. [CrossRef]
16. Zhou, D.; Jia, L.; Kamath, J. Scaling of counter-current imbibition processes in low-permeability porous media. *J. Petrol. Sci. Eng.* **2002**, *33*, 61–74. [CrossRef]
17. Zhou, D.; Jia, L.; Kamath, J. An Investigation of Counter-Current Imbibition Processes in Diatomite. In Proceedings of the SPE Western Regional Meeting, Bakersfield, CA, USA, 26–30 March 2001.
18. Li, J.S. The Effect of Surfactant System on Imbibition Behavior. Ph.D. Thesis, Institute of Porous flow and Fluid Mechanics of Chinese Academy of Sciences, Beijing, China, 2006.
19. Coates, G.R.; Xiao, L.; Prammer, M.G. *NMR Logging Principles and Applications*, 1st ed.; Halliburton Energy Services: Houston, TX, USA, 1999; ISBN 978-0967902609.
20. Al-Yaseri, A.Z.; Lebedev, M.; Vogt, S.J. Pore-scale analysis of formation damage in Bentheimer sandstone with in-situ NMR and micro-computed tomography experiments. *J. Petrol. Sci. Eng.* **2015**, *106*, 48–57. [CrossRef]
21. Morriss, C. Core analysis by low-field NMR. *Log Anal.* **1997**, *38*, 84–93.
22. Yao, Y.; Liu, D.; Che, Y. Petrophysical characterization of coals by low-field nuclear magnetic resonance (NMR). *Fuel* **2010**, *89*, 1371–1380. [CrossRef]
23. Baldwin, B.A.; Spinler, E.A. In situ saturation development during spontaneous imbibition. *J. Petrol. Sci. Eng.* **2002**, *35*, 23–32. [CrossRef]
24. Li, M.; Romerozerón, L.; Marica, F. Polymer Flooding Enhanced Oil Recovery Evaluated with Magnetic Resonance Imaging and Relaxation Time Measurements. *Energy Fuels* **2017**, *31*, 4904–4914. [CrossRef]
25. Song, Y.; Wang, S.; Yang, M. MRI measurements of CO_2-CH_4 hydrate formation and dissociation in porous media. *Fuel* **2015**, *140*, 126–135. [CrossRef]
26. Jiang, T.; George Hirasaki, A.; Miller, C.; Moran, K.; Fleury, M. Diluted Bitumen Water-in-Oil Emulsion Stability and Characterization by Nuclear Magnetic Resonance (NMR) Measurements. *Energy Fuels* **2007**, *21*, 1325–1336. [CrossRef]
27. Lai, F.P.; Li, Z.P.; Wei, Q.; Zhang, T.T.; Zhao, Q.H. Experimental Investigation of Spontaneous Imbibition in a Tight Reservoir with Nuclear Magnetic Resonance Testing. *Energy Fuels* **2016**, *30*, 8932–8940. [CrossRef]
28. Chen, T.; Yang, Z.; Ding, Y.; Luo, Y.T.; Qi, D.; Wei, L.; Zhao, X.L. Waterflooding Huff-n-puff in Tight Oil Cores Using Online Nuclear Magnetic Resonance. *Energies* **2018**, *11*, 1524. [CrossRef]
29. Oren, P.E.; Bakke, S.; Arntzen, O.J. Extending Predictive Capabilities to Network Models. *SPE J.* **1998**, *3*, 324–336. [CrossRef]

30. Ruth, D.; Mason, G.; Morrow, N.R. A Numerical Study of the Influence of Sample Shape on Spontaneous Imbibition. In Proceedings of the International Symposium of the Society of Core Analysts, Pau, France, 21–24 September 2003.
31. Li, H.; Guo, H.; Yang, Z. Tight oil occurrence space of Triassic Chang 7 Member in Northern Shaanxi Area, Ordos Basin, NW China. *Petrol. Explor. Dev.* **2015**, *42*, 434–438. [CrossRef]
32. Valvatne, P.H.; Blunt, M.J. Predictive pore-scale modeling of two-phase flow in mixed wet media. *Water Resour. Res.* **2004**, *40*, 187. [CrossRef]
33. Kleinberg, R.L.; Farooqui, S.A. T_1/T_2 ratio and frequency dependence of NMR relaxation in porous sedimentary rocks. *J. Colloid. Interface Sci.* **1993**, *158*, 195–198. [CrossRef]
34. Kleinberg, R.L.; Straley, C.; Kenyon, W.E.; Akkurt, R.; Farooqui, S.A. Nuclear magnetic resonance of rocks: T_1 vs. T_2. In Proceedings of the SPE Annual Technical Conference and Exhibition, Houston, TX, USA, 3–6 October 1993.

Article

Effect of Clay Mineral Composition on Low-Salinity Water Flooding

Shan Jiang, Pingping Liang and Yujiao Han *

Research Institute of Petroleum Exploration and Development, PetroChina, Beijing 100083, China; jssciences@163.com (S.J.); liangpp69@petrochina.com.cn (P.L.)
* Correspondence: yujiaohan@petrochina.com.cn; Tel.: +86-13167369300

Received: 15 October 2018; Accepted: 23 November 2018; Published: 28 November 2018

Abstract: Low-salinity water (LSW) flooding technology has obvious operational and economic advantages, so it is applied to practice in many oilfields. However, there are differences in the oil recovery efficiencies in different oilfields, the reasons for which need to be further studied and discussed. This paper studies the effect of different clay mineral compositions on low-salinity water flooding. For this purpose, three groups of core displacement experiments were designed with cores containing different clay mineral compositions for comparison. In the process of formation water and low-salinity water driving, the oil recovery and produced-water properties were measured. By comparing the two types of water flooding, it was found that the cores with the highest montmorillonite content had the best effect (5.7%) on low-salinity water flooding and the cores with the highest kaolinite content had the least effect (1.9%). This phenomenon is closely related to the difference in ion exchange capacity of the clay minerals. Moreover, after switching to low-salinity water flooding, the interfacial tension and wetting angle of the produced-water increased and the value of pH decreased, which are important mechanisms for enhancing oil recovery by low-salinity water flooding. This study reveals the influence of clay mineral composition on low-salinity water flooding and can provide more guidance for conventional and unconventional oilfield application of low-salinity water flooding technology.

Keywords: low-salinity water flooding; clay mineral composition; enhanced oil recovery; wetting angle; pH of formation water

---

1. Introduction

Compared with other enhanced oil recovery (EOR) technologies such as chemical flooding and thermal recovery, low-salinity water flooding is simple, economical and practical. Especially for unconventional reservoirs such as tight reservoirs, the general EOR methods such as polymer flooding cannot be applied because of the difficulty in injecting into such small pores, whose overpressure matrix permeability is less than 0.1×10^{-3} mD. On the other hand, low-salinity water flooding can enter the small pores at this scale, and the risk is low. Thus, low-salinity water flooding has great application potential in unconventional reservoirs. During the period of low oil prices, low-salinity water flooding has made great progress [1–3]. Martin [4] reported for the first time that decreasing the salinity of injected water improved the recovery of oil. However, this report did not get much attention until 1997 when Tang and Morrow [5] reported that the oil recovery was effectively improved by injecting low-salinity water and optimizing the composition of injected water. After that, numerous laboratory tests and field tests on low-salinity water flooding were carried out. Robertson, Lager and Seccombe [6–8] carried out oil field tests and achieved the desired results, confirming the feasibility of low-salinity water flooding. In 2006, Jerauld [3] found through numerical simulation that the injection of low-salinity water affected the relative permeability and formation pressure, thus changing the wettability of rock and finally improving the recovery. In 2008, Larger and others [9] proved through experiments that multicomponent ion exchange (MIE) occurs between low-salinity water injection, clay mineral surface and injected brine, thus enhancing the oil recovery. In 2010, Sorbie [10] proposed a mechanism for low-salinity water flooding, which is believed to be associated with rock porosity by low-salinity water flooding. At the same time, Rezaeidoust [11] also verified through application of low-salinity water flooding in Beihai sandstone reservoir that when the salinity of the injected water is low enough, the oil recovery can be improved. Many low-salinity water flooding experiments were carried out on limestone cores by Yousef and others [2,12,13] in 2011, and they found that continuous injection of low-salinity water can make rock more hydrophilic and improve recovery by improving pore throat connectivity. In 2015, Wu Jian [14] analyzed the oilfield experimental data and concluded that the main reason for the enhanced oil recovery effect of low-salinity water flooding is the microscopic transformation of reservoirs caused by migration of clay particles. Recently, Shehata [15] used zeta-potential measurements in low-salinity water flooding and found that chlorite and illite contributed to a smaller electrical-double-layer expansion compared to kaolinite, feldspars, montmorillonite, and muscovite. It has provoked people's attention to the effect of clay minerals in low salinity water flooding.

Clay minerals are widely distributed in China and the types and contents of clay minerals in different reservoirs vary greatly [16–18], so they have an important impact on oilfield development. According to Li and Zou [19,20], three main clay minerals (illite, kaolinite and montmorillonite) have very different crystal structures (Table 1), which may be the reason for their different effects on low-salinity water flooding.

Table 1. Crystallographic characteristics of kaolinite, illite and montmorillonite.

Property	Kaolinite	Illite	Montmorillonite
Layer	1:1	2:1	2:1
Grain size, μm	5–0.5	<0.5	2–0.1
Ion exchange capacity, meq/100 g	3–15	10–40	80–150
Surface area BET-N_2, m^2/g	15–25	50–110	30–80

Each unit of kaolinite crystal structure consists of Si-O tetrahedral sheet and A1-O/OH octahedral sheet. The head of Si-O tetrahedral sheet points to A1-O/OH octahedral sheet, and shares oxygen atoms with the A1-O/OH octahedral sheet. The layers are connected by hydrogen bonds, so that the interlayer spacing remains mainly unchanged and the expansion properties are small (Figure 1a). The interaction between atoms is strong, and desorption requires a greater chemical force.

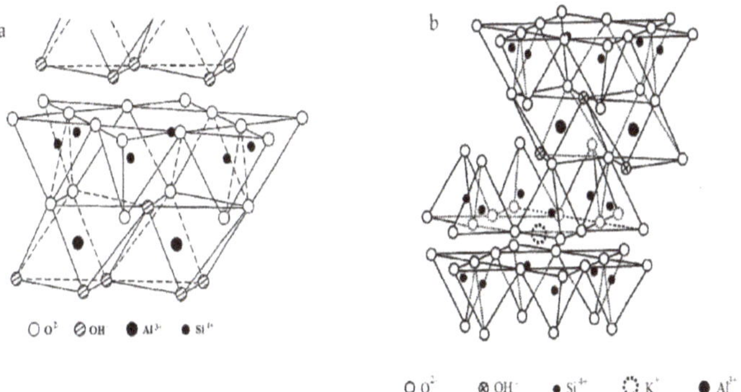

Figure 1. Structural diagram of clay minerals; (**a**) Kaolinite; (**b**) Illite.

Illite is made up of two Si-O tetrahedral sheets and one A1-O/OH octahedral sheet. The octahedral sheet is between two Si-O tetrahedral sheets. The head of Si-O tetrahedral sheets points to A1-O/OH octahedral sheet, and shares Oxygen atom with A1-O/OH octahedral sheet. Approximately 1/4 of Si^{4+} atoms in the Si-O tetrahedron are replaced by $A1^{3+}$, causing a lack of positive charge. Consequently, a layer of K^+ is formed between the two structural unit layers to balance the negative charge due to the displacement. During the evolution of mica into illite, K^+ is easily exchanged with other cations. Therefore, Ca^{2+} and Mg^{2+} are often found in the layers of illite, and the interlayer structure of illite is not stable (Figure 1b). This property is beneficial for ion exchange on its surface.

Each unit of the montmorillonite crystal structure consists of two Si-O tetrahedral sheets and one A1-O/OH octahedral sheet. The layers are bonded by the weak van der Waals forces. Montmorillonite can absorb water and other liquids, and can adsorb cations to balance interlayer charge. The interlayer water and exchangeable cations adsorbed between two structural layers can cause the lattice to expand, and the adsorbed cations allow montmorillonite to have significant ion exchange capacity. Therefore, montmorillonite generally has less chemical force between layers and contains more Na^+ or Ca^{2+} (Figure 2).

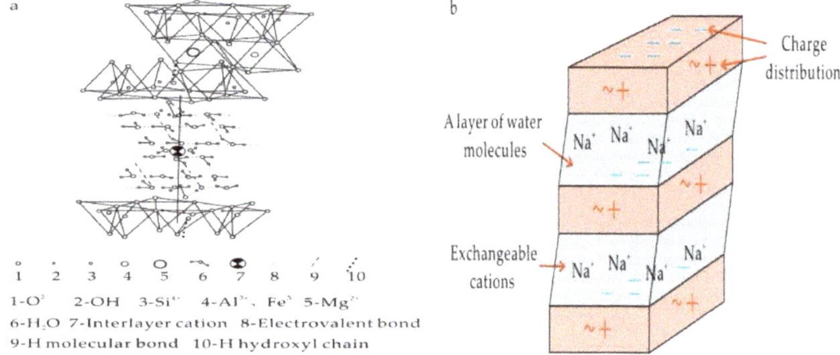

Figure 2. Structural diagram of montmorillonite. (**a**) Crystal structure of montmorillonite; (**b**) Crystal structure of montmorillonite after water swelling.

Great progress has been made in the study of the mechanism of low-salinity water flooding, but there is still a lot of controversy. Furthermore, the above studies did not quantitatively analyze the clay mineral composition of the cores used. The effect of low-salinity water flooding is probably related to clay mineral composition, but the influence of clay mineral composition on the EOR mechanism lacks the necessary basic experimental support. Therefore, three groups of experiments were designed in this study, and the influence of clay mineral composition on low-salinity water flooding was quantitatively studied based on the measurements of pH value, interfacial tension and wetting angle. The research results are of great significance for improving the recovery of clay-bearing mineral reservoirs.

2. Experiment

Three groups of high-temperature core displacement experiments were designed according to the experiment purposes. Each group of comparison experiments included the whole process of formation water flooding and low-salinity water flooding after the injection of 0.5 PV (1 PV = 1 pore volume) of formation water. The effects of clay mineral composition on low-salinity water flooding and the mechanism of low-salinity water flooding were comprehensively studied by statistical results of produced liquid and property tests of produced-water.

2.1. Core Preparation

The artificial cores were prepared in the laboratory as follows: First, the raw materials were prepared and mixed in a certain proportion. Then, they were placed in the mold for pressing and bonding. Finally, the cores were cut and stored for two weeks. Here, the clay mineral content of each core was designed as shown in Table 2.

Table 2. Clay mineral content of each group of cores.

Group	Clay Content, %	Montmorillonite Content, %	Kaolinite Content, %	Illite Content, %	Core Quantity
I	7	60	20	20	2
II	7	20	60	20	2
III	7	20	20	60	2

According to the designed content, montmorillonite, kaolinite and illite (Figure 3) were added to the raw materials to make the six artificial cores (Figure 4). Their physical parameters are listed in Table 3.

Figure 3. Clay minerals in the raw materials.

Figure 4. The 6 artificial cores for experiments.

Table 3. Physical parameters of the artificial cores.

Core Number	Length, m	Diameter, m	Porosity, %	Permeability, D
1	0.096	0.025	18.2	0.4435
2	0.096	0.025	17.8	0.4307
3	0.094	0.025	16.8	0.4266
4	0.097	0.025	18.4	0.4104
5	0.097	0.025	19.0	0.4637
6	0.096	0.025	18.5	0.4523

2.2. Fluid Preparation

The samples of the formation water and low-salinity water were provided by A block in Shengli Oilfield, and their ionic composition were tested (Table 4). According to the ionic composition, formation water was prepared by dissolving sodium chloride, magnesium chloride, and calcium chloride reagents in deionized water, and low salinity water was prepared by dissolving sodium chloride, magnesium chloride, calcium chloride and sodium bicarbonate reagents in deionized water.

Table 4. Ion content of formation water and low-salinity water.

Ion Composition	Na^+	Ca^{2+}	Mg^{2+}	Cl^-	HCO^{3-}	Total Salinity	pH
Ion content of formation water, g/m^3	8200	530	140	14000	550	23420	6.9
Ion content of low-salinity water, g/m^3	575	80	24	1100	-	1994	-

Field crude oil of the A block in Shengli Oilfield was selected for the experiments. The density of the crude oil is 0.857 kg/m^3 at 25 °C, the viscosity is 100 mPa·s, and the pH is 6.79. The experimental crude oil was subjected to a four-component separation test, and the crude oil components are given in Table 5.

Table 5. Contents of four components of crude oil.

Four-Component	Saturated Component	Aromatics	Colloid	Asphaltene
Content, %	43.45	12.97	3.80	0.54

2.3. Experimental Procedure

The instrument used in the experiment is a thermostatic displacement device (Figure 5).

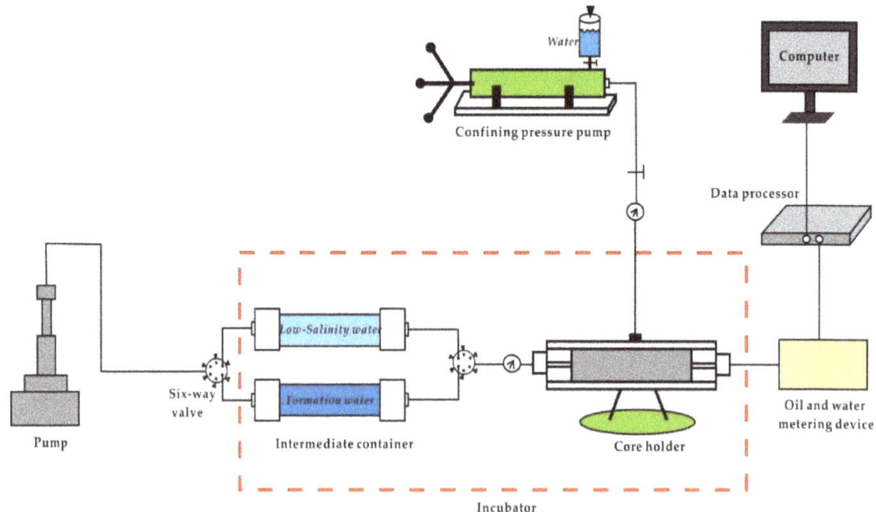

Figure 5. Schematic diagram of the experimental instrument.

The specific experimental procedures are as follows, and each step is in strict accordance with the experimental standards [21]. (1) The cores were first weighed, then they were evacuated for 7 h, and then saturated with formation water for 24 h at pressure of 20 MPa. (2) The samples were weighed and nuclear magnetic resonance (NMR) spectroscopy was performed; (3) Oil flooding was performed for 15 PV to saturate cores with crude oil, then the cores were weighed and placed in the oil for

240 h. (4) The pipeline was connected and the displacing equipment was started. The equipment was preheated for 2 h before every experiment to reach 90 °C. (5) For core No.1, core No. 3 and core No. 5, low-salinity water was injected after injecting 0.5 PV formation water. (6) For core No. 2, core No. 4 and core No. 6, continuous injection of formation water was carried out until the water cut reached 98%. In steps (5) and (6), during the experiment, the water output and oil output were measured and the pH, interfacial tension and wettability of produced-water were tested. The specific parameters of the comparison experiment are designed as Table 6.

Table 6. Design of the comparison experiment.

Group	Number	Main Component of Clay Mineral	Experimental Procedures and Parameters
I	1	Kaolinite	Injecting formation water at the beginning and injecting low-salinity water after 0.5 PV with flow rate of 0.05 cm^3/min
	2	Kaolinite	Injecting formation water with the flow rate of 0.05 cm^3/min
II	3	Illite	Injecting formation water at the beginning and injecting low-salinity water after 0.5 PV with the flow rate of 0.05 cm^3/min
	4	Illite	Injecting formation water with the flow rate of 0.05 cm^3/min
III	5	Montmorillonite	Injecting formation water at the beginning and injecting low-salinity water after 0.5 PV with the flow rate of 0.05 cm^3/min
	6	Montmorillonite	Injecting formation water with the flow rate of 0.05 cm^3/min

3. Results

3.1. Porous Structure of the Prepared Cores

The T_2 NMR spectrum showed good correspondence with the pore structure of the cores [22–25]. The T_2 spectrum curves of the six cores in saturated water state are shown in Figure 6. It can be observed that the pore structures of the six cores are relatively similar.

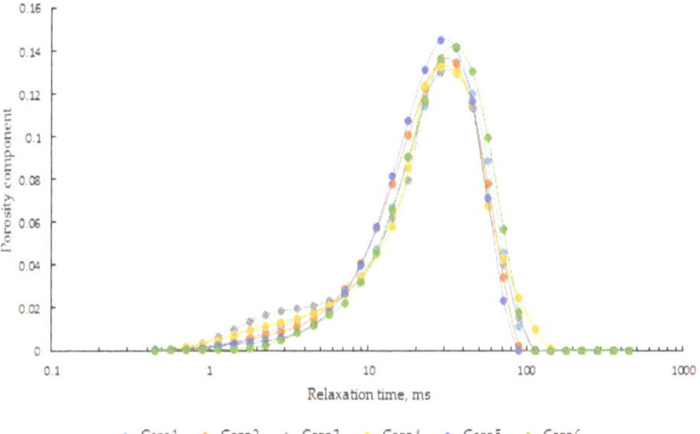

Figure 6. NMR curves of cores saturated with water.

3.2. Relationship between Clay Mineral Composition and Oil Recovery

Measurements were taken at ten points during the experiments, and the oil recovery results were calculated and plotted, as shown in Figure 7.

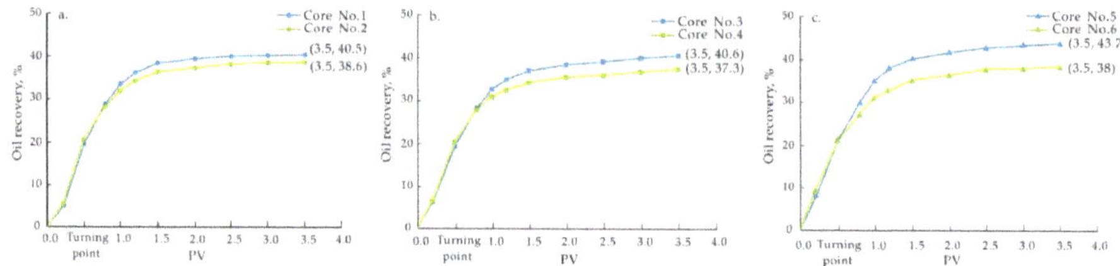

Figure 7. Comparison of oil recovery of the cores. (**a**) Cores No. 1 and No. 2; (**b**) Cores No. 3 and No. 4; (**c**) Cores No. 5 and No. 6.

As seen from the above graphs, injecting low-salinity water in cores containing clay minerals improved the oil recovery, but the degree of EOR was different. Cores with the highest content of montmorillonite had the most obvious effect of EOR, from 38% (Core No. 6) of the whole process of formation water flooding to 43.7% (Core No. 5), and the degree of recovery increased by 5.7%. On the other hand, cores with the highest content of kaolinite showed the poorest effect on low-salinity water flooding, from 38.6% (Core No. 2) of the whole process of formation water flooding to 40.5% (Core No. 1), and the degree of recovery increased by only 1.9%. Cores with the highest content of illite showed an intermediate effect of EOR (3.3%). The relative improvement of oil recovery for each group of cores was calculated, as shown in Table 7. It is obvious that cores with the highest content of montmorillonite are the most suitable for low-salinity water flooding.

Table 7. Difference in EOR.

Core	1–2	3–4	5–6
Difference in EOR, %	4.92	8.85	15.00

3.3. Change in Properties during Displacement Process

During the displacement process, five samples were taken from each core and each sample was subjected to measurements of pH value, interfacial tension and wetting angle. A total of 90 attribute tests were performed and the results are presented in Figure 8.

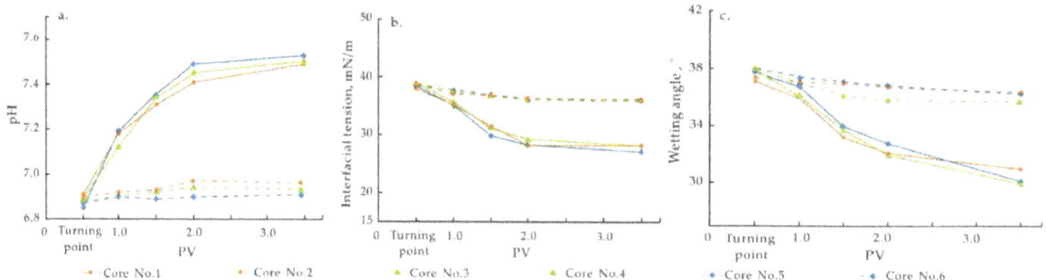

Figure 8. Results in attribute tests in Cores 1–6. (**a**) The change of pH; (**b**) The change of interfacial tension; (**c**) The change of wetting angle.

During the entire process of formation water flooding test, the changes in the three properties were not significantly affected by clay mineral composition, and all the cores showed a consistent pattern.

In the test involving injecting low-salinity water after injecting formation water, the values of different properties changed significantly, mainly reflected in the increase in pH value (0.5–0.6), and the decrease in interfacial tension (8–9 mN/m) and wetting angle (5–6.5°) of the produced-water. These results indicate the alkaline enhancement of displacing liquid system and the decrease in resistance of oil displacement. This effect is favorable for improving oil displacement efficiency, ultimately resulting in the improvement of oil recovery.

Groups with different clay mineral compositions showed various degrees of difference in the three properties between two diverse displacement tests. Taking these results in combination with the oil recovery data, a clear trend can be observed: cores with better effect of EOR on low-salinity water flooding will have higher pH value and lower interfacial tension and wetting angle.

4. Discussion

Based on the analysis of the above experimental data, combined with literature research, the mechanism of influence of clay mineral composition on low-salinity water flooding is discussed in this section. According to the NMR results in Section 3.1, the pore structure of the six cores is similar. Therefore, the effect of pore structure on the low-salinity water flooding in the experiment is excluded.

4.1. Ionic Mechanism of Low-Salinity Water Flooding

After injecting formation water, the organic acids and salt base in the crude oil and some cations in the formation water (such as Ca^{2+}) are adsorbed on the surface of the clay mineral due to the negative charge on the surface. At this point, the ion concentration, temperature, pressure, and pH are in a state of chemical equilibrium (Figure 9a).

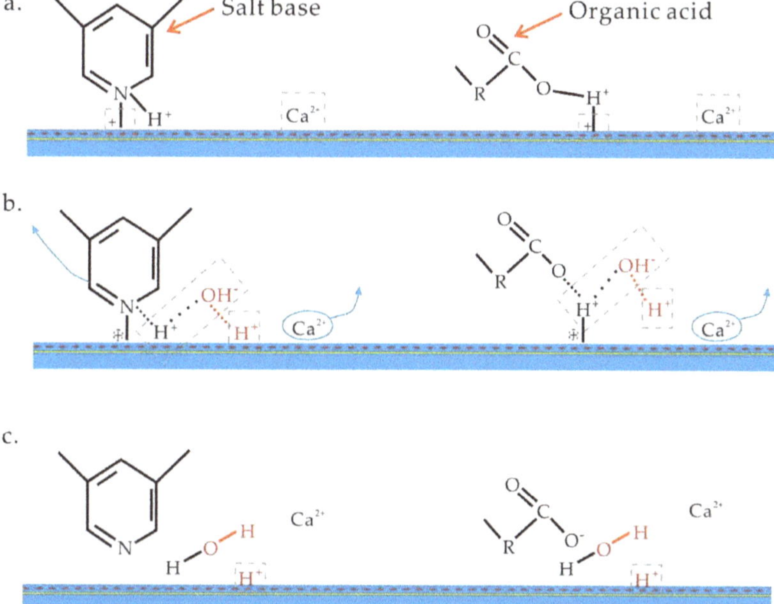

Figure 9. Ion exchange process during low-salinity flooding. (**a**) Initial state of clay surface; (**b**) Reaction occurring during low-salinity flooding; (**c**) Final state of clay surface.

After injecting low-salinity water, the original salt-rock interface equilibrium is destroyed. Then, the ion desorption reaction occurs on the rock surface, especially for Ca^{2+}, because the ion concentration of low-salinity water is much lower than that of the original formation water. Thus, in order to compensate for the loss of Ca^{2+} on the rock surface and maintain charge balance, H^+ in the water is adsorbed on the rock surface. In other words, the interaction between Ca^{2+} and H^+ occurs on the surface of clay minerals (Figure 9b).

As the amount of H^+ in the liquid system decreases and pH value of the system rises, the matrix and organic acids on the clay surface react with the OH^-, thereby forming a new interface of acid and salt and desorbing from the surface of the clay mineral (Figure 9c).

During this process, the amount of matrix and organic acids adhering to the clay surface is reduced by the reaction. On the one hand, this makes the rock surface more hydrophilic, and on the other hand, this reduces the resistance of oil displacement. Generally, this chemical reaction is similar to that of alkali flooding.

4.2. Analysis of Influence of Clay Mineral Composition on Low-Salinity Water Flooding

From the above analysis, it is evident that the surface ionic chemical reaction is closely related to the enhanced oil recovery mechanism of low-salinity water flooding. According to the statistics of the physical properties of the three clay minerals, it can be inferred that the fundamental reason for the different degrees of enhanced oil recovery is the differences in the degree of cation conversion occurring on the mineral surface for the three kinds of clay minerals.

Therefore, the different crystal structures of the three clay minerals determine the ion exchange capacity of their surface, which further influences the EOR effect of low-salinity water flooding. For kaolinite, the ion exchange capacity on its surface is weak, which is not conducive to the mechanism of low-salinity water flooding. As there are Ca^{2+} and Mg^{2+} ions inside the illite layer, and the interlayer structure of illite is unstable, its surface is favorable for ion exchange. At the same time, it is found that Ca^{2+} plays an extremely important role in EOR effect of low-salinity water flooding. Montmorillonite usually has less chemical force between layers and contains more Na^{+} or Ca^{2+}, which not only increases the surface area available for the ionic reaction, but also transfers and catalyzes the ionic reaction. Therefore, montmorillonite is more suitable for the application of water flooding to enhance the oil recovery.

5. Conclusions

Core flooding tests were designed to investigate the effects of different clay mineral compositions on the EOR effect of low-salinity water flooding. At the same time, the properties of the effluent were determined in different flooding stages, and the mechanism of EOR effect of low-salinity water flooding was analyzed. The following conclusions were drawn from the results of this work:

Compared to the entire process of water flooding test, the pH value increased by 0.5–0.6, the interfacial tension decreased by 8–9 mN/m, and the wetting angle decreased by 5–6.5° during the low-salinity water flooding. That is, after the low-salinity water was injected into the core, a similar effect to the alkaline flooding occurred, which increased the overall cleaning efficiency and ultimately improved oil recovery.

The composition of clay minerals had a significant influence on the effect of low-salinity water flooding. In particular, cores with the highest content of montmorillonite showed the most obvious effect on low-salinity water flooding (EOR of 5.7%), while cores with the highest content of kaolinite showed the poorest effect on low-salinity water flooding (EOR of 1.9%). The cores with the highest content of illite showed an intermediate effect (EOR of 3.3%). These results can be explained by the differences in the crystal structure of the clay mineral. The interlayer of montmorillonite is connected by van der Waals forces, which makes its surface have the highest ion exchange capacity. On the other hand, kaolinite has the largest crystal chemical force and the closest ion connection. Thus, the ion exchange capacity on its surface is weak. The performance characteristics are consistent with the experimental data, indicating that ion exchange is one of the essential mechanisms of EOR effect of low-salinity water flooding. Therefore, in formations with similar conditions but different clay mineral composition, the content of montmorillonite is the most important factor affecting the performance of low-salinity water flooding. If the content of montmorillonite is relatively high in the formation, low-salinity water flooding can achieve better results.

Author Contributions: Conceptualization, S.J. and P.L.; Methodology, S.J., Y.H.; Investigation, P.L.; Resources, P.L., Y.H.; Data curation, S.J.; Writing—Original Draft Preparation, S.J.; Writing—Review and Editing, Y.H.

Funding: This research was funded by Major National Science & Technology Program of China, grant number 2017ZX05035001.

Acknowledgments: This research received much help from the staff of Research Institute of Petroleum Exploration and Development, PetroChina. In the process of writing articles, our teachers (Dazhong Dong, Yuman Wang) gave us a lot of advice, which we really appreciate.

Conflicts of Interest: The authors declare no conflict of interest.

References

1. Wang, P.; Jiang, R.; Wang, G.C.; Liang, Y. Research Advance and Prospect of Low-Salinity Water Flooding. *Lithol. Reserv.* **2012**, *24*, 106–110.
2. McGuire, P.L.; Chatham, J.R.; Paskvan, F. Low-Salinity Oil Recovery: An Exciting New EOR Opportunity for Alaska's North Slope. In Proceedings of the SPE-93903-MS, SPE Western Regional Meeting, Irvine, CA, USA, 30 March–1 April 2005. [CrossRef]
3. Jerauld, G.R.; Lin, C.Y. Modeling Low-Salinity Waterflooding. In Proceedings of the SPE 102239, Annual Technical Conference and Exhibition, San Antonio, TX, USA, 24–27 September 2006. [CrossRef]
4. Martin, J.C. The Effects of Clay on the Displacement of Heavy Oil by Water. In Proceedings of the SPE1411, the 3rd Annual Venezuelan Region Meeting of AIME, Caracas, Venezuela, 14–16 October 1959.
5. Emadi, A.; Sohrabi, M. Visual investigation of oil recovery by Low-Salinity water injection: Formation of water micro-dispersions and wettability alteration. In Proceedings of the SPE 166435, SPE Annual Technology Conference and Exhibition, New Orleans, LA, USA, 30 September–2 October 2013. [CrossRef]
6. Robertson, E.P. Low-Salinity Waterflooding to Improve Oil Recovery-historical Field Evidence. In Proceedings of the SPE 109965, SPE Annual Technology Conference and Exhibition, Anaheim, CA, USA, 11–14 November 2007. [CrossRef]
7. Lager, A.; Webb, K.J.; Collins, I.R.; Richmond, D.M. LoSal Enhanced Oil Recovery: Evidence of Enhanced Oil Recovery at the Reservoir Scale. In Proceedings of the SPE 113976, SPE Symposium on Improved Oil Recovery, Tulsa, OK, USA, 20–23 April 2008. [CrossRef]
8. Seccombe, J.; Lager, A.; Jerauld, G.; Jhaveri, B.; Buikema, T.; Bassler, S.; Denis, J.; Webb, K.; Cockin, A.; Fueg, E. Demonstration of Low-salinity EOR at Interwell Scale, Endicott Field, Alaska. In Proceedings of the SPE 129692, SPE Improved Oil Recovery Symposium, Tulsa, OK, USA, 24–28 April 2010. [CrossRef]
9. Qi, X.; Kuang, Y. Influence of Clay Minerals on Surface Properties of Coal Slime. *Coal Sci. Technol.* **2013**, *7*, 126–128. [CrossRef]
10. Sorbie, K.S. A Proposed Pore-Scale Mechanism for How Low-salinity Waterflooding Works. In Proceedings of the SPE 129833, Improved Oil Recovery Symposium meeting, Tulsa, OK, USA, 24–28 April 2010. [CrossRef]
11. Rezaeidoust, A.; Puntervold, T.; Austad, T. A Discussion of the Low-salinity EOR Potential for a North Sea Sandstone Field. In Proceedings of the SPE 134459, SPE Annual Technical Conference and Exhibition, Florence, Italy, 19–22 September 2010. [CrossRef]
12. Zhang, Y.S.; Morrow, N.R. Comparison of Secondary and Tertiary Recovery with Change in Injection Brine Composition for Crude Oil/Sandstone Combinations. In Proceedings of the SPE/DOE Symposium on Improved Oil Recovery, Tulsa, OK, USA, 22–26 April 2006. [CrossRef]
13. Yousef, A.; Al-Saleh, S.; Al-Jawfi, M. New Recovery Method for Carbonate Reservoirs through Tuning the Injection Water Salinity: Smart Water Fooding. In Proceedings of the SPE 143550, SPE/EAGE Annual Conference and Exhibition, Vienna, Austria, 23–26 May 2011. [CrossRef]
14. Wu, J.; Chang, Y.W.; Li, J.; Liang, T.; Guo, X.F. Mechanisms of Low-salinity Waterflooding Enhanced Oil Recovery and Its Application. *J. Southwest Pet. Univ.* **2015**, *37*, 145–151. [CrossRef]
15. Shehata, A.M.; Nasr-El-Din, H.A. The Role of Sandstone Mineralogy and Rock Quality in the Performance of Low-Salinity Waterflooding. *SPE Reserv. Eval. Eng.* 2017. [CrossRef]
16. Zhang, X.F.; Liu, Z.B.; Liu, C.; Liu, H.; Zhang, R. Influences of the Clay Minerals on the Waterflooding Reservoir Physical Properties of Oilfield X in Liaodong Bay Depression. *Pet. Geol. Oilfield Dev. Daqing* **2017**, *36*, 62–67. [CrossRef]
17. Huang, J.; Zhang, C.L.; Xie, G.W.; Liu, X.G.; Ding, Y.Q.; Zhang, R.Y.; Lei, Q. Effects of authigenic clay minerals on reservoir properties and oilfield development—Take X reservoir formation in Jiyuan and Huaqing areas in Ordos basin as an example. *Petrochem. Ind. Appl.* **2017**, *36*, 90–96. [CrossRef]
18. Jiang, Y.Q.; Zhang, C.; Deng, H.B.; Wang, M.; Luo, M.S. Influences of Clay Minerals on Physical Properties of Low Permeability and Tight Sandstones. *J. Southwest Pet. Univ.* **2013**, *35*, 39–47. [CrossRef]

19. Li, L. Feasibility study on enhanced oil recovery by Low-Salinity Water Flooding in XM oilfield. Ph.D. Thesis, Southwest Petroleum University, Chengdu, China, 2015.
20. Zou, D.P.; Ke, S.Z.; Li, J.J.; He, Q.L.; Ma, X.R. Experimental Study on Resistivity Dispersion of Highly Clay Mineral Content Core. *Well Logging Technol.* **2018**, *42*, 261–266. [CrossRef]
21. Technical Regulation for Oilfield EOR Methods Screening. Available online: http://http://old.petrostd.com/read/view.aspx?id=3087 (accessed on 1 June 2016).
22. He, Y.D.; Mao, Z.Q.; Xiao, L.Z.; Ren, X.J. An improved Method of Using NMR T2 Distribution to Evaluate Pore Size Distribution. *Chin. J. Geophys.* **2005**, *48*, 373–378. [CrossRef]
23. Fleury, M.; Fabre, R.; Webber, J.B.W. Comparison of pore size distribution by NMR relaxation and NMR cryoporometry in shales. *SCA* **2015**, *25*, 25–36.
24. Yun, H.Y.; Zhao, W.J.; Liu, B.K.; Zhou, C.C.; Zhou, F.M. Researching rock pore structure with T2 distribution. *Well Logging Technol.* **2002**, *26*, 18–21. [CrossRef]
25. Zhang, C.M.; Chen, Z.B.; Zhang, Z.S. Fractal characteristics of reservoir rock pore structure based on NMR T2 distribution. *J. JPI* **2007**, *29*, 80–86.

Article

Hydrogeochemical and Isotopic Constraints on the Pattern of a Deep Circulation Groundwater Flow System

Xiting Long [1,2,3], Keneng Zhang [1,2], Ruiqiang Yuan [4,*], Liang Zhang [5] and Zhenling Liu [6]

1. Key Laboratory of Metallogenic Prediction of Nonferrous Metals and Geological Environment Monitoring Ministry of Education, School of Geoscience and Infophysics, Central South University, Changsha 410083, China; longxiting@csu.edu.cn (X.L.); ken@csu.edu.cn (K.Z.)
2. Hunan Key Laboratory of Nonferrous Resources and Geological Hazards Exploration, Changsha 410083, China
3. The 402 Team, The Bureau of Geology and Mineral Resources Exploration of Hunan, Changsha 410014, China
4. School of Environment and Resource, Shanxi University, Taiyuan 030006, China
5. Department of Biological Sciences, University of Windsor, 401 Sunset Avenue, Windsor, ON N9B 3P4, Canada; zhang.on.ca@gmail.com
6. Well Testing Sub-Company, BHDC, Langfang 065007, China; liuzhenling@cnpc.com.cn

* Correspondence: rqyuan@sxu.edu.cn; Tel./Fax: +86-0351-7010600

Received: 8 December 2018; Accepted: 21 January 2019; Published: 28 January 2019

Abstract: Characterization of a deep circulation groundwater flow system is a big challenge, because the flow field and aqueous chemistry of deep circulation groundwater is significantly influenced by the geothermal reservoir. In this field study, we employed a geochemical approach to recognize a deep circulation groundwater pattern by combined the geochemistry analysis with isotopic measurements. The water samples were collected from the outlet of the Reshui River Basin which has a hot spring with a temperature of 88 °C. Experimental results reveal a fault-controlled deep circulation geothermal groundwater flow system. The weathering crust of the granitic mountains on the south of the basin collects precipitation infiltration, which is the recharge area of the deep circulation groundwater system. Water infiltrates from the land surface to a depth of about 3.8–4.3 km where the groundwater is heated up to around 170 °C in the geothermal reservoir. A regional active normal fault acts as a pathway of groundwater. The geothermal groundwater is then obstructed by a thrust fault and recharged by the hot spring, which is forced by the water pressure of convection derived from the 800 m altitude difference between the recharge and the discharge areas. Some part of groundwater flow within a geothermal reservoir is mixed with cold shallow groundwater. The isotopic fraction is positively correlated with the seasonal water table depth of shallow groundwater. Basic mineral dissolutions at thermoneutral conditions, hydrolysis with the aid of carbonic acid produced by the reaction of carbon dioxide with the water, and hydrothermal alteration in the geothermal reservoir add some extra chemical components into the geothermal water. The alkaline deep circulation groundwater is chemically featured by high contents of sodium, sulfate, chloride, fluorine, silicate, and some trace elements, such as lithium, strontium, cesium, and rubidium. Our results suggest that groundwater deep circulation convection exists in mountain regions where water-conducting fault and water-blocking fault combined properly. A significant elevation difference of topography is the other key.

Keywords: deep circulation groundwater; groundwater flow; geothermal water; faults; isotopes

1. Introduction

There is a continuous heat-flow from the Earth's interior to the surface. Away from tectonic plate boundaries, geothermal gradient is about 25–30 °C/km of depth near the surface in most of the world [1]. For deep circulation groundwater, downward flow lowers crustal temperatures, while upward flow tends to raise temperatures. Deep circulation groundwater as part of the hydrologic cycle influences the distribution of heat and, thereby, the temperature field in the Earth's crust [2]. Deep circulation groundwater flow systems need more attention.

Deep circulation groundwater flow systems are usually connected to geothermal system, especially in neotectonic and volcanic areas. Geological structures relating to geothermal activities, such as faults, usually complicate the flow and chemistry of deep circulation groundwater. Generally, a geothermal system is mainly fed by meteoric water infiltrating at different altitudes including rain water and snowmelt [3–5]. Mountain regions are usually the recharge area [6,7]. Flow and geochemistry of geothermal water are usually structurally controlled [4,8]. Fault zone permeability influences the spatial distribution and behavior of hydrothermal and geothermal systems at all scales. Areas of spatial interaction between two groups of faults are structurally ideal places for concentrated hydrothermal activity [9].

Geothermal waters from metamorphic, granite, and sedimentary regions exhibit varying hydrogeochemical features [10]. Variant types of geothermal waters can be formed, such as HCO_3-Ca, HCO_3-Na, SO_4-Na (Ca), and Cl-SO_4-Na type [10–12]. Enhanced water–rock interaction increases concentrations of major and trace elements in geothermal waters. It was reported that the sodium and chloride concentrations of geothermal fluid reach up to 16,963 mg/L and 68,256 mg/L in a volcanic geothermal system, respectively [13]. It is possible that the deeply sourced geothermal fluids cause degradation of water quality of the shallow groundwater and surface water [14]. In addition, the geothermal water might show an altered water isotopic composition by a stronger oxygen shift in the deeper reservoir [7]. Although geothermal water is chemically different with cold water of the same area in most cases, the similarity of geothermal and cold water in chemical and isotopic compositions also exists [3].

Mixing is a common process during the upward flow of geothermal water. Geothermal waters might be a mixture of magmatic water, stream, deep geothermal fluids, shallow geothermal fluids, and cold water from the surface [13,15–19]. Mixing is also a process that controls outlet temperatures and causes dilution of geothermal water [5]. At the same time, mixing blurs the information of the deep circulation groundwater.

Deep circulation groundwater flow systems are structurally and geochemically complex. Isotopic investigations combined with geothermal applications represent powerful tools for the exploration of deep circulation groundwater flow systems [6,7,12]. In this study, a deep circulation hydrothermal system was surveyed based on hydrogeochemical and isotopic constraints to elucidate the origin of the geothermal fluids and the source of solutes and to discern the mixing and the hydrogeochemical alteration. The main goal is to gain a conceptual model and mechanisms for the deep circulation groundwater flow system.

2. Study Area

2.1. Geographical Settings

The study area, the Reshui River Basin, located around E 113°54′ and N 25°32′ with subtropical monsoon humid climate in southeast China. The multi-year average precipitation is 1670 mm. About 50% of annual precipitation happens during April to August. The coldest month is January with an average temperature of 6.5 °C, while the warmest month is July with a mean temperature of 27.8 °C. The annual mean air temperature is about 16.7 °C. The landscape is mainly hills and mountains with small basins distributed. Elevation changes from 200 m to 1700 m. Forests cover more than 70% of land surface. Many streams drain the area. The Shangyoujiang River has an annual mean flow of 16 m^3/s (Figure 1).

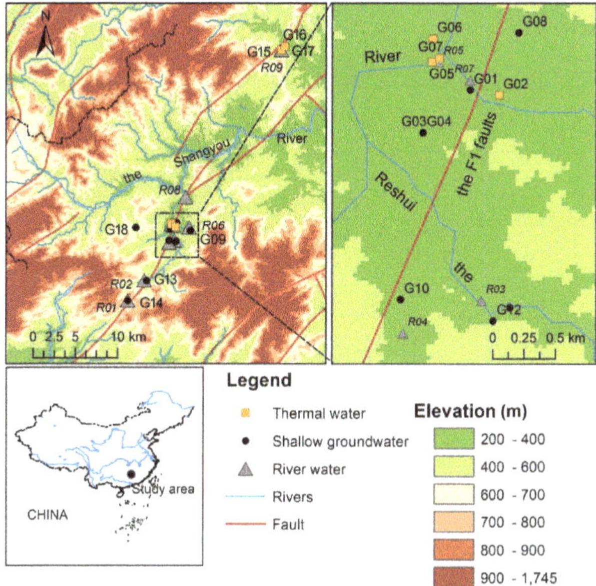

Figure 1. Elevation, rivers, main faults, and sampling sites in the study area.

2.2. Geological and Hydrogeological Settings

The study area is within the plate collision zone between the Cathaysian block and the Yangtze block. The exposing strata mainly include Sinian strata and Precambrian strata. Intrusive rocks are mainly monzonite granite formed during Triassic to Jurassic distributing on the east and south. There are four fault systems that dominated the geologic structure (Figure 2).

The F1 fault is an active about 200 km long fault system. Master faults strike NE and dip NW with an angle ranging 78–87°. The fault-throw is about several decameters. Many warm springs discharge waters controlled by the fracture system, which infers that the major fault of the system could penetrate deeply the crust. The F2 fault parallels to F1 located around 15 km on the southeast of the F1. The F2 is a NW dipping fracture system with an angle about 50°. The fault-throw changes from several meters to several decameters. The F3 is a SE dipping active normal fault with an angle 70–85°. The fracture zone is 7–12 m wide. The F4 is a NW trending thrust fault system with a maximum fault-throw about 20 m. The dipping angle of the F4 fault is around 66° towards SW.

The unconsolidated Quaternary alluvial sediments are distributed in the river valley. The sediment can be divided into two layers. The upper layer is sandy clay with a thick of 1–5 m. The second layer is gravels and sandy clay, which is 1–3 m thick. Mountains and hills are mainly outcrops of sandstone, slate, and granite (Figure 2). The thickness of weathering crusts ranges from 9 m to 24 m.

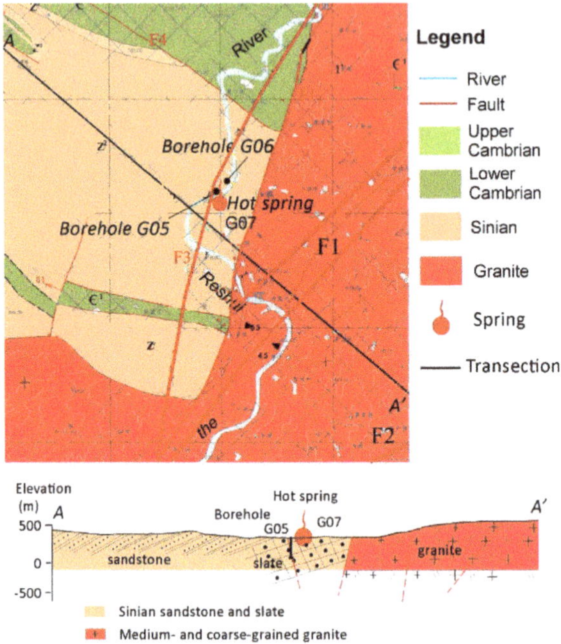

Figure 2. Geological sketch of the study area (revised from unpublished data).

There are three kinds of groundwater including: (1) pore water in unconsolidated sediments; (2) fissure water in weathering fracture and structural fracture of bedrocks outcrops; and (3) deep circulation groundwater. Groundwater occurred in Quaternary sediment exchanges with rivers seasonally and distributes in basins. The fissure water is recharged by precipitation flowing within fracture networks until discharged as spring or into Quaternary sediment on the base of slopes. The unconsolidated sediment and fracture networks of weathered crust are a connected system in which shallow groundwater flows. A part of the fissure water flows downwards into the deep through fault systems alimenting the deep circulation groundwater.

2.3. Thermal-Geological Features

The Reshui hot spring geothermal field has an area about 3 km^2. The spring locates on the riverside of the Reshui River that is a tributary of the Shangyoujiang River. A vertical stratum of slate forms the river bed where the hot spring discharges. The temperature of the spring water is usually around 90 °C with a maximum of 98 °C. Several boreholes were drilled around the hot spring. The production of hot water is up to 2500 m^3/d in total.

Boreholes with hot water (outlet temperature 84.6–92.2 °C) and the hot spring are distributed along the line trending 110–115°. Crustal derived granite locates on the east and the south of the hot spring. The geothermal field might be heated by magma activities and radioactive decay.

3. Method

Two field surveys were conducted on September 2016 and February 2017, respectively. Water samples of shallow groundwater, river water, and geothermal water were collected for major ions, trace elements, δD, and δ^{18}O (Figure 1). Samples of G07, G15, G16, and G17 were collected from geothermal springs. Samples of G02, G05, and G06 were collected from boreholes reaching the deep fissure network. The depths of the boreholes are 500, 200, and 150 m, respectively. The other groundwater samples were collected from open wells. Some of the wells are artesian on rainy seasons (summer). Before sampling, we measured water table depths and physicochemical parameters in situ (Horiba U-51 calibrated in advance), such as water temperature, pH, electrical conductivity (EC), oxidation-reduction potential (ORP). Water samples were collected in a 100 mL syringe and filtered immediately through 0.45 μm cellulose-ester membranes into three 60 mL and one 100 mL high density polyethylene (HDPE) bottles, which were filled to overflowing and capped. The samples in the 100 mL bottles were used for titration of bicarbonate on the day of sampling. The samples for cation analysis were acidified immediately (pH = 2).

Water samples were analyzed in the laboratory of the Institute of Geographic Sciences and Natural Resources Research, Chinese Academy of Sciences. The chemical compositions were characterized by ICP-OES (PerkinElmer, Optime 5300 DV, Waltham, MA, USA) for cations and Ion Chromatography (Shimadzu LC-10ADvp) for anions. The trace elements (Al, Ag, As, Ba, Be, Bi, Cd, Co, Cu, Cr, Cs, Fe, Ga, In, Li, Mn, Mo, Ni, Rb, Pb, Se, Sr, Sb, Tl, Ti, U, V, Zn) were analyzed by an inductively coupled plasma mass spectrometry (PerkinElmer, ICP-MS Elan DRC-e). Hydrogen and oxygen stable isotopic compositions of the water samples were analyzed by the isotope ratio mass spectrometer (Finnigan MAT-253, Silicon Valley, CA, USA) using the TC/EA (high-temperature conversion/elemental analyzer) method. The δ^{18}O and δD values were reported as per mill (‰) deviations from the international standard V-SMOW (Vienna Standard Mean Ocean Water). The δ^{18}O and δD measurements were reproducible to ±0.2‰ and ±1‰, respectively.

4. Results

4.1. Deep Circulation Geothermal Groundwater (TW) Features

Upward flow of the deep circulation groundwater usually carries heat of crust from the depth showing a high temperature. The water temperature is around 88 °C in the thermal spring (G07), which is located close to the F3 fault and the F1 fault. Long-term stress conditions would favor continuous fluid flow through vertical high-flux conduits in the faults [9]. While the borehole (G05) drilled into the F3 fault fractured zone with a depth about 200 m discovered the geothermal water directly with a water temperature of 93 °C. Given heat loss of water during ascending, the geothermal water temperature could be higher than 93 °C at the depth. The difference between G05 and G07 was induced by longer residence time and more heat loss of the geothermal water in fractures compared to the geothermal water in the borehole. The other geothermal water from springs or boreholes had water temperatures lower than 60 °C. It is suggested that the mixing of deep circulation geothermal groundwater with shallow cold groundwater in the subsurface. Based on the G05 and G07, the deep circulation geothermal water shows a pH value around 8.7 and an EC value about 340 μS/cm. Both were significantly higher than other water samples (Table 1 and Figure 3). The concentration of CO_2 degassing usually results in the geothermal water pH increase [19]. *Eh* (estimated by measured ORP values) of the geothermal water changes from 173 mV to 271 mV between September and February, which significantly lower than other water samples.

The hydrogeochemical type (Figure 4) of the deep circulation geothermal groundwater is $HCO_3{\cdot}SO_4$-Na (September rainy season) or $HCO_3{\cdot}SO_4$-Na (February dry season). The water isotopic composition is δ^{18}O −7.2‰ and δD −45‰ in average (Table 1). The concentrations of SiO_2 and F^- are about 150 mg/L and 10 mg/L, respectively. Specially, the trace elements, for example Li (370 μg/L), Rb (38 μg/L), Cs (36 μg/L), Mo (23 μg/L), As (15 μg/L), and Ga (3 μg/L) are significantly higher in the deep circulation geothermal groundwater than those in other water samples. Moreover, there is no obviously seasonal variance in hydrogeochemical features of the deep circulation geothermal groundwater.

Table 1. Results of hydrochemical measurements.

Sites	Sampling Date	Type	pH	EC (us/cm)	Eh (mV)	$\delta^{18}O$ (‰)	δD (‰)	Ca^{2+} (mg/L)	Mg^{2+} (mg/L)	Na^+ (mg/L)	K^+ (mg/L)	HCO_3^- (mg/L)	SO_4^{2-} (mg/L)	Cl^- (mg/L)	NO_3^- (mg/L)	F^- (mg/L)	SiO_2 (mg/L)	Fe (μg/L)	Al (μg/L)
G01	201702	SG	6.4	115	627	−6.1	−36	9.3	1.9	8.3	3.6	56.0	7.1	5.0	12.3	-	29.9	2.7	11.8
G02	201609	TW	6.8	144	475	−6.0	−36	17.7	2.3	8.2	2.8	99.5	2.0	1.6	1.0	0.2	46.05	2.8	60.8
G02	201702	TW	6.98	159	625	−5.6	−33	17.3	2.3	9.6	2.8	103.5	2.7	4.4	7.1	-	43.9	2.1	26.4
G03	201702	SG	6.14	46	676	−6.1	−36	4.0	1.1	3.5	2.0	27.0	1.3	2.5	7.1	-	18.3	1.6	4.4
G04	201702	SG	6.51	154	723	−6.1	−34	15.4	3.9	12.4	2.8	61.0	16.4	8.3	4.1	-	19.8	1.3	26.7
G05	201609	TW	8.84	342	176	−7.2	−46	4.9	0.1	53.2	3.2	126.0	32.9	2.4	6.1	10.1	150.2	23.0	47.6
G05	201702	TW	8.73	339	270	−7.2	−44	4.5	0.1	60.8	3.0	112.7	37.6	3.6	0.1	-	144.3	31.9	3.5
G06	201609	TW	7.87	312	298	−6.6	−41	14.5	3.8	39.5	4.2	130.1	23.5	6.4	4.1	6.8	126.5	3.6	42.2
G06	201702	TW	8.21	316	558	−6.8	−41	12.3	2.0	50.4	3.2	118.1	34.3	4.0	0.1	-	127.0	14.3	15.5
G07	201609	TW	8.44	339	170	−7.2	−46	4.8	0.1	52.0	2.9	123.5	33.6	4.6	0.9	10.2	153.5	40.8	14.4
G07	201702	TW	8.73	337	273	−7.2	−44	4.0	0.0	62.3	2.9	112.7	37.5	3.6	0.1	-	149.4	26.9	1.2
G08	201702	SG	6.22	63	664	−5.8	−34	3.5	1.4	7.0	1.7	40.5	3.1	1.8	4.6	-	33.3	3.0	-
G09	201609	SG	5.74	43	578	−5.9	−36	16.9	0.8	5.6	1.0	88.0	1.4	2.2	1.6	0.2	33.96	0.8	58.8
G09	201702	SG	6.22	43	644	−6.1	−37	3.1	0.6	6.3	0.8	35.0	0.8	0.4	0.8	-	30.4	1.9	-
G10	201609	SG	6.2	33	533	−6.1	−36	4.5	1.4	2.4	1.8	38.0	1.1	1.5	1.0	0.1	20.89	6.3	19.7
G10	201702	SG	7.15	33	631	−6.3	−37	2.7	1.4	2.6	1.3	30.0	0.8	0.2	1.3	-	21.3	17.8	11.8
G11	201609	SG	5.37	41	563	−5.3	−33	23.2	0.7	4.4	1.4	100.0	1.1	2.0	1.0	0.1	26.81	3.7	78.6
G11	201702	SG	7.82	39	615	−6.0	−35	8.0	0.5	4.7	1.2	47.0	1.4	1.3	5.2	-	26.9	5.1	5.0
G12	201702	SG	6.88	29	622	−6.2	−35	3.2	0.2	4.2	1.2	20.0	1.2	0.3	2.0	-	26.7	6.2	-
G13	201609	SG	6.68	37	503	−6.3	−38	9.9	0.5	3.3	2.0	40.0	2.0	2.0	7.6	0.1	27.87	30.1	52.4
G13	201702	SG	7.09	26	592	−5.8	−33	2.8	0.1	4.6	1.4	19.5	1.7	0.6	1.1	-	27.4	32.5	-
G14	201702	SG	6.16	22	580	−6.0	−34	1.8	0.1	2.7	1.7	14.0	0.8	0.4	0.6	-	22.5	2.7	-
G15	201702	TW	6.85	196	265	−6.7	−40	9.9	1.3	27.6	5.5	112.2	7.8	1.2	1.3	-	57.1	4.4	4.9
G16	201702	TW	7.66	259	520	−7.1	−43	7.7	0.7	50.2	2.5	137.1	10.5	2.2	1.4	-	68.1	4.8	2.2
G17	201702	TW	7.76	368	263	−7.6	−48	9.3	0.2	72.1	2.8	189.8	12.9	3.5	1.2	-	85.2	4.8	3.8
G18	201609	SG	6.8	56	503	−6.7	−39	11.9	1.3	3.2	1.2	63.0	2.1	1.9	2.1	0.3	24.78	6.3	39.5
R01	201702	RW	6.96	28	567	−6.0	−35	1.2	0.1	3.6	1.5	12.2	1.0	0.5	2.1	-	23.5	13.5	-
R02	201609	RW	7.03	26	476	−6.1	−35	2.4	0.2	2.6	1.0	16.0	1.9	2.2	2.0	0.2	19.47	58.1	34.7
R02	201702	RW	7.26	27	638	−6.1	−34	7.6	0.2	3.1	0.9	23.0	2.7	0.4	1.8	-	19.0	58.1	15.4
R03	201609	RW	7.16	30	517	−5.6	−35	3.6	0.3	2.9	1.2	10.0	0.1	5.2	3.4	0.6	19.59	20.7	42.9
R03	201702	RW	7.37	29	629	−6.4	−35	2.0	0.2	3.0	1.0	9.8	1.7	0.6	2.2	-	17.7	4.8	-
R04	201609	RW	6.84	34	448	−6.1	−36	9.8	1.1	2.7	1.5	38.8	0.1	2.3	0.1	0.1	19.29	19.6	65.5
R04	201702	RW	8.05	36	613	−6.2	−36	2.1	1.0	2.8	1.2	19.5	1.6	0.6	1.7	-	18.9	16.6	14.1
R05	201609	RW	7.12	46	420	−6.3	−37	4.4	0.7	3.2	1.6	22.0	5.0	3.5	0.5	0.2	19.73	8.1	19.3
R05	201702	RW	7.17	41	658	−6.1	−36	11.0	0.7	3.6	1.2	36.0	2.5	1.1	3.6	-	19.3	13.1	17.1
R06	201609	RW	7.31	26	476	−6.2	−37	2.0	0.2	2.9	1.2	11.0	2.5	5.1	1.8	0.6	21.2	13.7	18.3
R06	201702	RW	6.96	30	644	−6.0	−36	1.4	0.2	3.9	1.3	12.0	2.2	1.0	2.2	-	21.7	22.1	3.6
R07	201702	RW	6.99	47	625	−6.0	−34	3.4	0.6	4.5	1.7	17.8	2.4	1.4	3.5	-	21.6	6.3	6.7
R08	201702	RW	7.98	46	675	−6.0	−34	3.2	0.6	4.9	1.3	26.0	3.0	1.0	2.4	-	21.9	7.7	-
R09	201702	RW	8.63	56	527	−6.2	−36	3.6	1.0	6.0	1.8	31.2	2.1	1.0	2.5	-	26.3	9.6	5.0

Table 2. Calculated saturation indexes on September 2016.

Sites	Type	Talc	Chlorite	Chrysotile	Sepiolite	Willemite	Calcite	Aragonite	Dolomite	K-Mica	Quartz	Chalcedony	Goethite	Hematite	$Fe(OH)_3(a)$	K-Feldspar	Kaolinite	Ca-Montmorill	Illite	Gibbsite
G05	TW	9.49	7.49	4.84	1.22	6.38	0.62	0.51	−0.2	−1.96	0.4	0.14	5.3	12.85	−2.52	−2.16	-	−5.68	−5.43	−2.92
G07	TW	7.5	5.2	2.6	0.03	3.38	0.44	0.33	−0.5	0.35	0.54	0.28	4.99	12.23	−2.75	−1.41	−2.13	−3.46	−3.6	−2.13
G06	TW	7.9	-	2.95	0.75	2.56	0.4	0.28	0.64	-	0.66	0.36	6.18	14.56	−1.24	-	-	-	-	-
G02	TW	−5.41	-	−9.98	−6.15	−2.11	−1.23	−1.37	−3	-	0.86	0.43	7.23	16.46	1.33	-	-	-	-	-
G09	SG	−14.12	-	−18.47	−11.8	−7.33	−2.41	−2.56	−5.85	-	0.79	0.35	6.42	14.82	0.66	-	-	-	-	-
G11	SG	−16.92	-	−21.06	−13.71	−7.16	−2.59	−2.74	−6.41	-	0.68	0.24	6.16	14.31	0.4	-	-	-	-	-
G10	SG	−10.9	−19.39	−14.78	−9.93	−2.8	−2.8	−2.94	−5.75	6.94	0.52	0.09	6.33	14.66	0.42	−1.24	4.43	2.78	0.76	1.28
G18	SG	−7.32	−13.84	−11.37	−7.45	−3.39	−1.6	−1.74	−3.82	7.56	0.61	0.18	7.07	16.14	1.22	−0.58	4.57	3.32	1.55	1.26
G13	SG	−9.01	−15.7	−13.15	−8.58	−2.66	−1.98	−2.12	−4.92	9.47	0.66	0.22	7.09	16.18	1.22	0.22	5.77	4.74	2.89	1.82
R02	RW	−8.33	−13.98	−12.13	−8.32	−1.85	−2.58	−2.72	−5.87	8.72	0.46	0.04	7.01	16.03	1.04	−0.36	5.21	3.88	2.18	1.73
R04	RW	−7.71	−13.5	−11.54	−7.74	−2.6	−1.84	−1.98	−4.29	8.27	0.51	0.07	7.34	16.68	1.95	−0.46	4.95	3.62	1.49	1.56
R03	RW	−6.99	−12.96	−10.79	−7.44	−1.04	−2.47	−2.61	−5.65	7.09	0.46	0.04	7.17	16.37	1.2	−0.76	3.98	2.52	1.01	1.11
R05	RW	−6.19	−12.15	−10	−6.88	−2.32	−2.09	−2.24	−4.61	6.45	0.47	0.04	6.87	15.75	0.9	−0.91	3.5	1.97	0.58	0.87
R06	RW	−6.81	−13.59	−1.84	−7.54	−1.69	−0.74	−2.76	−6.78	6.49	0.53	0.1	6.97	15.96	1.1	−0.73	3.51	2.04	0.62	0.81

Table 3. Calculated saturation indexes on February 2017.

Sites	Type	Talc	Chrysotile	Sepiolite	Calcite	Aragonite	Quartz	Chalcedony
G05	TW	9.12	4.46	0.91	0.52	0.42	0.39	0.14
G07	TW	7.72	2.65	1.24	0.41	0.29	0.83	0.49
G06	TW	-	-	-	0.42	0.31	0.52	0.24
G02	TW	−4.68	−9.23	−5.56	−1.08	−1.23	0.87	0.44
G15	TW	−4.22	−8.86	−5.89	−1.23	−1.36	0.79	0.39
G16	TW	−0.85	−5.73	−3.21	−0.58	−0.72	0.98	0.56
G17	TW	−0.39	−5.34	−3.39	−0.14	−0.27	0.92	0.54
G01	SG	−9.53	−13.79	−8.63	−2.24	−2.39	0.77	0.32
G03	SG	−12.64	−16.48	−10.75	−3.16	−3.3	0.57	0.12
G04	SG	−9.36	-	−8.29	−1.98	−2.13	0.68	0.21
G08	SG	−10.63	−14.97	−12.17	−2.94	−3.09	0.8	0.36
G09	SG	−12.46	−16.77	−10.37	−3.12	−3.27	0.84	0.38
G10	SG	−6.74	−10.77	−6.46	−2.36	−2.51	0.73	0.26
G11	SG	−3.38	−7.58	−6.95	−1	−1.16	0.79	0.32
G12	SG	−10.11	−14.31	−8.84	−2.68	−2.83	0.78	0.32
G13	SG	−10.23	−14.5	−8.69	−2.6	−2.76	0.86	0.38
G14	SG	−15.2	−19.22	−12.43	−3.74	−3.89	0.66	0.2
R01	RW	−10.66	−14.74	−9.25	−3.22	−3.37	0.72	0.26
R02	RW	−8.86	−12.8	−7.87	−1.94	−2.09	0.69	0.21
R03	RW	−8.34	−12.22	−7.51	−2.75	−2.91	0.66	0.19
R04	RW	−2.1	−6.03	−5.85	−1.77	−1.92	0.69	0.22
R05	RW	−7.82	−11.78	−9.68	−1.68	−1.83	0.7	0.22
R06	RW	−10.44	−14.5	−8.88	−3.23	−3.38	0.75	0.28
R07	RW	−8.32	−12.32	−7.71	−2.59	−2.74	0.68	0.22
R08	RW	−2.72	−6.76	−6.4	−1.51	−1.66	0.72	0.26
R09	RW	2.4	−1.74	−3.17	−0.72	−0.88	0.75	0.29

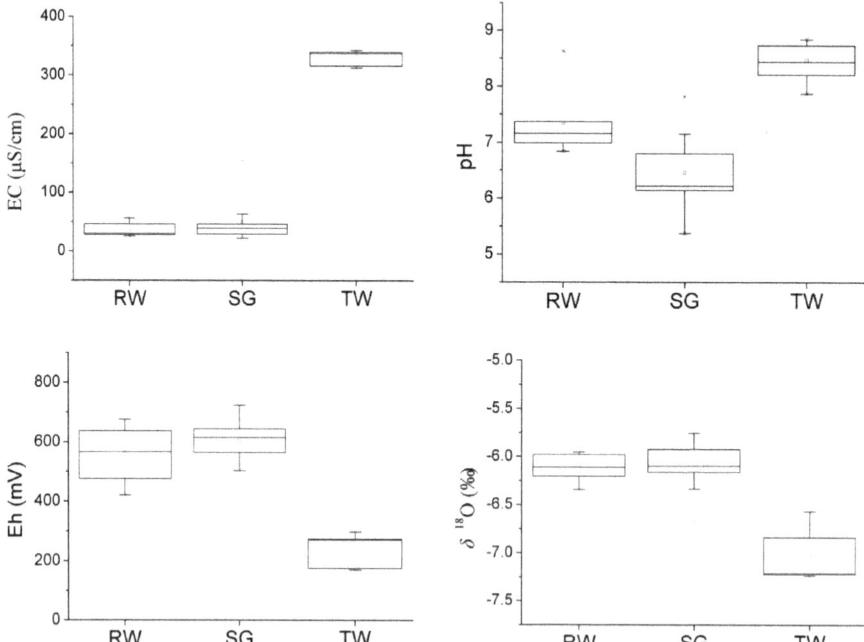

Figure 3. Hydrochemical comparison of geothermal water (TW) to river water (RW) and shallow groundwater (SG).

4.2. Shallow Groundwater (SG) Features

The circulation of deep geothermal water is deeply different with the shallow groundwater. The water table depth of the shallow groundwater is less than 0.5 m. Mean pH value is 6.1 in September and 6.7 in February for the shallow groundwater, while the average Eh values were 540 mV and 640 mV, respectively. EC value rarely changes in the shallow groundwater with a mean value about 40 μS/cm. Compared with the deep circulation groundwater, the shallow groundwater is weakly mineralized, oxidized, neutral, or slightly acidic.

The hydrogeochemical type of the shallow groundwater is HCO_3-Ca (September rainy season) or HCO_3-Na·Ca (February dry season). The mean water isotopic compositions are $\delta^{18}O$ −6.1‰ and δD −36‰. The concentrations of SiO_2 and F^- are about 26 mg/L and 0.2 mg/L (Table 1), respectively. The trace element concentrations are lower than the deep circulation geothermal water.

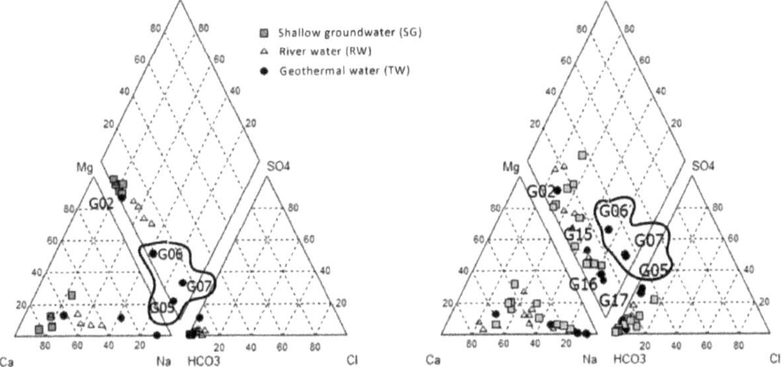

Figure 4. Piper plots in September rainy season (left) and February dry season (right).

4.3. River Water (RW) Features

The river water originates from the spring and outflowing shallow groundwater, which can be inferred from the water isotopes data (Figure 5). As a result, river water shows similar hydrogeochemical features to the shallow groundwater. Mean pH values are 7.2 in September and 7.5 in February for the river water, while the average Eh values are 470 mV and 620 mV, respectively. EC value rarely changes for the river water with a mean value about 38 μS/cm. The hydrogeochemical type of the river water is HCO_3-Ca·Na (September rainy season) or HCO_3-Na·Ca (February dry season). The mean water isotopic compositions are $\delta^{18}O$ −6.1‰ and δD −36‰. The concentrations of SiO_2 and F^- are about 20 mg/L and 0.3 mg/L, respectively. The trace element concentrations are generally lower than the deep circulation geothermal water. The river water is also weakly mineralized, oxidized, and neutral.

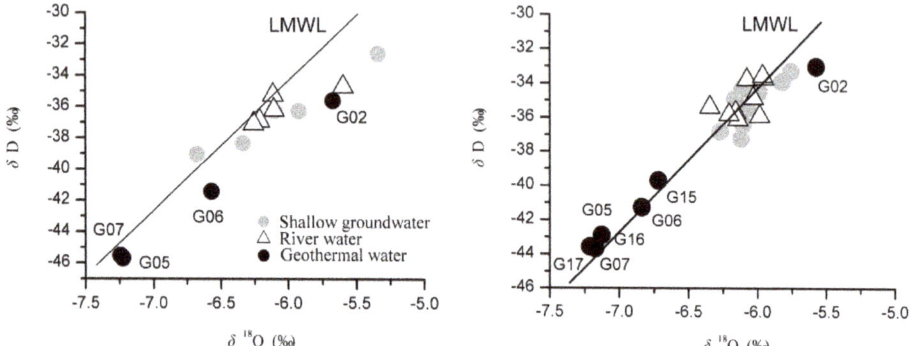

Figure 5. Isotopic compositions in different waters. The left is derived from the survey in September. The right is based on the result of February. The isotopic composition of the deep circulation geothermal water (G05 and G07) obviously depletes in heavy water isotopes indicating a different water source with other waters. The shallow groundwater is subject to evaporation in September due to very small depth of water table (<0.5 m), which tends to make an isotopic signal of the water locating below the local meteorological water line (LMWL).

5. Discussion

5.1. Mixing of Deep Circulation Groundwater (TW) with Shallow Groundwater (SG)

Mixing with shallow cool groundwater usually happens during the ascending of deep circulation groundwater. Mixing ratios should be considered accurately before assessing the deep circulation groundwater flow system based on hydrogeochemical and isotopic data. The thermal water from the borehole (G05) is withdrawn directly from the fracture zone of the F3 fault with a water temperature of 93 °C that is the highest among all water samples. The geothermal water (G05) and the geothermal spring (G07) are chemically equivalent featured by high contents of sodium, sulfate, chloride, fluorine, silicate, and some trace elements—such as lithium, strontium, cesium and rubidium. Moreover, the distinct physicochemical features exist between the geothermal water in G05 and G07 and the other water samples (Figure 3 and Table 1). According to the Figure 6 which identifies two end-members and the mixing line, geothermal groundwater from G05 and G07 are considered as the deep circulation water occurring in the geothermal reservoir without mixing with shallow cool groundwater (15 °C in February and 23 °C in September in average).

The shallow groundwater flows from east, west, and south to the Reshui geothermal field area following the topographic gradient with very shallow water table depth (within 0.5 m). In September, the end of rainy season, shallow groundwater receives significant recharge from precipitation. As a result, a part of wells even became artesian. In February, the period of low flow, water table depth fells with cease of the artesian wells. The seasonal change of water table depth strongly suggests a seasonal groundwater flow enhanced during rainy seasons. The seasonality of precipitation also caused differences in hydro-geochemistry of the shallow groundwater. As shown in Figure 6, the shallow groundwater has an identical hydrogeochemistry after a rainy season (September). However, a part of shallow groundwater gains more Na^+ and SO_4^{2-} during a dry season (February). Based on a two end-member mixing mechanism, a mixing line connects the deep circulation groundwater and the shallow groundwater (Figure 6). Therefore, we confirmed that (1) the geothermal water G06 is the mixture of the deep circulation geothermal groundwater and the local shallow groundwater; and (2) the geothermal water G15, G16, and G17 belong to another geothermal system (Figure 6).

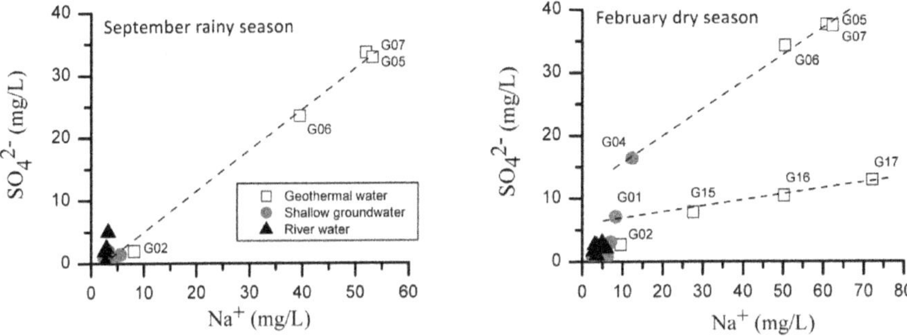

Figure 6. Plots of Na^+ versus SO_4^{2-} reveal the relationship between end-members. The left is derived from the survey in September. The right is based on the result of February.

Geothermal water of G02, G05, G06, and G07 belongs to the Reshui geothermal field, while the geothermal water of G15, G16, and G17 comes from another geothermal system named here the Nuanshui geothermal field. The two deep circulation geothermal water systems can also be verified by distinct hydrogeochemical features. The geothermal water of the Nuanshui geothermal field (G17) shows a little higher EC (370 µS/cm) and lower pH (7.8) than water in the Reshui system. The hydrogeochemical type of G17 is HCO_3-Na with obviously lower content of SO_4^{2-} compared to the G05 and G07 (Figure 6). Moreover, the concentrations of trace elements such as Li (417 µg/L), Sr (137 µg/L), Cs (101 µg/L), and As (22 µg/L) in G17 are higher than that of G05 and G07 (Figure 7).

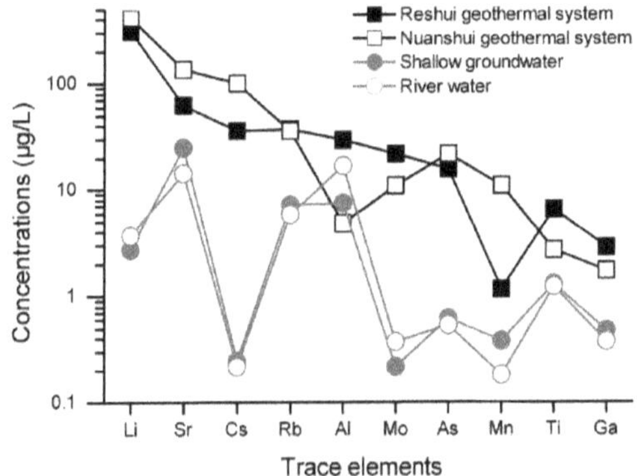

Figure 7. Trace elements concentrations in deep geothermal water (the mean of G05 and G07 represents the Reshui geothermal system, while G17 stands for the Nuanshui geothermal system), shallow groundwater (samples mean) and river water (samples mean). The other thermal water had obviously lower contents than G05, G07, and G17.

Based on the two end-member mixing method and tracer concentrations of Na^+ and SO_4^{2-}, the average mixing ratio of the deep thermal water is 71% in September (the rainy season) and 81% in February (the dry season) for G06. The mixing ratio decreases obviously during the rainy season, which is consistent with the enhanced shallow groundwater flow in rainy seasons. The borehole of G06 has a depth about 150 m where the upward deep geothermal water mixes with the shallow groundwater flow. When pumping, the water table depth drops down about 50 m. Once pumping ceases, the water table depth recovers quickly closing to the ground surface. The mixing ratio of the deep circulation geothermal groundwater only slightly changes from 5% to 6% for G02. The depth of borehole G02 is over 500 m drilled into thick diorite discovering thermal water with a temperature around 88 °C at the depth of 150 m. After the cease of pumping, shallow groundwater mixes into the borehole and dominated in the borehole tube.

5.2. Water Source of Deep Circulation Groundwater

The local meteorological water line (LMWL) is characterized as $\delta D = 8.42\ \delta^{18}O + 16.28$ based on water isotopes in precipitation of Guilin (1983–1998) [20] that are close to the study area with similar climate and geographical conditions. All samples of the study area are plotted around the LMWL indicating local precipitation origin of water (Figure 5). The deep circulation geothermal water, G05, G07, and G17 show the most depleted isotopic compositions. The shallow groundwater and the river water have similar isotopic compositions with an identical average isotopic composition ($\delta^{18}O$ −6.1‰ and δD −36‰ on September, $\delta^{18}O$ −6.0‰ and δD −35‰ on February). In addition, the shallow groundwater is significantly influenced by evaporation on September showing a more enriched isotopic composition.

The geothermal water of G06, G15, and G16 are distributed along the LMWL between G05, G07, and G17 and the average of shallow groundwater (Figure 5). It is the result of mixing of the ascending deep circulation geothermal water with the shallow groundwater. It should be noticed that the Reshui geothermal groundwater system and the Nuanshui geothermal groundwater system cannot be separated by isotope data due to the same precipitation isotopic input. The geothermal water of G02 shows an obviously enriched isotopic composition. The borehole G02 discovers deep circulation geothermal groundwater. The highest water temperature was observed around 88 °C, when the borehole was pumped. However, the borehole has been sealed for near a year long before our sampling with no pump installed. The water table depth was about 24.6 m with an around 500 m long water column in the borehole tube. The depth of the aquifer is about 150 m. We collected water samples from about 50 m below the ground surface. According the geochemical features (Table 1, Figures 4–6), the river water and shallow groundwater should have mixed into the borehole. Water in the tube is hard to flow and ready to be evaporated (Figure 5). Evaporation of waters also hinders the distinguishing of mixtures based on geochemical data.

The distinct distributions of isotopic compositions suggest different water sources. Water source areas can be identified by water isotopic compositions in surface water, groundwater and precipitation. Globally, the stable isotope lapse rate (change in stable isotope composition with elevation, namely, the altitude effect in precipitation isotopes) was reported to be in the range −0.15 and −0.5‰ per 100 m altitude increase for $\delta^{18}O$ [21]. Altitude is considered as the main geographic control on $\delta^{18}O$ in precipitation in Southern China with a lapse rate about −0.20‰ for every 100 m [22]. The average $\delta^{18}O$ in shallow groundwater and rivers is -6.1‰ with an elevation about 350 m in the lower part of the basin. At the same time, the mean value is −7.2‰ in the deep circulation geothermal groundwater. However, the $\delta^{18}O$ at the leeward slope of mountains is around 0.5‰ larger than that at the windward slope induced by continuous rain-out processes associated with orographic lifting at the windward side and sub-cloud evaporation at the leeward side [21]. The south China belongs to the eastern Asia monsoon region with rainy seasons controlled by south to north wet monsoon. It is suggested that the mean $\delta^{18}O$ should be revised as around −7.7‰ due to the leeward sampling positions. Therefore, the average recharge elevations are estimated about 1150 m for G05 and G07 of the Reshui geothermal system. The mean recharge elevation is close to the altitude of the halfway up to surrounding mountain tops, which suggested that the mixture of precipitation infiltration from different heights of mountains is the source of the deep circulation geothermal water. For the Reshui geothermal groundwater system, the water source area is located in the surrounding high mountains on the south.

5.3. Ion Origin and Phase Equilibrium of Minerals in Deep Circulation Groundwater

Simple solution and hydrolysis contributed to mineralization of the water. According to saturation indexes of minerals calculated by PHREEQCI [23], shallow groundwater and river water are supersaturated with respect to hematite, K-mica, goethite, kaolinite, Ca-montmorill, illite, gibbsite, quartz, and chalcedony (Tables 2 and 3), which are contained in widely distributed granite, slate, and quartz sandstone. These minerals contribute to constituents in water of the study area. Granite and slate covering most part of the watershed are usually hard to be weathered. However, carbonic acid formed by the reaction of carbon dioxide with the water takes apart in and promotes weathering progresses to derive mineral constituents such as metals and silica from granites. It is supported by a high percent of anionic milliequivalent of bicarbonate (average 82%) and the low pH value (average 6.4) in the shallow groundwater that means high activity of the hydrogen ion induced by the reaction of carbon dioxide with the water. Thus carbon dioxide takes an important role in the weathering of rock-forming minerals.

Compared with shallow groundwater and river water, the deep circulation geothermal water (G05 and G07) in the Reshui geothermal groundwater system is also supersaturated with respect to some altered minerals, such as talc, chrysotile, chlorite, and sepiolite (Tables 2 and 3), which is related with geothermal activities in depth supporting the deep circulation pattern [13]. Moreover, the deep circulation geothermal water shows distinct major ion chemistry. First, the content of sodium and sulfate are about ten times higher than that in shallow groundwater and river water. Second, the concentration of carbonate is around three times higher than that of shallow groundwater and river water. Third, there was no nitrate in the deep circulation geothermal water considering the rational measurement precision. The most widespread source of sodium is the weathering of feldspars in the study area. The deep circulation geothermal water in the Reshui geothermal system originated from the surrounding mountains on the south where feldspars are one of the common rock-forming minerals. Moreover, the weathering of feldspars would be promoted under geothermal condition with fresh replenishing of geothermal fluid. In hydrothermal environments, *S* is common and usually presents as sulfides. The sulfides would be oxidized to sulfates once exposed to the atmosphere or oxygenated. Therefore, substantial sodium and sulfate enter into the deep circulation geothermal water in the deep geothermal reservoir. However, the main source of nitrate in the water seems to be biologic activities in near-surface zones, forest litter, and the soil. During the deep circulation of the geothermal water, the nitrate could be consumed up by chemical and biochemical processes under an oxygen-deficient situation. That is the reason why nitrate is only observed in the shallow groundwater and the river water. Differences in saturation indexes of minerals and major ion chemistry suggest the different flow paths of the shallow groundwater and the deep circulation groundwater.

The deep circulation geothermal water (G05 and G07) also showed high contents of silicate and fluorine. The average concentrations of the silicate and fluorine are 150 mg/L and 10 mg/L in the geothermal water, respectively, which is consistent with a low temperature and large circulation depth geothermal water. At the same time, the concentrations are only 23 mg/L and 0.2 mg/L in the shallow groundwater and river water. Plagioclase, followed by K-feldspar and the ferromagnesian silicate minerals, are the major sources of dissolved silica in the groundwater. Large silica concentrations suggest that hydrolysis of primary silicates is a major process in the thermal water system. The very high saturation indexes (higher than 5) of hematite, K-mica, kaolinite, talc, and so on suggest the formation of secondary minerals, which is an indicator of a slowly flowing system.

The significant seasonal variation of phase equilibria of minerals is recorded for the shallow groundwater and river water that are supersaturated with respect to hematite, K-mica, goethite, kaolinite, Ca-montmorill, illite, gibbsite, quartz, and chalcedony in September and only quartz and chalcedony in February (Tables 2 and 3). Hematite, K-mica, and other clay minerals are chemical weathering productions of igneous rock by hydrolysis. Granite mountains are located on the south of the study area with a developed crust of weathering. In September, air temperature is high. Shallow groundwater ascends and even becomes artesian due to plenty rainfall during the rainy seasons when the weathering crust is flushed. Chemical weathering of the crust is intensive. As a result, the shallow groundwater and the river water are supersaturated with respect to hematite and clay minerals. In February, air temperature is low. Shallow groundwater descends and artesian springs disappear, because rainfall is infrequent. The crust would be exposed to dry air. The chemical weathering of the crust is greatly weakened. Chemical weathering changed seasonally following the seasonality of precipitation, which is presented by the seasonal variation of saturated minerals in the shallow groundwater and river water.

5.4. Circulation Depth and Passageway

Due to mixing with shallow cool groundwater, the geothermal water of G06 and G02 do not represent the true geothermal reservoir fluid and hence geothermal reservoir temperature estimated using geothermometers would deviate true value [3,17]. Geothermal water from G05 and G07 does not mix with the shallow cool groundwater. The geochemistry could represent the original result of the water-mineral equilibrium between the geothermal reservoir fluid and the host rocks in the geothermal reservoir. Chemical geothermometers give the last equilibration temperature for the reservoir. Many geothermometry techniques have been developed to predict reservoir temperatures in geothermal systems. All of these techniques are based on the assumption that temperature dependent water-mineral equilibrium is attained in the reservoir. The Na-K-Mg triangular diagram technology of Giggenbach [24] shows the geothermal water samples of G05 and G07 are located on the partial equilibrium zone (Figure 8). According to the saturation index of quartz (Tables 2 and 3), the ascending geothermal water is still saturated with silica. Therefore, geothermometers based on Na-K and SiO_2 can be used. However, the rates at which different species react vary, with silica adjusting faster than cations like Na^+ and K^+. Moreover, concentrations of SiO_2 in G05 and G07 are around 150 mg/L that is so high that SiO_2 loss (aggregation and precipitation) during sample storage has to be considered. The average reservoir temperature is estimated at 177 °C by the Na-K geothermometers [24,25], 173 °C [26], and 161 °C [27]. Therefore, the temperature of the geothermal reservoir is estimated as 170 °C, which is the average of results from above methods. When considering 13% SiO_2 loss during sample storage, the same result of reservoir temperature is reached by the SiO_2 geothermometer (no steam loss) based on quartz solubility [28]. Our result suggested that a cation geothermometer, such as Na-K, typically has a longer 'memory' during the storage than SiO_2 geothermometer when the thermal water has high SiO_2 concentrations.

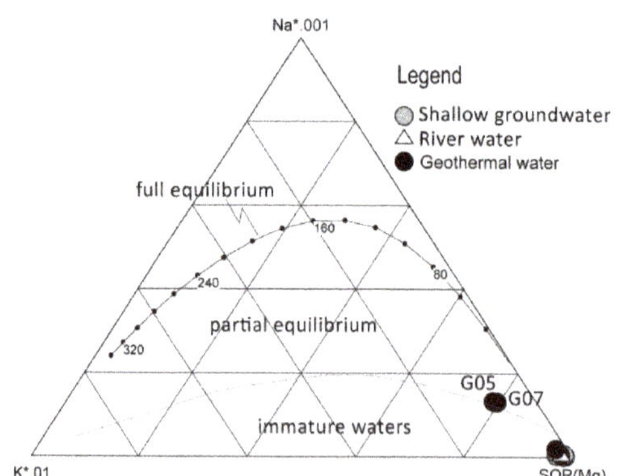

Figure 8. Ternary giggenbach's diagram (samples data in mg/L).

The equilibration temperature and the geothermal gradient of the region could offer a reliable estimate of the depth of geothermal reservoirs using the following formula [7]

$$D = (T - T_0)/K \times 1000 \tag{1}$$

where D is the reservoir depth (m); T is the reservoir temperature (°C); T_0 is the temperature of constant temperature zone (16.7 °C); K is geothermal gradient (°C/km). In South China, the average geothermal gradient is 24.1 °C/km, and the average heat flow is 64.2 mW/m^2 [29]. However, land plate boundaries and the active zones of the deep faults are coincident with anomalies of high heat flow values and high geothermal gradient. The Reshui geothermal field is within the plate collision zone with the measured heat flow values ranging 60–100 mW/m^2 that is obviously higher than the average. Referencing to the result of Yuan et al. [29], geothermal gradients of 35 °C/km and 40 °C/km are employed in this study. The depth of the geothermal reservoir is estimated ranging between 3.8–4.3 km using Equation (1), which indicates the circulation depth of the geothermal water.

The deep circulation geothermal water is recharged on the southern mountains. However, the geothermal reservoir is located 10 km northward away and 3.8–4.3 km in depth. The NE trending fault systems of F1, F2, and F3 penetrate the granite and connect the recharge area and the geothermal reservoir offering the passageway for the deep circulation groundwater.

5.5. Conceptual Model of the Deep Circulation Groundwater

Based on the above discussion, the pattern of the deep circulation geothermal water in the Reshui low-temperature geothermal field is reconstructed (Figure 9). The deep circulation geothermal water originates from precipitation infiltration in the southern mountains with an average elevation about 1150 m. The infiltrating water flows downward in the fracture network of the weathering crust, and then a part of the groundwater converges into the F1 (the dipping angle 78–87°) and F2 (the dipping angle 50°) fault fracture zones (trending NNE–SSW). Groundwater in the fault fracture zones moves downward and northward under the pressure of water from the recharge highlands. The deep circulation water is heated by deep heat sources to a temperature around 170 °C at a depth of 3.8–4.3 km. Simple solution and hydrolysis contribute to mineralization of the deep circulation water that is also supersaturated with respect to some altered minerals—such as talc, chrysotile, chlorite, and sepiolite—by hydrothermal alteration. Therefore, the hydrogeochemical features are distinct with the shallow groundwater. The alkaline deep geothermal water enriches in Na, F, Li, and other trace elements consistent with the granite reservoir nature [19]. The bicarbonate geothermal water with some sulfate and a little chloride is mainly heated by conduction partially by steam in the reservoir.

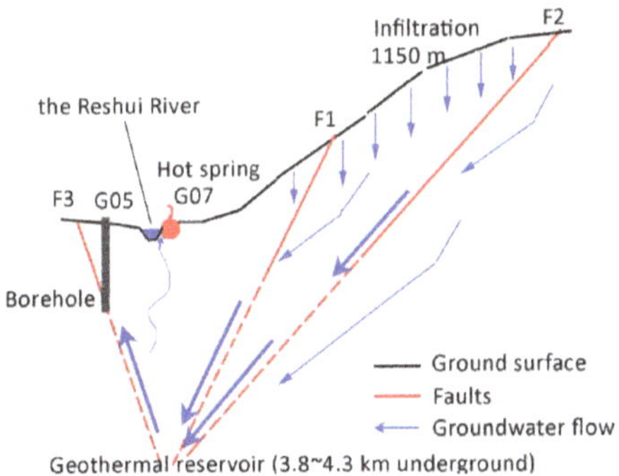

Figure 9. Sketch of the conceptual model of the geothermal groundwater flow system.

However, the deep circulation geothermal water encounters a 110° trending thrust fault F4 which hindered the movement of the deep circulation geothermal water. The deep circulation geothermal water subsequently flows upwards to approach the surface along the F3 fault fracture zone driving by high water pressure originating an elevation difference about 800 m between recharge and discharge areas of the deep circulation groundwater and reduced density after heating. A part of the ascending geothermal water directly discharges into the Reshui River valley forming the Reshui Spring (G07). The deep circulation geothermal water (G05) is derived from F3 fault. The other recharges into the overlying phreatic aquifer mixing with the local cool shallow groundwater (G06).

6. Conclusions

Deep circulation groundwater flow systems are usually connected to geothermal system, especially in neotectonic and volcanic areas. Based on hydrogeochemical and isotopic data, the deep circulation groundwater flow system was surveyed in the Reshui geothermal field where an 88 °C hot spring occurs. Precipitation, mostly falling in the highlands, is the water source of the deep circulation groundwater flow system. The weathering crust of the granite mountain on the south collects infiltration of precipitation, which is recognized as water source area of the deep circulation groundwater. The average recharge elevation is about 1150 m. A regional active normal fault system formed the passageway connecting the water source area on the surface and the deep geothermal reservoir. Groundwater from the surface circulates to a depth about 3.8–4.3 km where the water is heated up to around 170 °C by conduction and partially by steam from the geothermal reservoir. The ongoing geothermal water is obstructed by a thrust fault system. Then the geothermal water flows upwards and forms the hot spring forcing by the water pressure of convection derived from the 800 m elevation difference between the recharge (the south mountains) and discharge (the hot spring) areas. The geothermal water is equilibrium with hot rocks in the geothermal reservoir. A simple solution of minerals, hydrolysis with the aid of carbonic acid produced by the reaction of carbon dioxide with the water, and hydrothermal alteration in the geothermal reservoir contribute ions to the deep circulation geothermal water. The alkaline deep circulation geothermal water is chemically featured by high contents of sodium, sulfate, chloride, fluorine, silicate, and some trace elements—such as lithium, strontium, cesium, and rubidium—and by depleted water isotopic compositions. Although the hydrogeochemistry of deep circulation groundwater is greatly changed in the geothermal reservoir, the isotopic investigations combined with geothermal applications still represent powerful tools for the exploration of deep circulation groundwater flow system in a geothermal field.

Our results suggest that groundwater deep circulation convection exists in where water-conducting fault and water-blocking fault combined properly. A water-conducting fault with a bigger dipping angle would act as a passageway for deep circulation convection. Significant elevation difference in topography is the other key. Those conditions would be met in mountain areas in most cases. Groundwater would seldom circulate to a significant depth in plain regions, usually due to the thick and layered alluvial deposits.

Author Contributions: Conceptualization, R.Y. and K.Z.; Methodology, X.L.; Software, L.Z.; Validation, R.Y. and Z.L., K.Z. and L.Z.; Formal Analysis, R.Y.; Investigation, X.L. and R.Y.; Data Curation, X.L.; Writing-Original Draft Preparation, X.L and R.Y.; Writing-Review & Editing, R.Y.

Funding: This research was funded by the National Natural Science Foundation of China grant number 41301033.

Acknowledgments: The authors thank the 402 team in the Bureau of Geology and Mineral Resources Exploration of Hunan for research support, especially Xinping Deng for field survey assistance and data sharing.

Conflicts of Interest: The authors declare no conflict of interest.

References

1. Fridleifsson, I.B.; Bertani, R.; Huenges, E.; Lund, J.W.; Ragnarsson, A.; Rybach, L. The Possible Role and Contribution of Geothermal Energy to the Mitigation of Climate Change. In Proceedings of the IPCC Scoping Meeting on Renewable Energy Sources, Proceedings, Luebeck, Germany, 20–25 January 2008; pp. 59–80.
2. Kooi, H. Groundwater flow as a cooling agent of the continental lithosphere. *Nat. Geosci.* **2016**, *9*, 227. [CrossRef]
3. Mao, X.; Wang, Y.; Zhan, H.; Feng, L. Geochemical and isotopic characteristics of geothermal springs hosted by deep-seated faults in Dongguan Basin, Southern China. *J. Geochem. Explor.* **2015**, *158*, 112–121. [CrossRef]
4. Benavente, O.; Tassi, F.; Reich, M.; Aguilera, F.; Capecchiacci, F.; Gutiérrez, F.; Vaselli, O.; Rizzo, A. Chemical and isotopic features of cold and thermal fluids discharged in the Southern Volcanic Zone between 32.5° S and 36° S, Insights into the physical and chemical processes controlling fluid geochemistry in geothermal systems of Central Chile. *Chem. Geol.* **2016**, *420*, 97–113. [CrossRef]

5. Tassi, F.; Liccioli, C.; Agusto, M.; Chiodini, G.; Vaselli, O.; Calabrese, S.; Pecoraino, G.; Tempesti, L.; Caponi, C.; Fiebig, J.; et al. The hydrothermal system of the Domuyo volcanic complex (Argentina): A conceptual model based on new geochemical and isotopic evidences. *J. Volcanol. Geotherm. Res.* **2016**, *328*, 198–209. [CrossRef]
6. Battistel, M.; Hurwitz, S.; Evans, W.C.; Barbieri, M. The chemistry and isotopic composition of waters in the low-enthalpy geothermal system of Cimino-Vico Volcanic District, Italy. *J. Volcanol. Geotherm. Res.* **2016**, *328*, 222–229. [CrossRef]
7. Luo, L.; Pang, Z.; Liu, J.; Hu, S.; Rao, S.; Li, Y.; Lu, L. Determining the recharge sources and circulation depth of thermal waters in Xianyang geothermal field in Guanzhong Basin: The controlling role of Weibei Fault. *Geothermics* **2017**, *69*, 55–64. [CrossRef]
8. Avşar, Ö.; Kurtuluş, B.; Gürsu, S.; Kuşcu, G.G.; Kaçaroğlu, F. Geochemical and isotopic characteristics of structurally controlled geothermal and mineral waters of Muğla (SW Turkey). *Geothermics* **2016**, *64*, 466–481. [CrossRef]
9. Roquer, T.; Arancibia, G.; Rowland, J.; Iturrieta, P.; Morata, D.; Cembrano, J. Fault-controlled development of shallow hydrothermal systems: Structural and mineralogical insights from the Southern Andes. *Geothermics* **2017**, *66*, 156–173. [CrossRef]
10. Zhang, Y.; Tan, H.; Zhang, W.; Wei, H.; Dong, T. Geochemical constraint on origin and evolution of solutes in geothermal springs in western Yunnan, China. *Chem. Der Erde—Geochem.* **2016**, *76*, 63–75. [CrossRef]
11. Han, D.; Liang, X.; Currell, M.J.; Song, X.F.; Chen, Z.Y.; Jin, M.G.; Liu, C.; Han, Y. Environmental isotopic and hydrochemical characteristics of groundwater systems in Daying and Qicun geothermal fields, Xinzhou Basin, Shanxi, China. *Hydrol. Process.* **2010**, *24*, 3157–3176. [CrossRef]
12. Sayres, D.S.; Pfister, L.; Hanisco, T.F.; Moyer, E.J.; Smith, J.B.; St. Clair, J.M.; O'Brien, A.S.; Witinski, M.F.; Legg, M.; Anderson, J.G. Influence of convection on the water isotopic composition of the tropical tropopause layer and tropical stratosphere. *J. Geophys. Res.* **2010**, *115*, D00J20. [CrossRef]
13. Baba, A.; Yuce, G.; Deniz, O.; Ugurluoglu, D.Y. Hydrochemical and Isotopic Composition of Tuzla Geothermal Field (Canakkale-Turkey) and its Environmental Impacts. *Environ. Forensics* **2009**, *10*, 144–161. [CrossRef]
14. Newell, D.; Crossey, L.; Karlstrom, K.; Fischer, T.; Hilton, D. Continental-scale links between the mantle and groundwater systems of the Western United States; evidence from travertine springs and regional He isotope data. *GSA Today* **2005**, *15*, 4–10. [CrossRef]
15. Guo, Q.; Wang, Y.; Liu, W. O, H, and Sr isotope evidences of mixing processes in two geothermal fluid reservoirs at Yangbajing, Tibet, China. *Environ. Earth Sci.* **2010**, *59*, 1589–1597. [CrossRef]
16. Dotsika, E.; Poutoukis, D.; Kloppmann, W.; Guerrot, C.; Voutsa, D.; Kouimtzis, T.H. The use of O, H, B, Sr and S isotopes for tracing the origin of dissolved boron in groundwater in Central Macedonia, Greece. *Appl. Geochem.* **2010**, *25*, 1783–1796. [CrossRef]
17. Pasvanoğlu, S.; Chandrasekharam, D. Hydrogeochemical and isotopic study of thermal and mineralized waters from the Nevşehir (Kozakli) area, Central Turkey. *J. Volcanol. Geotherm. Res.* **2011**, *202*, 241–250. [CrossRef]
18. Liu, K.; Qiao, X.; Li, B.; Sun, Y.; Li, Z.; Pu, C. Characteristics of deuterium excess parameters for geothermal water in Beijing. *Environ. Earth Sci.* **2016**, *75*, 1485. [CrossRef]
19. Guo, Q.; Pang, Z.; Wang, Y.; Tian, J. Fluid geochemistry and geothermometry applications of the Kangding high-temperature geothermal system in eastern Himalayas. *Appl. Geochem.* **2017**, *81*, 63–75. [CrossRef]
20. Tu, L.; Wang, H.; Feng, Y. Research on D and O^{18} isotope in the precipitation of Guilin. *Carsol. Sin.* **2004**, *23*, 304–309.
21. Guan, H.; Simmons, C.T.; Love, A.J. Orographic controls on rain water isotope distribution in the Mount Lofty Ranges of South Australia. *J. Hydrol.* **2009**, *374*, 255–264. [CrossRef]
22. Liu, J.; Song, X.; Yuan, G.; Sun, X.; Liu, X.; Wang, S. Characteristics of $\delta^{18}O$ in precipitation over Eastern Monsoon China and the water vapor sources. *Chin. Sci. Bull.* **2010**, *55*, 200–211. [CrossRef]
23. Parkhurst, D.L.; Appelo, C.A.J. Description of input and examples for PHREEQC version 3—A computer program for speciation, batch-reaction, one-dimensional transport, and inverse geochemical calculations. In *U.S. Geological Survey Techniques and Methods, Book 6*; 2013; Chapter A43; 497p. Available online: http://pubs.usgs.gov/tm/06/a43 (accessed on 2 December 2016).
24. Giggenbach, W.F. Geothermal solute equilibria. Derivation of Na-K-Mg-Ca geoindicators. *Geochim. Cosmochim. Acta* **1988**, *52*, 2749–2765. [CrossRef]

25. Fournier, R.O. A Revised Equation for the Na/K Geothermometer. *Geotherm. Resour. Counc. Trans.* **1979**, *3*, 221–224.
26. Verma, M.P. Silica Solubility Geothermometers for Hydrothermal Systems. In Proceedings of the Tenth International Symposium on Water-Rock Interaction, WRI-10, Villasimius, Italy, 10–15 June 2001; Volume 1, pp. 349–352.
27. Verma, M.P.; Santoyo, E. New improved equations for Na/K, Na/Li and SiO_2 geothermometers by outlier detection and rejection. *J. Volcanol. Geotherm. Res.* **1997**, *79*, 9–24. [CrossRef]
28. Fournier, R.O. Chemical geothermometers and mixing models for geothermal systems. *Geothermics* **1977**, *5*, 41–50. [CrossRef]
29. Yuan, Y.; Ma, Y.; Hu, S.; Guo, T.; Fu, X. Present-Day Geothermal Characteristics in South China. *Chin. J. Geophys.* **2006**, *49*, 1005–1014. [CrossRef]

Article

Well-Placement Optimization in an Enhanced Geothermal System Based on the Fracture Continuum Method and 0-1 Programming

Liming Zhang, Zekun Deng, Kai Zhang *, Tao Long, Joshua Kwesi Desbordes, Hai Sun and Yongfei Yang

School of Petroleum Engineering, China University of Petroleum (East China), Qingdao 266580, China; zhangliming@upc.edu.cn (L.Z.); dzk9586@sina.cn (Z.D.); ltysdy@163.com (T.L.); joshuadesbordes@gmail.com (J.K.D.); sunhaiupc@sina.com (H.S.); yangyongfei@upc.edu.cn (Y.Y.)

* Correspondence: zhangkai@upc.edu.cn; Tel./Fax: +86-532-8698-1808

Received: 17 December 2018; Accepted: 18 February 2019; Published: 21 February 2019

Abstract: The well-placement of an enhanced geothermal system (EGS) is significant to its performance and economic viability because of the fractures in the thermal reservoir and the expensive cost of well-drilling. In this work, a numerical simulation and genetic algorithm are combined to search for the optimization of the well-placement for an EGS, considering the uneven distribution of fractures. The fracture continuum method is used to simplify the seepage in the fractured reservoir to reduce the computational expense of a numerical simulation. In order to reduce the potential well-placements, the well-placement optimization problem is regarded as a 0-1 programming problem. A 2-D assumptive thermal reservoir model is used to verify the validity of the optimization method. The results indicate that the well-placement optimization proposed in this paper can improve the performance of an EGS.

Keywords: enhanced geothermal system; well-placement optimization; fracture continuum method; 0-1 programming

1. Introduction

Development and utilization of renewable energy have been a hot topic in society in recent years because of increased energy consumption and pollution [1]. Due to its reproducibility and cleanness, geothermal energy has received extensive attention. Most geothermal energy is preserved in hot dry rock (HDR) with a temperature between 150 °C to 650 °C in a depth range of about 3–10 km [2].

The enhanced geothermal system (EGS) proposed in the 1970s is the representative technology for HDR development [3]. The connected fracture network is formed in HDR through hydraulic fracturing, and the fractured thermal reservoir, called an artificial thermal reservoir, can be injected with cold water to extract the thermal energy [4].

The cost of an EGS in reservoir development and management is expensive, especially in well-drilling. In the process of constructing an EGS, the cost of well-drilling would account for more than 50% of the total cost because the hard reservoir and high temperature in hot dry rock could damage the drill bit quickly [5]. The optimal well-placement and operation are important to the performance of the EGS [6]. Combining numerical simulation with optimization algorithms is an effective method to search for the optimal well-placement.

The seepage in EGS during heat extraction is affected by multiple factors such as multi-field coupling [7], geometrical parameters of porous media [8] and fractures. The effects of fractures on seepage and heat extraction cannot be ignored because the EGS is often developed by hydraulic fracturing [9]. Many methods have been applied to research the fluid flow in the fracture, including the equivalent porosity model (EPM) [10], dual-porosity model (DPM) [11], digital core [12], discrete fracture network (DFN) [13], stochastic continuum (SC) [14], fracture continuum method (FCM) [15] and lattice Boltzmann methods (LBM) [16].

The optimization algorithm has been used in groundwater resource management [17], porous media model building [18] and oil reservoir management [19] for many years. Optimal well-placement [20,21], or well pattern [22,23], has largely improved the performance of the reservoir. In addition, there are several optimization algorithms applied to the well location or well pattern of the reservoir, such as the adjoint gradient algorithm [24,25], genetic algorithm (GA) [26], particle swarm optimization (PSO) [27], and new unconstrained optimization algorithm (NEWUOA) [28].

Some research about the well-placement in an EGS has been proposed. Akin et al. [29] optimized a new injection well-placement using simulated annealing based on the Kizildere geothermal field. A trained artificial neural network replaced the commercial simulators to reduce the processing demand. Chen et al. [30] proposed that a suitable well layout can improve the heat extraction effect. In addition, they designed a five-spot well layout and confirmed its performance using a numerical simulation. Chen et al. [31] used a multivariate adaptive regression spline (MARS) to set a surrogate model to replace the numerical model and used the bound optimization by quadratic approximation (BOBYQA) to optimize the well-placement. Wu et al. [32] studied the relationship between well-placement and heat extraction based on the semi-analytical solution model with a single horizontal fracture. Guo et al. [33] proposed that more production wells are more effective in delaying the breakthrough of the cold front, and the well should be placed at a position with higher rock stiffness.

There is less research on well-placement and all studies used the traditional method to encode the well-placement. On the other hand, they did not fully consider the uneven distribution of lots of fractures. In order to improve the performance of heat extraction, an optimization framework based on 0-1 programming and genetic algorithms is used in EGS well-placement. The purpose of this work is to provide a valid method to determine where the best locations of wells are in an EGS with a complex fracture network. The framework for EGS well-placement optimization consists of two parts: coding the well-placement variable with a 0-1 variable instead of the traditional coordinates of well-placement and reducing the computational cost by the FCM model. The first part is used to decrease the possible well-placements and it also has the potential to do joint optimization for well-placement and the number of wells. The FCM model is used to simplify the fractured reservoir model to reduce the computational costs while preserving the effect of fractures on heat extraction. An assumptive model is used to verify the validity of the method, and GA is used to search the best well-placement in EGS.

2. Method

2.1. Fracture Continuum Method

For a geothermal reservoir with a large number of fractures, the discrete fracture network model needs to discretize each fracture, which is too computationally expensive, making it unsuitable for optimization problems that require multiple iterations. In this work, the FCM is used to describe the flow of fluids in the thermal reservoir, and to minimize the computational cost of numerical simulation while preserving the effect of the fracture network on the fluid flow.

The FCM model can be considered as a stochastic continuum model that preserves the characteristics of the fracture distribution. In the FCM model, the reservoir is divided into several sub-grids, and the permeability of each sub-grid is determined by the fractures passing through that sub-grid.

2.1.1. Backbone Network's Extraction

Matthäi [34] proposed that the disconnected fractures contribute little to the fluid flowing in the fractured reservoir when the permeability of the matrix is more than six orders of magnitude smaller than the fracture, and the effect of heat convection is much greater than heat conduction during heat extraction. Therefore, the disconnected fractures and the dead-end of fractures are eliminated and the backbone of the fracture network is extracted. Figure 1 shows an original fracture network and the backbone network extracted from the original network.

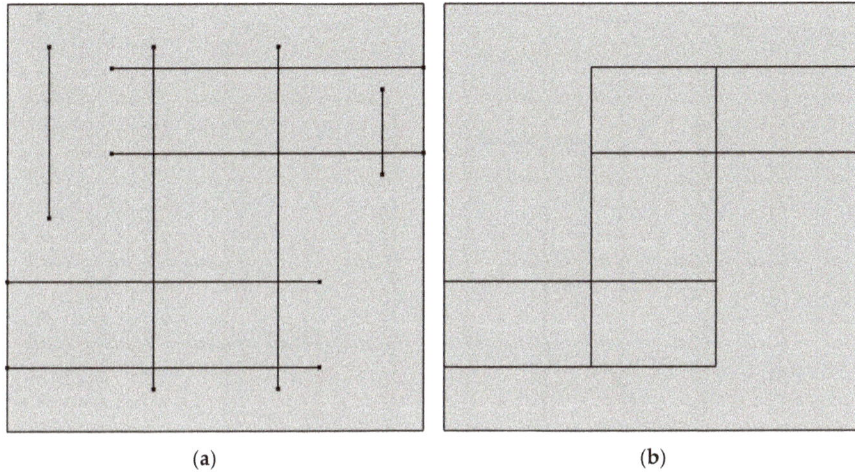

Figure 1. An original fracture network and the extracted backbone fracture network: (**a**) Original network. (**b**) Backbone network.

The permeability mapping in the next section is based on the backbone network, because the fractures that are not connected to each other are able to be connected in permeability mapping due to the dead end or disconnected fracture, which cause a higher permeability. Using the backbone network can largely avoid the higher permeability and retain the effect of the fracture on seepage.

2.1.2. Permeability Mapping Approach

The permeability mapping method is proposed in Ref [35]. An analytical method is used to calculate the permeability of each sub-grid. The permeability of the model in this work is expressed in tensor to preserve the effect of fracture direction on fluid flow. For a fracture with an angle θ to the x-axis, the permeability tensor of the fracture k_F in the two-dimensional coordinate system can be expressed as:

$$k_F = k_f \begin{bmatrix} \cos^2\theta & \sin\theta\cos\theta \\ \sin\theta\cos\theta & \sin^2\theta \end{bmatrix} \tag{1}$$

where θ is the angle between the fracture and the x-axis; k_f is the permeability of the fracture. The contribution of the fracture to the hydraulic conductivity of the sub-grid, which is crossed, can be estimated as T_f/Δ [36], where T_f is the hydraulic conductivity of the fracture and Δ is the size of sub-grid. The relationship of the fracture to the permeability of the sub-grid, which the fracture passes through, can be expressed as:

$$k_c = \frac{d_f}{\Delta} k_f \tag{2}$$

where k_c is the permeability contribution of fracture to the sub-grid; and d_f is the width of the fracture. Therefore, the permeability of each sub-grid can be calculated as:

$$k_{i,j} = k_m + \frac{d_f}{\Delta} k_f \begin{bmatrix} \cos^2\theta & \sin\theta\cos\theta \\ \sin\theta\cos\theta & \sin^2\theta \end{bmatrix} \tag{3}$$

where $k_{i,j}$ is the permeability of sub-grid(i, j); and k_m is the permeability of matrix. For cases where multiple fractures pass through the same sub-grid, the permeability sub-grid can be calculated as follows:

$$k_{i,j} = k_m + \sum_{n=1}^{N} k_c \begin{bmatrix} \cos^2\theta & \sin\theta\cos\theta \\ \sin\theta\cos\theta & \sin^2\theta \end{bmatrix} \tag{4}$$

where N is the number of fractures passing through the sub-grid(i, j); and $k_{i,j}$ is the permeability of the sub-grid(i, j). Figure 2 shows the process of fracture permeability mapping.

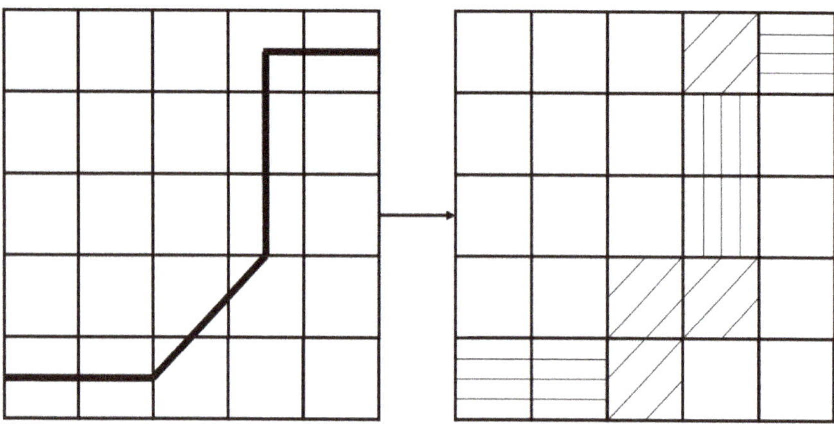

Figure 2. The schematic diagram of permeability mapping.

Considering the error between the FCM model obtained after mapping and the DFN model, the permeability of FCM needs to be corrected as follows:

$$k'_{i,j} = Ck_{i,j} \tag{5}$$

where $k'_{i,j}$ is the corrected permeability of the sub-grid(i, j); C is the permeability correction factor used to correct for the error in flow rate that occurs from mapping. In Ref [36] research C is calculated as $|\sin\theta| + |\cos\theta|$. In this work, the correction factor was calculated from the flow ratio between the DFN model and FCM model with uncorrected permeability.

2.2. Governing Equation

The fluid flowing in the thermal reservoir is described by Darcy's law. The mass balance equation in the porous media is as follows:

$$\frac{\partial\left(\varepsilon\rho_f\right)}{\partial t} + \nabla\cdot(\rho_f u) = Q_m \tag{6}$$

$$\frac{\partial\left(\varepsilon\rho_f\right)}{\partial t} = \rho_f S\frac{\partial P}{\partial t} \tag{7}$$

$$u = -\frac{k}{\mu}\nabla P \tag{8}$$

where ε is the porosity of the rock matrix; ρ_f is the density of the fluid; t is the time; u is the Darcy velocity; S is the storage coefficient; P is the pressure; k is the permeability of media; μ is the fluid dynamic viscosity; Q_m is the source-sink term.

In this work, the local thermal non-equilibrium theory is used to describe the heat exchange between the rock and the fluid flowing in the geothermal reservoir. The energy balance equations are as follows [37,38]:

$$(1-\varepsilon)\rho_s C_{p,s}\frac{\partial T_s}{\partial t} = \nabla\cdot[(1-\varepsilon)\lambda_s\nabla T] + q_{sf}\left(T_f - T_s\right) \tag{9}$$

$$\varepsilon\rho_f C_{p,f}\frac{\partial T_f}{\partial t} + \varepsilon\rho_f C_{p,f}u\nabla T_f = \nabla\cdot\left(\varepsilon\lambda_f\nabla T\right) + q_{sf}\left(T_f - T_s\right) \tag{10}$$

where ρ_s is the density of the matrix; T_s and T_f are the temperatures of the matrix and fluid respectively; $C_{p,s}$ and $C_{p,f}$ are the specific heat capacities of the matrix and fluid respectively; λ_s and λ_f are the matrix and fluid thermal conductivities respectively; q_{sf} is the interstitial convective heat transfer coefficient.

In Section 3 of this paper, the permeability of the FCM model needs to be corrected by using the results of the DFN model. The governing equations in the matrix are the same as that used in the FCM model.

The mass conservation equation in discrete fractures is written as:

$$d_f\frac{\partial \rho_f}{\partial t} + \nabla_T\cdot d_f\rho_f u_f = d_f Q_m \tag{11}$$

$$u_f = \frac{k_f}{\mu}\nabla_T P \tag{12}$$

where u_f is the Darcy velocity in the fracture. The porosity of fractures is assumed to be 100%, so the temperature of the rock is not considered in the energy balance equation of fractures. The energy balance equation for the fluid in the discrete fractures is written as [39]:

$$Q_{fe} = d_f\rho_f C_{pf}\frac{\partial T_{fr}}{\partial t} - d_f\rho_f C_{pf}u_f\nabla_t T_{fr} - d_f\lambda_f\nabla_t T_{fr} \tag{13}$$

where T_{fr} is the temperature of the fluid in fractures; Q_{fe} is a source term to describe the heat transfer between the matrix and fractures, which mainly results from the heat convection.

2.3. Well-Placement Optimization of EGS FCM Model

Generally speaking, there are two principles in EGS well-placement design [30]: longer major flow path and less preferential flow. However, it is difficult to find a long major flow path directly without preferential or short-circuit flow which is a notorious issue annoying EGS researchers and engineers [40]. Combinations of optimization algorithms and numerical simulations provide an idea for solving this problem.

2.3.1. Well-Placement Optimization Problem with 0-1 Programming

When designing an EGS well-placement, all wells, including injection wells and production wells, should pass through fractures because of the low permeability of the matrix. In the FCM model used in this paper, the thermal reservoir is divided into several equal-sized sub-grids, the parameters of each sub-grid represent the fractures' effect to this sub-grid. Therefore, all wells are located in the fractured sub-grids that have high permeability, which can ensure adequate connectivity between wells and reduce the number of potential well-placements. As shown in Figure 3, there are just 36 potential well-placements in a FCM model with 100 sub-grids.

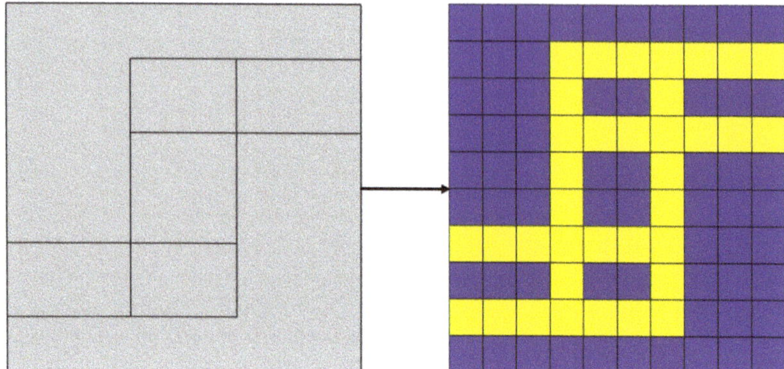

Figure 3. Schematic diagram of high permeability grid mapping from the fracture network.

However, this method of well-placement designing would bring some difficulties to the optimization. Complex fracture distribution in the reservoir results in uneven distribution of high permeability sub-grids, which makes it hard to deal with the constraints of well-placement. Transforming the well-placement optimization problem into a 0-1 programming problem can solve this difficulty.

In this work, the well-placement optimization problem of EGS is considered as a 0-1 programming problem. The one-well injection and multi-well production pattern are applied in this work. All wells are located in the high-permeability sub-grid and only one well at most on each sub-grid. For each grid, the grid without the well is recorded as 0, and the grid where the well is located in is recorded as 1. The coding form is shown in Figure 4, where the yellow grids represent the high-permeability sub-grids and the blue grids represent the low-permeability sub-grids.

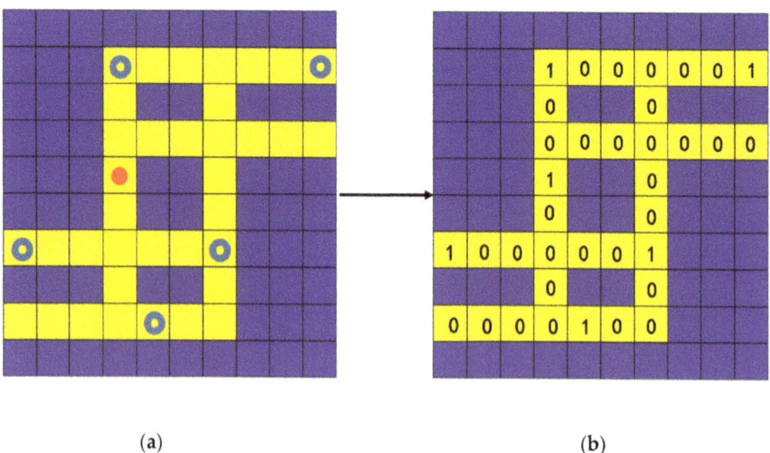

Figure 4. The well-placement and the optimal variables (**a**) The locations of wells (red point represents injection wells and blue represents production wells). (**b**) The optimal variables transformed from well locations.

This gives variables consisting of 0-1 to indicate the number and location of wells in the thermal reservoir. The high-permeability sub-grids have been numbered and the well-placement would be transferred to a one-dimensional vector consisting of binaries. The injection wells and production wells are not distinguished in coding, but the well closest to the center of the model is used as the injection well, and the other wells are used as production wells. The EGS well-placement problem based on 0-1 programming can be described as follows:

$$
\begin{array}{ll}
\max & f(x) \\
s.t. & \sum_{i=1}^{n} x_i \leq M \\
 & \sum_{i=1}^{n} x_i \geq N \\
 & x_i = 0 \quad or \quad 1
\end{array}
\tag{14}
$$

where x is a vector of 0s and 1s transformed from well-placement, $f(x)$ is the objective function, M is the upper limit of the number of wells, and N is the lower limit of the number of wells.

2.3.2. Genetic Algorithm

In this work, a GA is used to solve the EGS well-placement optimization problem. A GA is an optimization algorithm that searches for the best solution by simulating natural evolution. The iteration of a GA begins with a population of individuals. One individual represents a potential solution and the population represents a potential set of solutions to a problem. The individual is encoded by genes that represent the variables of the problem. The individual is evaluated by the fitness value determined by the user. The fitness value is the result of the objective function in most cases. The process of population regeneration consists of selection, crossover, and mutation. The role of selection is to eliminate individuals with low fitness, and crossover and mutation are used to generate new solutions to keep the diversity of the population. The best individual in the last generation is seen as the approximate optimal solution to the problem.

The strategies in GA adopted in this work are given below:

1. Initialization: N individuals are randomly generated before iterations, which is used as the first generation in GA.
2. Fitness calculation: the fitness (objective function) of each individual is calculated by a numerical simulation.
3. Selection: roulette is used to select parent individuals from the current population, which means that individuals with greater fitness are more likely to be selected, and the selected individuals enter the parents pool.
4. Crossover: do the single-point crossover of individuals in the parent pool based on crossover probability.
5. Mutation: single-point mutation is employed to make small random changes in the individuals in the parent pool
6. Elitist strategy: an elitist strategy is applied in the process of evolution. The individual with the best fitness in the current generation is retained to the next generation without crossover and mutation.
7. Stopping criteria: when the number of generations achieves the pre-set value, GA will stop.
8. Constraint: the constraint in this work is the number of wells. The first generation is initialized in the feasible region, and the infeasible solution generated in the iteration will be repaired.
9. Repair method: the production well closest to the center would be removed if the number of wells is above the upper bound of the number of wells, and the well would be added at random locations if the number of wells is below the lower bound.

The GA is written in MATLAB R2018b which is easy to combine with numerical simulation modules.

3. A Well-Placement Optimization Case

3.1. Computational Model

Natural fractures [41] or hydraulic fracturing fractures [42] in the reservoir can often be obtained from history matching or seismic inversion [43]. In this work, an assumptive model sized 200 m × 200 m is used as the original fracture network and the parameters of fractures are referenced from Ref [35]. Two hundred fractures, which are divided into two sets with different dip angles (i.e., 0° and 90°), are generated, and the fractures' lengths follow exponent distribution with the maximum length of 200 m, the minimum length of 20 m and exponent of 1. The backbone of the generated network is extracted using the method described in Section 2.1.1. The thickness of the permeable layer is 10 m. Figure 5 shows the original fracture network and the backbone network.

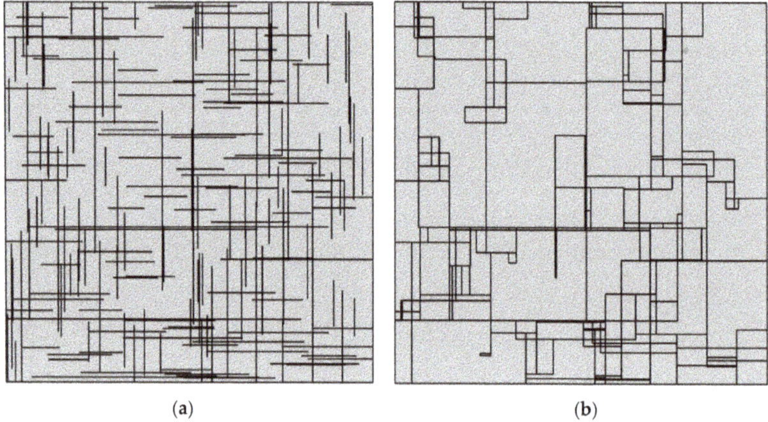

Figure 5. Fracture network: (**a**) original network (**b**) backbone network.

The model is divided into 20 × 20 square sub-grids with a side length of 10 m. The permeability mapping is based on the backbone network shown in Figure 5b. There is no permeability heterogeneity on the x-y and y-x components due to the direction of the fractures. The x-x component and y-y component of the permeability tensor are shown in Figure 6.

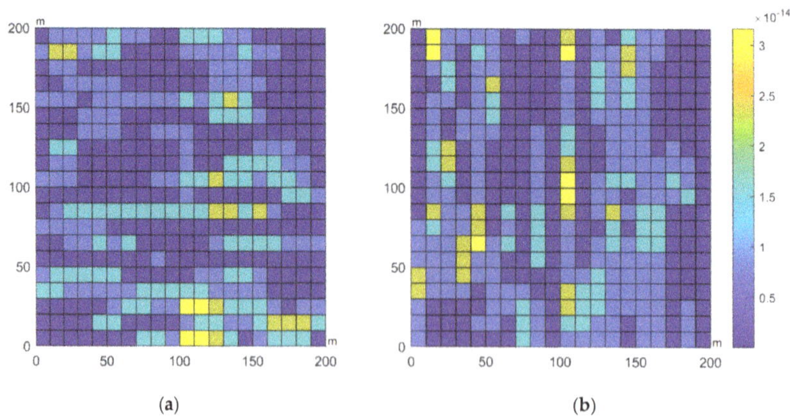

Figure 6. The permeability tensor (mm^2) of the FCM model. (**a**) x-x component (**b**) y-y component.

The mathematical model of the above governing equations including the matrix and fracture are discretized using the finite element method (the initial and boundary conditions are given in Section 3.2), which is solved using a commercial finite element software COMSOL Multiphysics 5.3 (COMSOL Co. Ltd., Stockholm, Sweden) and the GA written by MATLAB R2018b is linked to the software by LiveLink for MATLAB, which is developed by COMSOL. The parameters of GA are shown in Table 1.

Table 1. The parameters of genetic algorithm (GA).

Parameters	Value
Population size	400
Max generation	40
Crossover rate	0.6
Mutation rate	0.02
Number of wells	5

3.2. Model Parameters

The reservoir is initially saturated with water. All wells, including 4 production wells and 1 injection well work under constant pressure conditions. The working fluid for heat extracting is also water. The parameters are written in Table 2, which are referenced from some previous numerical studies [44,45]. The permeability of FCM has been corrected by the flow ratio of the DFN model and uncorrected FCM model.

Table 2. Model Parameters.

Parameters	Value
Matrix density (kg/m^3)	2700
Matrix porosity	0.01
Matrix permeability (m^2)	1×10^{-17}
Matrix heat capacity (J/(kg·K))	1000
Matrix heat conductivity (W/m·K)	3
Fracture permeability (m^2)	1×10^{-10}
Fracture width (m)	0.001
Water density (kg/m^3)	1000
Water viscosity (Pa·s)	0.001
Water heat capacity (J/(kg·K))	4200
Water heat conductivity (W/m·K)	0.6
Storage coefficient (1/Pa)	1×10^{-10}
Thickness of permeable stratum(m)	10
Correction factor	0.79

No-flow and adiabatic boundaries are around the reservoir. The adiabatic boundary is set to better observe the effect of well-placement on heat extraction in temperature distribution. The initial and boundary conditions can be found in Table 3.

Table 3. Initial and Boundary Conditions.

Conditions	Value
Initial pressure (MPa)	20
Initial temperature (°C)	200
Injection pressure (MPa)	30
Injection temperature (°C)	65
Production pressure (MPa)	20

In this work, we just consider a five-spot well-placement pattern with one injection well and four production wells. The injection well and production wells are not distinguished in the 0-1 code. The well closest to the center of the model would be used as an injection well while the other wells would be used as production wells. To facilitate analysis of single well performance, the production well number would be numbered clockwise from the well that is closest to that production well (0,0).

3.3. Objective Function

In this work, accumulative extracted thermal energy is used as the objective function of the well-placement optimization. Considering the adiabatic boundaries, it is equal to the decline in the thermal energy of the reservoir. It is expressed as follow [45]:

$$E = \iiint_{V_s} \rho_s c_{p,s}(T_i - T(t))dv \tag{15}$$

where E is the decline in thermal energy in the reservoir; T_i is the initial temperature; and $T(t)$ is the reservoir temperature in time t.

Besides E, the flow rate (Q), the accumulative extracted thermal energy (γ), the average production temperature ($\overline{T_{out}}$) and output thermal power (p) are also used to evaluate the performance of a geothermal reservoir with different well-placement. Q, γ, $\overline{T_{out}}$ and p are defined as:

$$Q = \int_L u(t)dl \tag{16}$$

$$\gamma = \int_0^{t_s} Q\rho_f C_{p,f}(T_{out} - T_{in})dt \tag{17}$$

$$\overline{T_{out}} = \frac{\int_L T(t)dl}{L} \tag{18}$$

$$p = Q\rho_f C_{p,f}(T_{out} - T_{in}) \tag{19}$$

where L is the length of the boundary of the well; γ is the accumulative extracted thermal energy; t_s is the simulation time; T_{out} is the average temperature of production water; T_{in} is the temperature of injection water; and L is the length of the outlet boundary.

3.4. Results and Discussion

Genetic algorithms do not require a given initial solution, which is different from traditional optimization algorithms. Therefore, two different five-spot injection/production patterns (named Case 1 and Case 2) are used as two comparisons of the optimization result. As in Section 2.3, the yellow grids represent the high-permeability sub-grids and the blue grids represent the low-permeability sub-grids. In Case 1 the production well 1 and 3 is located on a sub-grid without fractures, while in Case 2 all wells are set in the sub-grids passed through fractures just like the optimization result. Figure 7 shows the comparisons and the optimization result (named Case 3).

Figures 8 and 9 show the best individual of four different generations and the convergence process of the objective function. In Figure 9, the solid blue line indicates the best fitness in each generation, and the orange dotted lines indicate the average fitness of each generation.

The convergence process shows that the best fitness achieved a high value in previous generations, which may be due to the fact that the wells are always placed in a high-permeability sub-grid during the optimization process. The lower average fitness in previous generations illustrates that many low fitness individuals are generated during the population initialization and genetic manipulation of previous generations, and the rapid increase in average fitness indicates that the entire population is evolving, which can prove the validity of 0-1 programming and GA in geothermal well-placement optimization.

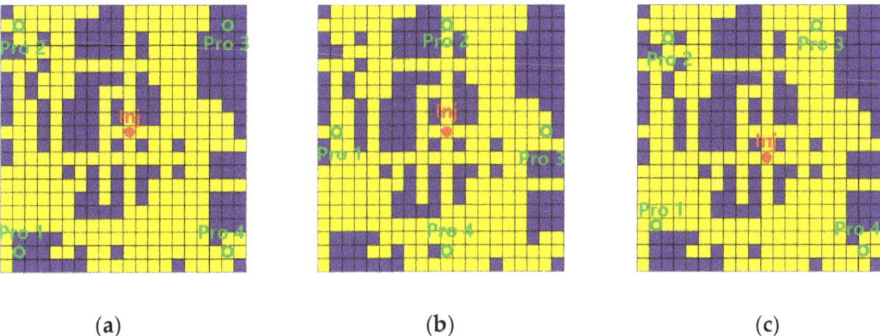

Figure 7. The well-placement of two comparisons and the optimization result (the green circle represents the production well and the red circle represents the injection well): (**a**) Case 1; (**b**) Case 2; (**c**) Case 3.

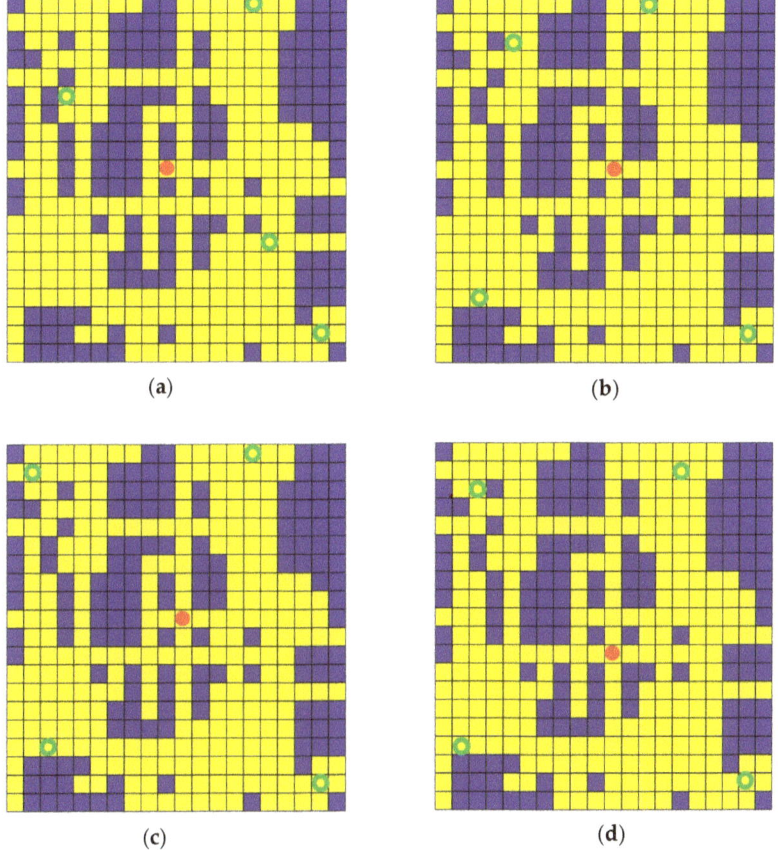

Figure 8. The best individual in (**a**) 1st generation (**b**) 10th generation (**c**) 20th generation (**d**) 40th generation.

The temperature variations of three cases are shown in Figure 10. The distribution of temperature is similar to each other at first because of the near location of the injection well. Gradually the cold front of each case migrates with the high-permeability sub-grid and the placement of production wells.

It is obvious that the migrations of the cold front in Case 2 and Case 3 are faster than Case 1. The cold front in Case 1 only migrated to the Pro2 and Pro4 that are located in high-permeability sub-grids caused by the connected fracture. From the temperature variation, it can be observed that the effect of fractures to seepage and heat extraction is preserved in the FCM model.

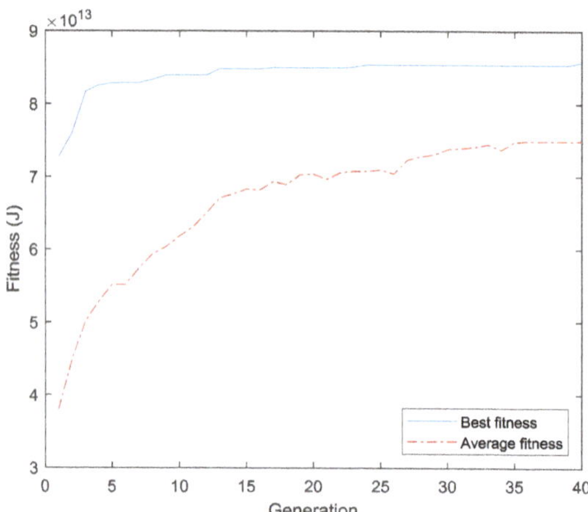

Figure 9. The convergence process of the GA in well-placement optimization.

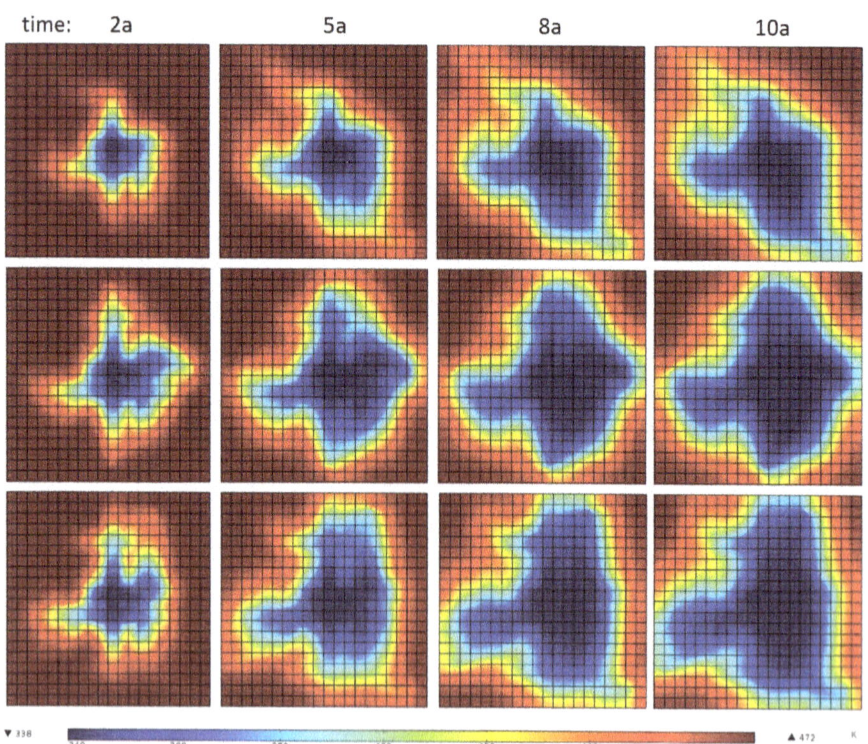

Figure 10. The temperature variations of three cases. The top is Case 1, the middle is Case 2 and the bottom is Case 3 (optimization result).

The difference in heat extraction between Case 2 and Case 3 is not as obvious as between Case 1 and other cases, but it can be found that the temperature of the northeast in Case 3 is lower than Case 2 and the low-temperature region in Case 3 is larger than Case 2 overall. Considering that there is no supply source, the heat extraction in Case 3 is more adequate. The accumulative extracted thermal energy of the three cases are plotted in Figure 11.

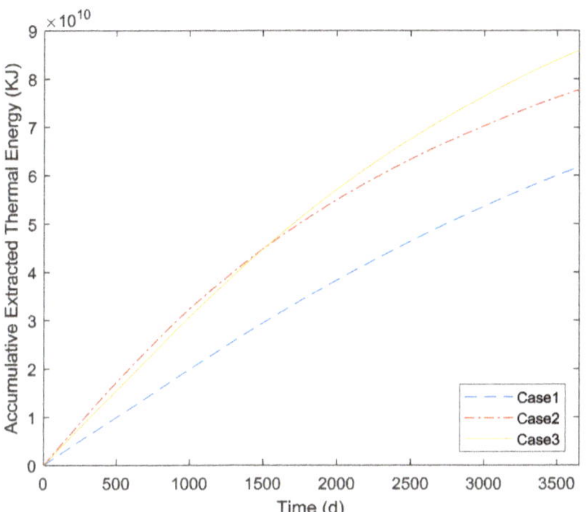

Figure 11. The accumulative extracted thermal energy in the three cases.

As shown in Figure 11, the final γ of Case 3 from GA is higher than the two five-spot patterns that set to compare, which can also prove the validity of the well-placement method applied in this work. It also can be found that the heat recovery rate of Case 2 is higher than that of Case 3 in the first 1500 days. Figure 12 shows the change in output thermal power in the three cases, which is consistent with Figure 11. As shown in Figure 12, the power of Case 2 is highest in the first 700 days, but it also has the fastest decline. After 2500 days, the power of Case 2 is the least in all three cases.

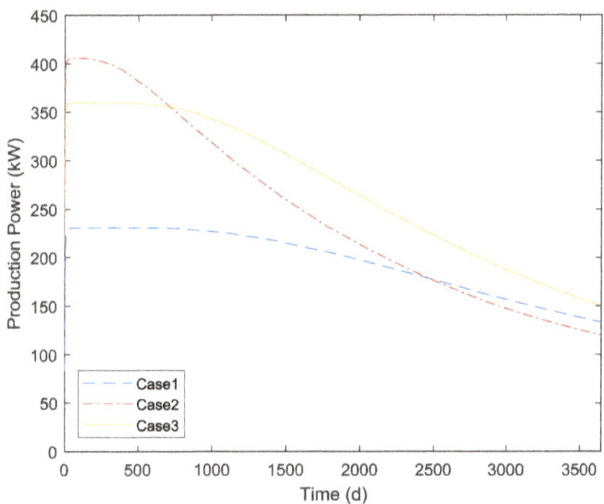

Figure 12. The production power in the three cases.

At the initial running stage, the higher heat recovery rate indicates a better flow connection between the production well and the injection well, which means there are more fractures connected with the production well and injection well, but it does not mean the final performance will be better. Preferential or short-circuit flow in the thermal reservoir is always a headache for geothermal development and management. High flow velocity may cause a rapid decrease in matrix temperatures beside the connected fractures, which decrease the efficiency of heat convection. The average temperature shown in Figure 13 and the flow rate shown in Figure 14 can prove it.

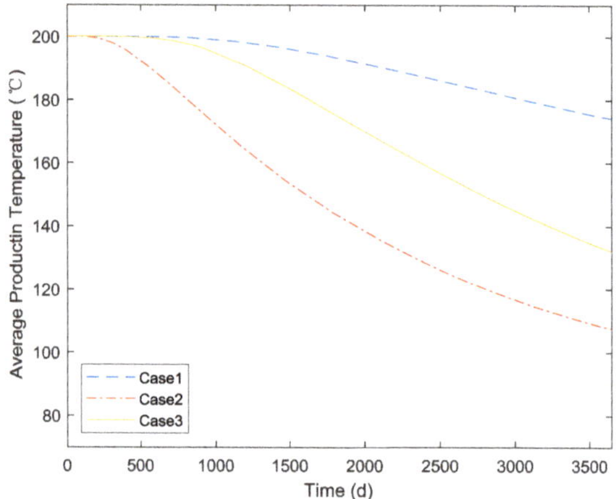

Figure 13. The average production temperature of the three cases.

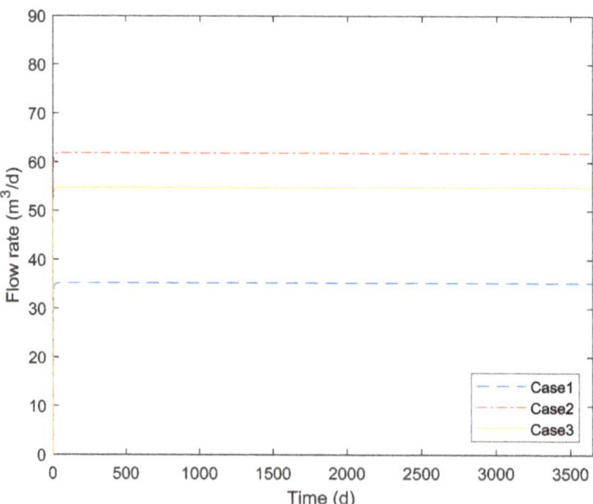

Figure 14. The production flow rate of the three cases.

The flow rate is rapidly stable because of the little storage coefficient. As shown in Figure 13, the average production temperature of Case 2 has a fast drop. It can be inferred from the temperature and flow rate that the preferential flow exists in Case 2.

Figures 15–17 show the accumulative energy, average temperature and the flow rate of each production well in the three cases. Consistent with the temperature distribution, the production wells, Pro1 and Pro3 in Case 1 contribute little to heat extraction, and the $\overline{T_{out}}$ of Pro3 shows the preferential flow in Case 2 mainly exists between the injection well and Pro3, and the optimization result has been improved in it.

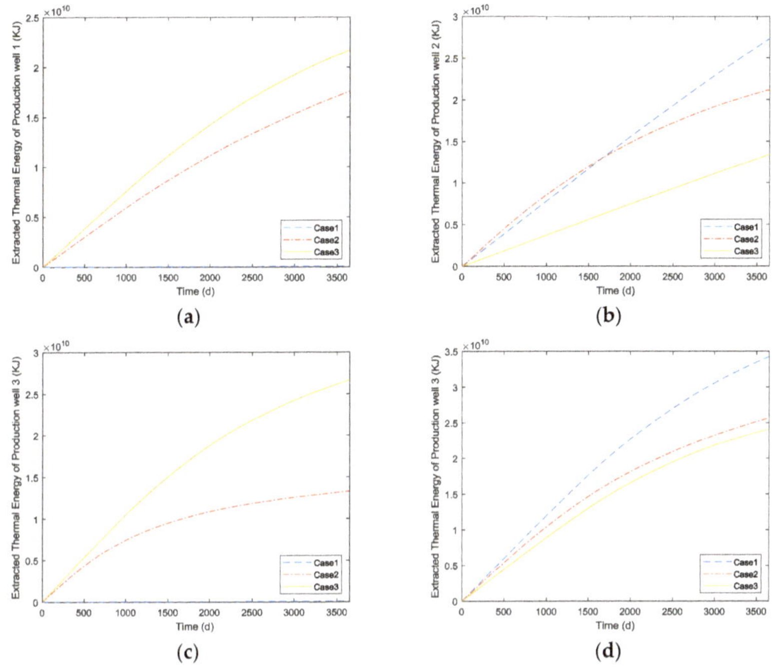

(a) (b) (c) (d)

Figure 15. The accumulative extracted thermal energy of the production well (**a**) Pro1 (**b**) Pro2 (**c**) Pro3 (**d**) Pro4.

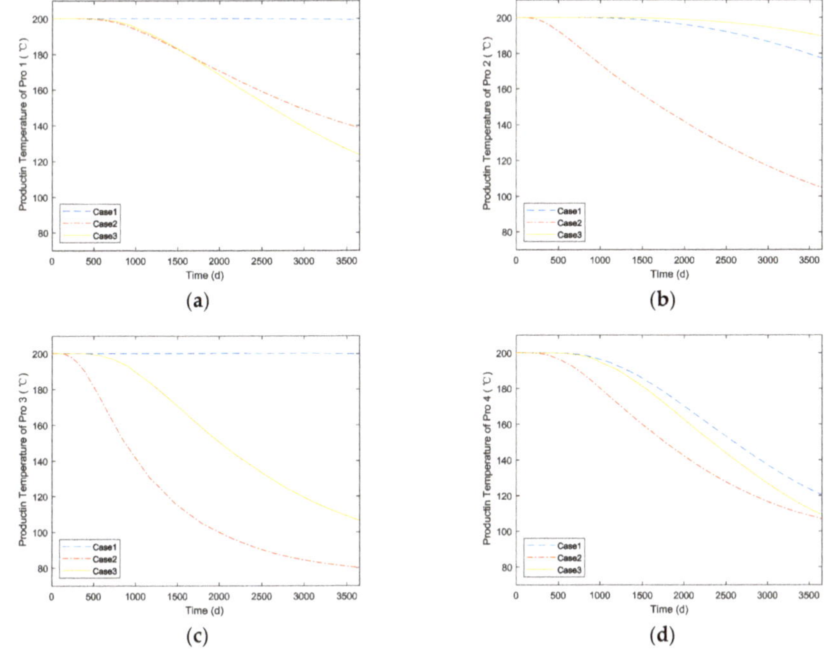

(a) (b) (c) (d)

Figure 16. The average temperature of the production well (**a**) Pro1 (**b**) Pro2 (**c**) Pro3 (**d**) Pro4.

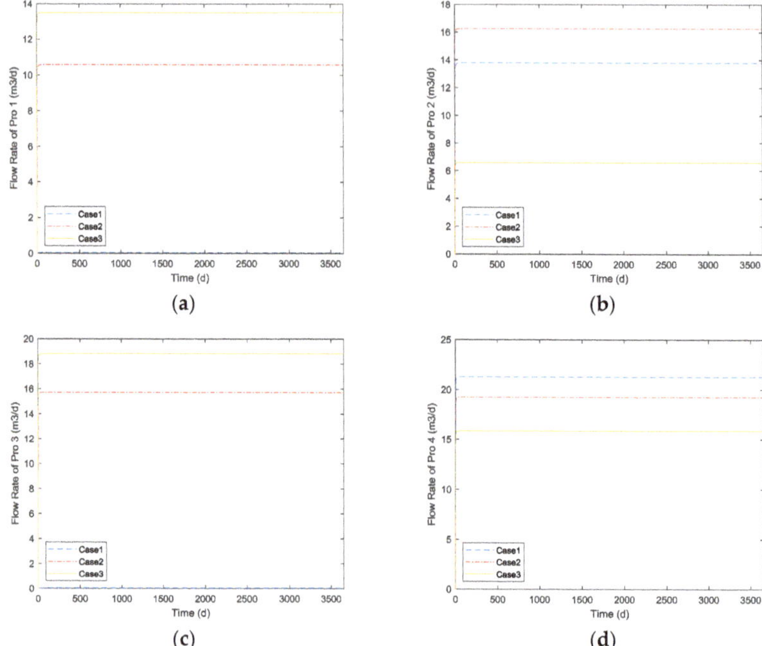

Figure 17. The flow rate of the production well (**a**) Pro1 (**b**) Pro2 (**c**) Pro3 (**d**) Pro4.

4. Conclusions and Future Work

A well-placement optimization framework is proposed in this paper. FCM is used to simplify the fractured thermal reservoir model and the GA is used to solve the well-placement optimization problem that was considered as a 0-1 programming problem.

1. The developed framework is efficient in the EGS well-placement optimization problem. The extracted thermal energy, which was the objective function, has increased in the convergence process of GA. And the optimization result shows better performance than comparison.
2. The FCM model can reflect the effect of fractures on seepage and heat transfer to a certain extent.
3. Regarding the well-placement optimization problem as a 0-1 programming problem can reduce the potential well-placements and improve the optimization effect. It also has the potential in joint optimization for well-placement and the number of wells.
4. In the well-placement design of EGS, the connectivity between the injection well and production well should be considered as the primary factor. The well in low-permeability contributes little to heat extraction.
5. Strong connectivity between wells does not mean better performance. Strong connectivity may lead to preferential flow and early heat breakthrough.

In this study, the framework only includes the 2-D model and the vertical well. In the future, this work will be generalized to the 3-D multi-field coupling model, a horizontal well, and the joint optimization of well-placement and the number of wells. A more advanced algorithm will be applied in EGS well-placement optimization, such as multi-objective optimization [46] and machine learning [47].

Author Contributions: Conceptualization, L.Z., Z.D. and K.Z.; Data curation, Z.D.; Formal analysis, L.Z. and T.L.; Funding acquisition, K.Z.; Methodology, L.Z. and K.Z.; Project administration, K.Z.; Software, Z.D.; Supervision, K.Z.; Validation, L.Z. and T.L.; Writing—original draft, Z.D.; Writing—review & editing, J.K.D., H.S. and Y.Y.

Funding: This work is supported by the National Natural Science Foundation of China under Grant 51722406, 61573018, 51874335 and 51674280, the Natural Science Foundation of Shan Dong Province under Grant JQ201808, the Key Research and Development Plan of Shan Dong Province under Grant 2018GSF116009, the National Science and Technology Major Project of China under Grant 2016ZX05025001-006, the Fundamental Research Funds for the Central Universities under Grant 18CX02097A, 17CX05002A and 17CX05003.

Conflicts of Interest: The authors declare no conflict of interest.

Nomenclature

The following terms are used in this manuscript:

k_F	fracture permeability tensor (m^2)
θ	angle between the fracture and the x-axis (°)
k_f	fracture permeability (m^2)
T_f	fracture hydraulic conductivity (m/s)
Δ	sub-grid size in FCM model (m)
k_c	permeability contribution of fracture to the sub-grid (m^2)
d_f	fracture width (m)
$k_{i,j}$	sub-grid (i, j) permeability (m^2)
k_m	matrix permeability (m^2)
N	fracture numbers
$k'_{i,j}$	corrected permeability of sub-grid (i, j) (m^2)
C	permeability correction factor
ε	matrix porosity
ρ_f	fluid density (kg/m^3)
t	time (s)
u	Darcy velocity (m/s)
S	matrix storage coefficient (1/Pa)
P	pressure (Pa)
k	porous media permeability (m^2)
μ	fluid dynamic viscosity
Q_m	source-sink term (1/s)
ρ_s	matrix density (kg/m^3)
T_s	matrix temperature (K)
T_f	fluid temperature (K)
$C_{p,s}$	matrix specific Heat capacity (J/kg/K)
$C_{p,f}$	fluid specific heat capacity (J/kg/K)
λ_s	matrix thermal conductivity (W/m/K)
λ_f	fluid thermal conductivity (W/m/K)
q_{sf}	interstitial convective heat transfer coefficient ($W/m^3/K$)
u_f	Darcy velocity in Fracture (m/s)
T_{fr}	fluid temperature in fracture (K)
E	The decline in thermal energy of the reservoir (J)
γ	accumulative extracted thermal energy (J)
L	the length of the boundary of well (m)
t_s	simulation runtime (s)
Q	mass flow rate in time t (m^3/s)
T_{out}	production water temperature In time T (K)
T_{in}	injection water temperature (K)
$\overline{T_{out}}$	average production temperature (K)
p	output thermal power (kW)
L	length of the outlet boundary (m)

References

1. Massachusetts Institute of Technology. *The Future of Geothermal Energy: Impact of Enhanced Geothermal Systems (EGS) on the United States in the 21st Century*; MIT: Cambridge, MA, USA, 2006.
2. Brown, D. The US hot dry rock program-20 years of experience in reservoir testing. In Proceedings of the World Geothermal Congress, Florence, Italy, 18–31 May 1995; pp. 2607–2611.
3. Office of Energy Efficiency and Renewable Energy; Lasala, R. *An Evaluation of Enhanced Geothermal Systems Technology*; Department of Energy: Washington, DC, USA, 2009.
4. Tenzer, H. Development of hot dry rock technology. *Geo-Heat Cent. Q. Bull.* **2001**, *32*, 14–22.
5. Polski, Y.; Capuano, L.; Finger, J.; Huh, M.; Knudsen, S.; Chip, M.A.; Raymond, D.; Swanson, R. *Enhanced Geothermal Systems (EGS) Well Construction Technology Evaluation Report*; Department of Energy: Washington, DC, USA, 2006.
6. Procesi, M.; Cantucci, B.; Buttinelli, M.; Armezzani, G.; Quattrocchi, F.; Boschi, E. Strategic use of the underground in an energy mix plan: Synergies among CO_2, CH_4 geological storage and geothermal energy. Latium Region case study (Central Italy). *Appl. Energy* **2013**, *110*, 104–131. [CrossRef]
7. Gan, Q.; Elsworth, D.; Cai, J. Heat transfer in enhanced geothermal systems: Thermal-Hydro-Mechanical coupled modeling. In *Petrophysical Characterization and Fluids Transport in Unconventional Reservoirs*; Cai, J., Hu, X., Eds.; Elsevier: Amsterdam, The Netherlands, 2019; pp. 201–205.
8. Qin, X.; Cai, J.; Xu, P.; Dai, S.; Gan, Q. A fractal model of effective thermal conductivity for porous media with various liquid saturation. *Int. J. Heat Mass Trans.* **2019**, *128*, 1149–1156. [CrossRef]
9. Abuaisha, M.; Loret, B.; Eaton, D. Enhanced Geothermal Systems (EGS): Hydraulic fracturing in a thermo–poroelastic framework. *J. Pet. Sci. Eng.* **2016**, *146*, 1179–1191. [CrossRef]
10. Bear, J.; Fel, L.G. A Phenomenological Approach to Modeling Transport in Porous Media. *Transp. Porous Media* **2012**, *92*, 649–665. [CrossRef]
11. Pruess, K.A. Practical method for modeling fluid and heat flow in fractured porous media. *SPE J.* **1985**, *25*, 14–26. [CrossRef]
12. Yang, Y.; Liu, Z.; Sun, Z.; An, S.; Zhang, W.; Liu, P.; Yao, J.; Ma, J. Research on stress sensitivity of fractured carbonate reservoirs based on CT technology. *Energies* **2017**, *10*, 1833. [CrossRef]
13. Ji, S.; Koh, Y. Appropriate Domain Size for Groundwater Flow Modeling with a Discrete Fracture Network Model. *Groundwater* **2017**, *55*, 51–62. [CrossRef]
14. Neuman, S.P.; Depner, J.S. Use of variable-scale pressure test data to estimate the log hydraulic conductivity covariance and dispersivity of fractured granites near Oracle, Arizona. *J. Hydrol.* **2015**, *102*, 475–501. [CrossRef]
15. Svensson, U. A continuum representation of fracture networks. Part I: Method and basic test cases. *J. Hydrol.* **2001**, *250*, 170–186. [CrossRef]
16. Yang, Y.; Liu, Z.; Yao, J.; Zhang, L.; Ma, J.; Hejazi, S.; Luquot, L.; Ngarta, T. Flow simulation of artificially induced microfractures using digital rock and lattice Boltzmann methods. *Energies* **2018**, *11*, 2145. [CrossRef]
17. Feyen, L.; Gorelick, S.M. Framework to evaluate the worth of hydraulic conductivity data for optimal groundwater resources management in ecologically sensitive areas. *Water Resour. Res.* **2005**, *41*, 147–159. [CrossRef]
18. Yang, Y.; Yao, J.; Wang, C.; Gao, Y.; Zhang, Q.; An, S.; Song, W. New pore space characterization method of shale matrix formation by considering organic and inorganic pores. *J. Nat. Gas Sci. Eng.* **2015**, *27*, 496–503. [CrossRef]
19. Chen, B.; Reynolds, A.C. Optimal Control of ICV's and Well Operating Conditions for the Water-Alternating-Gas Injection Process. *J. Pet. Sci. Eng.* **2016**, *149*, 623–640. [CrossRef]
20. Zhang, K.; Li, G.; Reynolds, A.C.; Yao, J.; Zhang, L. Optimal well placement using an adjoint gradient. *J. Pet. Sci. Eng.* **2010**, *73*, 220–226. [CrossRef]
21. Janiga, D.; Czarnota, R.; Stopa, J.; Paweł, W. Self-adapt reservoir clusterization method to enhance robustness of well placement optimization. *J. Pet. Sci. Eng.* **2019**, *173*, 37–52. [CrossRef]
22. Zhang, K.; Zhang, W.; Zhang, L.; Yao, J.; Chen, Y.; Lu, R. A study on the construction and optimization of triangular adaptive well pattern. *Comput. Geosci.* **2014**, *18*, 139–156. [CrossRef]
23. Zhang, K.; Zhang, H.; Zhang, L.; Li, P.; Zhang, X.; Yao, J. A new method for the construction and optimization of quadrangular adaptive well pattern. *Comput. Geosci.* **2017**, *21*, 499–518. [CrossRef]

24. Volkov, O.; Bellout, M.C. Gradient-based constrained well placement optimization. *J. Pet. Sci. Eng.* **2018**, *171*, 1052–1066. [CrossRef]
25. Zhang, L.; Zhang, K.; Chen, Y.; Li, M.; Yao, J.; Li, L.; Lee, J. Smart Well Pattern Optimization Using Gradient Algorithm. *J. Energy Resour.-Technol.* **2015**, *138*, 012901. [CrossRef]
26. Guyaguler, B.; Horne, R.N. Uncertainty Assessment of Well Placement Optimization. In Proceedings of the SPE Annual Technical Conference and Exhibition, New Orleans, LA, USA, 30 September–3 October 2001.
27. Jesmani, M.; Bellout, M.C.; Hanea, R.; Foss, B. Well placement optimization subject to realistic field development constraints. *Comput. Geosci.* **2016**, *20*, 1185–1209. [CrossRef]
28. Zhang, K.; Chen, Y.; Zhang, L.; Yao, J.; Ni, W.; Wu, H.; Zhao, H.; Lee, J. Well pattern optimization using NEWUOA algorithm. *J. Pet. Sci. Eng.* **2015**, *134*, 257–272. [CrossRef]
29. Akın, S.; Kok, M.V.; Uraz, I. Optimization of well placement geothermal reservoirs using artificial intelligence. *Comput. Geosci.* **2010**, *36*, 776–785. [CrossRef]
30. Chen, J.; Jiang, F. Designing multi-well layout for enhanced geothermal system to better exploit hot dry rock geothermal energy. *Renew. Energy* **2015**, *74*, 37–48. [CrossRef]
31. Chen, M.; Tompson, A.F.B.; Mellors, R.J.; Abdalla, O. An efficient optimization of well placement and control for a geothermal prospect under geological uncertainty. *Appl. Energy* **2015**, *137*, 352–363. [CrossRef]
32. Wu, B.; Zhang, G.; Zhang, X.; Jeffrey, R.G.; Kear, G.; Zhao, T. Semi-analytical model for a geothermal system considering the effect of areal flow between dipole wells on heat extraction. *Energy* **2017**, *138*, 290–305. [CrossRef]
33. Guo, X.; Song, H.; Killough, J.; Du, L.; Sun, P. Numerical investigation of the efficiency of emission reduction and heat extraction in a sedimentary geothermal reservoir: A case study of the Daming geothermal field in China. *Environ. Sci. Pollut. Res.* **2018**, *25*, 4690–4706. [CrossRef]
34. Matthäi, S.K.; Belayneh, M. Fluid flow partitioning between fractures and a permeable rock matrix. *Geophys. Res. Lett.* **2004**, *31*, L07602. [CrossRef]
35. Chen, B. Study on Numerical Methods for Coupled Fluid Flow and Heat Transfer in Fractured Rocks of Doublet System. Ph.D. Thesis, Tsinghua University, Beijing, China, 2009. (In Chinese)
36. Botros, F.E.; Hassan, A.E.; Reeves, D.M.; Pohll, G. On mapping fracture networks onto continuum. *Water Resour. Res.* **2008**, *44*, 134–143. [CrossRef]
37. Xu, C.; Dowd, P.A. Zhao, F.T. A simplified coupled hydro-thermal model for enhanced geothermal systems. *Appl. Energy* **2015**, *140*, 135–145. [CrossRef]
38. Saeid, S.; Al-Khoury, R.; Barends, F. An efficient computational model for deep low-enthalpy geothermal systems. *Comput. Geosci.* **2013**, *51*, 400–409. [CrossRef]
39. Chen, B.; Song, E.; Cheng, X. Plane-Symmetrical Simulation of Flow and Heat Transport in Fractured Geological Media: A Discrete Fracture Model with Comsol. In *Multiphysical Testing of Soils and Shales*; Laloui, L., Ferrari, A., Eds.; Proceeding of Springer Series in Geomechanics and Geoengineering; Springer: Berlin/Heidelberg, Germany, 2013.
40. Tenma, N.; Yamaguchi, T.; Zyvoloski, G. The Hijiori hot dry rock test site, Japan: Evaluation and optimization of heat extraction from a two-layered reservoir. *Geothermics* **2008**, *37*, 19–52. [CrossRef]
41. Zhang, L.; Cui, C.; Ma, X.; Sun, Z.; Liu, F.; Zhang, K. A Fractal Discrete Fracture Network Model for History Matching of Naturally Fractured Reservoirs. *Fractals* **2019**, *27*, 1940008.
42. Zhang, K.; Ma, X.; Li, Y.; Wu, H.; Cui, C.; Zhang, X.; Zhang, H.; Yao, J. Parameter prediction of hydraulic fracture for tight reservoir based on micro-seismic and history matching. *Fractals* **2018**, *26*, 1840009. [CrossRef]
43. Zhang, K.; Guo, Y.; Zhang, B.; Trumbo, A.M.; Marfurt, J.K. Seismic azimuthal anisotropy analysis after hydraulic fracturing. *Interpretation* **2013**, *1*, SB27–SB36. [CrossRef]
44. Sun, Z.; Zhang, X.; Xu, Y.; Yao, J.; Wang, H.; Lv, S.; Sun, Z.; Huang, Y.; Cai, M.; Huang, X. Numerical simulation of the heat extraction in EGS with thermal-hydraulic-mechanical coupling method based on discrete fractures model. *Energy* **2017**, *120*, 20–33. [CrossRef]
45. Song, X.; Shi, Y.; Li, G.; Yang, R.; Wang, G.; Zheng, R.; Li, J.; Lyu, Z. Numerical simulation of heat extraction performance in enhanced geothermal system with multilateral wells. *Appl. Energy* **2018**, *218*, 325–337. [CrossRef]

46. Zhang, L.; Wang, S.; Zhang, K.; Sun, Z.; Zhang, X.; Zhang, H.; Chipecane, M.T.; Yao, J. Cooperative Artificial Bee Colony Algorithm with Multiple Populations for Interval Multi-Objective Optimization Problems. *IEEE Trans. Fuzzy Syst.* **2018**. [CrossRef]
47. Wang, J.; Xu, C.; Yang, X.; Zurada, J.M. A Novel Pruning Algorithm for Smoothing Feedforward Neural Networks Based on Group Lasso Method. *IEEE Trans. Neural Netw. Learn. Syst.* **2018**, *29*, 2012–2024. [CrossRef] [PubMed]

MDPI
St. Alban-Anlage 66
4052 Basel
Switzerland
Tel. +41 61 683 77 34
Fax +41 61 302 89 18
www.mdpi.com

Energies Editorial Office
E-mail: energies@mdpi.com
www.mdpi.com/journal/energies

www.ingramcontent.com/pod-product-compliance
Lightning Source LLC
Chambersburg PA
CBHW051710210326
41597CB00032B/5433